PROGRESS IN

Nucleic Acid Research and Molecular Biology

Volume 53

PROGRESS IN
Nucleic Acid Research and Molecular Biology

edited by

WALDO E. COHN
Biology Division
Oak Ridge National Laboratory
Oak Ridge, Tennessee

KIVIE MOLDAVE
Department of Molecular Biology
and Biochemistry
University of California, Irvine
Irvine, California

Volume 53

ACADEMIC PRESS
San Diego   New York   Boston
London   Sydney   Tokyo   Toronto

This book is printed on acid-free paper. ∞

Copyright © 1996 by ACADEMIC PRESS, INC.

All Rights Reserved.
No part of this publication may be reproduced or transmitted in any form or by any means, electronic or mechanical, including photocopy, recording, or any information storage and retrieval system, without permission in writing from the publisher.

Academic Press, Inc.
A Division of Harcourt Brace & Company
525 B Street, Suite 1900, San Diego, California 92101-4495

United Kingdom Edition published by
Academic Press Limited
24-28 Oval Road, London NW1 7DX

International Standard Serial Number: 0079-6603

International Standard Book Number: 0-12-540053-5

PRINTED IN THE UNITED STATES OF AMERICA
96 97 98 99 00 01 BB 9 8 7 6 5 4 3 2 1

Contents

ABBREVIATIONS AND SYMBOLS ix
SOME ARTICLES PLANNED FOR FUTURE VOLUMES xi

Uracil Metabolism—UMP Synthesis from Orotic Acid or Uridine and Conversion of Uracil to β-Alanine: Enzymes and cDNAs 1

Thomas W. Traut and Mary Ellen Jones

I. Overview of Pyrimidine Metabolism 2
II. Allosteric Regulation by Dissociation of Enzyme Oligomers 13
III. UMP Synthase 18
IV. Uridine Kinase 40
V. β-Alanine Synthase 49
VI. Enzyme Evolution 59
VII. Conclusion and Future Prospects 70
References 72

The Importance of Being Modified: Roles of Modified Nucleosides and Mg^{2+} in RNA Structure and Function 79

Paul F. Agris

I. Biological Functions of Modified Nucleosides 81
II. The Large Array of Modified Ribonucleoside Structures and Chemistries 84
III. Site-selective Positioning of Modifications 101
IV. The Study of Modified-nucleoside Contributions to Structure and Function 105
V. Examples of Biophysical Contributions to Function by Modified Nucleosides 109
VI. Conclusions and Perspectives on the Importance of Being Modified 119
Appendix: Modified Nucleoside Symbols and Common Names 122
References 124

Chemical and Computer Probing of RNA Structure .. 131

N. A. Kolchanov, I. I. Titov, I. E. Vlassova
and V. V. Vlassov

I. Probing RNA Structure by Chemical and Enzymatic Approaches ... 133
II. Computer Analysis of the Secondary Structure of RNA 164
III. Concluding Remarks 190
   References 191

Transcriptional Activation of Thymidine Kinase, a Marker for Cell Cycle Control 197

Qing-Ping Dou and Arthur B. Pardee

I. Major Players in the G0 to S Transition 198
II. Transcriptional Regulation of Thymidine Kinase as a Model
   for Late G1 Events 202
III. Role of E2F Complexes in the Mouse Thymidine
   Kinase Transcription 204
IV. Role of New Complexes Yi1 and Yi2 in the Mouse
   Thymidine Kinase Transcription 208
   References 214

Eukaryotic Gene Expression: Metabolite Control by Amino Acids 219

Roney O. Laine, Richard G. Hutson
and Michael S. Kilberg

I. Amino-acid Control in Bacteria 219
II. Amino-acid Control in Yeast 220
III. Adaptive Regulation of Eukaryotic Plasma Membrane
   Amino-acid Transport 226
IV. Aminoacid-dependent Regulation of Asparagine
   Synthetase Expression 232
V. Identification and Characterization of Aminoacid-regulated
   Ribosomal-protein Genes 238
VI. Summary 245
   References 245

Molecular Recognition in the Assembly of the Segmented Reovirus Genome 249

Wolfgang K. Joklik and Michael R. Roner

I.	Basic Features of the Reovirus Multiplication Cycle	252
II.	The Nature of Reovirus Genome Segments	255
III.	The Nature of Reovirus Defective Interfering Particles	259
IV.	The Nature of Reovirus Genome Segment Reassortment and of the Reassortants That Are Generated	260
V.	The Functions of Reovirus Proteins	261
VI.	Reovirus Genome Segment Assortment Studied with Monoclonal Antibodies against Reovirus Proteins	269
VII.	The Infectious Reovirus RNA System	270
VIII.	The Interplay of Signals Required for the Insertion of Genetic Information into the Reovirus Genome	274
	References	277

Alu: Structure, Origin, Evolution, Significance, and Function of One-Tenth of Human DNA 283

Carl W. Schmid

I.	Identifying and Defining Repetitive Sequence Families	285
II.	Evolution of Alus (and Other Mammalian SINES)	296
III.	Genetic Effects of Alus and Other Repeats	301
IV.	Does Alu Have a Function?	302
V.	Prospectives and a Speculation	313
	References	314

Recent Advances in the Molecular Biology of Vitamin D Action 321

Hisham M. Darwish and Hector F. DeLuca

I.	The Vitamin D Receptor	322
II.	Vitamin D Receptor–DNA Interaction	325
III.	Transcriptional Regulation Mechanism by Vitamin D	329
IV.	The Molecular Biology of Vitamin D Hydroxylases	333
V.	Summary	338
	References	339

Regulation of Synthesis of Ribonucleotide Reductase and Relationship to DNA Replication in Various Systems ... 345

G. Robert Greenberg and John M. Hilfinger

I. Nature of Ribonucleotide Reductase Enzyme and Reaction	345
II. Time of Initiation of Bacteriophage T4 DNA Replication	349
III. Expression of T4 nrdA and nrdB Genes	351
IV. Expression of Escherichia coli nrdA and nrdB Genes	356
V. Phospholipid Synthesis and mRNA Breakdown Provide CMP for Synthesis of CDP and dCDP	359
VI. Control of Ribonucleotide Reductase Synthesis in Eukaryotes and Activity in Cell Division Cycle	361
VII. Effect of DNA Damage on nrdA and nrdB Expression	371
VIII. Ribonucleotide Reductase Regulation during Oogenesis	375
IX. Eukaryotic Animal DNA Viruses Encoding Ribonucleotide Reductases	379
X. Kinetics of Synthesis of T4-encoded Ribonucleotide Reductase and Relationship to T4 DNA Replication	383
XI. General Discussion	390
References	391

INDEX 397

Abbreviations and Symbols

All contributors to this Series are asked to use the terminology (abbreviations and symbols) recommended by the IUPAC-IUB Commission on Biochemical Nomenclature (CBN) and approved by IUPAC and IUB, and the Editors endeavor to assure conformity. These Recommendations have been published in many journals (1, 2) and compendia (3); they are therefore considered to be generally known. Those used in nucleic acid work, originally set out in section 5 of the first Recommendations (1) and subsequently revised and expanded (2, 3), are given in condensed form in the frontmatter of Volumes 9–33 of this series. A recent expansion of the one-letter system (5) follows.

SINGLE-LETTER CODE RECOMMENDATIONS[a] (5)

Symbol	Meaning	Origin of symbol
G	G	Guanosine
A	A	Adenosine
T(U)	T(U)	(ribo)Thymidine (Uridine)
C	C	Cytidine
R	G or A	puRine
Y	T(U) or C	pYrimidine
M	A or C	aMino
K	G or T(U)	Keto
S	G or C	Strong interaction (3 H-bonds)
W[b]	A or T(U)	Weak interaction (2 H-bonds)
H	A or C or T(U)	not G; H follows G in the alphabet
B	G or T(U) or C	not A; B follows A
V	G or C or A	not T (not U); V follows U
D[c]	G or A or T(U)	not C; D follows C
N	G or A or T(U) or C	aNy nucleoside (i.e., unspecified)
Q	Q	Queuosine (nucleoside of queuine)

[a] Modified from *Proc. Natl. Acad. Sci. U.S.A.* **83**, 4 (1986).
[b] W has been used for wyosine, the nucleoside of "base Y" (wye).
[c] D has been used for dihydrouridine (hU or H_2Urd).

Enzymes

In naming enzymes, the 1984 recommendations of the IUB Commission on Biochemical Nomenclature (4) are followed as far as possible. At first mention, each enzyme is described *either* by its systematic name *or* by the equation for the reaction catalyzed *or* by the recommended trivial name, followed by its EC number in parentheses. Thereafter, a trivial name may be used. Enzyme names are not to be abbreviated except when the substrate has an approved abbreviation (e.g., ATPase, but not LDH, is acceptable).

REFERENCES

1. *JBC* **241**, 527 (1966); *Bchem* **5**, 1445 (1966); *BJ* **101**, 1 (1966); *ABB* **115**, 1 (1966), **129**, 1 (1969); and elsewhere. General.
2. *EJB* **15**, 203 (1970); *JBC* **245**, 5171 (1970); *JMB* **55**, 299 (1971); and elsewhere.
3. "Handbook of Biochemistry" (G. Fasman, ed.), 3rd ed. Chemical Rubber Co., Cleveland, Ohio, 1970, 1975, Nucleic Acids, Vols. I and II, pp. 3–59. Nucleic acids.
4. "Enzyme Nomenclature" [Recommendations (1984) of the Nomenclature Committee of the IUB]. Academic Press, New York, 1984.
5. *EJB* **150**, 1 (1985). Nucleic Acids (One-letter system).

Abbreviations of Journal Titles

Journals	*Abbreviations used*
Annu. Rev. Biochem.	ARB
Annu. Rev. Genet.	ARGen
Arch. Biochem. Biophys.	ABB
Biochem. Biophys. Res. Commun.	BBRC
Biochemistry	Bchem
Biochem. J.	BJ
Biochim. Biophys. Acta	BBA
Cold Spring Harbor	CSH
Cold Spring Harbor Lab	CSHLab
Cold Spring Harbor Symp. Quant. Biol.	CSHSQB
Eur. J. Biochem.	EJB
Fed. Proc.	FP
Hoppe-Seyler's Z. Physiol. Chem.	ZpChem
J. Amer. Chem. Soc.	JACS
J. Bacteriol.	J. Bact.
J. Biol. Chem.	JBC
J. Chem. Soc.	JCS
J. Mol. Biol.	JMB
J. Nat. Cancer Inst.	JNCI
Mol. Cell. Biol.	MCBiol
Mol. Cell. Biochem.	MCBchem
Mol. Gen. Genet.	MGG
Nature, New Biology	Nature NB
Nucleic Acid Research	NARes
Proc. Natl. Acad. Sci. U.S.A.	PNAS
Proc. Soc. Exp. Biol. Med.	PSEBM
Progr. Nucl. Acid. Res. Mol. Biol.	This Series

Some Articles Planned for Future Volumes

Minute Virus of Mice cis-Acting Sequences Required for Genome Replication and the Role of the trans-Acting Viral Protein, NS-1
   CAROLINE R. ASTELL, QINGQUAN LIU, COLIN E. HARRIS, JOHN BRUNSTEIN, HITESH K. JINDALL AND PAT TAM

Structure and Transcription Regulation of Nuclear Genes for the Mouse Mitochondrial Cytochrome c Oxidase
   NARAYAN G. AVADHANI, A. BASU, C. SUCHAROV AND N. LENKA

The Large Ribosomal Subunit Stalk as a Regulatory Element of the Eukaryotic Translational Machinery
   JUAN P.G. BALLESTA AND MIGUEL REMACHA

Structure and Function of the Human Immunodeficiency Virus Leader RNA
   BENJAMIN BERKHOUT

High-Mobility-Group Chromosomal Proteins: Architectural Components That Facilitate Chromatin Function
   MICHAEL BUSTIN AND RAYMOND REEVES

Homologous Genetic Recombination in Xenopus: Mechanism and Implications for Gene Manipulation
   DANA CARROLL

The Internal Structure of the Ribosome
   BARRY S. COOPERMAN

S1-Nuclease-Sensitive DNA Structures Contribute to Transcriptional Regulation of the Human PDGF A-chain
   ZHAO-YI WANG AND THOMAS F. DEUEL

Hormonal and Cell-specific Regulation of the Human Growth Hormone and Chorionic Somatomammotropin Genes
   NORMAN L. EBERHARDT, SHI-WEN JIANG, ALLAN R. SHEPARD, ANDREW M. ARNOLD AND MIGUEL A. TRUJILLO

Eukaryotic Nuclear RNase P: Structures and Functions
   JOEL R. CHAMBERLIN, ANTHONY J. TRANGUCH, EILEEN PAGÁN-RAMOS AND DAVID R. ENGELKE

Mechanisms for the Selectivity of the Cell's Proteolytic Machinery
   ALFRED GOLDBERG, MICHAEL SHERMAN AND OLIVER COUX

Structure/Function Relationships of Phosphoribulokinase and Ribulosebisphosphate Carboxylase/Oxygenase
  FRED C. HARTMAN AND HILLEL K. BRANDES

The Nature of DNA Replication Origins in Higher Eukaryotic Organisms
  JOEL A. HUBERMAN AND WILLIAM C. BURHANS

Role of Translation Initiation Factor eIF-2B in the Regulation of Protein Synthesis in Mammalian Cells
  SCOT R. KIMBALL, HARRY MELLOR, KEVIN M. FLOWERS AND LEONARD S. JEFFERSON

Regulation and Function of Adenosine Deaminase in Mice
  MICHAEL R. BLACKBURN AND RODNEY E. KELLEMS

Enzymology of DNA Transfer by Conjugative Mechanisms
  WERNER PANSEGRAU AND ERICH LANKA

recA-independent DNA Recombination between Repetitive Sequences: Mechanisms and Implications
  XIN BI AND LEROY F. LIU

Experimental Analysis of Global Gene Regulation in Escherichia coli
  ROBERT M. BLUMENTHAL, DEBORAH W. BORST AND ROWENA G. MATTHEWS

The Elongation Phase of Protein Synthesis
  JOHN CZWORKOWSKI AND PETER B. MOORE

The Role of Ribosomal RNA in Translation
  JAMES OFENGAND

Bacterial and Eukaryotic DNA Methyltransferases
  NORBERT O. REICH

DNA Repair
  AZIZ SANCAR

Depletion of Nuclear Poly(ADP-ribose) Polymerase by Antisense RNA Expression: Influence on Genomic Stability, Chromatin Organization, DNA Repair, and DNA Replication
  CYNTHIA M.G. SIMBULAN-ROSENTHAL, DEAN S. ROSENTHAL, RUCHUANG DING, JOANY JACKMAN AND MARK E. SMULSON

Signals in Eukaryotic DNA Promote and Influence Formation of Nucleosome Arrays
  ARNOLD STEIN

SOME ARTICLES PLANNED FOR FUTURE VOLUMES

Transcriptional Regulation of Small Nuclear RNA Genes
 WILLIAM E. STUMPH

Bacillus subtilis as I Know It
 NOBORU SUEOKA

Uracil Metabolism—UMP Synthesis from Orotic Acid or Uridine and Conversion of Uracil to β-Alanine: Enzymes and cDNAs

THOMAS W. TRAUT[1] AND
MARY ELLEN JONES

*Department of Biochemistry and
Biophysics
University of North Carolina School of
Medicine
Chapel Hill, North Carolina 27599*

I. Overview of Pyrimidine Metabolism	2
A. Salvage vs. *de Novo* Pathways	4
B. The Central Position of the Liver	10
C. Is There Compartmentation of Nucleotide Pools?	12
II. Allosteric Regulation by Dissociation of Enzyme Oligomers	13
Models for Allosteric Regulation	16
III. UMP Synthase	18
A. Amino-acid Sequences of UMP Synthase and of Monofunctional OPRTases and ODCases	20
B. OPRTase Structure and Function	26
C. ODCase Structure and Function	29
D. Some Differences between UMP Synthase and the Monofunctional OPRTases and ODCases	33
IV. Uridine Kinase	40
A. Importance of Uridine Kinase and Possible Regulation of Gene Expression	40
B. Structure and Regulation of Uridine Kinase	44
C. Uridine Kinase DNA from cDNA and Genomic Libraries	46
V. β-Alanine Synthase	49
A. Structure and Dissociation of β-Alanine Synthase	52
B. The cDNA for β-Alanine Synthase from Rat Liver	57
VI. Enzyme Evolution	59
A. Motifs That Define Peptide Segments in Ligand-Binding Sites	60
B. Divergence of Kinases: Different Kinases with the Same Pattern of Motifs	60
C. Gene Fusion: Multifunctional Proteins	66
VII. Conclusion and Future Prospects	70
References	72

---

[1] To whom correspondence may be addressed.

I. Overview of Pyrimidine Metabolism

The importance of maintaining balanced pyrimidine nucleotide metabolism is evident in Fig. 1, where it is shown that many biosynthetic pathways are dependent on pyrimidines. The importance of the required synthesis of DNA and RNA needs no emphasis. Pyrimidine nucleotides are also important for the continued biosynthesis of various macromolecules such as phospholipids and polysaccharides, for which specific nucleotides are required as cofactors in activating the appropriate precursors, as exemplified by CMP-choline and UDP-glucose.

Even the conversion of uracil to β-alanine is a biosynthetic pathway, although historically this has been referred to as catabolism. A biosynthetic designation for this pathway comes from the importance of β-alanine, which functions as a neurotransmitter, as well as a building block for various dipeptides. These peptides include the ubiquitous pantothenate, the precursor for the acyl-carrier protein in fatty acid synthase and for coenzyme A; the inability of humans to synthesize pantothenate makes it a vitamin. The biochemical roles of carnosine and other dipeptides containing β-alanine are detailed in Section V.

In addition to the well-defined roles for uridine nucleotides as building blocks, or cosubstrates in the various forms of macromolecular synthesis

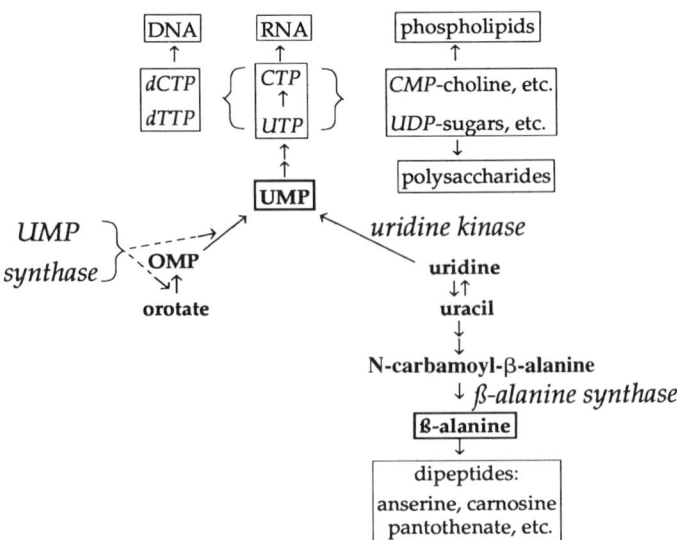

FIG. 1. Synthesis and interconversions of uracil compounds. Key pyrimidines are shown, as are the enzymes specific to this review.

indicated in Fig. 1, these compounds also appear to be involved in different physiological processes, wherein UMP or UDP or UTP serves as a specific agent in the activation of a complex process. Examples of these are shown in Table I, and include changes in calcium transport, blood vessel dilation, and sodium excretion. Although Seifert and Shultz (1) give more extensive information on this subject, in Table I are shown selected examples whereby an effort was made to define an effective concentration of the nucleoside or nucleotide for the process observed, such that this value is an estimate for the binding affinity of some protein for the uridine nucleotide. One study specifically measured the binding of [^{3}H]uridine or UTP to membrane proteins from rat liver (2), although the specific functions of these proteins remain unknown. The studies in Table I generally used isolated tissue or cells, immersed in a defined buffer solution, with the underlying assumption that the process observed *in vitro* might reflect the action of the same nucleotides in the blood.

Nucleotides presumably exist at very low concentrations in the extracellular environment, though no values have been published for their concentration in blood (3). Because measured concentrations for the different nucleosides in blood are in the range of 4–12 μM (3), it is a safe assumption that plasma nucleotide concentrations are certain to be below 100 μM. Some of the results in Table I can then support a possible role for these nucleotides *in vivo*. The values at the bottom of this table for platelet aggregation and phospholipase stimulation are not as good. Though the values for uridine show that this nucleoside is unlikely to have the role suggested in Table I under normal conditions, uridine could still be used for the effects shown under specific clinical conditions, where it would be administered as a therapeutic agent.

In mammals, on which most of the studies have been done, the two separate pathways for synthesis of UMP are generally active in most tissues

TABLE I
DIRECT PHYSIOLOGICAL ROLES FOR URIDINE AND URIDINE NUCLEOTIDES

Process affected	System	Compound (μM)				Ref.
		Urd	UMP	UDP	UTP	
Increase Ca^{2+} influx	Mouse tumor cells	—	—	—	5	248
Induce arterial constriction	Canine arteries	>1000	10	0.2	0.3	249
Induce natriuresis	Kidney tubules	—	>4	—	—	250
Platelet aggregation	Human blood	—	—	100	—	251
Stimulate phospholipase C	Rat pituitary	—	—	—	100	252
Induce hypothermia	Mice	≤20000	—	—	—	253

(exceptions are described below). Therefore the nucleosides uridine or cytidine, whether from the diet or the recycling of internal nucleotides, can be directly converted to UMP (or CMP) in an energetically efficient single reaction catalyzed by uridine kinase. Because such nucleosides may not always be sufficiently available, the *de novo* pathway is essential for synthesis of UMP; all other pyrimidine nucleotides can then be derived from UMP. The *de novo* pathway forms orotate from the normally available precursors aspartate, glutamine, and bicarbonate, and UMP synthase then converts orotate to UMP. UMP synthase contains two separate catalytic domains for the two sequential reactions, orotate phosphoribosyltransferase (OPRTase) and orotidylate decarboxylase (ODCase) shown in Fig. 1.

A. Salvage vs. de Novo Pathways

Early interest in these pathways led to many efforts to characterize the enzymes isolated from specific tissues, or to quantitate metabolic fluxes in whole cells or animals with exogenous labeled precursors. Such studies did not always provide consistent interpretations for the importance of the two pathways. Some of these difficulties were due to the experimental complexity of the studies. We now have sufficient data to separate out the variation of actual enzyme activities in different types of cells, as well as the variation in actual uptake of exogenous precursors.

The current data provide the following summary, for which details are given below. Though a few microorganisms have developed a special dependency on either the *de novo* or the salvage pathway, mammalian tissues generally have both pathways active, but not always at equal rates. During development, the *de novo* pathway is clearly important in almost all tissues. By adulthood, some tissues continue to have very active *de novo* UMP synthesis (liver, kidney, erythrocytes), and part of this UMP production may then be converted to uridine for export to the plasma, and thus supply other tissues for salvage. Many adult tissues show a significant decrease in the *de novo* pathway, and become more dependent on the salvage of uridine from blood. An important variable is the ability to transport orotate or uridine, which are frequently labeled to study either pathway under *in vivo* conditions. Only a few tissues demonstrate facilitated uptake of orotate, though most tissues have a facilitated uptake of uridine. In the adult rat, the most frequent experimental animal, a few tissues do not have, or have lost, the ability to use either exogenous orotate or uridine.

1. Variatioin in Enzyme Activity for *De Novo* and Salvage Enzymes

In normal mammalian tissues such as liver, lung, stomach, intestine, spleen, and colon, the assayed enzymatic activity for uridine kinase (the salvage pathway) commonly is 3- to 20-fold the activity of ODCase (*de novo* pathway) (4–7). It is likely that this ratio of enzyme activities reflects the ready availability of uridine to be salvaged, so that many tissues can rely on the salvage of continuously available uridine and cytidine. Many investigators have assayed either orotate PRTase or uridine kinase activities in samples from different tissues or organisms; however, values shown in Table II are from studies that assayed both of these activities, to facilitate comparison of the enzymes for the same cellular source and by the same experimental protocol. The direct salvage of uracil was also frequently assayed. The data in Table II show that four strategies are evident for the production of UMP, with some organisms relying largely on the *de novo* pathway. For *Crithidia*, *Trypanosoma*, or *Leishmania*, uridine kinase activity was marginal.

In mammals, there is a clear distinction between tissues that have more active OPRTase activity (liver) and tissues that have more active uridine

TABLE II
COMPARISON OF ACTIVITIES FOR UMP SYNTHESIS BY *de Novo* AND SALVAGE ENZYMES[a]

Synthesis	Organism	Relative enzyme activity[b]			Ref.
		Orotate PRTase	Uridine kinase	Uracil PRTase	
A. *De novo* > uridine salvage	*Schistomsoma mansoni*	10.6	1.0	5.7	162
	Crithidiae luciliae	100	(1.0)	116	236
	Trypanosoma brucei	140	(1.0)	12	13
	Leishmania mexicana amazonensis	450	(1.0)	13	13
	Rat liver	6.5	1.0	0.3	13
	Rat liver	6.8	1.0	—	12
B. Uridine salvage > *de novo*	Human colon	0.06	1.0	0.05	5
	Horse lymphocytes	0.1	1.0	—	254
	Mouse colon	0.16	1.0	0.03	5
	Rat liver	0.05	1.0	0.1	4
	Rat muscle	0.2	1.0	0.04	7
C. Uridine salvage ≅ *de novo*	Rat Heart	1.4	1.0	—	12
	Human lymphocytes	0.8	1.0	—	254
	Cow lymphocytes	1.0	1.0	—	254
	Rat brain	0.49	1.0	—	255
	Rat muscle	0.5	1.0	—	12
D. Uracil salvage > *de novo*	*Tritrichomonas foetus*	(1.0)	(1.0)	516	10

[a] Orotate phosphoribosyltransferase represents the *de novo* pathway, whereas uridine kinase and uracil phosphoribosyltransferase are two alternative salvage enzymes for the synthesis of UMP.

[b] Activities are normalized for each separate study; when the baseline value is in parentheses (1.0), this signifies the lower limit of detection.

kinase activity (colon, muscle, lymphocytes). For many mammalian tissues, the two pathways seem comparable, but with the salvage pathway enzymes having a higher specific activity, when assayed *in vitro*. Many organisms can salvage uracil directly by a PRTase activity, although this activity is generally low. Because this uracil PRTase activity is generally a fraction of the orotate PRTase activity (Table II), this may reflect that the OPRTase may catalyze the reaction with either substrate. For yeast, which has two separate genes for OPRTase (8), there is also a distinct uracil PRTase activity (9). Also for one parasite, *Tritrichomonas foetus*, the ability to salvage uracil by a PRTase activity is the most significant source of UMP (10), and therefore suggests that, for this organism, this is a distinct enzyme. A similar strategy for salvaging uracil has been shown for *Trichomonas vaginalis* (11).

Rat liver appears to depend more on the *de novo* pathway (12, 13), though one study found low OPRTase activity in rat liver (4). This most likely reflects a difference in the experimental methods and the difficulty of keeping enzyme activities stable in liver extracts. A higher activity for the *de novo* pathway in liver is more consistent with studies that show the importance of the liver for whole body pyrimidine metabolism, which will be detailed in Section I,B.

2. Variation in the Incorporation of Exogenous Orotate and Uridine

Transport into the cell could be the limiting feature in assaying the salvage or the *de novo* pathway, if all cells do not equally transport uridine and orotate. Reviews on this subject (14, 15) show that orotate and uridine are likely to be imported by different transport processes (14), with orotate permeating via a nonmediated process into some types of mammalian cells, such as rat hepatoma, mouse leukemia, and Chinese hamster ovary cells (15). Nucleosides such as uridine are generally taken up by facilitated transport in mammalian cells (16–18) and in bacterial cells (19). Specific examples illustrate such differences. Human erythrocytes incorporate orotate at 10 times the rate for uridine (20). For these cells, orotate uptake may be a facilitated process. Such differences are also evident in different bacterial species. Though *Salmonella typhimurium* utilizes uridine very readily (19), *Helicobacter pylori* utilizes exogenous orotate efficiently but uridine very poorly (21).

Many studies have now assayed the simple uptake, or the incorporation into nucleotides or nucleic acids, of labeled exogenous precursors. Such precursors include orotate, aspartate, bicarbonate, and ammonium for the *de novo* pathway, and uridine, cytidine, or deoxypyrimidine nucleosides for the salvage pathway. To facilitate comparison, data from such studies that assayed incorporation of orotate or uridine are shown in Table III, and were selected from studies that either used both orotate and uridine for the same

TABLE III
COMPARISON OF INCORPORATION OF EXOGENOUS OROTATE
OR URIDINE BY DIFFERENT TISSUES

		Incorporation of		
Use	Tissue	Orotate	Uridine	Ref.
A. Effective use of orotate	Human erythrocytes	Very good	None	20, 40
	Rat liver	Very good	Poor	22, 256, 257
	Rat kidney	Very good	Good	22, 257
	Rat lung	Good	Good	256
B. Effective use of uridine	Cat brain	—	Very good	26
	Rat bone marrow	—	Very good	22
	Rat brain (young)	—	Good	258
	Rat intestine	Very poor	Good	22, 23, 257
	Rat liver	—	Very good	36, 47
	Rat lung	—	Very good	22
	Rat kidney	—	Very good	47
	Rat spleen	—	Very good	22
	Pig lymphocytes	Poor	Very good	259
C. Poor uptake of orotate or uridine	Rat brain	Poor	Poor	260
	Rat brain	Very poor	—	257
	Rat heart	Very poor	Moderate	261
	Rat muscle	Very poor	Very poor	22, 257
	Rat thymus	Very poor	—	22

tissue, or that assayed one of these precursors for several different tissues. The results in Table III are shown as qualitative, because such studies routinely report results as "dpm" or "cpm" incorporated. Due to the variation in the specific activity of the tracer, plus other variables, a more quantitative comparison is not yet possible.

To facilitate the overall comparison, data from these studies are shown in three categories, depending on whether they use orotate effectively, use uridine effectively, or neither. Only four mammalian tissues incorporate orotate effectively. Two of these tissues, liver and kidney, function to maintain blood at stable concentrations of many compounds, and might be expected to remove orotate from blood. More mammalian tissues have maintained the ability to incorporate uridine actively, so that the salvage pathway could be a significant source for UMP synthesis. Another group of tissues appears to have low activities for transport of either orotate or uridine. As with measurements of enzyme activities in Table II, some rat tissues appear in more than one category of Table III. This may reflect differences in experimental technique, or possibly in the developmental stage of the rat.

This latter feature does not receive consistent attention, and there may be differences between young adults and older adult animals for some of the observed assays, as described in Section I,B. It is also worth comparing specific tissues in Tables II and III, because there are obvious discrepancies. For example, rat muscle has greater uridine kinase activity than OPRTase activity (7, 12), yet muscle appears inefficient at using exogenous uridine (22). A similar comparison is evident for adult rat brain.

Generally, these values for the assayed activities of the representative enzymes are in agreement with studies on the incorporation of [5-^{3}H]uridine in rats (22). These experiments showed very active incorporation by bone marrow, lung, kidney, and spleen. Rat intestinal mucosa appears to be almost totally dependent on salvage, because almost no OPRTase activity was detectable (23). For the tissues that actively use uridine, the salvage pathway contributed very significantly to the total pyrimidine nucleotide pool. Consistent with these studies are experiments using mice undergoing perfusion. In these experiments the incorporation of ^{15}NH$_4$Cl into total pyrimidine nucleotides (assayed by mass spectrometry) helped to define the contribution of the *de novo* pathway which produced about 2% of the pyrimidine nucleotides in liver, and about 6% in the intestine (24). The observed ratio of activities for uridine kinase and ODCase do not change much in cancer cells, because the need for increased nucleotides to support rapid DNA synthesis in actively dividing cells is generally satisfied by a fairly equal increase in the activities of both enzymes (4–7).

The remarkable increase in cancer cells of the potential activity for both *de novo* and salvage pathways was shown by human leukemia cells, which were compared to bone marrow cells as a control (25). The intracellular enzyme activities were estimated by following the incorporation into pyrimidine metabolites of NaH[^{14}C]O$_3$ (*de novo* pathway), or [^{3}H]uridine (salvage pathway). Incorporation by the salvage pathway was about 100–300 times that of the *de novo* pathway. Nevertheless, by calculating the total amount of pyrimidine nucleotides required for daily RNA and DNA synthesis, it was estimated that acute leukemia cells can still synthesize 70% of their daily requirement by the *de novo* pathway alone.

Some tissues may depend almost completely on uridine kinase for pyrimidine nucleotides. Early experiments showed that the brain of cats deprived of normal arterial blood and dependent on a perfusing solution required uridine or cytidine to maintain activity (26). In these studies, brain function was assayed by electrical activity and also by glucose uptake, oxygen utilization, etc. Especially interesting was the observation that the above measurements declined within an hour of perfusion with an artificial blood solution that had been extensively dialyzed. However, when uridine was then administered, all assayed activities increased toward the normal values, even when such perfusion was maintained for 3 hours. These studies showed that the brain of cats does not have sufficient *de novo* synthesis of UMP under such conditions, perhaps due to the lack of adequate available precursors, such as aspartate or glutamine, to initiate the *de novo* biosynthesis of UMP. But this requirement was completely fulfilled by the salvage of uridine.

These results emphasize the difficulty of interpreting data about cellular metabolic fluxes from studies in which uptake of an exogenous precursor may be the limiting feature. Because different mammalian tissues show both increased uridine kinase activity and an effective transport for incorporating uridine, the above studies suggest that mammals should maintain in the circulating blood a stable concentration of uridine that is adequate for the needs of tissues that manifest such a dependence. In a review of the physiological concentration of purines and pyrimidines, it appears that the average concentration of plasma uridine in seven different mammals is 5.3 ± 4.2 μM (3).

3. Changes in Enzyme Activities for UMP Synthesis with Development

A direct determination of OPRTase activity in extracts of mouse brain gave clear evidence for this enzymatic activity, and at a specific activity of about 40% of the activity in liver, suggesting that the brain of adult mice has a functional *de novo* pathway (27). The decline in activity appeared to be proportional to the stage of development, and by 90 days of age both OPRTase and ODCase had declined to about 25% of the activity in the fetal brain (28). These studies on uridine uptake by brain are consistent with a separate study that assayed enzyme activities from rat brain extracts obtained at birth, and through the first 18 days of growth (29). Both pathways appeared to be fully active in the brain of newborn rat pups, but by 18 days the enzyme activities of carbamoyl-phosphate synthase and aspartate carbamoyltransferase (*de novo* pathway) had already declined to one-third of their initial values, whereas uridine kinase activity had stabilized after only a modest decline. Furthermore, by following this activity from fetal brain to the neonate and then to the adult rat, it was shown that the *de novo* pathway is reduced to less than 1% of the activity observed in the fetus (30).

The greater activity of the *de novo* pathway in fetal brain has been confirmed in other studies (31, 32). Note however, that though these results on adult rat brain support the salvage of uridine as a major source for UMP synthesis, experiments on the incorporation of exogenous uridine by adult rat brain are not consistent with such an interpretation (see Table III). However, the uptake studies were done with slices of brain tissue, and isolated slices may be less effective in this activity.

Other studies on the importance of pyrimidines for neurobiological function were reviewed (33).

B. The Central Position of the Liver

To this point, we have focused on the synthesis of UMP by the two alternate pathways. When the additional steps involved in converting uracil to β-alanine are included, the importance of the liver for the metabolism of pyrimidines in the whole animal emerges. Three enzymes are required for the conversion of uracil to β-alanine (34). The first of these, dihydropyrimidine dehydrogenase, appears to be present in all tissues, but only liver and kidney have the next two enzymes, dihyropyrimidinase and β-alanine synthase (34, 35). Thus, though both liver and kidney can take up the circulating product from the first reaction in this pathway, dihydrouracil, the liver, due to its larger size and greater enzyme activities, is the major organ for converting available uracil and dihydrouracil from the whole organism to β-alanine.

The above pathway for the formation of β-alanine from uracil is completely consistent with the emerging model from studies showing that mammalian liver almost completely clears uracil and uridine from venous blood as it enters the liver. The liver then exports a defined amount of uridine back into the blood (36, 37). These observations are consistent with liver functioning as a chemostatic organ for maintaining stable blood uridine concentrations. The continued removal of uridine and uracil from blood entering the liver is consistent with the ability of the liver to use some fraction of these two compounds for the continuous formation of β-alanine.

In such experiments with isolated, perfused rat livers, infused [2-^{14}C]uridine at concentrations up to 15 μM was completely removed and totally converted to β-alanine + $^{14}CO_2$ (38). More specific work has shown that aspects of this recycling process are divided among different types of liver cells. The Kuppfer cells of the liver have unusually high levels of uridine phosphorylase activity, so that the uridine taken up from the blood is readily converted by them to uracil (39). Though these cells do not have all the enzymes for the conversion of uracil to β-alanine, the complete pathway is located in hepatocytes. Hepatocytes were the most effective liver cells in taking up uracil from the medium and converting it to β-alanine (39). Because ammonia is an additional product of the reaction catalyzed by β-alanine synthase, the liver is a most appropriate organ for this pathway. The liver also appears to be the major source for resupplying blood with a steady amount of uridine, although other tissues or cells may excrete uridine into the blood to an extent as yet undetermined.

This last function is a possible explanation for the observation that certain blood cells, such as human erythrocytes (40), have little uridine kinase activity but actively synthesize UMP from orotate. For such cells, the *de novo* pathway is not complete due to a deficiency of dihydro-orotate dehydrogenase because mitochondria are absent from erythrocytes. Because it requires a specific electron acceptor, dihydro-orotate dehydrogenase is located in the mitochondria of higher eukaryote cells *(41, 42)*. A possible function for UMP synthase in such cells may involve the salvage of circulating orotate. This compound is present at low concentrations in blood, with values of 0.2 μM for rat plasma *(43)* and <0.5 μM in human plasma *(44)*. By comparison, the blood concentration of the more standard salvage precursor, uridine, is about 5 μM *(3)*. Therefore, erythrocytes may take up and convert orotate to UMP and to uridine, and then secrete the uridine into the blood for use by other cells. By comparison, the orotate concentration in amniotic fluid of two normal pregnant women was 380 and 530 μM *(45)*. This study also found very high concentrations of uridine in amniotic fluid (730 and 960 μM), which was interpreted as a source of pyrimidine for the fetus.

The possible recycling of orotate to uridine was more specifically examined by an interesting protocol that assayed the incorporation by human lymphoblasts of [^{3}H]uridine or [2-^{14}C]orotate *(40)*. The lymphoblasts incorporated [^{3}H]uridine directly into uridine nucleotides, but showed very little incorporation of [2-^{14}C]orotate over 90 minutes. Human erythrocytes were then added to the lymphoblasts, and this coculture was continued for an additional 3 hours, with samples taken at two intervals and separated into the two cell populations, from which soluble metabolites were isolated. After the addition of erythrocytes to the lymphoblasts for coculture, ^{14}C label from the orotate was increasingly detected in the lymphoblasts. Because the erythrocytes excreted [^{14}C]uridine, formed from [^{14}C]orotate, the coculture results were interpreted as evidence that erythrocytes could take up and convert orotate to UMP and uridine, and excrete the final product, [^{14}C]uridine, which was then salvaged by the adjacent lymphoblasts.

That the blood uridine pool is fairly dynamic has been observed in various studies monitoring the fate of added [5-^{3}H]uridine. The labeled uridine is rapidly cleared from the blood, with a $t_{1/2}$ of 2–5 minutes in mice *(46, 47)* and 3 minutes in rats *(22)*. After administration, labeled uridine and cytidine are preferentially converted to nucleotides; uracil is not used for nucleotides, but almost completely converted to β-alanine *(46)*. Though cells may concentrate uridine to help expand the pyrimidine nucleotide pools, this process is naturally self-correcting because it has also been shown in such studies with human breast cancer cells that expansion of the UTP and CTP pools leads to feedback inhibition of both the *de novo* and salvage pathways *(48)*.

The above data were largely obtained from healthy animals, or people, on normal diets. We reemphasize the critical importance of the *de novo* pathway for conditions wherein there are not adequate nucleosides to maintain the body's needs by salvage alone. This is dramatically evident in patients with the genetic disorder that produces orotic aciduria. A key feature of such patients is megaloblastic anemia (44), in which bone marrow cells fail to synthesize DNA adequately. Studies already described showed that bone marrow cells have a very active salvage pathway (22), but if humans defective in UMP synthase have insufficient uridine to be salvaged, due to inadequate nutrition, they manifest the deleterious developmental patterns associated with this syndrome. Only when these patients are treated with supplemental uridine is the activity of uridine kinase (the salvage pathway) sufficient to maintain normal pyrimidine synthesis for the whole body (44). Because family members who share the (unsupplemented) diet of orotic aciduria patients have no symptoms, this supports the conclusion that the *de novo* pathway can generate all the pyrimidines needed by these people, either for direct use in cells where this is the principal pathway, or as a source of circulating pyrimidine nucleosides for other tissues that are more dependent on salvage.

An important feature of the liver's metabolic function is the circadian pattern that has been detected for pyrimidine synthesis in rodent liver (49, 50). In these experiments, rats or mice were maintained on fixed light–dark schedules, while receiving intraperitoneal injections of [^{14}C]orotate, [^{14}C]cytidine, or [^{14}C]uridine. Livers were then removed and extracts were analyzed for radioactivity in soluble nucleotides, as well as in RNA and DNA. These studies showed that the *de novo* pathway had a twofold increase in rate during the dark cycle, as measured by the incorporation of orotate into acid-soluble nucleotides (49). This was probably due to changes in enzyme activity, because a specific test was the assay of OPRTase activity in liver extracts prepared during the light and dark cycles. The specific activity of OPRTase during the dark cycle was twice the activity during the light cycle (50).

C. Is There Compartmentation of Nucleotide Pools?

Many of the studies described above used radioactive tracers to follow the uptake and incorporation of different precursors for pyrimidine nucleotides. In some cases, such studies yielded unexpected results, which led to the interpretation that separate pyrimidine nucleotide pools may exist, being formed separately by the salvage or *de novo* pathway, or being located in separate subcellular compartments. Experiments described in two earlier reviews (51, 52) show little conclusive support for compartmentation of nucleotides between the cytoplasm and the nucleus. One physical difficulty in maintaining separate pools between the nucleus and cytoplasm is that the pore size of the nuclear membrane is about 10 nm (51), so that even macromolecules such as enzymes may freely equilibrate between these two compartments.

Another review considers this subject in more detail with regard to the actual rates for uptake of precursors, or for their enzymatic conversion to nucleotides by *de novo* or salvage enzymes (3). This analysis found that frequently such separate pools are artifacts of the experimental approach. Compartmentation of separate pools between the cytoplasm and the nucleus was found to be unlikely, and only compartmentation between the cytoplasm and the mitochondrion had good support.

II. Allosteric Regulation by Dissociation of Enzyme Oligomers

In the past 30 years, we have acquired considerable information about a subset of enzymes whose distinctive characteristic is the ability to convert between oligomer and monomer, with a concomitant change in enzyme activity. Figure 2 shows the monomer and oligomer states for the three proteins that are the focus of this review; they are considered in more detail in Sections III–V. The numbers represent the number of subunits in a particular species. The native form is underlined. Superscript symbols designate the extent of enzyme activity as defined in the figure legend, and the effect of ligands on dissociation or association is also indicated. An enzyme species is defined as inactive when its measured activity is <5% of the optimum activity.

Any oligomeric protein can be converted to its subunits if sufficiently denaturing conditions are employed. The term *dissociating enzyme* has been

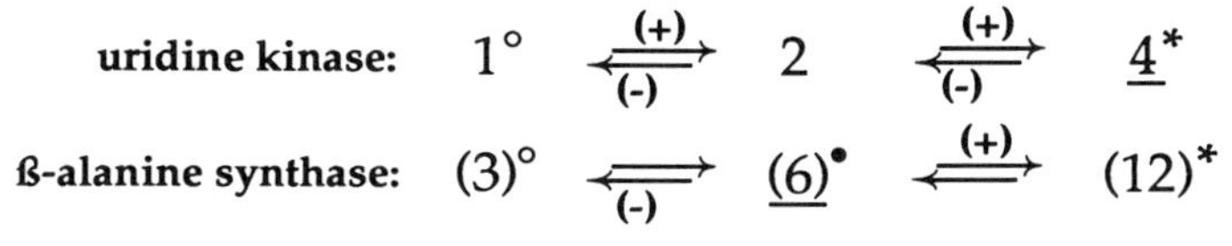

UMP synthase: $\underline{1}^\circ \xrightleftharpoons{(+)} 2^\bullet \xrightleftharpoons{(+)} 2^*$

uridine kinase: $\underline{1}^\circ \xrightleftharpoons[(-)]{(+)} 2 \xrightleftharpoons[(-)]{(+)} \underline{4}^*$

ß-alanine synthase: $(3)^\circ \xrightleftharpoons[(-)]{} \underline{(6)}^\bullet \xrightleftharpoons{(+)} (12)^*$

FIG. 2. Dissociating enzymes: a change in activity accompanies dissociation. Numbers indicate number of subunits; underlined numbers represent the native form of an enzyme (in the absence of ligands). Enzyme activity: *, most active; ●, partly active; ○, inactive; (+), positive effector, one of the substrates; (−), negative effector, an immediate product, or a more distant end product.

applied to enzymes that undergo a shift between oligomer and monomer in response to a physiologically relevant ligand, without the enzyme becoming denatured. For such enzymes, the equilibrium between monomer and oligomer is appropriately influenced by regulatory metabolite effectors. In a review of more than 70 different enzymes in nucleotide metabolism, it was found that 32% of these enzymes show allosteric behavior (53). Furthermore, 15% of these enzymes had oligomers that dissociated in response to varied concentrations of physiological ligands (53). In the pathways of carbohydrate metabolism and nucleotide metabolism, many enzymes have been reasonably well characterized, and it appears that almost one-half of the allosteric enzymes are also dissociating enzymes when tested *in vitro*.

Although the great majority of these dissociating enzymes contains only 2 or 4 subunits, there also are examples of very large polymers with more than 100 subunits. The term *native* is used here to designate the quaternary structure that is stable in the absence of any ligands. This is a useful reference state, and should not be assumed to be the form of the enzyme favored under *in vivo* conditions, where the enzyme will always be in the presence of varying concentrations of substrates and effectors.

Isolated enzymes studied *in vitro* can be made to change their aggregation state by appropriate changes in pH, temperature, salt concentration, or various denaturants. However, at least in most multicellular organisms and *in vivo*, an enzyme would not usually be subject to wide variations in environmental conditions. Regulation of an enzyme under physiological conditions is primarily related to (1) enzyme concentration and (2) effector ligand concentration. If either the concentration of enzyme or of an appropriate effector ligand is known to vary significantly *in vivo*, this parameter is likely to be significant for the physiological regulation of these enzymes. However, studies done *in vitro* across any concentration range can provide useful information about the enzyme and its protein structure.

The process of reversible association and dissociation can readily be investigated with isolated enzyme preparations using techniques such as molecular-sieve chromatography, sedimentation, or light scattering. Because enzymes may exist as mixtures of monomers and various oligomers, with the different aggregation states that equilibrate either rapidly or slowly, the observed profile of migrating enzyme peaks depends on whether the rate of equilibration between monomers and oligomers is faster or slower than their rate of migration (separation) in the experimental apparatus. The approach consists of obtaining a molecular mass (or migration profile) of native enzyme in the absence of any effector ligands, and in the presence of increasing concentrations of effector ligand. For a rapidly equilibrating system, one observes a migration profile that represents a statistical average for the entire population during the experiment. This is most obvious for simple monomer–dimer systems.

An example of such an enzyme is UMP synthase. In the absence of ligands, the enzyme sediments as a monomer, whereas in the presence of the saturating substrate, orotidine-5′-phosphate (OMP), the enzyme sediments as a dimer (54, 55). Very sharp symmetric peaks are also observed at intermediate ligand concentrations, representative of the average sedimenting position of mixtures rapidly interconverting between monomer and dimer during the time course of the experiment. As shown in Fig. 2, UMP synthase can exist in two distinct conformational forms of the dimer; this is further described in Section III.

When such experiments are initially done in the absence of ligands, the buffer is sometimes ignored as a possible ligand. This feature may be significant, because even buffer components can be ligands and must therefore be carefully considered. As an example, when the native molecular weight of UMP synthase was first measured in phosphate buffer, it produced an unusually high M_r for the species assumed to be a monomer (27). Subsequent studies on this enzyme demonstrated that the native monomer could be titrated with increasing phosphate concentration, leading to a concomitant increase in M_r until the enzyme became completely converted to the dimer (54). As detailed by Traut (56), inorganic phosphate is an allosteric ligand and alters the molecular weight for at least 10 enzymes. An important result of those studies was that UMP synthase behaves as a rapidly equilibrating mixture of monomers and dimers that produces quite symmetric sedimentation peaks at a position appropriate for the statistically average M_r for a given mixture of monomers and dimers.

In some cases, as with β-alanine synthase (Fig. 2), the native oligomer species can associate to larger oligomers or dissociate as a function of different effectors. The native enzyme without ligands is a hexamer (57); in the presence of the substrate (N-carbamoyl-β-alanine), the enzyme is converted to the active dodecamer, whereas in the presence of the product (β-alanine), it dissociates to the inactive trimer (58).

For a slowly equilibrating enzyme, one may observe two or more distinctly migrating peaks as the separate species separate more rapidly than they can reequilibrate. About half the enzymes in a recent survey can form three or more different aggregates; some of these show at least two forms at the same time (56). It is a significant diagnostic feature of slowly equilibrating enzymes that multiple peaks are produced during sedimentation or molecular-sieve chromatography. This is also observed with ion-exchange chromatography, because oligomerization can change the exposed surface with ionic charge.

Models for Allosteric Regulation

1. CONFORMATIONAL CHANGE STABILIZED BY THE LIGAND

Classic models for explaining enzyme regulation by allosteric ligands are based on the assumption of a change between at least two conformational states, and this model assumes an equilibrium between inactive and active forms of the enzymes (59, 60). The binding of positive effector ligands stabilizes the active form, whereas inhibitors stabilize the inactive form. When enzymes are formed as oligomers with two or more subunits, binding of a ligand, such as the substrate, to one subunit can communicate a conformational change to an adjoining subunit, leading to an observed change in activity, designated as cooperativity (61). Consistent with such a simple conformational model, dissociation and reassociation is a process that can lead to a different conformation (Fig. 3A): if the intrinsic conformation of the dissociated monomer is inactive, then interaction between subunits in the associated tetramer could serve to stabilize an active conformation. Because substrate (S) binds preferentially to the tetramer, it would stabilize this active oligomer.

An alternative mechanism for utilizing dissociation and association to regulate enzyme activity comes from the ability of some enzymes to form a functional ligand binding site between two subunits. As diagrammed in Fig. 3B, the sites for the two substrates, A and B, in the dissociated subunit are spatially too far apart to permit the chemical interaction for catalysis to occur. However, in the dimer the two substrate binding sites are now appropriately positioned to facilitate binding of A next to B. For simplicity, Fig. 3B was drawn to emphasize the positions of the A and B sites, but this clearly also involves some conformational changes. Therefore, any process that

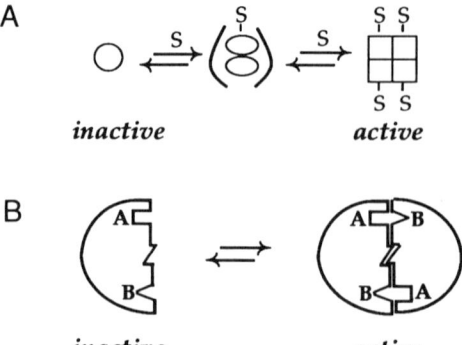

FIG. 3. Models for loss of activity by dissociation due to conformational change (A) or by disruption of the catalytic site (B). The dimeric species in (A) is a transient intermediate.

shifts the equilibrium in Fig. 3B directly leads to a change in enzyme activity. The two mechanisms illustrated by Figs. 3A and 3B are not mutually exclusive, and some enzymes appear to include aspects of both.

2. ENZYMES WITH LIGANDS BINDING BETWEEN SUBUNITS

Though Fig. 3B may appear speculative, there is evidence emerging for the binding of ligands between subunits in a few crystal structures. This feature is slightly more complex than illustrated in Fig. 3B, which illustrates a complete catalytic site between subunits, but shows each ligand bound to only one subunit. More realistic would be the binding of a ligand in a pocket between two subunits, with residues from each subunit required for proper binding. Eight examples of a substrate binding at a catalytic site between subunits are summarized in Table IV. There is one example of a substrate binding at a regulatory site between subunits (hexokinase), and several examples of an allosteric effector binding between subunits.

Of the enzymes in Table IV, those shown to dissociate and reassociate in response to physiologically relevant effectors include phosphofructokinase (62–69), hexokinase (70–72), and glycogen phosphorylase (73–75). For the other enzymes in Table IV, no systematic studies have explored if these enzymes undergo reversible dissociation of the oligomer under physiological conditions. Nevertheless, such structural studies demonstrate that ligands do bind between subunits in dimers or larger multimers, and thereby directly demonstrate that a simple physical mechanism for ligand binding between subunits occurs in a variety of enzymes. Because most of the enzymes in

TABLE IV
LIGAND BINDING BETWEEN SUBUNITS, DETERMINED IN CRYSTAL STRUCTURES

Enzyme	Source	Site (ligand)		Ref.
		Catalytic	Regulatory	
Aspartate carbamoyl transferase	*Escherichia coli*	Carbamoyl-P	—	262
Cytidine deaminase	*Escherichia coli*	Cytidine	—	263
β-Galactosidase	*Escherichia coli*	Disaccharide	—	264
Glutamine synthetase	*Salmonella typhimurium*	NH_4^+	—	265
Glycogen phosphorylase	Rabbit muscle	—	AMP	266, 267
Hexokinase	Yeast	—	ATP	72
Orotate phosphoribosyl-transferase	*Salmonella typhimurium*	P-Rib-PP	—	76
Phosphofructokinase	*Bacillus stearothermophilus*	F-6-P	ADP; PEP	268–271
Phosphofructokinase	*Escherichia coli*	F-6-P	ADP	272

Table IV are allosterically regulated, it is then evident how a simple change in the alignment of subunits with their neighbors can form or disrupt the specific ligand binding site.

It is also relevant to the present review that one of the enzymes in Table IV, orotate phosphoribosyltransferase, is the bacterial enzyme that corresponds to one of the two catalytic domains of UMP synthase. The structure available for this enzyme predicts that it can only be active as a dimer, because residues from both subunits appear necessary to form a functional catalytic site at the dimer interface (76). Because kinetic studies show that UMP synthase is only active as the dimer (77), as described in Section III,D, the protein structure that is emerging for the bacterial OPRTase may serve as a model for the structure and function of this domain in the mammalian UMP synthase.

III. UMP Synthase[2]

This bifunctional protein contains OPRTase and ODCase activities (Fig. 4). The oldest organisms having UMP synthase are the protista, such as *Dictyostelium discoideum* (78). UMP synthase has an OPRTase domain at the N-terminal end of the protein, which is attached to the ODCase domain that occupies the C-terminal end. Therefore, a review of UMP synthase is simultaneously a review of the new literature for monofunctional OPRTases and ODCases.

The discovery that UMP synthase is a bifunctional protein began with the discovery of orotic aciduria (44, 79), a human genetic disease. The mutant gene causes the simultaneous loss of both OPRTase and ODCase activities in all but one patient. The proofs that both enzyme activities were part of a single protein came from the simultaneous cosedimentation of the two enzyme activities in sucrose gradients (80) followed by the isolation of a pure protein having both enzyme activities from mouse Ehrlich ascites cells (81) and from human placenta (82). In addition, Suttle *et al.* (83) demonstrated the amplification of the UMP synthase gene in cultured cells. They constructed a cDNA that produced a human protein having both enzyme activities.

Patients with orotic aciduria have two mutant autosomal genes, one from each parent, on chromosomes 3, at position q13 (44, 84). These patients are usually detected because of a failure to thrive and a marked megaloblastic

[2] This review covers discoveries since 1980. It is not exhaustive but highlights the major and dramatic progress that has been made on the structure and function of UMP synthase and its two constituent enzyme centers in this bifunctional protein and in monofunctional OPRTases and ODCases.

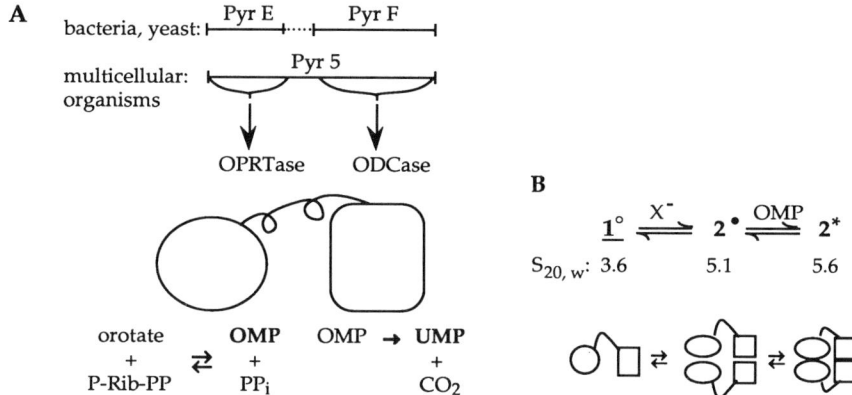

FIG. 4. Domain structure of UMP synthase. (A) Two separate genes in microorganisms code for two proteins that are homologous to the two domains of the single-gene product for UMP synthase in multicellular eukaryotes. (B) Association of the inactive UMP synthase monomer to form the active dimer by sedimentation in the presence of allosteric effectors (see Fig. 2 for notation); conformational changes and interactions of domains are suggested.

anemia that can be followed by death if the child does not receive effective uridine therapy, either as a dietary supplement or by intramuscular injections. Megaloblastic anemia is a disease that occurs whenever DNA synthesis in the bone marrow is slow. It occurs in patients with normal genes when their diet lacks sufficient B_{12} or folic acid, and it is fully cured in these patients when the missing vitamin is added to their diet. However, for children and adults with two defective UMP synthase genes, these two agents have no effect (nor does Fe^{2+} feeding). Only dietary uridine, which can be converted to UMP by the salvage pathway, will suffice to relieve the anemia and the other reversible results of this genetic disease. Lymphopenia is frequently observed in these patients as well as low serum IgG, IgA, and borderline IgM concentrations (85, 86). Both IgG and IgM levels respond to uridine, but IgA and the lymphopenia do not (86). Failure to thrive prior to uridine feeding and various growth abnormalities have also been observed.

The only other mammalian species known to have orotic aciduria is cattle, in which at least one mutant UMP synthase gene (87) was originally selected by a breeding program because heterozygous cows with one good and one defective gene were high milk producers. Attempts by appropriate matings to produce a calf having a defective gene for UMP synthase on both autosomal chromosomes have failed so far and it appears likely that the homozygous embryos are reabsorbed sometimes before 40 days into the pregnancy (87). The uniformity of the time period when the embryo loss occurs has led investigators to suggest that a homozygous bovine fetus is at greater risk than a homozygous human embryo, because in cattle there are two membranes that nutrients from maternal blood must cross to enter the fetal calf blood stream, whereas in human placentas there is only a single membrane separating the maternal and fetal blood streams. However, if some homozygous human mutant embryos are similarly reabsorbed it could explain why this hereditary disease is rare (44). Another reason could be that a child needs to enter a rather sophisticated medical center to be properly diagnosed and treated.

These patients, before treatment, excrete approximately between 1 and 10 mmol of orotic acid per day; the range presumably represents the degree to which their particular mutation causes a loss of the enzyme activities of UMP synthase. For comparison, normal healthy humans excrete about 10 μmol of orotic acid per day, presumably because the orotic acid formed by the first four enzymes of the *de novo* UMP biosynthetic pathway is effectively converted to UMP in the cell where it was synthesized. Because orotic acid does not enter most cells readily, as described in Section I,A, investigators who have studied its metabolism in whole cells in detail have permeabilized cells with dextran sulfate 500 and then incubated them briefly with orotic acid (or PRPP) followed by immediate resealing of the cells prior to the experiment, to avoid protein losses (88).

A. Amino-acid Sequences of UMP Synthase and of Monofunctional OPRTases and ODCases

The known sequences include six UMP synthases derived from humans, cattle (*Bos taurus*), *Drosophila melanogaster*, tobacco, and two protista organisms, *Dictyostelium discoideum* and *Naegleria gruberi*. UMP synthases has also been sequenced in plants, for example, wheat (89). The sequence for monofunctional OPRTase includes six bacteria and eight eukaryotic species (mainly fungi). There currently are 25 monofunctional ODCase sequences known; 6 bacteria and 19 simple eukaryotes, mainly fungi. In addition there are 6 OPRTase and ODCase domains as parts of the UMP synthases listed above and an ODCase domain from mouse UMP synthase.

We have summarized these data in Table V, where the top line gives the amino-acid sequence for human UMP synthase predicted from the cDNA (83). Certain of the amino acids are underlined, and these amino acids are important in the binding of substrates and/or in the catalysis of the two enzyme reactions. In a few cases, two amino acids are listed, e.g., Y/F. In this case, these are the only two amino acids at this position in all sequences, and the first-listed amino acid (e.g., Y of Y/F) is in the human protein. Underlined amino acids in Table V are residues with defined functions. Below the amino-acid sequences in Table V are three lines with symbols showing a conserved position (+); dots (·) allow proper spacing within

TABLE V
AMINO-ACID SEQUENCE FOR HUMAN UMP SYNTHASE[a]

Amino-acid sequence

OPRTase

Domain	1	2	3	4	5	6	7	8	9	10	11	12	13	14	15	16	17	18	19	20	21	22	23	24	25	26	27	28	29	30
	M	A	V	A	R	A	A	L	G	P	L	V	T	G	L	Y	D	V	Q	A	F	K	E/L	G	D	F/Y	V	L	K	S/V
All species	.	.	.	.	.	.	.	.	.	.	.	.	.	.	.	.	.	.	.	.	.	.	.	.	.	.	.	.	.	.
All eukaryotes	.	.	.	.	.	.	.	.	.	.	.	.	.	.	.	.	.	.	.	.	.	.	.	.	.	+	.	.	.	+
All UMP synthases	.	.	.	.	.	.	.	.	.	.	.	.	.	.	.	.	.	.	.	+	.	+	+	.	+	+	.	.	+	+

Domain	31	32	33	34	35	36	37	38	39	40	41	42	43	44	45	46	47	48	49	50	51	52	53	54	55	56	57	58	59	60
	G	L	S	S	P	I	Y/F	I	D/N	L	R	G	I	V	S	R	P	R	L	L	S	Q	V	A	D	I	L	F	Q	T
All species	.	.	+	+	.	.	+	.	+	.	.	.	.	.	.	.	.	.	.	.	.	.	.	.	.	.	.	.	.	.
All eukaryotes	.	.	+	+	.	.	+	.	+	.	.	.	.	+	.	.	.	.	.	.	.	.	.	.	.	.	.	.	.	.
All UMP synthases	.	.	+	+	.	.	+	.	+	.	.	.	.	+	.	.	.	.	.	+	.	.	.	.	.	.	.	.	.	.

Domain	61	62	63	64	65	66	67	68	69	70	71	72	73	74	75	76	77	78	79	80	81	82	83	84	85	86	87	88	89	90
	A	Q	N	A	G	I	S	F	D	T	V	C	G	V	P	Y	T	A	L	P	L	A	T	V	I	C	S	T	N	Q
All species	.	.	.	.	.	.	.	.	.	.	.	+	.	.	.	.	.	.	.	.	.	.	.	.	.	.	.	.	.	.
All eukaryotes	.	.	.	.	.	.	.	.	.	.	.	+	.	.	.	.	.	.	.	.	.	.	.	.	.	.	.	.	.	.
All UMP synthases	.	.	.	.	.	.	.	.	.	.	+	+	+	+	+	+	+	.	.	+	.	.	.	.	.	.	.	.	.	.

Domain	91	92	93	94	95	96	97	98	99	100	101	102	103	104	105	106	107	108	109	110	111	112	113	114	115	116	117	118	119	120
	I	P	M	L	I	R	R	K/S	E	T	K	D	Y	G	T	K	R	L	V	E	G	T	I	N	P	G	E	T	C	L
All species	.	.	.	.	.	.	.	+	.	.	+	.	.	.	.	.	.	.	.	.	+	.	.	.	.	.	.	.	.	.
All eukaryotes	.	.	.	.	.	.	.	+	+	.	+	.	.	.	.	.	.	.	.	.	+	.	.	.	.	.	.	.	.	.
All UMP synthases	.	.	+	.	.	.	.	+	+	.	+	.	.	.	.	.	.	.	.	.	+	.	.	.	.	.	.	.	.	.

Domain	121	122	123	124	125	126	127	128	129	130	131	132	133	134	135	136	137	138	139	140	141	142	143	144	145	146	147	148	149	150
	I	I	E/D	D/V	L	V	I/S	S	G/C	S	S	V	L	E	T	V	E	V	L	Q	K	E	G	L	K	V	T	D	A	I
All species	.	.	+	+	.	+	+	.	+	.	.	.	.	.	.	.	.	.	.	.	.	.	.	.	.	.	.	.	.	.
All eukaryotes	.	.	+	+	.	+	+	.	+	.	+	.	.	.	.	.	.	.	.	.	+	.	.	.	.	.	.	.	.	.
All UMP synthases	.	.	+	+	.	+	+	.	+	.	+	.	.	.	.	.	.	.	.	.	+	.	.	.	.	.	.	.	+	.

(continued)

TABLE V (Continued)

Domain	Amino-acid sequence[a]
	V L L D R E Q G G K D K L Q A H G I R L H S V C T L S K M L180
All species	. . . .
All eukaryotes	. . . . + ...
All UMP synthases	+ . . . + ...
	E I L E Q Q K K V D A E T V G R V K R F I Q E N V F V A A N210
All species	. . . .
All eukaryotes	. . . .
All UMP synthases	. . + ... + ...

ODCase

Domain	Amino-acid sequence[a]
	H N G S P L S I K E A P K E L S F G A R A E L P R I H P V A240
All species	. . . .
All eukaryotes	. . . . + ... + ... +
All UMP synthases	. . . . + ... + . . + ... +
	S K L L R L M Q K K E T N/L/V C L S A D V S L A R E L L Q L A270
All species	. . . .
All eukaryotes	. . . . + . . . + + ... + ... + . . .
All UMP synthases	. . + ... + . . . + + . . . + + ... + . . +
	D/E A L G P S/I/V C M L K T H V D I L N D F T L D V M K E L I T300
All species	+ . . . . + ...
All eukaryotes	+ . + . . + ... + + + . + + . + ... + . + . .
All UMP synthases	+ . + . . + + . . . + + + . + + . + ... + . . + . . .

	L	A	K	C	H	E	F	L	I	F	E	D	R	**K**	F	A	D	I	G	N	T	V	K	K	Q	Y/F	E	G	G	I$_{330}$
All species	.	.	.	.	.	.	.	.	.	.	.	+	+	+	.	.	.	.	.	.	.	.	.	.	.	.	.	+	+	.
All eukaryotes	.	+	.	+	+	+	.	.	.	.	+	+	+	+	+	+	+	+	+	+	+	+	+	+	+	+	+	+	+	.
All UMP synthases	.	+	.	+	+	+	.	.	.	.	+	+	+	+	+	+	+	+	+	+	+	+	+	+	+	+	+	+	+	.
	F	K	I	A	S	W	A	D	L	V	N	A	H	V	V	P	G	S	G	V	V	K	G	L	Q	E	V	G	L	P$_{360}$
All species	.	.	.	.	.	.	.	.	.	.	.	.	.	.	.	.	.	.	.	.	.	.	.	.	.	.	.	.	.	.
All eukaryotes	.	.	+	.	.	.	.	.	.	.	.	.	.	.	.	.	+	.	.	.	.	.	.	.	.	.	.	.	.	.
All UMP synthases	.	.	+	.	.	.	+	.	.	.	.	.	.	.	.	.	+	.	.	.	.	.	.	.	.	.	.	.	.	.
	L	H	R	G	C	L	L	I	A	E	M	S	S	T	G	S	L	A	T	G	D	Y	T	R	A	A	V	R	M	A$_{390}$
All species	.	.	.	.	.	.	.	.	.	.	.	.	.	.	.	.	.	.	.	.	.	.	.	.	.	.	.	.	.	.
All eukaryotes	.	.	.	.	.	.	.	.	.	.	.	.	.	.	.	.	.	.	.	.	.	.	.	.	.	.	.	.	.	.
All UMP synthases	.	.	.	.	.	.	.	.	.	+	.	.	+	+	.	.	.	.	.	.	.	.	.	.	.	.	.	.	.	.
	E	E	H	S	E	F	V	V	G	F	I/V	S	G	S	R	V	S	M	K	P	E	F	L	H	L	T/S/P/Y	G	V	Q$_{420}$	
All species	.	.	.	.	.	.	.	.	.	.	.	.	.	.	.	.	.	.	.	.	.	.	.	.	.	+	.	.	.	
All eukaryotes	.	.	.	.	.	.	.	.	.	.	+	.	.	.	.	.	.	.	.	.	.	.	.	.	.	+	+	+	+	
All UMP synthases	.	.	.	.	.	.	.	.	+	+	+	.	.	.	.	.	.	.	.	.	.	.	.	.	.	+	+	+	+	
	L	E	A	G	G	D	N	L	G	Q	Q	Y	N/S/T	P	Q	E	V	I	G	K	R	G	S	D	I	I	I	V	G$_{450}$	
All species	.	.	.	.	.	.	.	.	.	.	.	.	.	.	.	.	.	.	.	.	.	.	.	.	.	.	.	.	+	
All eukaryotes	.	.	.	.	.	+	.	.	+	.	+	.	+	.	.	.	.	.	.	.	.	.	+	.	.	.	.	.	+	
All UMP synthases	.	.	.	.	.	+	.	.	+	.	+	.	+	.	.	.	.	.	.	.	.	.	+	.	.	.	+	+	+	
	R	G	I	I	S	A	A	D	R	L	E	A	A	E	M	Y	R	K	A	A	W	E	A	Y	L	S	R	L	G	V$_{480}$
All species	.	.	.	.	.	.	.	.	.	.	.	.	.	.	.	.	.	.	.	.	.	.	.	.	.	.	.	.	.	.
All eukaryotes	+	+	.	.	.	.	.	.	.	.	.	.	.	.	.	+	.	.	.	.	.	+	.	+	.	.	.	.	.	.
All UMP synthases	+	+	+	.	.	.	.	.	.	.	.	.	.	.	.	+	+	.	.	.	.	+	.	+	.	.	.	.	.	.

[a] Known functional amino acids are in bold type; identical amino acids occurring in sequences of all proteins studied with OPRTase or ODCase activity are indicated by "+."

the 10 amino-acid positions. The bold symbols clearly cluster in a few important areas. The relatedness of sequences is greater the closer the species are related to one another. This does not show up in Table V for bacteria because of the way the table is constructed. However, it does show up in the other two categories; for example, the amino-acid sequences of UMP synthases from *D. discoideum*, *D. melanogaster*, and wheat (F. Lacroute, personal communication) share 47, 47, and 56% identity with human UMP synthase (when each sequence is compared solely to the human sequence). In contrast, when the sequences of the monofunctional OPRTase and ODCase of either *S. cerevisiae* or *S. typhimurium* are compared to the OPRTase and ODCase domains of UMP synthase, *S. cerevisiae* OPRTase and ODCase have 30 and 50% identical amino acids in the same regions as human UMP synthase, whereas *S. typhimurium* OPRTase and ODCase have only 21 and 28%, respectively.

Some of these cDNAs have been used to construct vectors that allow the overproduction of *S. typhimurium* OPRTase (90), the human OPRTase domain (91, 92), yeast ODCase (93, 94), the mouse and human ODCase domains (91, 92), and intact human UMP synthase (95). These vectors have made these previously rare proteins amenable to the structure/function studies described in this section.

The monofunctional OPRTases are approximately 23,000 Da in mass. The monofunctional ODCases vary considerably in size, but most are about 28,000 to 30,000 Da. Human UMP synthase has a mass of 52,133 Da calculated from the amino-acid sequence predicted by the cDNA. It also takes into account the post-translational modification of the amino terminus of this protein, namely, the removal of the terminal methionine followed by acetylation of the formerly penultimate alanine (96). Unpublished studies by L. Livingstone, M. E. Perry, and M. E. Jones of lymphocytes grown in tissue culture with ^{32}P demonstrate that UMP synthase is not phosphorylated under conditions where the trifunctional protein CAD (dihydroorotate synthase) is phosphorylated (97). Because denatured UMP synthase migrates by size-exclusion chromatography as if its mass were 52,000 Da, it would appear that the modification of the N-terminal sequence may be the only major post-translational modification affecting protein size.

UMP synthase from either mouse or human tissues (81, 82) undergoes post-translational modifications that do not shorten the protein detectably, but do change the net charge on the exterior surface of the protein so that there are several isoelectric species in the pure protein preparations. These isozymes occur in concentrations that exceed the 5% of extraneous proteins in the "pure" UMP synthase preparations. All isozymes react with antibody against pure UMP synthase, whereas the extraneous proteins do not We think that these isozymes may represent "aging" of the original gene product in such a manner that the major and most basic protein band loses basic or neutral groups to produce successively smaller amounts of several more acidic isozymes.

The only three-dimensional structure known is that of S. *typhimurium* OPRTase with OMP bound at the active site (76). Several crystal forms of apo-OPRTase of *E. coli* have been reported (98). Yeast ODCase has also been crystallized (99) with the mononucleotide of barbituric acid, an analog of the substrate OMP, and a tightly bound inhibitor of ODCases (K_d of 10^{-12} M) (100).

Until more three-dimensional structures are solved, it is very hard to decide where the OPRTase domain of UMP synthase ends and where the ODCase domain begins, and whether there is a large or small "linker" region. Table V shows that, for human UMP synthase, the last essential amino acid of the OPRTase domain is presumably R155. R155 of human UMP synthase is equivalent to R156 in *S. typhimurium* OPRTase, which interacts with the carboxyl group of orotic acid. After this point, as Table V indicates (and as the analysis of all sequences by the Genetics Computer Group program "PILEUP" confirms), there is not much amino-acid identity among the known OPRTases and OPRTase domains. A secondary structure analysis may reveal that the latter sequences, despite the lack of amino-acid identities, do produce similar secondary structures. Indeed, in four of the bacterial species, including two from bacilli, it is hard to find an arginine residue that one can clearly equate to R156 of the OPRTase of *S. typhimurium*, even though an equivalent R is in all eukaryotic OPRTases and in all UMP synthases. The ODCases and the ODCase domain of UMP synthase have very few identical amino acids near the C-terminal end of the OPRTase domain, as indicated by the lack of identities between R155 of human UMP synthase and the aspartate, D269, which is definitely in the ODCase domain of all species that have been sequenced. R230 is invariant in the ODCases of all eukaryotes. The sequence from L183 through A229 is devoid of any sequence of identities for all known species, either for OPRTases, ODCases, or UMP synthases.

Fragments of human or mouse UMP synthase having only one enzyme activity support the idea that the linking region may be located in this area. The first ODCase domain of UMP synthase was obtained in pure form by controlled proteolysis with either trypsin or elastase of pure mouse UMP synthase (101). The fragment had a mass of 28,500 Da by SDS gel electrophoresis and had no OPRTase activity. In fact, the proteolysis conditions led to an early and steady loss of OPRTase activity but no accompanying loss of ODCase activity (101). This 28,500-Da ODCase domain was established to be at the carboxyl-terminal end of the protein by the first plasmid that had the 3' end of a cDNA fragment for mouse UMP synthase (102). Expression in

E. coli JFS116 (*102*) or yeast (*103*) of a vector having this cDNA fragment attached to a linker sequence produced an ODCase that allowed growth of mutants lacking this enzyme. The arginine codon equivalent to R230 was the first amino acid of the ODCase domain in the vector sequence. The protein expressed from this plasmid had a mass of 28,500 Da and had ODCase but no OPRTase activity.

More recently, human UMP synthase cDNA has been cleaved in two laboratories to give defined DNA segments. In one, a fragment was prepared producing an OPRTase that terminated at the codon for P222 (*91*). This vector, having the 5' end of the cDNA, produced only 4–9% of the normal OPRTase activity expected when expressed in an OPRTase$^-$ *E. coli* strain; the protein had no ODCase activity and did not allow growth of an ODCase$^-$ cell line. The second vector, coding from amino acid A221 through amino acid V480 of human UMP synthase, had only ODCase activity and was overproduced in an ODCase$^-$ *E. coli* mutant that thereafter had 3- to 24-fold more ODCase activity than the normal strain. More recently, we (*92*) produced two baculovirus vectors: the first contains the amino acid from M1 of the human sequence through G213; the second has a vector with a methionine codon inserted preceding the human UMP synthase sequence I218 through V480. The first vector produces a protein with only OPRTase activity, and the second a protein with only ODCase activity. All of this work places the location of the linker region but does not establish the minimal peptide length for functional OPRTase or ODCase domains.

B. OPRTase Structure and Function

The purification (*90*) and crystallization (*76, 104*) of *S. typhimurium* OPRTase has been achieved using a vector containing the open reading frame for this protein. Expression of the vector produces *S. typhimurium* OPRTase in large quantities. Studies of the kinetics of the various purine and pyrimidine phosphoribosyl transferases have given differing results (see *90*). These reactions all occur with anomeric inversion at C-1 of the ribosyl moiety. There are two possible mechanisms: (a) a direct attack of N-1 nitrogen of orotate at C-1 of PRPP and the displacement of pyrophosphate to give the b anomer; (b) the formation of an enzyme-bound PRPP whereby the pyrophosphate is eliminated to form a carboxonium-type intermediate followed by an orotate attack on the intermediate to give OMP.

For the enzyme from *S. typhimurium*, the authors found that the "pingpong" kinetics previously reported for this reaction are incorrect. However, repurification of the substrate, commercial OMP, to remove inorganic pyrophosphate was necessary to eliminate the orotate–OMP exchange reaction observed prior to the OMP repurification. The enzyme also had to be repurified because it apparently binds OMP tightly enough throughout purification so that a PRPP–PP$_i$ exchange occurred without addition of either orotate or OMP. Denaturation of the OPRTase with urea, followed by both gel filtration of the protein to remove OMP and renaturation of the protein after gel filtration, eliminated the second exchange reaction. As a result of having pure enzyme and pure OMP, it was demonstrated (90) that the OPRTase reaction proceeds by purely random kinetics, a result compatible with a direct attack of orotate on PRPP.

In a second study (105) it was demonstrated that Mg^{2+} is probably required to bind PRPP (and PP$_i$) but is not required to bind orotate and OMP to the enzyme. The authors also found that the PRPP–PP$_i$ exchange appears to be rate-limiting and therefore the release of either of these two ligands may control the rate of the reaction. The K_{eq} for the forward reaction is 0.12, 0.07, or 0.1 (90, 106, 107) using assays of the concentration of one or more of the reactants at equilibrium. Only the value of 0.1 was obtained with a pure OPRTase (90). These results indicate that the reaction favors the accumulation of orotate and PRPP.

Two other studies come to this same conclusion (108, 109), but have much larger K_{eq} values. Both of these studies were done with impure yeast enzyme. In one case, NMR was used to measure the concentration of the P-containing reactants; the K_{eq} obtained was 0.7 (109). In the second case, the K_{eq} observed was 0.49 (108); however, this value was calculated from rate constants and K_m values that assumed a ping-pong mechanism, which appears to be an incorrect assumption (90).

The crystal structure of S. typhimurium OPRTase with bound OMP is by far the most signal addition to our knowledge of the OPRTase reaction, and for that matter for this class of enzymes in general (76). Some of the results found in the X-ray studies were anticipated from an earlier study of inactivation of OPRTase by modifying lysine residues with 2,4,6-trinitrobenzene sulfonate (110). The authors noted that of the 25 basic residues (12 lysines and 13 arginines) in S. typhimurium OPRTase, only 5 basic residues were invariant in the amino-acid sequences of all species then known, from S. typhimurium to humans. These were S. typhimurium K26, R99, K100, K103, and R156. Indeed the inactivation of OPRTase with 2,4,6-trinitrobenzene sulfonate occurred by modification of K26, K100, K103, and K100 + K103. The modification of these three lysine residues was reduced when protective ligands were available, with PRPP > OMP > orotate + PP$_i$ > orotate > PP$_i$ > no substrate addition. By studying each of these modifications alone and observing which substrate or substrates protected a given lysine residue from modification, the authors concluded that both PRPP and PP$_i$ protect K100 and K103, although PP$_i$ was not as effective as PRPP. Lysine 26 was not protected by PP_i, but was also protected by OMP and by PRPP, suggesting that this residue recognizes the 5' phosphate of both OMP and PRPP as well as the orotate ring.

The X-ray structure (76) reveals that OPRTase with OMP bound is a dimer, with 1 mol of OMP bound to each monomeric unit. The orotate of OMP is in a solvent-inaccessible crevice formed by a "hood" structure (Table V) whose equivalent human structure is composed of the 17-amino-acid sequence from F/L23 to D/N39 (*S. typhimurium* F20 to N37). The orotate ring is stacked under F24 of the *Salmonella* protein (which is equivalent to Y27 in the human sequence). Hydrogen bonds to the main peptide chain anchor the orotate N3, orotate O4, and the orotate-carboxylate forms a fourth hydrogen bond to the main-chain N of K26 of *S. typhimurium* (K29 of the human protein). The amino-acid sequence alignments for OPRTase structure show a K in 15 of the 20 OPRTase sequences at this position. Replacements are one N, one S, one T, and two As. It is interesting and perhaps significant that a PILEUP analysis of sequence elements shows that K26 precedes an S in all OPRTase sequences except one, where the amino-acid equivalent to S27 is a V. The S27 as well as K26 may also be important and the KS, XS (where X = N, S, T, or A), or KV arrangements may all work well. The orotate C5 is tightly packed in the hood crevice near R156 (R155 in the human sequence), precluding most bulky substituents; however, it does not exclude F^- because 5-F orotate can be converted to a mononucleotide by the OPRTase of UMP synthase (*111*).

The ribose oxygen 2' and 3' groups are in a solvent-accessible region and form H-bonds to the amino group of K26 (K29 of the human protein). The ribosyl ring-O and the 5'-O are hydrogen-bonded to T128 (T127 in the human protein), which is well conserved. A pocket formed from V126–T131 of the *S. typhimurium* sequence holds the 5' phosphate; T128 to T131 (probably S131 or S132 in the human sequence) interacts with the 5' phosphate by the T side-chain oxygens. The whole human sequence from E123 through S131 is important because backbone N-atoms of six of these residues interact with phosphate oxygens. The exact role of the ED or DD sequence, which is so readily recognized in all PRTases, was not revealed by this X-ray structure. The authors suggest it may be a site where Mg^{2+} is bound and that Mg^{2+} chelates with PP_i, or the PP_i group of PRPP. In the structure of OPRTase obtained with OMP, no Mg^{2+} was included, so that ED residues were not seen to bind the ligand.

One should read the original paper (76) to have a full appreciation of the multiple interactions. Experiments showing how the various residues that bind substrates in the "hood" region and in the residue 123–131 sequence are elegant and extensive. The excellent fit between the amino acids that are rigorously conserved in the 19 known OPRTase sequences from a very diverse number of species agrees well with the importance of these residues as viewed in the OPRTase X-ray structure, and suggests that while one waits for appropriate crystals, such amino acids can be worth examining by biotechnological or chemical modification studies.

C. ODCase Structure and Function

The use of biotechnology to prepare large amounts of pure yeast ODCase (93, 94) has made it possible to use techniques such as crystallography (99), NMR (40), and $^{13}C/^{12}C$ isotope studies (112) to solve the mechanism of this interesting decarboxylation reaction (112–115).

The decarboxylation mechanism was correctly deduced in 1976 (116) from studies of the nonenzymatic rates of decarboxylation of a number of orotate derivatives. It was proposed that the noncovalent mechanism illustrated below, in which the substrate, OMP (A), is protonated by the enzyme. This forms a zwitterionic transition state intermediate (B), which is decarboxylated to yield CO_2 and a nitrogen ylide (C). Loss of the proton from the enol

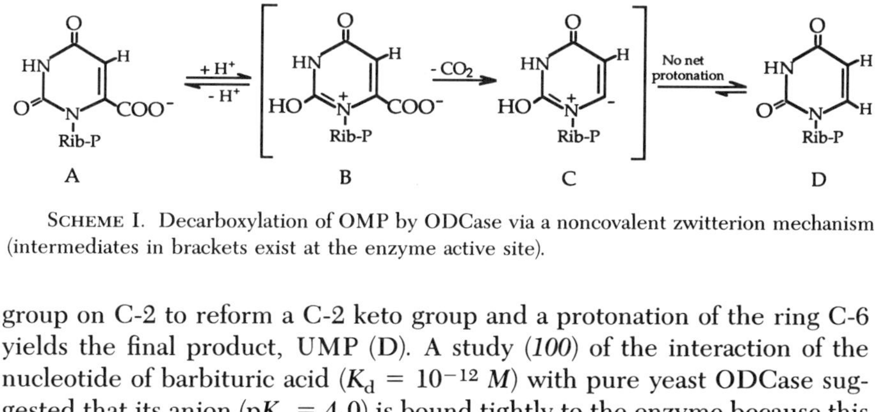

SCHEME I. Decarboxylation of OMP by ODCase via a noncovalent zwitterion mechanism (intermediates in brackets exist at the enzyme active site).

group on C-2 to reform a C-2 keto group and a protonation of the ring C-6 yields the final product, UMP (D). A study (100) of the interaction of the nucleotide of barbituric acid ($K_d = 10^{-12}$ M) with pure yeast ODCase suggested that its anion ($pK_a = 4.0$) is bound tightly to the enzyme because this nucleotide resembles the zwitterion of OMP (B above). Barbituric monoribonucleotide could therefore be considered a transition-state analog.

The ODCase reaction is unusual among enzymatic decarboxylations because there is no cofactor requirement (i.e., neither thiamin nor pyridoxal phosphate is required) and no metal need be added, nor is the enzyme a metalloprotein (114). The affinity of ODCase for the zwitterion in the transition state is estimated to be near 10^{-24} M (115), a remarkable affinity. The proof that the zwitterion mechanism is correct for ODCase was achieved by using pure yeast ODCase (94) in two separate studies (112, 113). The first was to see if an addition of an enzyme nucleophile to the C-5 ring-carbon of OMP was followed by decarboxylation with subsequent elimination of the enzyme, as suggested by Silverman and Groziak (*117*). When OMP with ^{13}C in C-5 in its orotate ring was synthesized (*113*), NMR studies showed that the enzyme does not catalyze a transition in ring C-5 from trigonal (sp^2) to tetrahedral (sp^3), which would be essential for this mechanism. The replacement of the proton on ring C-5 with deuterium also did not affect the rate of the reaction (*113*).

The second approach was more positive, namely, a study of ^{13}C/^{12}C isotope effects as one changes the temperature, pH, and solvent for the reaction (*112*). Such studies can clearly distinguish the mechanism of Silverman and Groziak (*117*) and that proposed by Beak and Siegel (*116*). In the first case, the ^{13}C isotope effects would be small, but for the zwitterion mechanism the ^{13}C isotope effect would be large. The data were clear cut; the enzyme reaction occurs by the zwitterion mechanism. In addition, the changes in k^{12}/k^{13} vs. pH and in log V_{max}/K_m vs. pH showed that the acidic (below pH 7) and basic (above pH 7) reactions differ. At acid pH, the enzyme-bound transition-state intermediate forms, but catalysis cannot proceed when a catalytic amino acid is protonated. At basic pH, the enzyme binds substrate but cannot form the zwitterion transition-state intermediate, which has lost the essential proton and, therefore, can no longer donate this proton to C-2 of the orotate ring. The pK_a values of the two amino acids participating in these reactions are so near one another (near pH 7) that a single peak (not a plateau) is observed in the graph of V_{max}/K_m vs. pH.

It seemed possible that the protonated catalyst that becomes uncharged might be a lysine residue. Because K93 of the yeast ODCase (K314 in the human UMP synthase sequence) is in the most highly conserved peptide segment of the protein (Table V), we chose to replace the codon for this residue with as many other codons as possible (*118*). All amino acids except glutamate, phenylalanine, and proline were placed in this position, and all 17 analog proteins produced were inactive. We chose to use for further study the cysteine analog protein, i.e., K93 → C93 (referred to as K93C), because Smith and Hartman (*119*) and Planas and Kirsch (*120*) had shown that a cysteine residue replacing a lysine residue can be treated with Br-ethylamine to yield a partially active protein containing an S-(2-aminoethyl)cysteine, an isostere of lysine. Indeed the K93C analog ODCase can react with Br-ethylamine if the yeast ODCase is denatured with urea, treated with Br-ethylamine, and then renatured after the interval needed for the chemical modification.

The K93C protein had no apparent activity even when the activity test was made so sensitive that a 10^{-6} reduction in normal ODCase activity could be detected. We can, therefore, state that, *if* K93C has any activity, it must be less than 10^{-6} of normal activity. The chemically modified analog of K93C recovered 10^4 or more activity so that yeast ODCase with an S-(2-

aminoethyl)cysteine residue had 1% of the ODCase activity of the wild-type protein. When a similar chemical reaction was done with 2-bromoethanol, 2-bromoacetamide, or trimethyl (2-bromoethyl) ammonium bromide, no detectable activity resulted (*118*). Although the degree to which the K93C cysteine group was modified has not been directly assayed, this result suggests that the amine group of the ethylamine added to C93 in K93C is necessary for reactivation.

Dilute yeast ODCase at 0.5 μM is a monomer, whereas concentrated yeast ODCase at 35 μM is a dimer. These concentrations are the initial values of enzyme samples that are layered onto sucrose gradients for sedimentation studies; during the sedimentation the enzyme becomes diluted 5- to 10-fold as it moves through the gradient. This demonstrates the ability of the protein subunits to associate at increasing concentrations of subunits. If the substrate, OMP, is added to dilute yeast ODCase in the sedimentation buffer, a dimer is formed (*94, 118*). This dimerization process is similar to that described for UMP synthase, which is illustrated in Fig. 4. If the product (UMP) or an OMP analog (such as 6-azaUMP) is added to the dilute yeast ODCase solution, the dimer is also formed (*94,118*). We could use this test to establish whether the K93C analog protein binds these nucleotides. Indeed K93C does bind nucleotides, but the specificity for the nucleotides 6-aza-UMP and UMP differs from the normal (K93) enzyme. The normal protein (at 0.3 μM) forms a dimer in the presence of 0.5 μM 6-azaUMP and 80 μM UMP, whereas K93C requires 120 μM 6-azaUMP and only 0.5 μM UMP.

These changes fit well with the studies by Levine *et al.* (*100*) on the nature of the binding of UMP, 6-azaUMP, and barbituric monoribonucleotide (BMP) with yeast ODCase. These authors proposed that the anion of OMP, 6-azaUMP, and UMP is bound to ODCase. K93 could be the protonated residue of yeast ODCase that aids the positioning of the pyrimidine ring. With wild-type yeast ODCase, 6-azaUMP would bind more tightly than UMP, because the pK_a for the formation of their anions is 7.0 and 9.5, respectively (Fig. 5). With the K93C analog, the cysteine has no attracting positive charge, so the pyrimidine base could be held mainly by the hydrophobic pocket surrounding K93; and because UMP at pH 7.5 is mainly uncharged, it would now bind more favorably to this pocket than the negatively charged 6-azaUMP. The tight-binding BMP could bind equally well in the wild-type pocket if its pyrimidine base were rotated either way (Fig. 6). There is a spectral change in the absorbance of a mixture of equimolar 6-azaUMP and yeast ODCase with changing pH that suggests that ODCase K93 not only positions a nucleotide but enolizes the C-2 oxygen of the 6-azaUMP (*114*). However, further studies of X-ray analyses are needed to solve this point.

Because the sequences of all ODCases and all ODCase domains of UMP

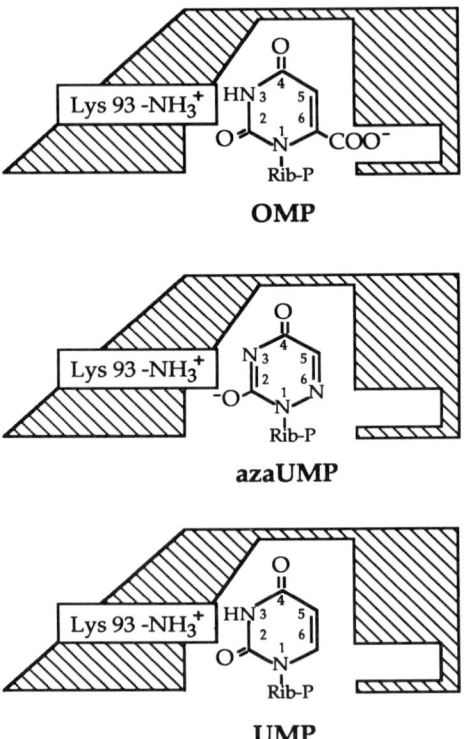

FIG. 5. Schematic representation for proposed role of Lys-93 in catalysis and inhibitor binding to the ODCase catalytic site. Lys-93 is postulated to donate the catalytic proton to O-2 of the substrate OMP. Lys-93 is also proposed to be the key protein functional group that attracts the anionic forms of 6-azaUMP (pK_a = 7.0, one tautomeric form of anion shown) and UMP (pK_a = 9.5, neutral form shown).

synthase are highly conserved (indeed an ODCase identifying sequence), from E311 through V322 of the human UMP synthase, this may well be a site similar to the two major binding sites of the OPRTases and the OPRTase domains where there are a series of interactions with OMP to position the orotate ring properly. In the case of K314, it is not only useful in positioning the pyrimidine ring, but is also necessary to donate the proton to form the zwitterion transition-state intermediate (*112*).

Yeast ODCase has been crystallized with 6-azaUMP (*99*), and an apoenzyme crystal was also formed but not studied in detail.

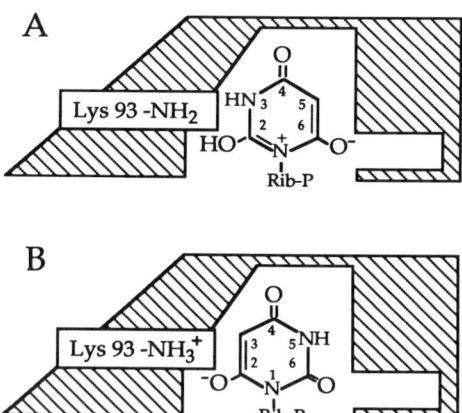

FIG. 6. Schematic representation of alternate forms of barbituric monoribonucleotide binding at ODCase active site. (A) BMP binds with the unchanged C-2–C-4 side of pyrimidine facing Lys-93 and becomes protonated, leading to the zwitterion resembling catalytic intermediates. (B) Charged portion of BMP aligns with Lys-93, yielding a favorable charge interaction.

D. Some Differences between UMP Synthase and the Monofunctional OPRTases and ODCases

As mentioned earlier, the construction of a recombinant baculovirus vector containing the cDNA for human UMP synthase has allowed this protein to be overexpressed in cabbage looper larvae infected with the virus (95). The overexpression led to a crude larval homogenate that was 180-fold enriched in UMP synthase over the concentration in human placental extracts (our previous source for human UMP synthase). The insect system has the advantage that the protein produced by the larvae is modified, as it would be in human tissues, by removal of the N-terminal methionine, which is replaced by an acetyl group. Such modification is thought to make proteins more resistant to proteolysis.

UMP synthase has both similarities and differences from monofunctional OPRTases and ODCases. It has sequence identities with the monofunctional proteins, as illustrated in Table V. This includes the two amino-acid sequences demonstrated by the interaction of OMP with S. typhimurium OPRTase to be essential for binding OMP, PRPP, and PP$_i$. It also has the same sequence in the ODCase domain surrounding K314 as yeast ODCase has surrounding K93. If the human K314 codon is converted to a cysteine codon, human UMP synthase also loses ODCase activity, but it retains full OPRTase activity (B. Han, D. Pasek and M. E. Jones, unpublished). The individual OPRTase and ODCase domains have also been purified to homogeneity (92).

1. NATURE OF DIMERS FOR OPRTASES, ODCASES, AND UMP SYNTHASES

Salmonella typhimurium OPRTase exists as a dimer in the crystal with bound OMP (76); we do not know if the apoenzyme is a dimer, or what concentration of this protein may be required to form the dimer. Yeast ODCase is a monomer in dilute solution, but forms a dimer of the expected size when nucleotides that bind to the OMP binding site of the ODCase domain are added to this dilute solution (94, 118). When the concentration of the ODCase is increased approximately 100-fold and no nucleotide is present to promote dimerization, the concentrated enzyme forms a dimer, which must be more stable than the monomer by itself.

Mouse UMP synthase has three molecular-size species that have been observed by size-exclusion chromatography (54) and by sedimentation through sucrose gradients (54, 107). The $S_{20,w}$ values for the three forms are 3.6, 5.1, and 5.6 (54), as shown in Fig. 4. These forms of UMP synthase are not spherical and have f/f_0 values of 1.30, 1.47, and 1.51, and Stokes radii of 3.05, 4.45, and 5.10 nm for the 3.6-S, 5.1-S, and 5.6-S species, respectively (54). We believe these forms represent a monomeric protein (3.6-S; $M_r \sim$ 50,000), and a simple dimer (5.1-S; $M_r \sim$ 100,000). The third species was a "dense" dimer at 5.6-S, with an apparent M_r of $\sim$118,000; this species was considered to be a conformational variation of the 5.1-S dimer. When the UMP synthase is dilute, the monomer is converted to the dimer by the addition of many anions (55). The characteristic that the anions share is that they are all competitive inhibitors of OMP at the ODCase binding site. The equilibrium between the 3.6-S, 5.1-S, and the 5.6-S forms is more rapid than the methods used to separate the various species, so that single sharp peaks with intermediate S-values were observed when the ligand concentration was lower than that needed to fill the two ODCase ligand-binding sites, i.e., low concentrations of AMP or XMP can yield a 4.7-S peak (54), but saturating XMP gave the usual 5.1-S species, and the AMP concentration curve was still rising toward 5.1-S when solubility limitations interfered with obtaining further values. As a result, a ligand concentration curve must be done to establish whether the ligand will form only the 5.1-S species, or whether at higher concentrations the 5.6-S species will be formed. The anions that can form the 5.1-S dimer include chloride, acetate, orthophosphate, PRPP (with and without Mg^{2+}), UMP, 6-azaUMP, and OMP. Mg^{2+} ion itself does not cause dimerization. With excess OMP, only the 5.6-S species is formed, unless the OMP is not fully saturating, so that one sees both the 3.6-S and the 5.6-S species. Subsaturation with BMP also yields this mixture of 3.6-S and 5.6-S species, or the 5.6-S species alone when BMP was saturating. If the concentration of OMP used converts all of the UMP synthase to the 5.6-S form, and then 50 to 200 mM orthophosphate, a competitive anion, is added, the P_i apparently displaces OMP (or more likely UMP formed from OMP during the sedimentation interval) and shifts the $S_{20,w}$ value back to 5.1-S (54).

Because the concentration of anion effector needed to produce the 5.1-S species can be accurately predicted by the [concentration of effector]/K_i for a given effector as a competitor of OMP binding to ODCase, it is clear that the ODCase domain determines the formation of the 5.1-S dimer. Because the concentrations of 6-azaUMP and UMP needed to form the 5.6-S dimer are very much higher than the K_i values vs. OMP at the decarboxylase site, it is tempting to suggest that nucleotides that give the 5.6-S species bind to both the OPRTase and ODCase domain to give this unique conformation of UMP synthase. The K_m values for OMP binding to the OPRTase and ODCase sites, respectively, are 0.07 and 0.23 µM (55).

Kinetic studies sought to determine if enzyme activity is associated with one form or with all three of the conformational forms of UMP synthase (77). These studies found that the UMP synthase dense dimer is essential for ODCase activity, and that a lag of almost 10 seconds occurred before the 3.6-S monomer form of UMP synthase had full activity after the addition of OMP (77). The same type of kinetic experiments found that the monomeric UMP synthase appears to have OPRTase activity. However, the fact that *S. typhimurium* OPRTase may use catalytic residues from each of the two monomers at a single active site may mean that with UMP synthase dimerization would be necessary for OPRTase activity also. This suggests that the earlier kinetic studies (77) may not have been sufficiently rapid to detect the dimerization of the UMP synthase in the studies of OPRTase activity.

2. STABILITY OF OPRTASES, ODCASES, AND UMP SYNTHASES

Although the monofunctional OPRTase is a rather stable protein, the monofunctional ODCase is so hydrophobic that it completely loses activity when frozen at −20°C unless there is 20–50% glycerol present. OPRTase is rather hydrophobic also; although it can be stored frozen at −20°C without glycerol for a longer interval, it also decays slowly when frozen (121). In general, human UMP synthase and its isolated domains are best preserved if glycerol is present when they are frozen (82).

The bifunctional protein (52,133 Da) stabilizes the ODCase domain toward elevated temperature; the ODC domain by itself (28,500 Da) is heat-labile in the sense that a plot of velocity vs. time for converting substrate to product is linear for less than 2 minutes at 37°C, when enzyme activity ceases. The activity of the domain is linear for about 5 minutes at 25°C and for a full 20 minutes or longer at 0°C (*101*). Also, for storage at 4°C, the isolated ODCase domain must either be stored in 5 mM P_i buffer plus 50% glycerol or in high 300 mM P_i buffer to be held at −70°C; the 5 mM P_i buffer plus 50% glycerol is best (*101*). The conclusion is that the interactions of domains within the intact UMP synthase monomer offer stability to the ODCase domain because none of the precautions, other than 20–50% glycerol for storage at −70°C, is necessary.

3. PROTEOLYSIS OF UMP SYNTHASES

The proteolysis of UMP synthase is retarded by conditions that favor dimerization of the monomers of this bifunctional protein (*101*). These include high UMP synthase concentrations or the presence of other proteins (such as those in the crude placental extracts) or of nucleotides that favor dimer formation (*101*).

It was rather surprising to find that 6-azaUMP apparently stabilizes cellular UMP synthase in a human orotic aciduria mutant cell (*122*). Krooth and co-workers (*123*) had found that cells from orotic aciduria patients grown in tissue culture with 6-azauridine (which is converted to 6-azaUMP after entering into the cells) plus uridine or cytidine to supply pyrimidine nucleotides by the salvage pathway led to a "normal level" of UMP synthase (as assayed *in vitro*) in the mutant cell lines. A reexamination of these data (*122*) led to the observation that 6-azauridine treatment did indeed increase the UMP synthase activities of both mutant and normal cell lines; in addition, it changed the conditions for the stability of the mutant enzyme in the activity assay. Without 6-azaUMP the ODCase activity of mutant cell line extracts increases from 2 to 7 pmol UMP formed from OMP/min/mg protein with time, or with a higher concentration of the cell extract (i.e., cell protein).

However, the same cells grown in 60 μM 6-azauridine had a specific activity of 100 pmol of UMP formed/min/mg protein and this value was independent of the protein or time interval used as long as OMP, the substrate, was maintained in excess. The ODCase activity of the UMP synthase from mutant cell lines was more heat-labile in cells grown in the absence of 6-azauridine. All of these experiments point to the mutant enzyme being unstable unless cells are grown with 6-azauridine, which is converted to 6-azaUMP inside of the cell and which causes UMP synthase to dimerize.

Pulse/chase experiments were used to measure the intracellular rates of synthesis and decay of UMP synthase; the UMP synthases of mutant or normal cell lines were synthesized at the same rate but the turnover of the mutant enzyme was much increased: i.e., $t_{1/2}$ = 9 hours in the mutant cells grown in the absence of azauridine vs. 215 hours in normal cells grown under the same conditions. The time intervals used to determine the decay rate for the mutant cell lines gave a normal first-order decay curve, but for the normal cell lines the rate was so slow that first-order decay was not observed, suggesting that 215 hours is a low value and that a longer decay interval might be observed if longer time intervals had been used to study the UMP synthase decay in normal cells.

In summary, this particular mutant cell line synthesizes UMP synthase at a normal rate (122). This observation is in agreement with data by Winkler and Suttle that showed that this mutant cell line has the same UMP synthase mRNA concentrations as the normal cell line (124). The mutant gene produces a protein with a ratio of the specific activity of ODCase to the specific activity of OPRTase of eight, whereas the value for the normal gene product is two or less. The mutant enzyme had this ratio of eight whether it was grown with or without azauridine; therefore the mutant gene apparently codes for a protein with low OPRTase activity.

4. KINETICS OF THE ODCASE DOMAIN OF UMP SYNTHASE

Three preparations of the ODCase domain of mouse (101–103) and one of the ODCase domain of human UMP synthase (95) have been prepared (see Table VI). One mouse ODCase domain (101) and the human domain preparation (95) were pure, but the protein produced from vector carrying the cDNA for the mouse ODCase domain inserted into either a yeast (103) or an E. coli (102) vector were only tested in crude yeast or E. coli cell extracts. Nonetheless, the results deserve some comparison. The ODCase domains vary in length. The first to be obtained was made by proteolysis of pure mouse UMP synthase with trypsin (101). This protein (referred to as ODCase domain I) migrated by SDS gel electrophoresis as a protein of M_r ~28,500. The pure human ODCase domain (III) was prepared by constructing a recombinant baculovirus vector. The vector was constructed to place a human cDNA fragment coding for amino acids I218 through the C-terminal amino acid V480 behind the vector promoter and a vector ATG codon; this should yield a protein of 28,672 Da.

The mouse cDNA vector for the ODCase domain began at an arginine residue equivalent to the human R230 through V480 and contains five N-terminal amino acids, MTGSG, from the yeast plasmid pSDL5 (103). This ODCase domain II binds both OMP and 6-azaUMP about a third as well as the normal mouse UMP synthase, and it binds OMP a third as tightly as the mouse ODCase domain I (Table VI). The mouse and human sequences for the ODCase domain align very well with no gaps. Using the numbering for the human sequence, the ODCase domain III construct contains an additional 12 amino acids at the N-terminus of the domain, in comparison to the mouse ODCase domain II (Table VI).

The pure ODCase domain should have a specific activity about 1.8 times that of UMP synthase due to the difference in the polypeptide chain

TABLE VI

COMPARISON OF THE CHARACTERISTICS OF THE ODCASE ACTIVITY OF UMP SYNTHASE ODCASE DOMAINS WITH THE INTACT PROTEIN

Protein	Size (M_r)	Pure	No. of dimers	Relative[a] S.A.	K_m OMP (μM)	K_i 6-azaUMP (μM)
ODCase domain I–mo[b]	~28,500	Yes	—	2 → 1[c]	0.2	—
ODCase domain II–mo[b] R230–Q481[d]	28,886	No	1	—	0.6	0.007[e]
ODCase domain III–hu[b] I218–V480	28,672	Yes	—	2	?	—
UMP synthase–mo[b]	~52,000	Yes	2	1.0	0.2	0.003

[a] All values are normalized to the value for UMP synthase, defined as 1.0.

[b] mo, mouse; hu, human.

[c] During proteolysis, no ODCase activity was lost; however, the domain was reisolated from the peptide fragments, and it had the same specific activity as the pure mouse UMP synthase.

[d] The mouse sequence is incomplete, and residue numbers are by comparison to the human sequence.

[e] Determined at 20°C; all other kinetic values determined at 37°C.

length between the domain and the bifunctional protein. This was true for ODCase I prior to its repurification from the protein digestion mixture; no ODCase activity was lost during the trypsin digestion, but during the removal of small peptides and the reconcentration of the ODCase domain a 50% loss of ODCase specific activity occurred. The domain III generated from a baculovirus vector, and purified over a monoclonal antibody column, has the expected specific activity, i.e., approximately double that of UMP synthase (95).

5. HUMAN MUTANTS

A rare opportunity arose to study the genes and enzyme activities of a child with orotic aciduria and her parents (44, 86). The parents have very different lesions. The mother's codon for a proline (P) at position 92 had been changed to a serine. The mother's red cell OPRTase and ODCase levels (Table VII) remained in proportion to those of control subjects (ODCase/OPRTase = 1.4), but both activities were reduced to about 15% of the normal level. The father had the genetic change in the ODCase domain in amino-acid 286 from an isoleucine to an asparagine,[3] and the ratio of ODCase/OPRTase was distorted from the normal 1.4 to 1.7, suggesting that proteolysis of UMP synthase may have produced a modest accumulation of the ODCase domain (101). The result of having these two mutant genes plus the

---

[3] R. F. Reed, T. Lin and D. P. Suttle, unpublished.

TABLE VII
UMP Synthase Activity in Erythrocytes from Parents and Child with Mutant UMP Synthase Genes

Subject	Genotype	Amino acid sequence changes	Specific activity[a]		Ratio ODC/OPRT
			OPRT	ODC	
Control (2 persons)	RR	None	0.530	0.780	1.4
Mother	Rr	P92 → S (OPRTase)	0.080	0.116	1.4
Father	Rr	I286 → N (ODCase)	0.260	0.440	1.7
Child	rr	P92 → S + I286 → N	0.038	0.008	0.2

[a] In units of nanomoles per hour per milligram protein.

lack of a normal gene in the child is very dramatic, with a massive loss of ODCase activity and a significant loss of OPRTase activity (Table VII). Neither of the amino-acid changes appear to be in the known substrate binding sites; however, P92 is an invariant residue for all UMP synthases. I286, which is changed to an asparagine, is adjacent to an aspartate residue that is invariant in the ODCase structure of all eukaryotes; however, this change only reduces the OPRTase and ODCase activities for the father to about 50%. Both parents can have three dimer pairs—nm, nn, and mm (n, normal; m, mutant). In the child the pairs are m_1m_1, m_1m_2, and m_2m_2, where m_1 is the maternal mutation and m_2 is the paternal mutation. It will be interesting to see why these changes lead to such a dramatic drop in the activities of the child's ODCase and OPRTase. The data suggest that dimer formation may be grossly affected in the child; i.e., m_1 and m_2 do not form very stable dimers. If so, one or both of the mutations may be located in an area where the two monomers must meet to form a dimer, or must so distort the protein that dimer formation cannot occur.

6. Abzymes and Thermodynamics

Because ODCase is a decarboxylase that needs no cofactor, it is an ideal enzyme for which to develop a ligand that will cause the formation of an abzyme. The haptens used are shown here (125):

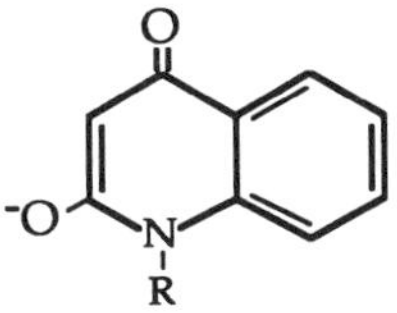

$R_1 = (CH_2)_3COOH$
$R_2 = H$

An enol was selected to aid the positioning of a positively charged amino acid adjacent to the enolate. The fused phenyl ring was designed to create a hydrophobic space within the antibody binding site.

A recent report demonstrates that the uncatalyzed chemical half-time for the decarboxylation of orotic acid is 78 million years in neutral aqueous solution at room temperature (115). ODCase, either from yeast (94) or human UMP synthase (95), enhances this rate by 10^{17}, thereby making ODCase a most proficient enzyme (115).

IV. Uridine Kinase

A. Importance of Uridine Kinase and Possible Regulation of Gene Expression

The importance of uridine kinase as one of the two sources for the formation of UMP led to the initial studies on this enzyme (126, 127). Interest in the enzyme increased significantly as researchers explored the possibility of blocking this enzyme activity as part of cancer chemotherapy, and many of the early studies have been reviewed (128). It has become generally established that enzymes directly related to the synthesis of DNA are not expressed in a constant fashion, but are subject to regulatory factors that alter gene expression as a function of the cell cycle. Because RNA is synthesized fairly continuously by all cells, there is a constant requirement for nucleotides. Therefore, enzymes such as uridine kinase were considered to be constitutive enzymes, and regulation would only occur as a consequence of ligands interacting directly with the enzyme. Two types of data have emerged in recent years that suggest that, at least in specific tissues or physiological situations, uridine kinase is subject to more complex regulation: (1) induction or repression of enzyme expression and (2) expression of tissue-specific isozymes.

1. ALTERED EXPRESSION OF URIDINE KINASE

These studies were done before the advent of modern molecular biology, and therefore observed changes in enzyme activity extracted from cells or tissues exposed to some test stimulus. An early example of such experiments are the various studies with mammalian lymphocytes exposed to phytohaemagglutinin (129–132). Increased expression of uridine kinase protein is a likely explanation for these studies, because there was a significant increase (20 to 30-fold) in enzyme activity after cells were stimulated with phytohaemagglutin for at least 20 hours. These results were obtained with lymphocytes from humans (129–132), as well as lymphocytes from horse, pig, sheep, and cattle (*132*). That induction by phytohaemagglutin represented true induction was verified with human lymphocytes. After an exposure time of 72 hours, the activity of uridine kinase was increased about 50-fold, and, as assayed with an antibody in a Western blot, the increase in uridine kinase protein was about the same (*133*).

The likelihood that serum growth factors initiate the induction of uridine kinase was supported by studies using BALB/c 3T3 cells. When mitotically quiescent cells were exposed to fresh calf serum for 6 to 9 hours, the measured activity of uridine kinase extracted increased about fourfold (*134*). In efforts to define the growth factor, the serum was replaced by either platelet-poor plasma or by platelet-derived growth factor. When added alone, either one of these caused two- to threefold increase in uridine kinase activity after 6 hours, but this enzyme level then declined even with continued exposure to the growth factor. Only when these two components were jointly added was the induction of uridine kinase enzyme activity maintained at a level greater than threefold for at least 12 hours.

Later studies with these mouse fibroblasts showed that exposure to serum for 9 hours led to an eightfold increase in enzyme activity, as well as a fourfold increase in enzyme proteins, as assayed by a Western blot with antibody to the pure mouse enzyme (*133*). Such stimulation by growth factors may also account for the results observed in a study where the activity of uridine kinase was examined in human skeletal muscle (*135*). When muscle cells were isolated from biopsy samples, and then grown in culture for 7 days to form fresh myotubes, there was an increase of 100- to 200-fold in the measured uridine kinase activity.

The activity of uridine kinase was also increased after rats received injections of D-galactosamine, known to trap the uridylate pool in the liver (*136*). When protein samples from the hippocampus were tested at 2 and 4 hours postinjection, the activity of the enzyme was increased more than 50%. In this experiment, it is not clear what mechanism produced the increase in activity. The more moderate size of this increase, compared with the above studies that employed phytohaemagglutinin, would be consistent with a decrease in feedback inhibition by CTP and UTP as these nucleotide pools were decreased due to the effects of D-galactosamine. However, although such a decrease in feedback inhibition should alter the activity *in vivo*, it is not clear that the isolated enzyme would show any effects when tested *in vitro*, so that this experimental result may represent an increase in the amount of enzyme.

More likely to reflect a true physiological stimulant, studies with rats have demonstrated the effect of administered insulin on the ability of skeletal muscle to take up uridine from blood, and to synthesize UTP (*137, 138*). In these experiments, diaphragm muscle was excised and rapidly placed in buffer with appropriate concentrations of [³H]uridine and insulin. After tissues were so incubated for 15 minutes, the intracellular uridine nucleotide pool increased by 20% for tissue exposed to insulin (*138*). When muscle tissue was exposed to insulin for only 5 minutes, extracts prepared from them showed an increase in uridine kinase activity of about 15%. This is not a large increase, but it was statistically very significant because the experiment used more than 20 samples for the different conditions. The possibility of a greater increase in uridine kinase activity with longer exposure time of the muscle to insulin was not tested.

The physiological processes listed in Table I were mentioned previously (Section I). In an effort to further define the vasodilator effects of uridine nucleotides, experiments were done with intact rats exposed to minoxidil, a pyrimidine analog. The coadministration of this compound abolishes the effect on arterial constriction by uridine, but not by UMP (*139*). These results suggested that the blocking effect of minoxidil was not at the site of the membrane receptor protein itself, but is more likely an effect on the enzyme that converts uridine to UMP. Additional studies assayed the synthesis of UMP from uridine by plasma membrane preparations, and found that the addition of minoxidil reduces the uridine kinase activity by 24%. These results are consistent with the values in Table I showing that uridine nucleotides are most effective in altering the vasodilation in such studies.

For the experiments described above, the types of cells or tissues varied, as did the experimental parameter that was used to initiate changes in uridine kinase activity. When the magnitude of increase in enzyme activity for some of these experiments was 4- to 5-fold or 100- to 200-fold, this probably represented true induction of the expression of uridine kinase. In those experiments where the magnitude of this increase was lower, the cause of the induction in activity is not as certain.

2. ARE THERE ISOZYMES OF URIDINE KINASE?

The importance of UMP synthesis led to early efforts to isolate and characterize uridine kinase. Different investigators observed multiple enzyme peaks or bands by chromatography (*140–149*), isoelectric focusing (*148–150*), or electrophoresis (*151, 152*). Up to four apparently distinct species were observed by chromatography or electrophoresis, and two species were seen by isoelectric focusing. These data are summarized in Table VIII, and although the same numerals for isozymes are shown for the different studies, isozyme II (as an example) is not necessarily the same enzyme species in different studies, because in some studies this number was based on the size of the enzyme and in other studies it was based on ionic charge.

One of the initial studies with uridine kinase from rat liver used size-exclusion chromatography, thus the authors were able to interpret the two

TABLE VIII
POSSIBLE ISOZYMES OF URIDINE KINASE

Tissue or cell	Isozyme forms	Analytic method	Ref.
Rat			
Liver	I, II	Size exclusion	140
	II	Ion exchange	143
	I, II, III, IV	Electrophoresis	151
Kidney	II, III, IV	Electrophoresis	151
	I, II, III, IV	Size exclusion/affinity	146
Spleen	III, IV	Electrophoresis	151
Brain	IV	Electrophoresis	151
Fetal rat			
Liver	I, II	Ion exchange	143
	I	Electrophoresis	151
Brain	I	Electrophoresis	151
Rat hepatoma	I	Size exclusion	142, 273
	I, II	Size exclusion	142, 273
	I, II	Size exclusion	145
	I, II	Isoelectric focusing	150
	I, II, III, IV	Electrophoresis	152
Mouse			
Ehrlich ascites cells	I, II	Size exclusion	141
	I, II, III, IV	Size exclusion	153
	I, II, III, IV	Ion exchange	153
Lymphoid cells	I, II	Size exclusion	144
L1210 leukemia	I, II	Isoelectric focusing/size exclusion	149
Human			
Lymphoma	I, II	Isoelectric focusing/size exclusion	148
Leukemia	I	Size exclusion	147
Tumors	I, II	Size exclusion	147

observed species as being monomers and tetramer forms of the enzyme (*140*). These authors partly purified the two separate species, and showed that they had very similar kinetics. This size relationship was also seen by electrophoresis of the enzyme from fetal rat liver or rat hepatoma, where three of the four species were interpreted as monomer, dimer, and tetramer forms (*151, 152*). Two isozymes were still judged necessary to interpret all four species observed (*151, 152*). Although most of the reports in Table VIII concluded that the separate species were genuine isozymes of uridine kinase, these isozymes were generally not purified and characterized.

In contrast to the isozymes observed above, work with the completely purified uridine kinase from mouse Ehrlich ascites cells showed this enzyme to exist as a single protein species as measured by two-dimensional electro-

TABLE IX
MECHANISMS FOR THE PRODUCTION OF ISOZYMES

1. Two or more isogenes → multiple isozymes
2. One gene → one protein + post-translational modification → multiple isozymes
3. One gene → one dissociating oligomeric protein → multiple M_r forms

phoresis (153). Nevertheless, this pure form of uridine kinase could still migrate as four distinct species by either size-exclusion or ion-exchange chromatography (153). These results could easily result from a tetrameric enzyme that partly dissociated to produce additional dimer and monomer forms, which easily appear as distinct forms when the column flow rate separates these different forms more rapidly than they can reassociate to form the tetramer.

A possible interpretation for all the data on isozymes comes from the mechanisms for isozyme formation listed in Table IX. The existence of multiple genes coding for multiple isozymes is the classic pattern. Additional features can cause a single-gene product to exist in multiple forms. Post-translational modification of a protein can lead to N-terminal modification, acylation, deamidation, phosphorylation, etc. Some of these clearly lead to changes in surface charge. Also, uridine kinase is a tetramer that readily dissociates and reassociates as a function of enzyme concentration or physiological effectors (Fig. 2); this can also produce multiple forms of the enzyme. The three mechanisms listed in Table IX are not mutually exclusive; a single enzyme can be observed in multiple forms due to the action of all three mechanisms. The important conclusion is that more caution and better data are desirable in making such interpretations about isozymes. The likelihood that such isozymes of uridine kinase are real and of physiological significance is enhanced by observations that some forms of uridine kinase appear only in specific tissues, or in certain developmental states. The current information for uridine kinase, summarized in Table VIII, supports the hypothesis that at least two genes for uridine kinase exist. However, both gene products may not be observed at all stages of development, or in all tissue at a given stage of development.

B. Structure and Regulation of Uridine Kinase

The pure mouse enzyme isolated from Ehrlich ascites cells has a subunit M_r of 31,000, and is normally a stable tetramer (153). CTP and UTP, the end products of the pyrimidine nucleotide pathway (Fig. 1), are allosteric feedback inhibitors of uridine kinase. They cause the active tetrameric enzyme to dissociate to the inactivate monomer (153, 154). The intrinsic activity of these different oligomeric species was measured by kinetic studies in which the enzyme was incubated under conditions known to produce either the monomer or the tetramer. The substrate ATP was then added to initiate the reaction (155). Such experiments showed that the dissociated monomer initially had no detectable activity, but within about 30 seconds it had been converted to a fully active form. By comparison, the tetrameric species had activity at all timepoints after the addition of substrate.

The enzyme shows sensitivity to temperature by different experimental methods. An Arrhenius plot for the measured activity from 0 to 40°C is biphasic, showing that below and above room temperature the enzyme has different changes in the enthalpy for the reaction (155). Consistent with this observation are results demonstrating that the oligomer state of the enzyme also changes over this temperature range, with the enzyme being a stable tetramer at ≥22°C, but existing in a mixture of various oligomeric states at 4°C. All these data are consistent with the model of uridine kinase as a stable, active tetramer that can be dissociated to inactive monomers by inhibitors at 37°C, or by low temperature in the absence of ligands. These multiple M_r species observed at 4°C are important for the interpretation of data about isozymes, described in Section IV, because in such studies chromatography is routinely performed at 4°C.

A very interesting feature of how uridine kinase may be regulated comes from the kinetic studies with the inhibitors CTP and UTP. These nucleotides are competitive inhibitors of the substrate ATP (155), and thus should bind at the catalytic site. However, the catalytic site shows no pattern in its preference for a phosphate donor, because both purine and pyrimidine nucleotides can be substrates and because both ribo- and deoxyribonucleotides can be substrates (155). The only exceptions are CTP and UTP, the allosteric inhibitors of uridine kinase. One possible explanation for this unusual lack of activity with CTP and UTP is that they do not bind at the catalytic site in the normal position where ATP, and other NTPs, would bind. Instead, CTP and UTP may bind backward, as bisubstrate analogs that occupy the nucleoside site (because cytidine and uridine are the favored nucleoside substrates), and also bind to the phosphate-binding part of the ATP site.

Although this explanation is plausible, it is inconsistent with the observed results showing dUTP to function as a phosphate donor. It is expected that dUTP can also bind in the bisubstrate position, because the corresponding nucleoside, deoxyuridine, is also a substrate for uridine kinase, though with a lower activity (156). Also, dUTP does not produce dissociation of the enzyme tetramer, as happens with CTP and UTP. These results led to the hypothesis that, on the enzyme, a distinct regulatory site exists that is specific for binding CTP and UTP (155). Because UTP and CTP should bind part of the time in the correct orientation at the ATP site, it is the binding at this proposed regulatory site that leads to allosteric changes in the protein, which cause enzyme dissociation and complete loss of activity.

Although the mammalian uridine kinase has highest activity with ATP as the phosphate donor, it showed moderate to fairly good activity with the other NTPs (except the inhibitors UTP and CTP), for enzyme obtained from mouse (*155, 157*), or from rat (*158*). The nucleotides that had the lowest activities were GTP and dGTP (*155, 157*), and this may be due to tight binding, because the enzyme demonstrates an affinity for the guanine nucleotides that is almost 10-fold that for ATP or dATP (*155*). This differential affinity may reflect some ancestral physiological rationale, because kinetic studies with the enzyme from other organisms show two different patterns. Some bacteria show either a marked preference or an apparent requirement for GTP as the phosphate donor: examples include *E. coli* (*159*) and *Mycobacterium microti* and *Mycobacterium avium* (*160*). By comparison, a pattern similar to the mammalian affinities for NTPs was observed with uridine kinase from *Bacillus stearothermophilus* (*158*) or from *Tetrahymena* (*161*). In *Schistosoma mansoni*, a parasite dependent on salvage, the uridine kinase activity uses ATP effectively (*162*).

C. Uridine Kinase DNA from cDNA and Genomic Libraries

1. THE cDNA FOR URIDINE KINASE FROM MOUSE BRAIN

A cDNA for uridine kinase from mouse brain was isolated, sequenced, and characterized (*163*). It is only 95% complete, missing the first 18 codons, which are found in a pseudogene for uridine kinase (described below). It has over 1 kb of untranslated 3′ sequence, containing the standard poly(A) site. When the cDNA sequence (missing the first 18 codons) was inserted after an ATG start codon in a pET expression plasmid transformed into *E. coli*, the cells produced uridine kinase protein that assembled into a native tetramer and had uridine kinase activity. The translated sequence, with the first 18 amino acids from the pseudogene, codes for a protein of 278 amino acids (M_r = 31,239), in good agreement with the observed M_r of 31,000 for the native mouse enzyme on SDS gel electrophoresis.

The combined translated amino-acid sequence for the cDNA plus the first 18 codons from the pseudogene has a high identity (38 and 37%) to the sequences for uridine kinase from *E. coli* and from yeast. The sequence also has good to moderate identity with three nucleotide kinases, for which crystal structures exist that identify a common ATP-binding nucleotide fold (*164–166*). The sequence alignment in Fig. 7 [by the Pileup program (*167*)] shows that the three uridine kinase sequences align directly with the three nucleotide kinases at the positions for two of the three peptide segments involved in the ATP-binding site, which are defined by their consensus motifs (shown in bold in Fig. 7) and identified as *Kinase-1a* and *Kinase-3a* (*168*); these are detailed further in Section VI,B. A *Kinase-2* site is at the

```
                        1                                                         50
Urd-K_mouse     ..........  ..........  ..........  ..masagggg  sestalargr
Urd-K_yeast     mshriapske  rsssfisild  detrdtlkan  avmdgevdvk  ktkgkssryi
Urd-K_Ecoli     ..........  ..........  ..........  ..........  ........mt

UMP-K_yeast     ..........  ..........  ..........  ..........  mtaattsqpa
AMP-K_pig       ..........  ..........  ..........  ..........  ........me
GMP-K_yeast     ..........  ..........  ..........  ..........  ..........

                       51           Kinase-1a                                    120
Urd-K_mouse     rpqprpflIG  VsGGtaSGKS  TvcEKImell  gqnevdrrqr  klviLsqDcf
Urd-K_yeast     ppwttpyIIG  igGaSGSGKt  svaaKIVssi  n.......vp  wtvLislDnf
Urd-K_Ecoli     dqshqcvIIG  iaGaSaSGKS  liastLyrel  reqvGdeh..  .igvipeDcy

UMP-K_yeast     fspdqvsvIf  VlGGpGaGKg  TqcEKLVkdY  ....sfvhlS  agdLLRAeqg
AMP-K_pig       eklkkskIIf  VvGGpGSGKg  TqcEKIVqkY  ....GythlS  tgdLLRAevs
GMP-K_yeast     .....srpIv  isGpSGtGKS  TllkKLfaeY  pdsfGfsvsS  ttrtpRAgev

                      101            Kinase-2                                    150
Urd-K_mouse     Yk...vLtaE  qkakalkgqy  nfDhPdAfdn  dLmhktlkni  veGKtvevPt
Urd-K_yeast     Yn...pLgpE  drarafkney  dfDePnAinl  dLaykcilnl  KeGKrtniPv
Urd-K_Ecoli     YkdqshLsmE  erv.....kt  nyDhPsAmdh  sLllehlqal  KrGsaidlPv

UMP-K_yeast     ra...gsqyg  elikncikeg  qivpqeitla  lLrnaisdnV  KanKh.....
AMP-K_pig       .s...gsarg  kmlseimekg  qlvpletvld  mLrdam akV  dtsK......
GMP-K_yeast     ngkdynfvsv  defksmiknn  efiewaqfsg  nyygstvasV  Kqvsksgktc

                151 Kinase-2                                                     200
Urd-K_mouse     YdfVthsrlp  e.ttVvypad  VVlfEGILvf  ytqeirDmfh  lrlFVDtdsD
Urd-K_yeast     YsfVhhnrvp  dkniViygas  VVviEGIyaL  ydrrllDlmd  lkiyVDadlD
Urd-K_Ecoli     YsyVehtrmk  e.tvtvepkk  ViilEGILLL  tdarlrDeln  fsiFVDtplD

UMP-K_yeast     kflIDgfprk  mdQaisferd  IVe..skfiL  ..........  ...FfDcpeD
AMP-K_pig       gflIDgypre  vkQgeeferk  Igq..ptLLL  ..........  ...yVDagpe
GMP-K_yeast     ildID.....  m.QgVksvka  IpelnarflLf  .........  ...iappsve

                      201            Kinase-3a                                   250
Urd-K_mouse     vrLsRRvL.R  D..vqRGRDl  EqiltQytaf  VkPaFeeFcl  PTkkyADvIi
Urd-K_yeast     vcLaRRLs.R  D..ivsRGRDl  dgcIqQwekf  VkPnavkFvk  PTmknADaIi
Urd-K_Ecoli     icLmRRik.R  D..vneRGRsm  dsvmaQyqkt  VrPmFlqFie  PskqyADiIv

UMP-K_yeast     i.mleRLLeR  gktsGRRdDn  iEsIkKR...  .....FntFke  tsmpvieyfE
AMP-K_pig       t.mtkRLLkR  getsGRvdDn  EEtIkKR...  .....letyyk  aTepviafyE
GMP-K_yeast     d.L.kk...R  l..eGRgtet  EEsInKR...  .....ls....  aaqaelayaE

                      251                                                        300
Urd-K_mouse     PRGvdNmVAi  NliVqhIqdi  Ln........  ..gdlckrhr  ggpngrnhkr
Urd-K_yeast     PsmsdNatAv  NliinhIksk  LelksnehLr  eliklgssps  qdvlnrniih
Urd-K_Ecoli     PRGgkNriAi  dilkakIsqf  fe........  ..........  ..........

UMP-K_yeast     tkskvvrVrc  drsVedvykd  vqdairdsL.  ..........  ..........
AMP-K_pig       kRGivrkVna  egsVddvfsq  vcthl.dtLk  ..........  ..........
GMP-K_yeast     t.GahdkViv  Nddldkayke  Lkdfifaek.  ..........  ..........

                      301              327
Urd-K_mouse     tfpepgdhpg  vlatgkrshl  esssrph
Urd-K_yeast     elpptnqvls  lhtmllnknl  ncadfvf
```

FIG. 7. Alignment of the three sequences for uridine kinase and the three nucleotide kinases. Sequences in bold type code for the peptides shown to bind ATP in the crystal structures for the three nucleotide kinases (*164–166*).

same position for the three nucleotide kinases, but at a slightly different position for the three uridine kinases.

2. A Pseudogene for Uridine Kinase

The initial efforts to isolate the complete cDNA for uridine kinase were unsuccessful. Several different cDNA libraries were explored with the same result: a cDNA that appeared to be missing the first 18 codons. This led to the interpretation that, because the 5' end of uridine kinase has a (G + C) content above average, this might produce local secondary structure at the 5' end of the mRNA that could not be traversed by the reverse transcriptase used in the synthesis of cDNA libraries. Genomic libraries from mouse brain were then screened. Using DNA primers identified from the cDNA sequence already established, PCR was used to isolate additional sequences from the genomic library. One additional sequence, with 94% identity to the cDNA, was obtained from a mouse genomic library; this was considered to be a pseudogene for uridine kinase, because it lacked introns, and had four stop codons that result in an interrupted coding region.

Sequencing of this genomic DNA band revealed an additional 130 base pairs of 5' uridine kinase DNA sequence that contained a possible translational start methionine conforming to the consensus translational start sequence, RNNATGG, proposed by Kozak (169). Comparison of the sequence of this clone to the sequence of the cDNA showed the presence of a T insertion at nucleotide position 130, and 44 base changes in the genomic DNA relative to the cDNA at various positions along the overall alignment. Overall, there was a 94% identity between the two nucleotide sequences. It is highly likely that this genomic clone represents a uridine kinase pseudogene. The pseudogene contains the standard features identified for DNA elements that appear to have originated from an RNA template before being inserted into host DNA by a virus (170).

3. The Number of Genes for Uridine Kinase

Southern analysis of mouse kidney genomic DNA was carried out to identify the presence of the uridine kinase gene(s) (163). Separate samples of mouse genomic DNA were digested with *Eco*RI, *Bpm*I, *Hin*dIII, or *Bam*HI to yield various possible fragment patterns. Probing of the Southern blot with a digoxigenin–dUTP-labeled uridine kinase cDNA probe revealed the presence of two or three bands in each digest, with two bands being more distinctly labeled by this probe, at a size of about 13 kb. These gene fragments were distinct from the pseudogene, because probing with the isolated pseudogene labeled a completely different band at 5.6 kb, which appears to be distinct from the two very similarly sized bands labeled by the cDNA probe.

To ascertain if there were multiple uridine kinase RNA transcripts, Northern blots of mouse grain poly(A)$^+$ RNA were initially screened using a digoxigenin–dUTP-labeled uridine kinase cDNA probe that contained approximately 92% of the coding region for mouse uridine kinase. On probing a Northern blot containing total mouse brain poly(A)$^+$ RNA, only one poly(A)$^+$ RNA species was recognized. The size of this poly(A)$^+$ RNA species was approximately 1.9 ± 0.1 kb. The size of this band was in very good agreement with the combined DNA sequence for the cDNA plus the extra 5′ part of the pseudogene, of 1940 bases (163), and is interpreted as representing the complete cDNA.

The most direct interpretation of the data from the southern blots is that mouse genomic DNA contains one or possibly two copies of a gene coding for uridine kinase, and one copy of a uridine kinase pseudogene. The Northern blot data showed that brain mRNA contains only a single band, so that if two or more functional uridine kinase genes exist, only one appears to be expressed in brain. These results are consistent with the earlier studies already described on isozymes (Table VIII). Two (or more) genes for uridine kinase may exist, but certain tissues appear to express only one gene product as mRNA and protein.

Previous studies on uridine kinase isolated from mouse brain show a single subunit band at about 28 kDa, as measured by a Western blot with antibody to the pure mouse uridine kinase (133). Because the predicted protein size from the mouse brain DNA (M_r 31,239) corresponds very well with the size of the native enzyme from mouse ascites cells, this suggests that the uridine kinase previously isolated from mouse brain may have undergone more extensive post-translational processing. Although two functional uridine kinase genes may exist, because the Northern blot showed that brain mRNA contains only a single band, only one gene appears to be expressed in brain. This may also be true for mouse ascites cells, because the protein purified from these cells appeared to be a single protein species (single-gene product) as determined by two-dimensional electrophoresis (153).

V. β-Alanine Synthase

β-alanine synthase is a trivial name used to stress the important product; the formal name of this enzyme is N-carbamoyl-β-alanine amidohydrolase, and the trivial name is β-ureidopropionase (EC 3.5.1.6). It is the only enzyme catalyzing the biosynthesis of β-alanine in animals. The reaction is via hydrolytic cleavage of N-carbamoyl-β-alanine to yield NH_3 + CO_2 + β-alanine. This enzyme is important, because β-alanine has only recently been recognized for its pleiotropic physiological roles (Fig. 8); it functions in the

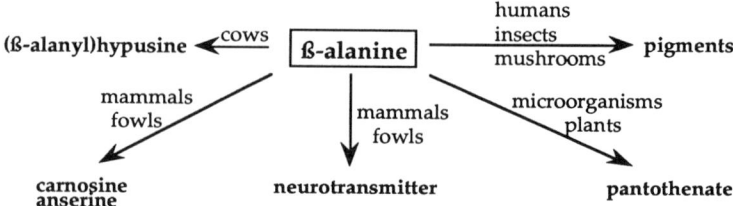

FIG. 8. Physiological roles for β-alanine show this compound to be used as a neurotransmitter and as a building block in many organisms for the synthesis of various dipeptides (carnosine, anserine) or dipeptide-like compounds (pantothenate, pigments).

brain and visual system as a neurotransmitter (171–177), and in the activation of ion channels in multiple systems (178). Of the four naturally occurring β-amino acids, β-alanine is the only one that is rapidly transported across the blood/brain barrier (179–181) by a high-affinity transporter (176–179, 182, 183). Furthermore, there are receptors in brain that respond uniquely to β-alanine (184–186). Pharmacological studies show that β-alanine is a depressant (171), and that it may have effects similar to γ-aminobutyric acid (GABA) (171, 174). In addition, β-alanine depresses release of thyrotropin-stimulating hormone (174) and facilitates wound healing (187). Injection of β-alanine into the ventricle of rat brain leads to significant increases of lutenizing hormone in plasma (188).

The bulk of the body's β-alanine exists in the form of dipeptides: carnosine (β-alanyl-histidine) is abundant in mammalian muscle and brain, whereas anserine (β-alanyl-methyl-histidine) is more prevalent in avian tissues (189). The high concentrations of these dipeptides in muscle (7–10 mM) suggest a physiological role, which is not yet well defined, although there is emerging support for the action of carnosine as an antioxidant (190). Pantothenate, a precursor of coenzyme A, is probably the best known of these compounds. The formation of coenzyme A may be the major role for β-alanine in microorganisms. This requirement has been shown for S. typhimurium in studies on a strain mutated in this pathway. This resulted in the strain being temperature-sensitive for the conversion of uracil to β-alanine, and requiring either N-carbamoyl-β-alanine (NCβA), β-alanine, or pantothenate for growth (191).

Abnormal β-alanine concentrations appear to have serious consequences for humans. Several cases of hyper-β-alaninemia have been reported (189, 192, 193), always resulting in early infant death. At least a dozen cases have now been reported of decreased β-alanine formation resulting from lesions in the first two enzymes of the pathway for uracil catabolism (194–199). Such cases are not always fatal, but usually show severe neural dysfunction and seizures. These clinical studies are consistent with the above-mentioned neurochemical functions attributed to β-alanine.

All these clinical findings support the hypothesis that β-alanine levels are carefully regulated in the body. The results predict that the enzyme that produces β-alanine should be under regulatory control. Consistent with this hypothesis, β-alanine synthase (the irreversible step in this pathway) shows allosteric responses to substrate, product, and analogs, as assayed by kinetic analysis and by the extent of subunit assembly (57, 58, 200). In contrast, the first two activities in the conversion of uracil to β-alanine are reversible, and these enzymes show no allosteric properties (201, 202). In adult rats, significant β-alanine synthase activity is reliably detected only in liver and kidney (58), though β-alanine is used largely in brain and muscle. This compartmentation is logical, because liver and kidney salvage uracil from blood (46), converting it to β-alanine, and are the two tissues that can handle the ammonia produced by this reaction.

The levels of β-alanine synthase activity also undergo dramatic changes during development. In the fetal rat, and in the adult, liver is the major source of β-alanine synthase activity. Enzyme activity is also easily detected in fetal brain. However, young adult rats show a dramatic increase in enzyme activity in liver of 3000-fold, compared to 20-day fetal liver, whereas in brain the enzyme activity declines to barely detectable levels. Interestingly, 13 different tumor lines appear to have lost tissue-specific regulation of the expression of β-alanine synthase and exhibit high levels of this enzyme activity (203).

This finding has important implications for cancer chemotherapy. β-alanine synthase catalyzes the only irreversible step in the degradation of the anticancer drug, 5-fluorouracil (5-FU) to fluoro-β-alanine (111, 204, 205). 5-FU is catabolized very rapidly in liver or kidney. In other normal tissues, deficient in this pathway, 5-FU is converted to F-dUMP, the active form of the drug as the inhibitor of thymidylate synthase. The finding that many tumors have the three enzymes that catabolize 5-FU (203) means that the drug will have a short $t_{1/2}$ in the very cancer tissues for which the drug is intended. This limitation in the effectiveness of 5-FU could be overcome by application of a potent inhibitor of β-alanine synthase. In such a chemotherapeutic strategy, normal nonmitotic tissues could be spared from β-alanine deficiency by dietary supplements of this amino acid.

Because of the clinical interest in 5-FU and F-β-alanine, experiments have been done to follow the tissue distribution of [^{3}H]F-β-alanine after it was administered intravenously to rats (206). The bulk of the injected F-β-alanine was rapidly taken up by the kidney, but within 15 minutes brain tissue had accumulated this drug to a concentration of about 20 μM, and maintained a concentration of 15 μM 24 hours after the injection. Because it may well be that both β-alanine and F-β-alanine bind to the same membrane receptors and transporters, the results from this study are consistent with the earlier studies on the uptake of β-alanine by brain.

A. Structure and Dissociation of β-Alanine Synthase

Historically, β-alanine synthase has not received adequate attention because the enzyme was difficult to purify, and there was no sensitive and convenient assay. We developed a convenient and sensitive radioactive assay (34). This procedure facilitated the separate enzymatic assay for each of the three sequential enzymatic activities in the conversion of uracil to β-alanine, and enabled the work on an effective purification protocol for the enzyme from rat liver (57). The importance of this purification protocol was that the pure enzyme maintained the many allosteric properties that had previously been measured with partially purified enzyme preparations (58). These results were in contrast to a purification protocol for the same enzyme that involved a heating step during the purification (207). Because both laboratories had purified the enzyme from rat liver, the significant differences in the observed allosteric properties for the pure enzyme suggested that the heat treatment (207) may have had deleterious consequences.

This was specifically shown by our results that heating the enzyme for only 10 minutes at 50°C produced significant changes. When the enzyme was heated in the standard buffer, without any stabilizing ligands, there was a significant loss in V_{max}. When the enzyme was heated briefly in the presence of the ligand propionate, V_{max} was essentially the same, but there was a dramatic decrease in the affinity for substrate (57). Such variant results point out the difficulty in devising an effective purification protocol. The purification procedure that uses the heat step is faster, and has a better yield, and therefore appears to be a desirable procedure. However, if preliminary studies with less pure (more "native") enzyme had not shown the allosteric features of β-alanine synthase, they would not as easily be discovered with pure enzyme that had been subjected to heat.

The enzyme has also been completely purified from calf liver, with a heat step during the purification (208). The calf enzyme had an oligomeric structure and subunit M_r completely consistent with that of the rat liver enzyme, but showed no cooperativity toward the substrate N-carbamoyl-β-alanine, and had unusually poor affinity for this substrate (K_m of 500 μM).

β-Alanine synthase appears to be regulated by dissociation and reassociation of enzyme oligomers. It is unusual in having an unliganded oligomer (hexamer) that is an intermediate between being a trimer, or in the fully associated and active dodecamer form (Fig. 2). The dissociation and association response of the enzyme to different ligands is shown in Fig. 9. The native hexameric enzyme occurs in the absence of ligands, but readily dissociates to trimers in response to β-alanine, or associates to form dodecamers on binding the substrate N-carbamoyl-β-alanine, or analogs of this substrate such as propionate. The distinct end points in these two processes are consistent with the trimer being a table protomer, and the process shown in Fig. 9 is described by $\alpha_6 \rightarrow 2\alpha_3$ or by $2\alpha_6 \rightarrow \alpha_{12}$, so that α_{12} is formed by four α_3 protomers. The fact that various M_r values that are not a multiple of α_3 were observed in Fig. 9 is explained by the likelihood that the given oligomers associate or dissociate in a rapid equilibrium, leading to a single enzyme band with an M_r representing the statistical average M_r.

The unliganded hexamer has only modest intrinsic catalytic activity (58), but in the presence of substrate or product it associates to the most active or dissociates to the least active quaternary states (Fig. 2). As with the studies on UMP synthase and uridine kinase already discussed, these kinetic experiments to determine the enzyme activity associated with a particular oligomeric species were done under conditions in which activity could be assayed within a few seconds of adding the substrate, and with the preexisting oligomeric state of the enzyme determined by the concentration of enzyme and ligands. β-Alanine synthase is also unusual in maintaining a stable trimeric structure that functions as a protomer. Whereas most dissociating enzymes can be converted to the monomer by appropriate ligands under physiological conditions, this was achieved with β-alanine synthase only

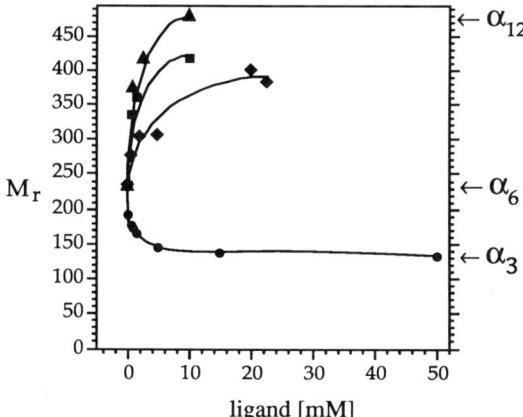

FIG. 9. Ligand-induced change in the oligomer assembly of β-alanine synthase. The molecular weight was measured for the enzyme in the presence of the ligands: NCβA (■), β-alanine (●), N-amidino-β-alanine (♦), and propionate (▲). The size of the three major oligomeric species is shown. M_r is in kDa.

under nonphysiological conditions. At pH 8.0 and 4°C, the enzyme showed two forms, a trimer and a monomer, by size-exclusion chromatography. Both species of enzyme exhibited activity. Because rapid kinetic studies had shown the trimer to be inactive, these last results suggest that even the fully dissociated monomer can reform into an active oligomer on binding the substrate, as diagrammed in Fig. 2.

This enzyme also has the interesting property of exhibiting an activation constant for the substrate that is significantly lower than the K_m for that substrate. β-Alanine synthase has a K_m of 8 μM for the substrate, but an S_{act} of 9 nM. This value was obtained from the kinetic experiment shown in Fig. 10 (57), where the enzyme shows distinct positive cooperativity at the lowest concentrations of substrate, but no cooperativity above 15 nM NCβA, at which concentration the enzyme has achieved the fully active form, with a Hill coefficient of 1.0. Thus, this enzyme becomes 50% activated at a substrate concentration of 9 nM. This strong affinity shown for the activation response suggests a separate regulatory site for the substrate, a feature that may appear to be unnecessary because the substrate must bind at the catalytic site, and could presumably achieve conformational effects on binding there. However, there is an expanding set of well-characterized allosteric enzymes that share this more complex feature of a separate regulatory site to mediate conformational effects by a substrate. Among these enzymes are hexokinase (72, 209, 210), phosphofructokinase (211, 212), and purine nucleoside phosphorylase (213).

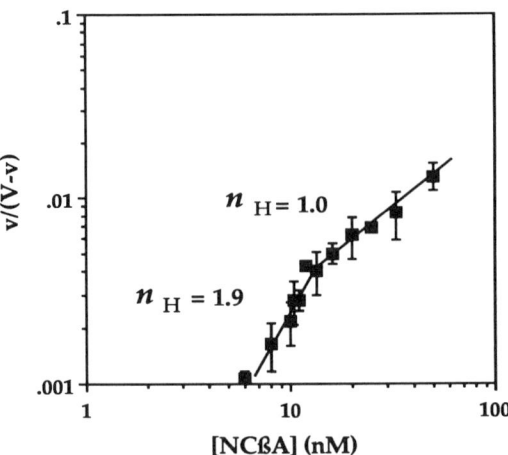

FIG. 10. Hill plot for β-alanine synthase. Pure enzyme was used to test activity at concentrations of the substrate, NCβA, far below the K_m, which is 8 μM.

The interaction of ligands with β-alanine synthase is summarized in Fig. 11, which shows the structures of the different compounds, the effect they produce on dissociation or association, and their affinity for the enzyme. The two alternate substrates bind with high affinity, and each causes the enzyme to associate toward the fully active dodecamer. The two alternate products bind fairly poorly, and β-alanine causes the enzyme to dissociate to the inactive trimer, whereas 2-Me-β-alanine does not produce the allosteric response leading to association or dissociation. Because γ-aminobutyrate also inhibits the enzyme and produces dissociation, it appears that a positively charged amino group is necessary, although not sufficient, for binding at the site on the enzyme that leads to oligomer dissociation.

This interpretation is not inconsistent with the data shown for the substrate analog N-amidino-β-alanine. This analog has a positively charged guanidino group; the pK_a for this group is expected to be about 11–12, so that at pH 8, less than 10^{-3} of this ligand would be in the unprotonated, neutral form. But, compared with the two substrates, the unprotonated form is expected to bind at the same site, and consistent with this interpretation is the unusually high K_i for N-amidino-β-alanine. If it is the neutral species,

		ΔM_r	(K_m) or K_i
			μM
substrates:			
N-carbamoyl-ß-alanine: (ureidopropionate)	$NH_2 - \overset{O}{\overset{\|}{C}} - NH - CH_2 - CH_2 - COO^-$	↑	(8 ± 2)
2-Me-N-carbamoyl-ß-alanine: (ureidobutyrate)	$NH_2 - \overset{O}{\overset{\|}{C}} - NH - CH_2 - \underset{\underset{CH_3}{\|}}{CH} - COO^-$	↑	(6.1)
substrate analogs:			
N-Amidino-ß-alanine: (guanidinopropionate)	$^+NH_2 = \overset{NH}{\overset{\|}{C}} - NH - CH_2 - CH_2 - COO^-$	↑	11,000
propionate	$CH_3 - CH_2 - COO^-$	↑	90
products:			
ß-alanine	$^+NH_3 - CH_2 - CH_2 - COO^-$	↓	1,100
2-Me-ß-alanine (ß-aminobutyrate)	$^+NH_3 - CH_2 - \underset{\underset{CH_3}{\|}}{CH} - COO^-$	0	3,900
product analog:			
γ-aminobutyrate:	$^+NH_3 - CH_2 - CH_2 - CH_2 - COO^-$	↓	1,600

FIG. 11. Ligands that bind to β-alanine synthase. Alternate ligand names are shown in parentheses. Under "ΔM_r" is shown whether the ligand leads to aggregation of the native hexamer to form the active dodecamer (↑), or dissociation to form the inactive trimer (↓).

present at very low concentrations, that is actually binding to the enzyme, the apparent K_i for this unprotonated species would be in the neighborhood of 10 μM, very similar to the affinities shown by the two natural substrates. Of additional interest are the results with propionate, a compound that would normally be designated as a product analog. But when one considers that the affinity of propionate is much more like that of the substrates, and that the allosteric results produced by propionate are very much like the substrates (Figs. 9 and 11), then propionate behaves as a substrate analog.

The results with 2-Me-β-alanine (β-aminobutyrate) appear inconsistent with the other ligands. Because many of the ligands are either substrates or products, it is expected that they bind at the same site. The extra methyl group on 2-Me-N-carbamoyl-β-alanine does not hinder binding of this substrate, or the allosteric response of β-alanine synthase to this substrate, thus the product made from this ligand should bind effectively and have effects comparable to the alternate product, β-alanine. However, previous studies made no effort to distinguish between the L and D forms for the chiral center of β-aminobutyrate. Because it is likely that one enantiomer is bound preferentially, this could well explain the apparent increase in the K_i of β-aminobutyrate in comparison to β-alanine (Fig. 11); these studies remain to be completed. Alternatively, allosteric regulation could be mediated at a separate regulatory site, and thereby respond more specifically to β-alanine, whose synthesis needs to be controlled. In support of this hypothesis are the results showing that the enzyme shows two types of affinity for the substrate N-carbamoyl-β-alanine (Fig. 10), which could also be the result of two separate binding sites.

Although controlling the physiological levels of β-alanine must be important, the *in vitro* kinetic studies suggest that β-alanine synthase will always be in the active conformation; i.e., the affinity of the substrate at the postulated regulatory site is so much lower (S_{act} = 9 nM) than the K_m (8 μM). Thus, under most conceivable physiological concentrations of NCβA [normal ~10 μM (57)], the enzyme should be active. Therefore, the interesting question: How is β-alanine synthase regulated? One hypothesis is that a second (competing) regulatory ligand normally functions in opposition to the substrate. Such a competing ligand has not yet been identified, but these studies have elucidated basic properties of the enzyme that will make identification of such a putative inhibitory ligand possible.

Because this enzyme has now been purified from several sources, and partially characterized, it is possible to begin an assessment of what features are commonly conserved, and what features may be unique to different organisms, as shown in Table X. The subunit M_r for the mammalian enzyme is about 42,000, although there is some variability. This value is in agreement with the predicted M_r from the sequence for the rat liver enzyme (214). The

TABLE X
PHYSICAL AND KINETIC PROPERTIES OF β-ALANINE SYNTHASE FROM DEFINED SOURCES

Source	Subunit M_r	Native oligomer	Affinity for NCβA K_m (μM)	Affinity for propionate K_i (μM)	Cooperativity for NCβA	Ref.
Rat liver	42,000	6	8	90	Positive	57, 58
Rat liver[a]	54,000	6	170	300	Positive	200, 207
Calf liver[a]	38,500	6	500	35,000	None	208
Pseudomonas putida	45,000	2	3740	4690	None	216
Euglena gracilis	?	≥24	38	160	None	215

[a] Purification involved a heat step. NCβA, N-carbamoyl-β-alanine.

subunit for the enzyme from *Pseudomonas putida* is very similar in size. The mammalian enzymes are stable hexamers; the enzyme from *Euglena gracilis* is very large, and if it has a subunit of comparable size, it would form the large oligomers shown in Table X. Only the enzyme from rat liver shows allosteric kinetic properties (57, 58, 200).

The loss of affinity for ligands due to heat treatment in the purification has already been described. It is therefore interesting that the enzyme from *E. gracilis* is the only other example that shows very high affinity for both the substrate NCβA and the substrate analog propionate (215). The most different enzyme in Table X is β-alanine synthase from *P. putida*, which has an unusual dimeric form and is active on a large number of substrates (216). This feature may require less discrimination at the catalytic site, and thus explain the very poor affinity for both the substrate and substrate analog shown in Table X. It is quite remarkable that two enzymes with the same catalytic reaction differ in K_m value for the same substrate by almost 500-fold.

B. The cDNA for β-Alanine Synthase from Rat Liver

The cDNA for β-alanine synthase was cloned from rat liver cDNA libraries in λgt11, using affinity-purified polyclonal antibodies against β-alanine synthase protein. This cDNA codes for 1432 bp, and the open reading frame codes for 393 amino acids (M_r = 44,042) (214), in good agreement with the M_r of ~42,000 for the native enzyme on SDS gel electrophoresis (57). Chemical analysis of the native enzyme showed two zinc atoms per subunit, and the sequence of β-alanine synthase contains two putative zinc-binding site motifs. Comparison of the amino-acid sequence to sequences in the protein data base showed that it has about 20% identity to aspartate carbamoyltransferase, ornithine carbamoyltransferase, urease, and leucine aminopeptidase, enzymes that bind comparable ligands or have a similar mechanism.

```
ACT   1 ANPLYQKHIISIN.DLSRDDL.................NL.............VLATAAK   29
        |.| :|.       :::  .|..|||           ||            |:  |
βAS   2 AGPEWQSLEQCLEKHLPPDDLSQVKRI....LYGKQTRNL.............DLPRKA    43
        .:|::  |  ::  ::| |...:.|.::  .|    |  :  .||.:       :|.:|.
LAP  16 DEPQFTSAGENFNK.LVSGKLREILNISGPPLKAGKTRTFYGLHEDFPSVVVVGLGKKT.   73

ACT  30 LKANPQPELLKHKVIASCFFEASTRTRLSFETSMHRLGASVVGFSDSAN.....TSL...   81
        |.|..:.::  ....:  ::  :|:|..  :  : ...  :::...||           ..
βAS  44 LEAASERNF....ELKGYAFGAAKEQQRCPQIVRVGLVQNRIPLPTSAPVAEQVSAL...   96
        :  ...: |:.         .:  ||  ...  |.||    :::.   |...|.:...|
LAP  74 AGIDEQENW....HEGKENIRAAVAAG.CRQIQDLEIPSVEVDPCGDAQAAAEGAVLGLY  128

ACT  82 ....GKKGETLADTISV....ISTYVDAIVMR.............HPQEGAARLATEFSG  120
            |:.|.|:|:.    :   ::   :|     |.              :.:... |.| |:.
βAS  97 ....HKRIEEIAEVAAMCGVNIICFQEAWNMPFAFCTREKLPWTEFAESAEDGLTTRFCQ  152
        |.   .:.    |  : |       |||||....|..   :.|::      :...|::    .|:|.
LAP 129 EYDDLKQKRKVVVSAKLHGSED...QEAWQRGVLFASGQNLARRLMETPANEMTPTKFAE  185
                                                            | catalytic →

ACT 121 ................NVP...VLNAGDG.........................SNQH  134
        |           :||:  : :
βAS 153 ......KLAKKHNMVVISP...ILERDRD.........................HGGV  176
        | |.. .. |.|.|       | |.. ::                             ...:
LAP 186 IVEENLKSASIKTDVFIRPKSWIEEQEMGSFLSVAKGSEEPPVFLEIHYKGSPNASEPPL  245
                                                            | catalytic →

ACT 135 PTQTLLDLFTIQETQGRLDNLHVAMVGDLK..........YGRTVHSLTQA.LAKFDGN  182
        ..|  :  :   .  |:   .  |::.||||.             |.  .|.:  |.  :::::.|
βAS 177 LWNTAVVISNSGLVMGKTRKNHIPRVGDFNE.....STYYMEGNLGHPVFQTQFGRIAVN  231
        ::    . .:. ::||  |.   |:::|:.           |..    ..:|:  |.  .::.:.  |
LAP 246 VFVGKGITFDSGGISIKAAANMDLMRADMGGAATICSAIVSAAKLDLPINIVGLAPLCEN  305

ACT 183 RFYFIAPDALAMPQYILDMLDEKGIAWSLHSSIEEVMVE.....VDILY............  226
        :::     ..:|         .::|    :::  ::|   |:.     ..|:|         :  ::
βAS 232 .ICYGRHHPL...NWLMYSVNGAEIIFNPSATIGEL........SESMWPIEARNAAIAN  279
        :.  |: :.         |  |:  :.                     |  ...|       |||..
LAP 306 .MPSGKANKP...GDVVRARNGKTIQVDNTDAEGRLILADALCYAHTFNPKVIINAATLT  361
                                           catalytic →  |

ACT 227 ......MTRVQKERLDPSEYANVKAQ.....FVLRASDLHNAKANMKVLHPLPR......  269
              :.||..|:.  |.|:.....:.           :..       .  |  ::     ...|.|
βAS 280 HCFTCALNRVGQEHY.PNEFTSGDGKKAHHDLGYFYGSSYVAAPDGSRTPGLSRNQDGLL  338
        .:..||....|     .:.|.                :.  :..|.   .:..|  |  :   :.:.|
LAP 362 GAMDIALGSGATGVFT.NSSW..........MNKLFEASIETGDRVWRMPLFEHYTRQVI  410
                         catalytic →  |

ACT 270 VDEI....ATDVDKT.........PHAWY.FQQAGN.GIEARQALLAL.......VLNRDLVL    310
        |.|:     .   ::         :.::    |..  .|.  :::||:      |:         :::.  ||||
βAS 339 VTEL....NLNLCQQ.........INDFWTFKMTGRLEMYARELAEAVKPNYSPNIVKEDLVLAPSSG  393
        .:|    |:: ..           :.|     |.|  |::::              |    :   ....:|
LAP 411 DCQLADVNNIGKYRSAGACTAAAFLKEFVTHPKWAHLDIAGVMTNKDEVPYLRKGMAGRPTRFSQDSA  478
```

FIG. 12. Alignment of the sequence for β-alanine synthase (βAS) with the sequences for aspartate carbamoyltransferase (ACT) and leucine aminopeptidase (LAP). For ACT and LAP with a defined crystal structure, the catalytic region is indicated, as are key residues important for catalysis.

Because two of these other sequences are for enzymes for which high-resolution crystal structures have been determined, the sequence of β-alanine synthase was separately aligned with each of these. The alignment was done with a standard algorithm for this procedure (217), and the two results are superimposed in Fig. 12. The percent sequence identity of β-alanine synthase to the other two sequences in Fig. 12 is below the threshold commonly used to establish common ancestry for proteins. However, a common ancestry is supported by the unusual oligomeric structure of these three enzymes. Each of them is composed of a trimeric protomer, which is very stable and is dissociated only by denaturing conditions. For each of the three enzymes, this protomer in turn acts as a unit for further aggregation to form the active oligomer: six catalytic subunits are required in the active form of aspartate carbamoyltransferase and leucine aminopeptidase, and twelve for β-alanine synthase. Thus, the greater-than-average level of sequence identity, combined with the shared and highly unusual oligomeric structure, will support the interpretation of a common ancestry. This facilitates analysis of β-alanine synthase by helping to identify the likely catalytic region of the protein, and, especially where critical amino-acid residues in one of the other sequences are shared by β-alanine synthase, this gives a first estimate for those residues to be tested experimentally.

The common ancestry suggested for the three enzymes is not contradicted by the fact that they do not share a common chemical reaction mechanism, and common ancestry of the other two proteins was already suggested by the authors for their structures (218). An interesting feature for both of those structures is that the catalytic centers are oriented toward each other, forming a tunnel between each adjacent pair of subunits within the interior of the trimer. Especially interesting is that these tunnels extend so that they are superimposed from one trimer onto the other trimer in the native hexamer of leucine aminopeptidase. That is, when this hexameric oligomer is viewed from one end, three tunnels are evident that go all the way through the oligomer (218). To the extent that β-alanine synthase shares such an oligomeric structure, this architecture would help to explain how modest changes in the alignment of subunits could lead to profound conformational changes. These would account for the dramatic allosteric consequences that have been observed for this enzyme (57, 58) (see Figs. 1 and 10).

VI. Enzyme Evolution

Our work on sequencing the cDNA for uridine kinase led naturally to comparing the amino-acid sequence for this protein to other kinases or nucleotide-binding enzymes. The goal of such sequence comparisons is the possibility of identifying established structural and functional elements in the new sequence. This requires that the reference protein has one or more established crystal structures, and that a convincing identity can be observed for the two sequences being compared. A few results from this analysis are shown in Fig. 7, and this work has led to the identification of consensus peptide segments widely found to form the ATP-binding site of many kinases, and also the nucleoside monophosphate-binding site of enzymes that phosphorylate these substrates (168).

A. Motifs That Define Peptide Segments in Ligand-Binding Sites

The initial identification in a few kinases of peptide segments involved in binding ATP (219) has led to a widely expanding field for defining fairly specific amino-acid sequences that code for polypeptide segments having the same ligand-binding function in many different proteins. Two proteins may have a very similar ligand-binding pocket because (1) they are descended from a common ancestor, and retain much of the original structure and function, or (2) they have arisen by some genomic recombinatorial mechanism that has placed similar polypeptide regions, specifying the same ligand-binding pocket, into otherwise different proteins. The success of the above model is due to the wide availability of supporting data, and to the underlying assumption (supported by data) that the amino-acid sequence defining a specific structure does not usually change that much over evolutionary time. The outcome of this approach is that specific sequence motifs, amino-acid sequences defining a specific structure or function, are frequently able to identify a ligand-binding site on a novel protein.

B. Divergence of Kinases: Different Kinases with the Same Pattern of Motifs

1. Types of ATP-Binding Sites

Using the available structures for 16 different kinases in the protein data base, it was possible to identify a family of such motifs that were required to form different types of binding sites for ATP, as well as for AMP and GMP (168). Although a single universal ATP-binding site might appear to be optimal, this is in fact not possible because function dictates the type of protein structure needed. Table XI briefly summarizes the known functions for ATP, and it is important to see that even when ATP is used as a substrate, the specific details of the catalytic mechanism suggest that the local architecture at the catalytic site be appropriate for one of the five catalytic mechanisms involving ATP. ATP may also bind at a noncatalytic site, where it acts as an inhibitor or an activator, and specific structures for each of these are known.

TABLE XI
Types of ATP-binding Sites

Function	Examples	Consensus segments
A. Catalytic site		
1. Direct transfer of γ-P to acceptor	Most kinases	*Kinase-1, Kinase-2, Kinase-3*
2. Transfer of γ-P via P-enzyme	NDP kinase	Insufficient data
3. Transfer of γ-P to ADP → ATP	Phosphoglycerate kinase	*Kinase-1, Kinase-3, Kinase-2*
4. Transfer of β,γ-PP to acceptor	Diphosphokinases	Insufficient data
5. Transfer of AMP to acceptor	Aminoacyl-tRNA synthetases	*HIGH, MSK, Kinase-3*
B. Regulatory site		
1. ATP as activator	Aspartate carbamoyltransferase	*Regulatory-1, Regulatory-2*
2. ATP as inhibitor	Phosphofructokinase	Insufficient data

The analysis of the data bases for protein structures, as well as for protein sequences, led to the identification of the motifs shown in Table XII (168). The *Kinase* motifs identify the peptide segments for an NTP-binding site, and were obtained from enzymes that bind ATP or GTP. The NMP motifs were obtained from proteins that bind AMP or GMP as substrates. Five of these sequence motifs in Table XII have one or two invariant residues, but two motifs have not even one invariant residue. Therefore, a simple search with one motif sequence against the sequence for any selected protein will frequently identify more than one such motif in the target sequence.

The motifs were named for their function, as well as for the likely position (N- to C-terminal) in the kinase, as shown in Fig. 13. This figure illustrates the linear structures of the proteins for various kinases in pyrimidine

TABLE XII
Consensus Sequences for Peptide Segments Involved in Nucleotide-binding Sites

Segment	Consensus[a]
Kinase-1a	(G, A, S, N) X$_4$ (G, A, C, S) **K** (G, S, T, V, A, P) (T, S, A, D, G, N, M)
Kinase-2	(V, G, I, L, N, T, A, Y, K,) (A, F, L, I, G, D, E, T, C, K, P) (A, L, I, G, V, S, P, E, F, H, T) (L, G, V, I, T, D, F, Q, M, Y, K) **D**
Kinase-3a	(E, A, G, P, F) (T, S, G, F) X$_3$ (**Y**, **R**)
Kinase-3b	**N K** X (**D**, **W**)
NMP-1	(L, I, V, K, E, S) (S, L, Y, A) (**T**, S, A, V, L) (G, S, T, R, N, K) (**D**, R, F, H) (L, M, T, P) (L, F, P, **P**, E, Q) **R**
NMP-2	(K, E, G, M, V, C, S, P, Q) (**L**, I, F, Y, K, Q, S) (V, L, I, K, D, E, T, Q) X (D, E, L, I, G, N, S, T) (D, E, Q, A)
NMP-3a	G (**F**, Y, T, S,) (P, S, T) (**R**, V) X$_{3-4}$**Q**

[a] Within parentheses, residues are listed in decreasing frequency of occurrence at that same position. Residues in bold type are invariant, or occur in a majority of sequences.

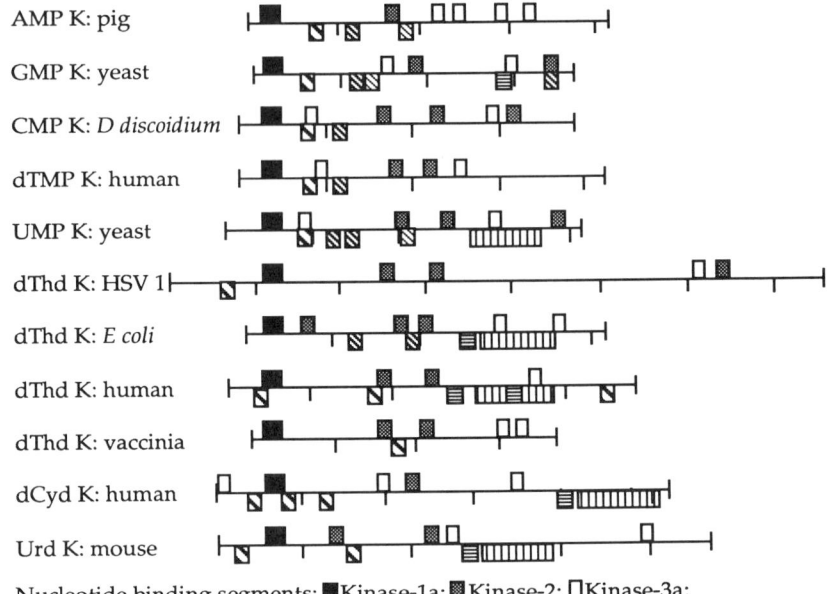

Nucleotide binding segments: ■Kinase-1a; ▨Kinase-2; ▫Kinase-3a;
◩NMP-1; ◪NMP-2; ◨NMP-3; ◩ + ⊞ = regulatory site.

Fig. 13. Defined position of motifs for binding NTPs or NMPs (K, kinase). Intervals of 50 amino acids are indicated.

and purine metabolism, because five of these proteins have a crystal structure defining the ATP-binding site. Appropriate symbols are used to distinguish the different motifs. In these sequences the *Kinase-1a* motif is never duplicated, and also occurs at or near the N-terminus. The schemes were therefore aligned in Fig. 13 so as to place the *Kinase-1a* motif at the same position.

Although there are frequently multiple copies for the other motifs, it becomes evident that the *Kinase-2* motif is normally near the middle of the protein, and the *Kinase-3a* motif is nearer the C-terminus. Because this spatial pattern emerged in the data set, the position and distance from *Kinase-1a* of the *Kinase-2* and *Kinase-3a* motifs in many sequences were determined, leading to the result that the *Kinase-2* motif most commonly is located at an average of 61 amino acids (range, 49–80) from *Kinase-1a*, although a few enzymes have a separation of about 145 amino acids (*168*). The same analysis showed that the position for *Kinase-3a* is normally at a distance of 126 amino acids from *Kinase-1a*.

The particular spatial positioning of the three motifs needed to form an NTP-binding site was also evident in additional kinases that function in various other metabolic pathways (*168*). Although it may seem less likely that all of these enzymes have diverged from a single common ancestor, the alternative mechanism to explain such common sequence and structure features is that convergent evolution frequently led to the same structure for enzymes with the same function of transferring the γ-P of an NTP to an acceptor (function A1 of Table XI). A valuable outcome of this analysis is that although individual sequence motifs, by themselves, may not be sufficiently specific to locate a functional peptide in a protein sequence, when two or three different motifs are used in a search, and when the identified locations are then screened for the distance positions established for the data set above, the likelihood of accurately finding the elements of an NTP-binding site becomes very high.

2. Types of Nucleoside-Monophosphate-Binding Sites

The same analysis of protein structures and sequences led to the identification of three motifs required to form a binding site for a nucleoside monophosphate (*168*). These *NMP* motifs are shown in Table XIII, and are numbered for their spatial position along the protein. Kinases that phosphorylate the different nucleotides AMP, GMP, and UMP have very similar *NMP-1* motifs, with an invariant arginine that bonds to the phosphate on the acceptor nucleotide substrate (*164, 166, 220–222*), and sometimes with the purine base of AMP. Also, adenylate kinases and uridylate kinase have very

TABLE XIII
Sequences for Peptide Segments Required to Bind a Nucleoside Monophosphate

Enzyme and source (NMP)	Segment [a]		
	NMP-1	*NMP-2*	*NMP-3*
Adenylate kinase:			
Pig (AMP)	[37]L S **T G** D **L L R**	[62]**E** G **K G Q** **L V P** L E T **V**	[91]**L** I **D G** Y**P R E** V K Q
Escherichia coli (AMP)	[29]L S **T G** D**M L R**	[53]**M**D A **G** **K L V T** **D E L V**	[82]**L L D G** **F P R T** I P Q
Bovine (AMP)	[34]L S **S G** D **L L R**	[58]I D **Q G** **K L I P** D D V M	[86]**L L G G** **F P R T** L P Q
Yeast (AMP)	[35]**L A T G** D**M L R**	[59]**M**D **Q G** **G L V S** D D I M	[89]I **L D G** **F P R T** I P Q
Guanylate kinase			
Yeast (GMP)	[34]**S** S T T **R T P R**	[57]**E** F K S **M I** K N **N E F I**	[68]I **E W A** **Q F S G** **N**Y Y G S
Uridylate kinase			
Yeast (UMP)	[45]L **S** **A G** D L **L R**	[70]I **K E G** **Q** I **V P** **Q E** **I T**	[101]**L** I **D G** **F P R K** M D Q
OPRTase			
Salmonella typhimurium (OMP)	[28]**L K** **S G** **L S** **S P** I Y	[114]**P G E T** **C L** I I **E D V V** T S G S	[149]**A** I V **L** **L D R E** Q G G

[a] Residues in bold were shown to bind to the ligand at the active site. Numbers give the amino-acid residue position in the sequence, and residues at consensus positions are underlined.

similar *NMP-3* motifs, which interact more with the base. The kinases all have the same reaction mechanism, and transfer a second phosphate onto the α-P of their NMP substrate. By comparison, the established binding site for OMP in the structure of OPRTase shows a somewhat similar *NMP-1* motif at the appropriate position (76), with a comparable basic amino acid, lysine, bonding to the phosphate. Furthermore, OPRTase has an *NMP-3* motif at the appropriate position that has some similarity to the corresponding *NMP-3* motif for the adenylate and uridylate kinases. OPRTase has an *NMP-2* motif that is not at a comparable position in the kinases, and that has only modest similarity to the *NMP-2* motif of the kinases. This motif is at the center of the OPRTase catalytic site and involves residues critical to the reaction, so that it is expected that a difference should be evident here. This particular *NMP-2* motif, shown for the OPRTase in Table XIII, has previously been found in about 60 sequences for various phosphoribosyltransferases and phosphorylases (223). Here we have evidence that enzymes with somewhat different catalytic mechanisms share a consensus sequence that defines part of the catalytic site. Although the mechanism for the reaction may vary, the unifying feature is that all these enzymes bind a nucleoside or nucleotide at this catalytic site.

The observation that a nucleotide-binding site on a phosphoribosyltransferase has some resemblance to a nucleotide-binding site on various kinases was not expected. These enzymes have very different reaction mechanisms, and it is not expected that they all evolved from a common ancestor, though this is likely for the kinases in Table XIII. However, an interesting relationship has already been observed for OPRTase with ODCase, the other domain on the bifunctional UMP synthase. Alignment of the two sequences for the separate yeast proteins showed an identity greater than 24% for a central region in both sequences (*114*), and this alignment was used to suggest the possible binding site for OMP in the OPRTase. This prediction was quite good, because the recent crystal structure for OPRTase shows that this peptide sequence is involved in binding to the pentose and the phosphate of the ligand OMP (76). Such sequence alignments thereby demonstrate the ability of this approach to help identify ligand-binding sites.

The conserved structure for the three nucleotide kinases of Table XIII has been described. This led to the alignment of the sequence for uridine kinase to the sequence for uridylate kinase (Fig. 14), and the possibility of predicting a comparable tertiary structure for uridine kinase. The alignment of the sequence motifs that define the ATP-binding site is evident in Fig. 7. In Fig. 14 can be seen the three NMP motifs for uridylate kinase, which define the binding site for the acceptor substrate, UMP, as determined in the crystal structure for this enzyme (*164*). The likelihood that both kinases in Fig. 14 have evolved from a common ancestor suggests the possibility of

```
                          1                                    Kinase-1a            50
Urd-K_mouse      MASAGGGGSE STALARGRRP QPRPFLIGVS GGTASGKSTV CEKIMELLGQ
                 ::  ::::.::           . .   :| |.  ||.::||:|  |||::. :
UMP-K_yeast      ........MT AATTSQPAFS PDQVSVIFVL GGPGAGKGTQ CEKLVKDY..

                 51                                                       98
Urd-K_mouse      NEVDRRQRKL VILSQDCFYK VLTAEQKAKA LKGQYNFDHP DAFDN...DLM
                 .  .......   .:|:::      ..    :.|.::.  :|:  .:  :..  .:.:     :|:
UMP-K_yeast      ..SFVHLSAG DLLRAEQGRA ...GSQYGEL IKNCIKEGQI VPQEITLALL
                      UMP-1                                        UMP-2

                 99          Kinase-2                                     148
Urd-K_mouse      HKTLKNIVEG KTVEVPTYDF VTHSRLPETT VVYPADVVLF EGILVFYTQE
                 ::::.: |.:  :..  . ..|      .:|  .:.:  :.:..|:|.    :  ||:|..:|
UMP-K_yeast      RNAISDNVKA NKH.KFLIDG ..FPRKMDQA ISFERDIVES KFILFFDCPE
                                          UMP-3

                 149                            Kinase-3a                198
Urd-K_mouse      IRDMFHLRLF VDTDSDVRLS RRVLRDVQRG RDLEQILKQY TAFVKPAFEE
                 ....:               ||.  .|...:  .|:  .|..:.:|:.  .:..|.:...
UMP-K_yeast      DIMLE..... .......RLL ERGKTS.GRS DDNIESIKKR FNTFKETSMP

                 199                                                      248
Urd-K_mouse      FCLPTKKYAD VIIPRGVDNM VAINLIVQHI QDILNGDLCK RHRGGPNGRN
                 ...  .::::::  |:..|....:  .::...||:.  ...:.
UMP-K_yeast      VIEYFETKSK VVRVRCDRSV EDVYKDVQDA IRDSL..... ..........
```

FIG. 14. Alignment of the sequence for uridine kinase (Urd-K) with the sequences for uridylate kinase (UMP-K). For uridylate kinase with a defined structure (164), the catalytic region is indicated, as are key residues important for binding the substrates.

locating a uridine-binding site by comparison to the position for the UMP-binding site. However, although the alignment shows a high identity for the ATP-binding site at the *Kinase-1a* motif, and a moderate identity at the *Kinase-3a* motif, there is only marginal sequence identity for the three NMP motifs. Therefore more data are desirable to establish the uridine-binding site, although the current alignment may still serve as an initial hypothesis, which can then be tested by appropriate experiments to modify this site on the protein by covalent modification or by site-directed mutagenesis.

Comparison of Figs. 7 and 14 also points out the limits to the information that may be obtained from such sequence alignments. The crystal structure for uridylate kinase was obtained with free ATP bound, and thus did not show the binding for *Kinase-2* (164), which normally coordinates to the Mg^{2+} ion when Mg-ATP is used. Although the expected *Kinase-2* segment was not observed in the structure, the most likely position for this motif was readily identified when the sequence for uridylate kinase was aligned with the two nucleotide kinases for which *Kinase-2* had been established (Fig. 7). As is also evident in Fig. 7, the three uridine kinase sequences all have a *Kinase-2* motif at a common position that is different from the nucleotide kinases. However, when only uridine kinase and uridylate kinase are aligned, as in Fig. 14, an alternative *Kinase-2* motif in the uridine kinase sequence now aligns directly with the *Kinase-2* motif of uridylate kinase. This last alignment again emphasizes the inherent ambiguity of small sequence motifs that contain only one invariant residue, such as *Kinase-2*.

C. Gene Fusion: Multifunctional Proteins

Except for dihydro-orotate dehydrogenase, the enzymes for the *de novo* UMP pathway all function in the cytoplasm of all higher eukaryotes. Perhaps this common spatial location has facilitated the process by which the other enzyme activities in the *de novo* pyrimidine pathway have come to be associated into multifunctional proteins in eukaryotic organisms (53). Such proteins contain two or more domains, with each domain functioning as a separate catalytic center, and it is now reasonably established that this protein architecture has in most cases evolved from gene fusion. For example, in some bacteria the two separate genes coding for OPRTase (*pyr E*) and ODCase (*pyr F*) are juxtaposed in the bacterial DNA, where this spatial configuration permits the joint regulation of these genes.

Such regulons containing *pyr E* and *pyr F* have now been identified in *B. subtilis* (224) and in *Bacillus caldolyticus* (225). Given such a favorable gene localization, it would then be possible for a simple mutation to remove a termination signal at the end of each gene, such that the RNA polymerase would continue to read through subsequent genes. This would produce a larger mRNA and a larger protein product. The short noncoding nucleotide regions between genes would then become expressed, and presumably lead to connector or linker peptide regions between catalytic domains.

An obvious requirement is that all genes after the initial gene be in the same reading frame. This organization of genes leading to fusion and multifunctional proteins is observed in different organisms and for different metabolic pathways (226). One of the best examples for coordinating an entire pathway by such architecture is fatty acid synthase (227). In *E. coli* there are eight separate genes that produce eight proteins, which become associated in a multienzyme complex. Yeast has clustered all these genes into two larger genes and produces an enzyme complex with the two different protein subunits that are now specified. Mammals have completed this gene consolidation by having a single gene, equivalent to the eight bacterial genes, and having all functions organized on a single multidomain protein.

A clear exception to this gene fusion model is the example from the bacterial thermophile *Thermotoga maritima*, which expresses a 70-kDa protein having catalytic centers for phosphoglycerate kinase (PGK) and triosephosphate isomerase (TIM) (228). This organism has separate genes for the two enzymes, and the genes are directly adjacent. A cDNA from a library contained the entire triosephosphate isomerase sequence, but lacked a start codon, although it had at the 5' end an extra 19 codons from the phosphoglycerate kinase sequence. These data were interpreted in support of the polymerase undergoing a frame shift near the end of the PGK gene, and then continuing in a new and correct reading frame through the TIM sequence. This novel mechanism has the effect of producing the normally observed fusion protein.

Additional structural support for the multidomain architecture of different multifunctional proteins defined above comes from studies on such isolated proteins in which selected proteolysis was used to cleave connecting loops between the domains, with the object of defining the remaining catalytic activity associated with the separated domains that are often obtained by careful proteolysis. For the multifunctional protein containing most of the *de novo* pyrimidine pathway, called CAD or dihydro-orotate synthase, tryptic digestion of the 240-kDa protein produced fragments with sizes comparable to the individual gene products in bacteria, and also had the same enzyme activity as the individual bacterial enzymes (229–231). Trypsin cleavage of UMP synthase (subunit M_r of 52,000) also produced a separate domain of M_r 28,500 that contained only the ODCase activity, but the OPRTase domain did not survive this treatment (101).

An additional approach to characterize further such multifunctional proteins has used the current technology of making cDNAs coding for a defined amount of protein, and then expressing them in appropriate hosts to obtain abundant quantities of such protein constructs. For dihydro-orotate synthase this approach has been used successfully to produce only the catalytic domain coding for dihydroorotase (232, 233), or the amidotransferase domain (234), and also for the regulatory domain (235). For UMP synthase, this experimental approach has led to the cloning and expression of the two separate cDNAs that code for the human OPRTase and ODCase domains (91, 92). When these cDNAs were expressed in *E. coli*, the ODCase was expressed very effectively, whereas expression of the OPRTase was less efficient (91). When two similar cDNAs were expressed via the baculovirus system in cabbage looper larvae, both domains were obtained in good yield and with good enzyme activity (92). Thus, the studies cited above have led to the current model that multifunctional proteins could have evolved by gene fusion, and that the separated catalytic domains fold separately to maintain the structure and function originally coded for each catalytic domain.

Although the appearance of multifunctional proteins is widespread, and is evident in both prokaryotes and eukaryotes, the absolute importance of such a protein architecture remains unclear, because this pattern is not sufficiently systematic. It is not used consistently for the same enzyme cluster in different organisms. And though many mammals use this architecture, even this single group does not use this clustering strategy as systematically as it might. By using gene fusion for covalently joining enzymes that are normally sequential in a specific metabolic pathway, several possible advantages occur, as shown in Table XIV. The first of these functions must be important, because it is presumed to be the reason why separate genes coding for enzymes in the same pathway are commonly clustered in bacteria into regulons and thus become subject to coordinated regulation. However, this function can obviously be served by the simple placement of the genes into a local lineup without an obligatory fusion event.

The second function in Table XIV is not as obvious. The specific catalytic activity for different enzymes can easily vary by many orders of magnitude, and it is frequently assumed that it is the absolute activity of each enzyme in a pathway that should be at a similar level. This appears desirable, because a metabolic bottleneck may arise when a "fast" enzyme is followed by a "slower" enzyme, leading to accumulation of the intermediate metabolite. However, the reverse pattern is observed when a slow enzyme is followed by a more rapid activity. A specific example of this comes from studies with OPRTase and ODCase from *H. pylori*, in which the ODCase activity is 50-fold the OPRTase activity (21). A possible explanation for these results is that the equilibrium for the OPRTase reaction does not favor the synthesis of OMP (55).

The OPRTase reaction can be made to continue in the biosynthetic direction if OMP does not accumulate, by having this product rapidly removed by conversion to UMP. Thus, when these two catalytic centers are on separate proteins that have never been shown to interact, then OMP synthesized by the OPRTase must diffuse into the bulk solvent. Because the OPRTase has a high affinity for OMP [a K_m of 3.1 µM for the enzyme from *S. typhimurium* (90)], the back reaction of OMP to orotate could then hinder sufficient biosynthesis of UMP. Therefore, this may require appropriately higher levels of ODCase activity to maintain the necessarily low concentration of OMP required for biosynthesis to continue at a steady rate. However, if the kinetic properties of the second enzyme in any metabolic sequence are appropriately optimized, then it may not need to be expressed in excess to the

TABLE XIV
PROPERTIES ASSOCIATED WITH CLUSTERING OF MULTIPLE CONSECUTIVE ENZYME CENTERS INTO A MULTIFUNCTIONAL PROTEIN

1. Induction or repression is uniform
2. Separate enzymes are present at equimolar concentrations
3. Increased stability of catalytic centers when domains are stabilized by local interactions
4. Kinetic interaction between adjacent enzyme centers is facilitated

preceding catalytic center. It is significant that the high ratio of ODCase activity to OPRTase activity observed in bacteria is not found in any eukaryotic organisms. These organisms generally show the frequently observed ratio of about 1–3 for these two catalytic activities, whether they are protozoans such as *Trypanosoma* (13) or *Crithidia* (236), invertebrates such as *Schistosoma* (237), plants (238), or mammals (27, 28, 55, 81, 82, 239–241). Comparison of these results suggests that the coordinate expression of two sequential catalytic domains in a single protein is more likely when the kinetic features of each catalytic domain have been optimized to facilitate steady and continuous product synthesis.

If neighboring domains can interact when joined into a multifunctional protein, this may enhance the stability of such domains, relative to their condition in an isolated state (Table XIV, number 3). An example of this property comes from studies of the ODCase domain from mouse UMP synthase, which is sensitive to temperature. Activity with the isolated domain was constant at 0°C and fairly stable at 25°C for 20 minutes, but this same ODCase domain showed a significant loss of activity within 2–3 minutes at 37°C (101). However, when tested in the bifunctional human UMP synthase, the ODCase activity produced an almost linear Arrhenius plot in the temperature range from 4 to 45°C (96). These results demonstrate a greater sensitivity to thermal inactivation at *physiological* temperatures for the mammalian ODCase activity in the isolated domain, but not in the intact bifunctional protein. This observed difference in stability suggests that some form of protein–protein interaction is enhanced in the architecture of the native bifunctional enzyme.

The two catalytic centers OPRTase and ODCase are joined in a single multifunctional protein in all multicellular eukaryotes, and the initial catalytic centers for the synthesis of dihydro-orotate are joined in a multifunctional protein, thus an additional benefit that may be inherent in such an architecture for the covalent joining of sequential catalytic domains is the fourth function in Table XIV. Interactions between such catalytic centers are facilitated, and most commonly this refers to the preferential transfer of a common metabolite from the catalytic site where it is produced, to the next catalytic site, where it will function as a substrate. When such a transfer of a common ligand is very efficient, it may be referred to as *channeling*.

Kinetic studies with these two multifunctional proteins gave results showing a preferential flow of common metabolites between adjacent catalytic centers. With the first multifunctional protein, dihydro-orotate synthase, or CAD, this was observed for the transfer of carbamoyl phosphate between the carbamoyl-phosphate synthase and aspartate carbamoyltransferase sites (242, 243), as well as for the transfer of N-carbamoyl aspartate between the aspartate carbamoyltransferase and dihydroorotase sites (242, 243). This latter channeling result was duplicated for the same multifunctional protein obtained from Syrian hamster cells (244). For the same enzyme domains in the multifunctional protein from yeast, channeling of carbamoyl phosphate was also observed (245).

For UMP synthase, the preferential transfer of OMP between the two catalytic centers was observed for the mammalian enzyme (107, 246), although the use of a crude enzyme fraction introduced side reactions involving a nucleotidase and a uracil phosphorylase that caused some cycling of OMP → orotidine → orotate → OMP, which complicated the interpretation of channeling (246). Channeling of OMP between the UMP synthase centers was also observed for the enzyme from *Crithidia lucilae*, for which a channeling factor of 50 was calculated by comparing the amount of [6-^{14}C]UMP synthesized from endogenous [6-^{14}C]OMP produced by the OPRTase, vs. the amount of UMP produced from [7-^{14}C]OMP added exogenously to compete with the endogenous OMP in the same experiment (247). For such channeling studies, control experiments with the separate yeast enzymes can be informative. For the UMP synthase activities, almost no channeling was observed with the yeast enzymes, using the same experimental protocol (247).

VII. Conclusion and Future Prospects

Recent results on the three enzymes that directly lead to the synthesis of UMP, or the disposal of uracil, emphasize the importance of balanced metabolism, because all of these enzyme activities are allosterically regulated. Regulation by effector ligands involves the dissociation or reassociation of subunits with the oligomer; models have been presented to show how such conformational states are stabilized, and to illustrate how this could help to form or disrupt the catalytic site. In an increasing set of recently characterized proteins, this latter feature is more frequently seen to be the result of a catalytic site being situated at the junction of two subunits in an oligomer. The monofunctional OPRTase from *S. typhimurium* is an example of this architecture. This structural feature has the benefit of facilitating small movements between subunits to alter the actual shape of a ligand-binding site, without requiring major conformational distortion in the subunit itself.

DNA sequencing suggests that many mammalian or human proteins may be coded by two or more genes, and that such gene duplication may provide a simple safety factor in the event of a mutation in one gene, or provide a more sophisticated regulatory strategy if the genes lead to slightly altered isozymes to be expressed in unique tissues. For the enzymes discussed in this review, the current evidence supports the existence of only a single gene for UMP synthase and for β-alanine synthase, and perhaps two genes for uridine kinase. Although the metabolism of uracil compounds is required in all tissues, the evolutionary strategy for its steady metabolism has largely relied on unique enzymes that are sensitive to allosteric regulation.

Analysis of the sequence motifs associated with ligand-binding sites, as defined by protein structures for many kinases, shows the usefulness of sequence data in helping to identify such binding sites. The use of these consensus motifs, combined with their spatial position along the sequence, has made the identification of ATP- or GTP-binding sites much more favorable. In addition, binding sites for AMP, GMP, and UMP also show a similarity in the sequence motifs for the peptide segments of the catalytic site where they are bound. It is not yet clearly resolved whether such common consensus motifs for the same binding site represent divergence from a common ancestor, or perhaps convergence toward an optimum structure, but the wealth of sequence data makes it more likely for any new sequence to be used in finding corresponding structure/function examples already in the data base.

Future studies will address the interesting possibility that these allosteric proteins contain a separate regulatory site, at which either a substrate or product may bind to produce conformational changes. These separate sites have been hypothesized to explain some very unusual differences in the affinity demonstrated for binding at the catalytic site, measured by activity, vs. the affinity for some ligand as measured by a separate biophysical property, such as oligomer dissociation. Of additional importance will be experiments to determine the benefit of the bifunctional architecture of UMP synthase. Although more primitive organisms have two genes coding for separate proteins, is the gene fusion that joined the two domains in higher eukaryotes a neutral event, or has it endowed the mammalian bifunctional protein with additional (and now necessary) features? The most likely benefit of this architecture is union of the two sequential catalytic centers provided for interaction between these two domains, such that now they are more stable, they can be jointly regulated, and they can more efficiently transport the intermediate metabolite (OMP). By lowering the steady-state concentration of OMP in the cell, this latter feature would have the benefit of diminishing the back reaction for the OPRTase, while also increasing OMP decarboxylation to produce UMP. The recent synthesis of plasmids to express UMP synthase, or its domains, will enable the overproduction of these proteins for structural studies by crystallography, and for determination of catalytic function in response to site-directed mutagenesis.

References

1. R. Seifert and G. Schultz, *Trends Pharm. Sci.* **10**, 365 (1989).
2. R. Tauber, L. Richter and W. Reutter, *FEBS Leett.* **141**, 198 (1982).
3. T. W. Traut, *MCBchem* **140**, 1 (1994).
4. G. Weber, et al., *Adv. Enz Regul* **16**, 3 (1978).
5. J. E. Denton et al., *Cancer Res.* **42**, 1176 (1982).
6. N. K. Ahmed, R. C. Haggitt and A. D. Welch, *Biochem. Pharm.* **31**, 2485 (1982).
7. G. Weber, et al., *Cancer Res.* **43**, 1019 (1983).
8. J. De Montigny, L. Kern, J. C. Hubert and F. Lacroute, *Curr. Genet.* **17**, 105 (1990).
9. L. Kern, J. de Montigny, R. Jund and F. Lacroute, *Gene* **88**, 149 (1990).
10. C. C. Wang, R. Verham, S.-F. Tzeng, S. Aldritt and H.-W. Cheng, *PNAS* **80**, 2564 (1983).
11. C. C. Wang and H.-W. Cheng, *Mol. Biochem. Parasitol.* **10**, 171 (1984).
12. S. Matsushita and B. L. Fanburg, *Circul. Res.* **27**, 415 (1970).
13. D. J. Hammond and W. E. Gutteridge, *BBA* **718**, 1 (1982).
14. J. Hochstadt, *CRC Crit. Rev. Biochem.* **9**, 259 (1974).
15. R. M. Wohlhueter, S. R. McIvor and P. G. W. Plagemann, *J. Cell. Physiol.* **104**, 309 (1980).
16. P. G. W. Plagemann and D. P. Richey, *BBA* **344**, 263 (1974).
17. P. G. W. Plagemann, R. M. Wohlhueter and J. Erbe, *BBA* **640**, 448 (1981).
18. P. G. W. Plagemann and C. Woffendin, *BBA* **981**, 315 (1989).
19. J. C. Williams, C. E. Lee and J. R. Wild, *MGC* **178**, 121 (1980).
20. E. H. Harley and P. Berman, in "Advances in Experimental Medicine and Biology" (C. H. M. M. De Bruyn, H. A. Simmonds and M. M. Müller, eds.), Vol. 165A, p. 103. Plenum, New York, 1984.
21. G. L. Mendz, B. M. Jimenez, S. L. Hazell, A. M. Gero and W. J. O'Sullivan, *J. Appl. Bacteriol.* **77**, 1 (1994).
22. J. D. Moyer, J. T. Oliver and R. E Handschumacher, *Cancer Res.* **41**, 3010 (1981).
23. A. Raisonnier, M.-E. Bouma, C. Salvat and R. Infante, *EJB* **118**, 565 (1981).
24. D. W. Zaharevitz, E. A. Napier, L. W. Anderson, J. M. Strong and R. L. Cysyk, *EJB* **175**, 193 (1988).
25. Y. Sugiura, S. Fujioka and S. Yoshida, *Jpn. J. Cancer Res.* **77**, 664 (1986).
26. A. Geiger and S. Yamasaki, *J. Neurochem.* **1**, 93 (1956).
27. P. Reyes and M. Guganig, *JBC* **250**, 5097 (1975).
28. P. Reyes and C. Intress, *Life Sci.* **22**, 577 (1978).
29. M. E. Weichsel, Jr., N. J. Hoogenraad, R. L. Levine and N. Kretchmer, *Pediatr. Res.* **6**, 682 (1972).
30. G. C. Tremblay, U. Jimenez and D. E. Crandall, *J. Neurochem.* **26**, 57 (1976).
31. K. Lipp, N. van der Meulen, G. Wenske and F. Linneweh, *Biol. Neonate.* **33**, 62 (1978).
32. S. Bhasin and G. E. Shambaugh III, *Am. J. Physiol.* **243**, E234 (1982).
33. G. P. Connolly, *Paths Pyrim.* **2**, 37 (1994).
34. T. W. Traut and S. Loechel, *Bchem* **23**, 2533 (1984).
35. G. Weber, S. F. Queener and J. A. Ferdinandus, *Adv. Enz. Regul.* **9**, 63 (1971).
36. T. Gasser, J. D. Moyer and R. E. Handschumacher, *Science* **213**, 777 (1981).
37. A. Monks and R. L. Cysyk, *Am. J. Physiol.* **242**, R465 (1982).
38. A. Holstege, H.-M. Gengenbacher, L. Jehle and W. Gerok, *J. Hepatol.* **14**, 335 (1992).
39. A. Holstege, H.-G. Leser, J. Pausch and W. Gerok, *EJB* **149**, 169 (1985).
40. P. Berman and E. Harley, in "Advances in Experimental Medicine and Biology" (C. H. M. M. De Bruyn, H. A. Simmonds and M. M. Müller, eds.), Vol. 165A, p. 367. Plenum, New York, 1984.

41. J.-J. Chen and M. E. Jones, *ABB* **176**, 82 (1976).
42. V. Hines, L. D. Keys III and M. Johnston, *JBC* **261**, 11386 (1986).
43. P. Cortes, F. Dumler, D. L. Paielli and N. W. Levin, *Am. J. Physiol.* **255**, F647 (1988).
44. D. P. Suttle, D. M. O. Becroft and D. R. Webster, *in* "The Metabolic Basis of Inherited Disease" (C. R. Scriver, A. L. Beaudet, W. S. Sly and D. Valle, eds.), 6th Ed., Vol. I, p. 1095. McGraw-Hill, New York, 1989.
45. S. Ohba et al., *J. Inher. Metabol. Dis.* **16**, 872 (1993).
46. J. D. Moyer, N. Malinowski and O. Ayers, *JBC* **260**, 2812 (1985).
47. J. W. Darnowski and R. E. Handschumacher, *Cancer Res.* **46**, 3490 (1986).
48. J. M. Karle, K. H. Cowan, C. A. Chisena and R. L. Cysyk, *Mol. Pharm.* **30**, 136 (1986).
49. J. Seifert, *ABB* **201**, 194 (1980).
50. D. N. M. Naguib, S.-J. Soong and M. H. El Kouni, *Biochem Pharmacol.* **45**, 667 (1993).
51. G. Siebert, *Biochem Soc. Trans.* **6**, 5 (1978).
52. J. D. Moyer and J. F. Henderson, *CRC Crit. Rev. Biochem.* **19**, 45 (1985).
53. T. W. Traut, *CRC Crit. Rev. Biochem.* **23**, 121 1988).
54. T. W. Traut and M. E. Jones, *JBC* **254**, 1143 (1979).
55. T. W. Traut, R. C. Payne and M. E. Jones, *Bchem* **19**, 6062 (1980).
56. T. W. Traut, *Crit. Rev. Biochem. Mol. Biol.* **29**, 125 (1994).
57. M. M. Matthews, W. Liao, K. L. Kvalnes-Krick and T. W. Traut, *ABB* **293**, 254 (1992).
58. M. M. Matthews and T. W. Traut, *JBC* **262**, 7232 (1987).
59. J. Monod, J. Wyman and J. Changeux, *JMB* **12**, 88 (1965).
60. D. E. Koshland, Jr., *in* "The Enzymes" (P. D. Boyer, ed.), 3rd Ed., Vol. I, p. 342. Academic Press, New York, 1970.
61. J. Monod, J. Wyman and J.-P. Changeux, *JMB* **12**, 88 (1965).
62. R. P. Aaronson and C. Frieden, *JBC* **247**, 7502 (1972).
63. E. Hofmann, et al., *Adv. Enz. Regul.* **13**, 247 (1975).
64. L. K. Hesterberg and J. C. Lee, *Bchem* **20**, 2974 (1981).
65. L. K. Hesterberg and J. C. Lee, *Bchem* **21**, 216 (1982).
66. D. Kotlarz and H. Buc, *EJB* **117**, 569 (1981).
67. M. A. Luther, H. F. Gilbert and J. C. Lee, *Bchem* **22**, 5494 (1983).
68. G. D. Reinhart, *JBC* **258**, 10827 (1983).
69. V. Guixe and J. Babul, *ABB* **264**, 519 (1988).
70. J. Gazith, I. T. Schulze, R. H. Gooding, F. C. Womack and S. P. Colowick, *Ann. N.Y. Acad. Sci.* **151**, 307 (1968).
71. J. P. Shill, B. A. Peters and K. E. Neet, *Bchem* **13**, 3864 (1974).
72. T. A. Steitz, W. F. Anderson, R. J. Fletterick and C. M. Anderson, *JBC* **252**, 4494 (1977).
73. E. Helmreich, M. C. Michaelides and C. F. Cori, *Bchem* **6**, 3695 (1967).
74. C. Y. Huang and D. J. Graves, *Bchem* **9**, 660 (1970).
75. B. Metzger, E. Helmreich and L. Glaser, *PNAS* **57**, 994 (1967).
76. G. Scapin, C. Grubmeyer and J. C. Sacchettini, *Bchem* **33**, 1287 (1994).
77. T. W. Traut and R. C. Payne, *Bchem* **19**, 6068 (1980).
78. M. Jacquet, R. Guilbaud and H. Garreau, *MGG* **211**, 441 (1988).
79. C. M. Huguley, Jr., J. A. Bain, S. Rivers and R. Scoggins, *Blood* **14**, 615 (1959).
80. W. T. Shoaf III and M. E. Jones, *Bchem* **12**, 4039 (1973).
81. R. W. McClard, M. J. Black, L. R. Livingstone and M. E. Jones, *Bchem* **19**, 4699 (1980).
82. L. R. Livingstone and M. E. Jones, *JBC* **262**, 15726 (1987).
83. D. P. Suttle, B. Y. Bugg, J. K. Winkler and J. J. Kanalas, *PNAS* **85**, 1754 (1988).
84. M. B. Qumsieh, M. B. Valentine and D. P. Suttle, *Genomics* **160** (1989).
85. R. Girot, M. Hamet and J. L. Perignon *N. Engl. J. Med.* **308** (1983).
86. C. S. Alvarado et al., *J. Pediatr.* **113**, 867 (1988).

87. J. L. Robinson, M. R. Drabnik, D. B. Dombrowski and J. H. Clark, *PNAS* **80**, 321 (1983).
88. J.-J. Chen and M. E. Jones, *JBC* **254**, 2679 (1979).
89. M. Minet, M.-E. Dufour and F. Lacroute, *Plant J.* **2**, 417 (1992).
90. M. B. Bhatia, A. Vinitsky and C. Grubmeyer, *Bchem* **29**, 10480 (1990).
91. T. Lin and D. P. Suttle, *Som. Cell. Mol. Genet.* **19**, 193 (1993).
92. M. J. Yablonski, B.-D. Han, T. W. Traut, D. A. Pasek and M. E. Jones, *FASEB J.* **9**, A 1287 (1995).
93. F. Lue, D. I. Chasman, A. R. Buchman and R. D. Kornberg, *Mol. Cell. Biol.* **7**, 3446 (1987).
94. J. B. Bell and M. E. Jones, *JBC* **266**, 12662 (1991).
95. B.-D. Han, L. R. Livingstone, D. A. Pasek, M. J. Yablonski and M. E. Jones, *Bchem.* **34**, 10835 (1995).
96. M. J. Yablonski, D. A. Pasek, B. D. Han, M. E. Jones and T. W. Traut (unpublished).
97. E. A. Carrey, D. G. Campbell and D. G. Hardie, *EMBO J.* **4**, 3735 (1985).
98. N. Aghajari, K. F. Jensen and M. Gajhede, *JMB* **241**, 292 (1994).
99. J. B. Bell, M. E. Jones and C. W. Carter, Jr., *Proteins* **9**, 143 (1991).
100. H. L. Levine, R. S. Brody and F. H. Westheimer, *Bchem* **19**, 4993 (1980).
101. E. E. Floyd and M. E. Jones, *JBC* **260**, 9443 (1985).
102. C.-A. Ohmstede, S. D. Langdon, C.-B. Chae and M. E. Jones, *JBC* **261**, 4276 (1986).
103. S. D. Langdon and M. E. Jones, *JBC* **262**, 13359 (1987).
104. G. Scapin, J. C. Sacchettini, A. Dessen, M. Bhatia and C. Grubmeyer, *JMB* **230**, 1304 (1993).
105. M. B. Bhatia and C. Grubmeyer, *ABB* **303**, 321 (1993).
106. I. Lieberman, A. Kornberg and E. S. Sims, *JBC* **215**, 403 (1957).
107. T. W. Traut and M. E. Jones, *JBC* **252**, 8374 (1977).
108. J. Victor, L. B. Greenberg and D. L. Sloan, *JBC* **254**, 2647 (1979).
109. A. Tavares, C. S. Lee and W. J. O'Sullivan, *BBA* **913**, 279 (1987).
110. C. Grubmeyer, E. Segura and R. Dorfman, *JBC* **268**, 20299 (1993).
111. J. P. Sommadossi *et al.*, *JBC* **257**, 8171 (1982).
112. J. A. Smiley, P. Paneth, M. O. O'Leary, J. B. Bell and M. E. Jones, *Bchem* **30**, 6216 (1991).
113. S. S. Acheson, J. B. Bell, M. E. Jones and R. Wolfenden, *Bchem* **29**, 3198 (1990).
114. K. Shostak and M. E. Jones, *Bchem* **31**, 12155 (1992).
115. A. Radzicka and R. Wolfenden, *Science* **267**, 90 (1995).
116. P. Beak and B. Siegel, *JACS* **98**, 3601 (1976).
117. R. B. Silverman and M. P. Groziak, *JACS* **104**, 6434 (1982).
118. J. A. Smiley and M. E. Jones, *Bchem* **31**, 12162 (1992).
119. H. B. Smith and F. C. Hartman, *JBC* **263**, 4921 (1988).
120. A. Planas and J. F. Kirsh, *Bchem* **30**, 8268 (1991).
121. A. Yoshimoto, T. Amaya, K. Kobayashhi and K. Tomita, in "Methods in Enzymology" (P. A. Hoffee and M. E. Jones, eds.), Vol. 51, p. 69. Academic Press, New York, 1978.
122. M. E. Perry and M. E. Jones, *JBC* **264**, 15522 (1989).
123. L. Pinsky and R. S. Krooth, *PNAS* **57**, 1267 (1967).
124. J. K. Winkler and D. P. Suttle, *Am. J. Human Genet.* **43**, 86 (1988).
125. J. A. Smiley and S. J. Benkovic, *PNAS* **91**, 8319 (1994).
126. P. Reichard and O. Sköld, *Acta Chem. Scand.* **11**, 17 (1957).
127. O. Sköld, *JBC* **235**, 3273 (1960).
128. A. Cihak and B. Rada, *Neoplasma* **23**, 233 (1976).
129. Z. J. Lucas, *Science* **156**, 1237 (1967).
130. P. Hausen and H. Stein, *EJB* **4**, 401 (1968).

131. K. Ito and H. Uchino, *JBC* **251**, 1427 (1976).
132. G. J. Peters, A. Oosterhof and J. H. Veerkamp, *Int. J. Bchem.* **15**, 51 (1983).
133. N. Cheng and T. W. Traut, *JCB* **35**, 217 (1987).
134. W. Wharton and W. J. Pledger, *In Vitro* **17**, 706 (1981).
135. A. E. M. Jacobs, A. Oosterhof and J. H. Veerkamp, *BBA* **970**, 130 (1988).
136. S. Staak, N. Popov and H. Matthies, *Biomed. Bchem. Acta* **48**, 325 (1988).
137. E. S. Haugaard, K. B. Frantz and N. Haugaard, *PNAS* **74**, 2339 (1977).
138. N. Haugaard, A. Torbati, T. Smithgall and G. Wildey, *MCBchem* **93**, 13 (1990).
139. G. Macdonald, T. Walker, R. Assef and K. Duggan, *Clin. Exp. Pharmacol. Physiol.* **17**, 287 (1990).
140. G. Krystal and T. E. Webb, *BJ* **124**, 943 (1971).
141. G. Krystal and P. G. Scholefield, *Can. J. Biochem.* **51**, 379 (1973).
142. R. C. Keefer, H. P. Morris and T. E. Webb, *Cancer Res.* **34**, 2260 (1974).
143. M. C. Fulchignoni-Lataud, M. Tuilie and J. M. Roux, *EJB* **69**, 217 (1976).
144. N. Greenberg, D. E. Schumm, P. E. Hurtubise and T. E. Webb, *Cancer Res.* **37**, 1028 (1977).
145. N. Greenberg, D. E. Schumm and T. E. Webb, *BJ* **164**, 379 (1977).
146. J. Vesely and J. Smrt, *Neoplasma* **24**, 461 (1977).
147. M. Otal-Brun and T. E. Webb, *Cancer Lett.* **6**, 39 (1979).
148. N. K. Ahmed and A. D. Welch, *Cancer Res.* **39**, 3102 (1979).
149. N. Ahmed and D. R. Baker, *Cancer Res.* **40**, 3559 (1980).
150. I. G. Dubinina, V. G. Shkurko and G. I. Vornovitskaya, *Biokhimya* **46**, 1503 (1982).
151. J. Absil, M. Tuilie and J. M. Roux, *BJ* **185**, 273 (1980).
152. M.-C. Fulchignoni-Lataud and J. M. Roux, *J. Cell. Physiol.* **118**, 34 (1984).
153. R. C. Payne, N. Cheng and T. W. Traut, *JBC* **260**, 10242 (1985).
154. R. C. Payne and T. W. Traut, *JBC* **257**, 12485 (1982).
155. N. Cheng, R. C. Payne and T. W. Traut, *JBC* **261**, 13006 (1986).
156. N. Cheng, R. C. Payne, W. E. Kemp, Jr. and T. W. Traut, *Mol. Pharm.* **30**, 159 (1986).
157. E. P. Anderson, in "Methods in Enzymology" (P. A. Hoffee and M. E. Jones, eds.), Vol. 51, p. 314. Academic Press, New York, 1978.
158. A. Orengo and G. F. Saunders, *Bchem* **11**, 1761 (1972).
159. P. Valentin-Hansen, in "Methods in Enzymology" (P. A. Hoffee and M. E. Jones, eds.), Vol. 51, p. 308. Academic Press, New York, 1978.
160. P. R. Wheeler, *J. Gen. Microbiol.* **136**, 189 (1990).
161. W. Plunkett and J. G. Moner, *BBA* **250**, 92 (1971).
162. M. H. El Kouni and F. N. M. Naguib, *Int. J. Parasitol* **20**, 37 (1990).
163. P. A. Ropp and T. W. Traut (unpublished).
164. H.-J. Müller-Dickmann and G. E. Schulz, *JMB* **236**, 361 (1994).
165. T. Stehle and G. E. Schulz, *JMB* **211**, 249 (1990).
166. D. Dreusicke, A. Karplus and G. E. Schulz, *JMB* **199**, 359 (1988).
167. J. Devereux, "The GCG Sequence Analysis Software Package Version 7.2." Genetics Computer Group, Inc., Madison, WI, 1992.
168. T. W. Traut, *EJB* **222**, 9 (1994).
169. M. Kozak, *JBC* **266**, 19867 (1991).
170. E. F. Vanin, *Annu. Rev. Genet.* **19**, 253 (1985).
171. K. Krnjevic, *Brit. Med. Bull.* **21**, 10 (1960).
172. M. Sandberg and I. Jacobson, *J. Neurochem.* **37**, 1353 (1971).
173. G. Toggenburger, D. Felix, M. Cuenod and H. Henke, *J. Neurochem.* **39**, 176 (1982).
174. A. Sanchez-Herranz, R. M. Del Rio and E. Montoya, *Horm. Metab. Res.* **17**, 588 (1985).

175. N. Tremblay, R. Warren and R. W. Dykes, *Neuroscience* **26**, 745 (1988).
176. L. M. Orensanz, E. Ambrosio, I. Fernandez and M. T. Montero, *Neurochem. Res.* **13**, 1133 (1988).
177. M. Kihara, Y. Misu and T. Kubo, *Life Sci.* **42**, 1817 (1986).
178. D. Choquet and H. Korn, *Neurosci. Lett.* **84**, 329 (1988).
179. Y. Tsukuda, Y. Nagata, S. Hirano and T. Matsutani, *J. Neurochem.* **10**, 241 (1963).
180. L. Agullo, B. Jimenex, C. Aragon and C. Gimenez, *EJB* **159**, 611 (1986).
181. H. S. Sidhu and J. D. Wood, *Neuropharmacology* **25**, 555 (1986).
182. F. Zafra, M. C. Aragon, F. Valsidivieso and C. Gimenez, *Neurochem. Res.* **9**, 695 (1984).
183. O. M. Larrson, R. Griffiths, I. C. Allen and A. Shousboe, *J. Neurochem.* **47**, 426 (1986).
184. I. Parker, K. Sumikawa and R. Miledi, *Proc. R. Soc. Lond. Ser. B* **233**, 201 (1988).
185. H. Akagi and R. Miledi, *Neurosci. Lett.* **95**, 262 (1988).
186. H. Akagi and R. Miledi, *Science* **242**, 270 (1988).
187. K. Nagai, T. Suga, K. Kawasaki and S. Mathuura, *Surgery* **100**, 815 (1986).
188. J. G. Ondo, K. A. Pass and R. Baldwin, *Neuroendocrinology* **21**, 79 (1976).
189. C. R. Scriver, T. L. Perry and W. Nutzenadel, in "Metabolic Basis of Inherited Disease," Vol. 4, p. 570. Academic Press, New York, 1983.
190. A. R. Pavlov, A. A. Revina, A. M. Dupin, A. A. Boldyrev and A. I. Yaropolov, *BBA* **1157**, 304 (1993).
191. T. P. West, T. W. Traut, M. S. Shanley and G. A. O'Donovan, *J. Gen. Microbiol.* **131**, 1083 (1985).
192. C. R. Scriver, S. Pueschel and E. Davis, *N. Engl. J. Med.* **274**, 635 (1966).
193. P. J. Schechter, P. J. Lewis and J. W. Newberne, *Lancet* **I**, 737 (1984).
194. S. K. Wadman et al., in "Advances in Experimental Medicine and Biology," Vol. 165A, p. 109. Plenum, New York, 1984.
195. S. K. Wadman et al., *J. Inherit. Metab. Dis. Supp* **2**, 113 (1985).
196. J. A. J. M. Bakkeren et al., *Clin. Chim. Acta* **140**, 247 (1984).
197. M. Tuchman et al., *N. Engl. J. Med.* **313**, 245 (1985).
198. B. Wilcken, J. Hammond, R. Berger, G. Wise and C. James, *J. Inherit. Metab. Dis. Supp* **2**, 115 (1985).
199. J. P. Braakhekke, et al., *J. NeuroSci.* **78**, 71 (1987).
200. M. Kigugawa, S. Fujimoto, C. Mizota and N. Tamaki, *FEBS Lett.* **229**, 345 (1988).
201. B. Podschun, G. Wahler and K. D. Schnackerz, *EJB* **185**, 219 (1989).
202. K. P. Brooks, E. A. Jones, B.-D. Kim and E. G. Sander, *ABB* **226**, 469 (1983).
203. F. N. M. Naguib, M. H. El Kouni and S. Cha, *Cancer Res.* **45**, 5405 (1985).
204. A. Holstege, J. Pausch and W. Gerok, *Cancer Res.* **46**, 5576 (1986).
205. G. C. Daher, B. E. Harris and R. B. Diasio, *Pharm. Therap.* **48**, 189 (1990).
206. R. Zhang, S.-J. Soong, S. Liu, S. Barnes and R. B. Diasio, *Drug Metab. Disp.* **20**, 113 (1992).
207. N. Tamaki, N. Mizutani, M. Kikugawa, S. Fujimoto and C. Mizota, *EJB* **169**, 21 (1987).
208. G. Waldman and K. D. Schnackerz, *Biol. Chem. Hoppe-Seyler* **370**, 969 (1989).
209. J. P. Shill and K. E. Neet, *JBC* **250**, 2259 (1975).
210. P. A. Lazo, A. Sols and J. E. Wilson, *JBC* **255**, 7548 (1980).
211. D. Blangy, H. Buc and J. Monod, *JMB* **31**, 13 (1968).
212. R. A. Poorman, A. Randolph, R. G. Kemp and R. L. Heinrikson, *Nature* **309**, 467 (1984).
213. P. A. Ropp and T. W. Traut, *JBC* **266**, 7682 (1991).
214. K. L. Kvalnes-Krick and T. W. Traut, *JBC* **268**, 5686 (1993).
215. C. Wasternack, G. Lippmann and H. Reinbotte, *BBA* **570**, 341 (1987).
216. J. Ogawa and S. Shimizu, *EJB* **223**, 625 (1994).
217. S. B. Needleman and C. D. Wunsch, *JMB* **48**, 443 (1970).

218. S. K. Burley, P. R. David, R. M. Sweet, A. Taylor and W. N. Lipscomb, *JMB* **224**, 113 (1992).
219. J. E. Walker, M. Saraste, M. J. Runswick and N. J. Gay, *EMBO J.* **1**, 945 (1982).
220. U. Egner, A. G. Tomasselli and G. E. Schulz, *JMB* **195**, 649 (1987).
221. K. Diedrichs and G. E. Schulz, *JMB* **211**, 249 (1990).
222. C. W. Müller and G. E. Schulz, *JMB* **224**, 159 (1992).
223. A. R. Mushegian and E. V. Koonin, *Prot. Sci.* **3**, 1081 (1994).
224. C. G. Lerner, B. T. Stephenson and R. L. Switzer, *J. Bacteriol.* **169**, 2202 (1987).
225. S. Y. Ghim, P. Nielsen and J. Neuhard, *Microbiology* **140**, 479 (1994).
226. H. Bisswanger and E. Schmincke-Ott, eds., "Multifunctional Proteins." Wiley, New York, 1980.
227. A. D. McCarthy and D. G. Hardie, *Trends Biochem. Sci.* **9**, 60 (1984).
228. H. Schurig *et al.*, *EMBO J.* **14**, 442 (1995).
229. J. N. Davidson, P. C. Rumsby and J. Tamaren, *JBC* **256**, 5220 (1981).
230. M. I. Mally, D. R. Grayson and D. R. Evans, *PNAS* **78**, 6647 (1981).
231. D. R. Grayson, L. Lee and D. R. Evans, *JBC* **260**, 15840 (1985).
232. L. Crofts, Y. Peide, A. Woodhouse, E. M. Algar and R. I. Christopherson, *Prot. Expr. Purif.* **1**, 45 (1990).
233. N. K. Williams, Y. Peide, K. K. Seymour, G. B. Ralston and R. I. Christopherson, *Prot. Eng.* **6**, 333 (1993).
234. H. I. Guy and D. R. Evans, *JBC* **269**, 7702 (1994).
235. X. Liu, H. I. Guy and D. R. Evans, *JBC* **269**, 27747 (1994).
236. S. Tampitag and W. J. O'Sullivan, *Mol. Biochem. Parasitol.* **19**, 125 (1986).
237. M. H. Iltzsch, J. G. Niedzwicki, A. W. Senft, S. Cha and M. H. El Kouni, *Mol. Cell. Parasitol.* **12**, 153 (1984).
238. H. D. Doremus, *ABB* **250**, 112 (1986).
239. G. K. Brown, R. M. Fox and W. J. O'Sullivan, *Biochem. Pharm.* **21**, 2469 (1972).
240. P. R. Kavipurapu and M. E. Jones, *JBC* **252**, 5589 (1976).
241. G. K. Brown and W. J. O'Sullivan, *Bchm* **16**, 3235 (1977).
242. R. I. Christopherson and M. E. Jones, *JBC* **255**, 11381 (1980).
243. R. I. Christopherson, T. W. Traut and M. E. Jones, *Curr. Top. Cell Regul.* **18**, 59 (1981).
244. M. I. Mally, D. R. Grayson and D. R. Evans, *JBC* **255**, 11372 (1980).
245. M. Belkaïd, B. Penverne and G. Hervé, *ABB* **262**, 171 (1988).
246. T. W. Traut, *ABB* **268**, 108 (1989).
247. S. Pragobpol, A. M. Gero, C. S. Lee and W. J. O'Sullivan, *ABB* **230**, 285 (1984).
248. G. R. Dubyak and M. B. De Young, *JBC* **260**, 10653 (1985).
249. Y. Shirasawa, R. P. White and J. T. Robertson, *Stroke* **14**, 347 (1983).
250. G. Macdonald, R. Assef, S. Watkins and J. Burrell, *Clin. Exp. Pharmacol. Physiol.* **14**, 253 (1987).
251. J. F. Mustard and M. A. Packham, *Pharmacol. Rev.* **22**, 97 (1970).
252. R. Limor, I. Schvartz, E. Hazum, D. Ayalon and Z. Naor, *BBRC* **159**, 209 (1989).
253. G. J. Peters *et al. Cancer Chem. Pharmacol.* **20**, 101 (1987).
254. W. J. M. Tax, G. J. Peters and J. H. Veerkamp, *Int. J. Biochem.* **10**, 7 (1979).
255. B. W. Potvin, H. J. Stern, S. R. May, G. F. Lam and R. S. Krooth, *Biochem. Pharmacol.* **27**, 655 (1978).
256. H. Witschi, *Cancer Res.* **32**, 1686 (1972).
257. R. E. Handschumacher and J. Coleridge, *Biochem. Pharmacol.* **28**, 1977 (1979).
258. M. Petricevic, C. W. Denko and L. Messineo, *Int. J. Biochem.* **15**, 751 (1983).
259. D. P. Ringer, B. A. Howell, J. L. Etheredge, J. A. Clouse and D. E. Kizer, *FEBS Lett.* **224**, 59 (1987).

260. A. F. Hogans, G. Guroff and S. Udenfriend, *J. Neurochem.* **18**, 1699 (1971).
261. J. Olivares and A. Rossi, *J. Mol. Cell Cardiol.* **20**, 313 (1988).
262. E. Kantrowitz and W. N. Lipscomb, *Science* **241**, 669 (1988).
263. L. Betts, S. Xiang, S. A. Short, R. Wolfenden and C. W. Carter, Jr., *JMB* **235**, 635 (1994).
264. R. H. Jacobson, X.-J. Zhang, R. F. DuBose and B. Matthews, *Nature* **369**, 761 (1994).
265. S.-H. Liaw and D. Eisenberg, *Bchem* **33**, 675 (1994).
266. V. L. Rath, C. B. Newgard, S. R. Sprang, E. J. Goldsmith and R. J. Fletterick, *Proteins* **2**, 225 (1987).
267. D. Barford and L. N. Johnson, *Nature* **340**, 609 (1989).
268. P. R. Evans and P. J. Hudson, *Nature* **279**, 500 (1979).
269. P. R. Evans, G. W. Farrants and P. J. Hudson, *Philos. Trans. R. Soc. London Ser. B* **293**, 53 (1981).
270. P. R. Evans, G. W. Farrants and M. C. Lawrence, *JMB* **191**, 713 (1986).
271. T. Schirmer and P. R. Evans, *Nature* **343**, 140 (1990).
272. Y. Shirakihara and P. R. Evans, *JMB* **204**, 973 (1988).
273. R. C. Keefer, D. J. McNamara, D. E. Schumm, D. E. Billmire and T. E. Webb, *Biochem. Pharmacol.* **24**, 1287 (1975).

The Importance of Being Modified: Roles of Modified Nucleosides and Mg^{2+} in RNA Structure and Function

PAUL F. AGRIS

Department of Biochemistry
North Carolina State University
Raleigh, North Carolina 27695

I. Biological Functions of Modified Nucleosides	81
II. The Large Array of Modified Ribonucleoside Structures and Chemistries	84
III. Site-selective Positioning of Modifications	101
IV. The Study of Modified-nucleoside Contributions to Structure and Function	105
V. Examples of Biophysical Contributions to Function by Modified Nucleosides	109
A. Stabilization of Domain Structure and Induced Dynamics	112
B. Modification Influences Local Nucleoside Conformation and Dynamics	116
VI. Conclusions and Perspectives on the Importance of Being Modified . .	119
Appendix: Modified Nucleoside Symbols and Common Names	122
References	124

Nucleic acids consist of over 100 modified nucleosides, in addition to 8 major ribo- and deoxyribonucleosides. The "Importance of Being Modified," as with the Wilde[1] play from which the title of this article is taken, has multiple meanings, the significance of which is only fully understood when the contribution of each and every "player" is investigated. Nucleoside modifications contribute an astonishing array of chemistries and structures to nucleic acids. The importance of these chemistries and structures to DNA and RNA function is only beginning to be understood. In general, evolutionary selection, conservation, and enhancement of nucleic acid modifications, in species ranging from bacteria to humans, have been an enigma ever since the first modified base, 5-methylcytosine, was identified in 1948 (1). During the subsequent half-century, more than 100 modified nucleosides have been

---
[1] Oscar Wilde (1854–1900), Irish poet and playwright, whose most famous of plays is probably "The Importance of Being Earnest."

discovered and their structures determined. Discovery and structural determination of the naturally occurring modified nucleosides continue, and are focused particularly on the nucleic acids of organisms living in extremes of temperature, salt, and light conditions. At least three books (2–4), one compendium of information comprising an additional three volumes (5), and many reviews (6–14) have been written about them. However, little has been said about the physicochemical contributions of the modified nucleosides and their effects on function.

Until recently, few biochemical reactions and interactions of RNA were known for which a modified nucleoside was absolutely required. Yet, the importance of modified nucleosides to biological function is irrefutable. Nucleosides are modified after polymerization of the nucleic acid. In order to accomplish site-specific modifications of the bases in DNA and the bases and ribose of RNA, a very large investment in genetic information, energy, and material resources has evolved. For instance, there are many more genes for modifying enzymes than there are genes for the RNAs being modified. Nucleic-acid modifications have been linked to control of gene expression at both the level of transcription (15) and the level of translation (13, 16, 17). An antibiotic resistance can be conferred by a loss of modification in the *E. coli* 16-S rRNA (18). The product of the *PET56* nuclear gene of *Saccharomyces cerevisiae* is required for mitochondrial rRNA methylation of a highly conserved nucleotide at the peptidyltransferase center of the large subunit. A mutation reducing transcription of the gene by 80% also significantly reduces formation of functional large subunits (19). The presence of another modification, pseudouridine, at the peptidyltransferase center in both yeast cytoplasmic and mitochondrial rRNA implicates modified nucleosides in the biosynthesis of the peptide bond (20, 20a). The activity of HIV AIDS reverse transcriptase may be modulated by the modified nucleosides found in the anticodon domain of the tRNA^{Lys} primer of the enzyme (21). A relationship exists between cellular differentiation and the degree to which specific modifications appear in particular tRNA (22–27).

Although a few modifications have been found in DNA, 93 modified nucleosides exist in the various RNAs (29). For the most part, the chemistry and structure that modified nucleosides, individually and in combination, uniquely contribute to DNA or RNA function have yet to be explained. However, *there are 10 physicochemical contributions that can be attributed to modified nucleosides* (Table I). Any particular modification may contribute one or more of the properties listed. Of particular interest is the increasingly documented relationship between the presence of modified nucleosides in RNAs, particularly tRNAs, and the site and affinity of Mg^{2+} binding to RNA. Transfer RNAs display the greatest variety and largest numbers of modified

TABLE I
PHYSICOCHEMICAL CONTRIBUTIONS OF MODIFIED NUCLEOSIDES

1. Introduction of transient charges dependent on protonation
2. Introduction of positive and/or negative charges
3. Alteration and restriction of nucleoside conformation
4. Inhibition of noncanonical base-pairings
5. Disruption of canonical base-pairs
6. Enhancement of base-stacking interactions
7. Reordering of water
8. Facilitation of, or direct coordination with, metal ions
9. Restriction of phosphodiester bond conformation and/or nearest-neighbor nucleoside sugar pucler
10. Interaction between modifications of modified nucleosides producing new conformations and chemistries

nucleosides. The modified nucleosides of tRNA are well documented and the most highly studied.

This review begins with a summary of the more evident relationships between modified nucleosides and tRNA function. *In vivo* and *in vitro* syntheses of modified nucleoside-containing RNAs have been developed for physicochemical and biochemical studies. The different approaches to RNA synthesis are compared. Chemical and structural contributions of modified nucleosides are emphasized and examples cited for each of the 10 listed in Table I. By coupling physicochemical approaches to those of biochemistry and molecular biology, patterns of function attributable to modified nucleosides have begun to emerge. An understanding of the chemistry, structure, and function of modified nucleosides of tRNAs is applicable to the study of modified nucleosides of other RNAs and DNA. The structure–function relationships of the simplest of nucleoside modifications may provide insights into the "RNA World" and could be applied to the design of nucleic acids with new or altered functions.

I. Biological Functions of Modified Nucleosides

Most modified nucleosides for which specific functions have been demonstrated are in tRNA (see *13*). Determination of modified nucleoside functions, for the most part, has remained a challenge to conventional approaches. Experimental approaches to the functions of modified nucleosides have been limited to a comparison of completely modified RNA, purified from wild-type organisms, to the corresponding RNA from modification mutants or completely unmodified RNA from *in vitro* transcription. Two commonly used techniques when applied to tRNA research have been extremely rewarding but, unfortunately, have often yielded disappointing results in the study of modifications. *Escherichia coli* and yeast strains deficient in the synthesis of a specific tRNA modification, more often than not, have yielded little information as to the modification's function. The removal of any one modification, especially one common to many tRNA species, such as 5-methyluridine (m^5U_{54} or rT_{54}) at position 54, or N^6-isopentenyladenosine at position 37 (i^6A_{37}) (Fig. 1), has little effect on bacterial or yeast cell-growth (30–42).[2] RNAs from relaxed strains of *E. coli* cultured in the absence of methionine, the precursor to methyl groups, lack a number of methylated nucleosides (35), whereas RNAs from bacteria and yeast strains carrying mutations in specific modification enzymes are missing particular methylations (30, 31), or thiolations (36–38), but all other modifications are still present. Aminoacylations of species of tRNAs that are produced *in vitro* and are missing specific modifications, with surprisingly few but extremely interesting exceptions, are only moderately different from those of the wild type. However, translational efficiency, codon choice, and control of frame shifting can be affected significantly (13).

Totally unmodified tRNA species are produced in large quantities for studies of function by an easily performed *in vitro* T7 polymerase transcription of chemically synthesized DNA templates or cloned genes. Application of this technology resulted in identification of nucleoside determinants for some aminoacyl-tRNA synthetases (aaRS) as a "code" solely within the domain of the acceptor stem of the cognate tRNA (39–41). However, most aaRS require, in addition, contributions from the anticodon and other domains of the cognate tRNA (17, 42–44). Anticodon domains are particularly rich in modified nucleosides (see Section III). A few interesting tRNA/aaRS systems have exhibited significant differences *in vitro* with respect to the aminoacylation kinetics of the transcript and that of fully modified native tRNA. Although first observed 20 years ago (16), more recent studies, employing mutants and the ligation of modified and unmodified tRNA portions, proved that the anticodon wobble position $mnm^5s^2U_{34}$ of *E. coli* tRNA^{Glu} is a recognition determinant of glutamyl-tRNA synthetase (ERS)[3] (17, 17a). The same modification appears in tRNA^{Lys}, and the unmodified transcript of *E. coli* tRNA^{Lys} has $\frac{1}{140}$th the aminoacylation activity of the fully modified native tRNA (45).

An early observation in the study of yeast FRS recognition of cognate and

---

[2] The symbols and abbreviations used for the modified nucleosides are listed in the Appendix at the end of this article.

[3] Standard single-letter code for amino acids: E, glutamic acid; F, phenylalanine; I, isoleucine; K, lysine.

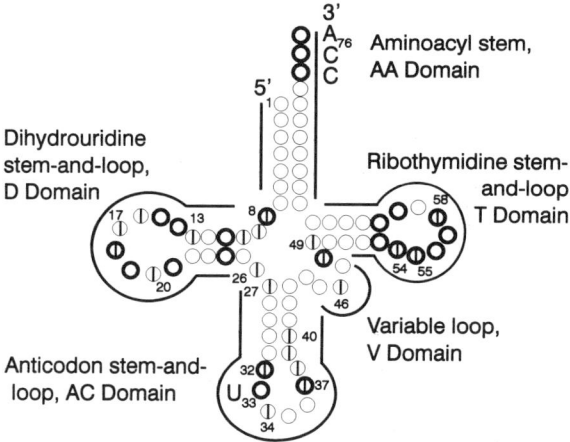

FIG. 1. Cloverleaf depiction of tRNA secondary structure. The stem-and-loop domains of the molecule are outlined. Invariant and semiconserved residues are those encircled in bold. Residues that are frequently modified have a line through the circle.

noncognate tRNAs, yet to be fully explored, is the importance of the wybutosine (yW_{37}; see Appendix for complete list of nucleotide symbols) and N^6-isopentenyladenosine (i$^6A_{37}$) modifications at position 37 of yeast tRNA^{Phe} and tRNA^{Tyr}, respectively (46). *Escherichia coli* IRS exhibits a reduction in activity when a lysine-modified cytidine, lysidine (k$^2C_{34}$), at wobble position 34 in a minor tRNA^{Ile} species is replaced by C (47) and when A is substituted for N^6-threonylcarbamolyadenosine (t$^6A_{37}$) at position 37, 3′ adjacent to the anticodon of tRNA$_I^{Ile}$ (48). Modifications of yeast tRNA^{Asp} act as negative determinants by decreasing the amount of misacylation (49). In particular, the 1-methylguanosine at position 37, 3′ to the anticodon (m$^1G_{37}$), decreases misacylation by arginyl-tRNA synthetase (50). Although the presence of m$^5U_{54}$ is important for formylation of initiator tRNAs in *E. coli* and yeast, in strains lacking m$^5U_{54}$ synthesis protein synthesis is correctly initiated by unformylated Met-tRNA$_i^{Met}$ (51, 52). A modification unique to eukaryotic initiator tRNA^{Met} at position 64, 2′-O-ribosyladenosine-[phosphate] (53), is a probable negative determinant, deterring recognition by elongation factor (54, 55). However, the individual chemical, structural, and functional relationships of the s^2-, mnm^5-, yW, i^6-, k^2-, t^6-, m^1-, ribosyl-, and other modifications are not resolved for aminoacyl-tRNA synthetase recognition of cognate tRNA and discrimination of noncognate tRNAs.

Unmodified tRNA transcripts and chemically synthesized unmodified anticodon domains have very much weaker ribosome associations (10^{-2}) than do native tRNAs (56). The importance of tRNA modifications to ribosome binding and codon recognition is exemplified by (1) 5-substituted 2-thiouridines at wobble position 34 ($s^2U^*_{34}$), which specifically recognize codons ending in A (12, 36–38); (2) inosine, I_{34}, which reads the third base of the codons ending in U, C, and G, but rarely codons ending in A (57); (3) the lysine-modified cytidine, lysidine (k^2C_{34}), at wobble position 34 in $tRNA^{Ile}_{minor}$, which ensures recognition of the Ile codon AUA instead of the Met codon AUG (47); (4) 2-methylthio-N^6-isopentenyladenosine ($ms^2i^6A_{37}$) and its cis-hydroxy isomer ($ms^2io^6A_{37}$) at position 37, 3'-adjacent to the anticodon, which maintain codon context and reading fidelity (58, 59); (5) yW_{37}, which stabilizes anticodon–codon interaction, possibly by stacking with the anticodon bases (60), or by reducing the free energy of interaction with the ribosome (61); and (6) m^1G_{37}, which prevents shifts in the translational reading fame (62).

Additional evidence of modified nucleoside importance to translation comes from studies of strains of *Schizzosaccharomyces pombe* depleted of those $tRNA^{Ser}$ isoacceptors that read the UCA codon (36). The $tRNA^{Ser}$ isoacceptors have either a 2-thio- or 5-methoxycarbonylmethyluridine at wobble position 34. In these strains, the major $tRNA^{Ser}$ with inosine-34 was incapable of rescuing the reading of Ser codons ending in A. Suppressor $tRNA^{Ser}_S$ having 2-thiouridine derivatives at wobble position 34 are ineffective at reading UGA when in thiodeficient *S. pombe* strains (37, 38). Undermodification is prevalent in viral-infected, transformed, and normal cells that are growing rapidly. Retroviruses may take advantage of frame-shifting caused by undermodification in the required expression at the *gag–pro* and *pro–pol* junctions of these viruses (63). The results of these and many other investigations firmly establish the importance of anticodon-domain modifications in ribosome-mediated codon-binding, and support the "extended anticodon" hypothesis (64), but do not explain the physicochemical contributions of the modified nucleosides.

II. The Large Array of Modified Ribonucleoside Structures and Chemistries

The modified nucleosides of RNA have evolved to alter RNA chemistry and structure, and their ability to accomplish this is as varied as are their 93 individual chemistries and structures (29). Nucleoside modifications are as "simple" as methylations (65), thiolations (66), and glycoyl bond substitutions (i.e., pseudouridylation) (67, 68), and as complex as hypermodifications that add amino acids (69) or heterocyclic rings (70). Even simple methylations significantly change the character of a nucleoside by increasing either hydrophobicity or hydrophilicity (71) or by inhibiting participation in canonical

Watson–Crick base-pairing (72). Chemical identification of newly found modified nucleosides has outpaced determination of their contributions to biological function, because sensitive HPLC and MS analytic technologies (73–77) have developed faster than methods for analyzing structure/function relationships of modified nucleosides.

The chemistries and structures that the 93 ribonucleoside modifications contribute to RNA are presented here in a manner analogous to the conventional textbook presentation of the 20 common amino acids. The many different nucleoside modifications are representative of the same three broadly based types of chemical contributions as relate to the amino-acid side-chains: charge (Tables IIA and IIB); increased polarity, but without charge (Tables IIC and IID); and increased hydrophobicity (Tables IIE–IIG). Structures of the modified nucleosides are presented in Tables IIA–IIG in the predominant anti glycosyl bond conformation and with the charges contributed by the modifications at pH 7. Of the 93 modifications, 24 impart charges to the mononucleosides under physiological conditions (Tables IIA and IIB). Simple methylation of one of the heterocyclic ring nitrogens of either the purine or pyrimidine base can result in a positively charged quaternary nitrogen. Approximately, 25% of all tRNAs have m^1A^+ at position 58 of the tRNA TψC stem-and-loop domain, and approximately 40% have m^7G^+ at position 46 of the "extra arm" or variable loop (Fig. 1). (In order to emphasize the chemical contributions of modified nucleosides in this review, the standard abbreviations of the modified nucleosides have been altered to include the charge, i.e., m^1A^+, instead of m^1A.) However, neither of these mononucleosides carries a net charge under physiological conditions (71).

The two relatively common purine modifications, m^1A^+ and m^7G^+, exemplify the physical chemical contribution listed first in Table I, the *introduction of transient positive or negative charges*. The m^1A tautomer prevalent at pH 7 has no net charge and $m^7G^{\pm}$ is a zwitterion (71). NMR analyses of native, but $^{13}CH_3$-enriched, yeast and *E. coli* tRNA^{Phe} demonstrate that the two methylated nucleosides are positively charged in the tRNAs because they have been protonated in the tertiary structure of the molecule (78). Protonation occurs through tertiary structure hydrogen bonding, $m^1A^+_{58}$ to T_{54} and $m^7G^+_{46}$ to G_{22} and C_{13} (78). Thus, the purpose of methylations at A_{58} and G_{46} in tRNAs is to produce site-specific electrostatic charges within the tertiary structure of the tRNA and not to add methyl groups to the molecular structure. The positive charges are transient because of their dependence on the weak tertiary interactions that can easily be disrupted by interaction with a protein (78).

The study of other charged nucleosides has not been as thorough. With H-bonded tertiary interactions similar to those of m^1A^+, the nucleoside m^3C^+ would be positively charged, as depicted in Table IIA. Thus, methyla-

TABLE IIA
MODIFIED NUCLEOSIDES THAT CONTRIBUTE CHARGE[a]

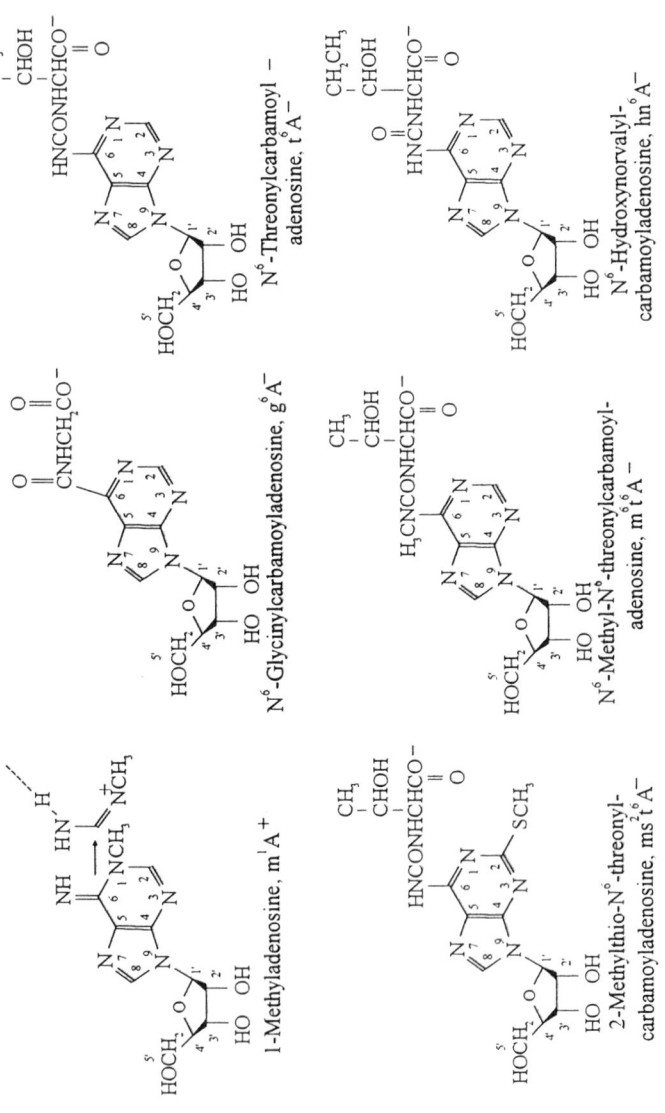

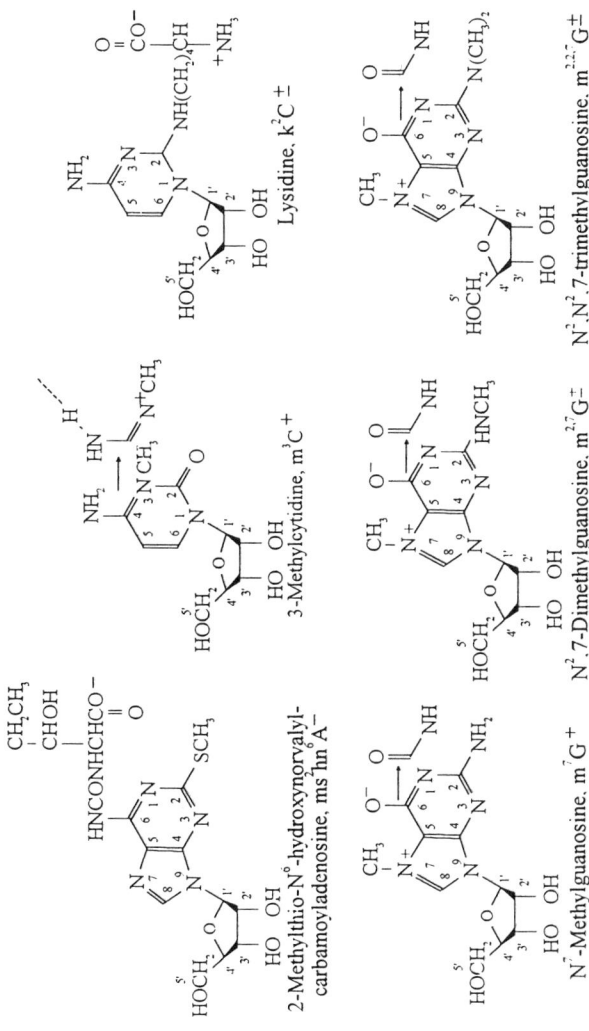

[a] Standard abbreviations of modified nucleoside names have been adapted for this article by the inclusion of charge.

TABLE IIB
Modified Nucleosides That Contribute Charge[a]

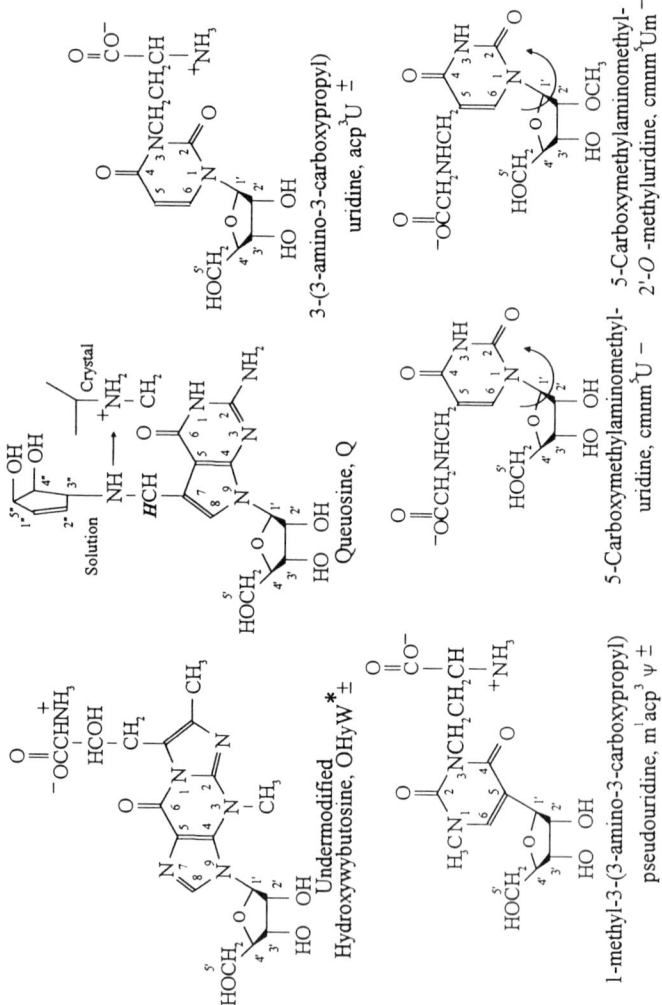

5-Carboxymethylaminomethyl-2-thiouridine, cmnm^{5}s^{2}U–

5-Carboxymethyluridine, cm^{5}U–

Uridine-5-oxyacetic acid, cmo^{5}U–

5-(carboxyhydroxymethyl)uridine, chm^{5}U–

2′-O-ribosyladenosine [phosphate], Ar[p]–

2′-O-ribosylguanosine [phosphate], Gr[p]–

---

^{a} Standard abbreviations of modified nucleoside names have been adapted for this article by the inclusion of charge.

TABLE IIC
Modified Nucleosides That Increase Polarity

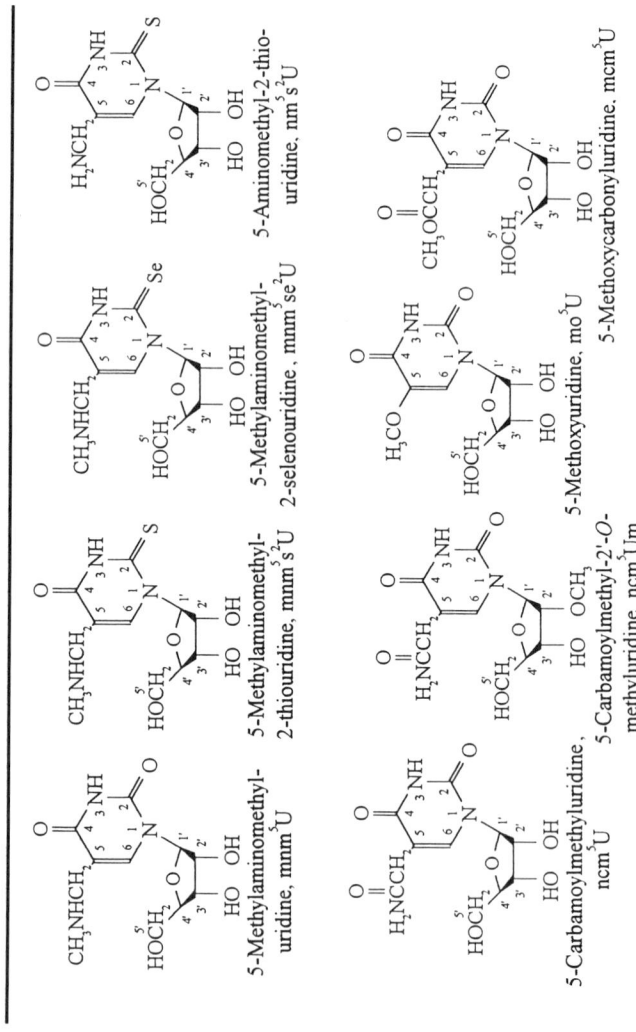

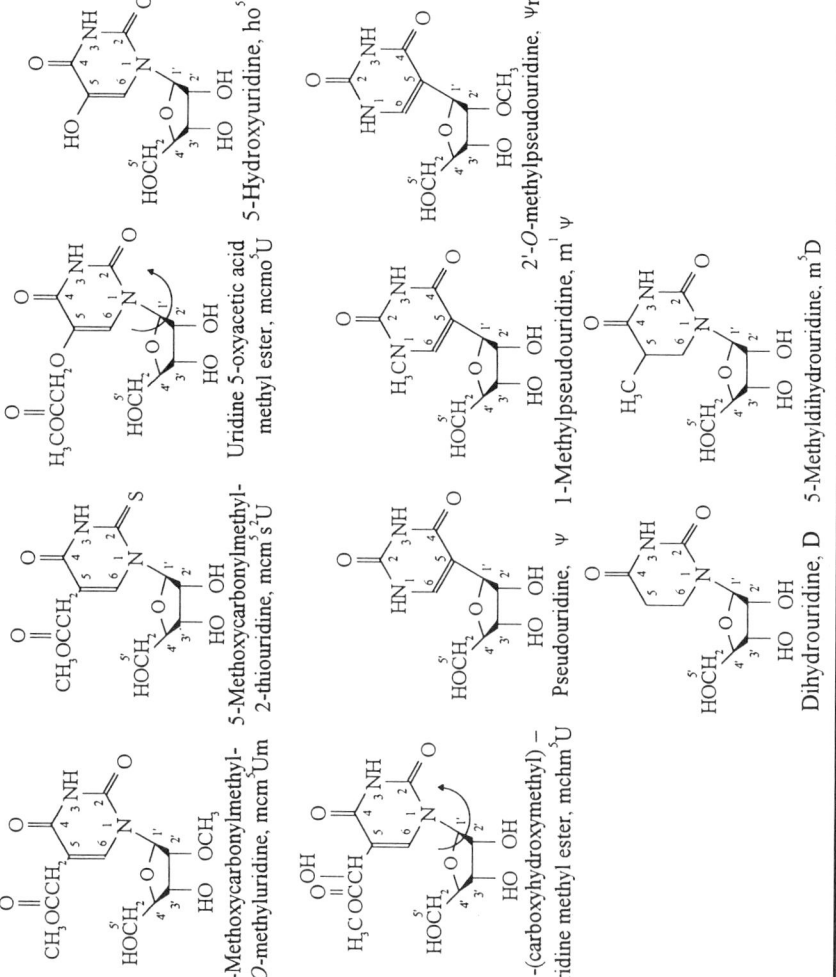

TABLE IID
MODIFIED NUCLEOSIDES THAT INCREASE POLARITY

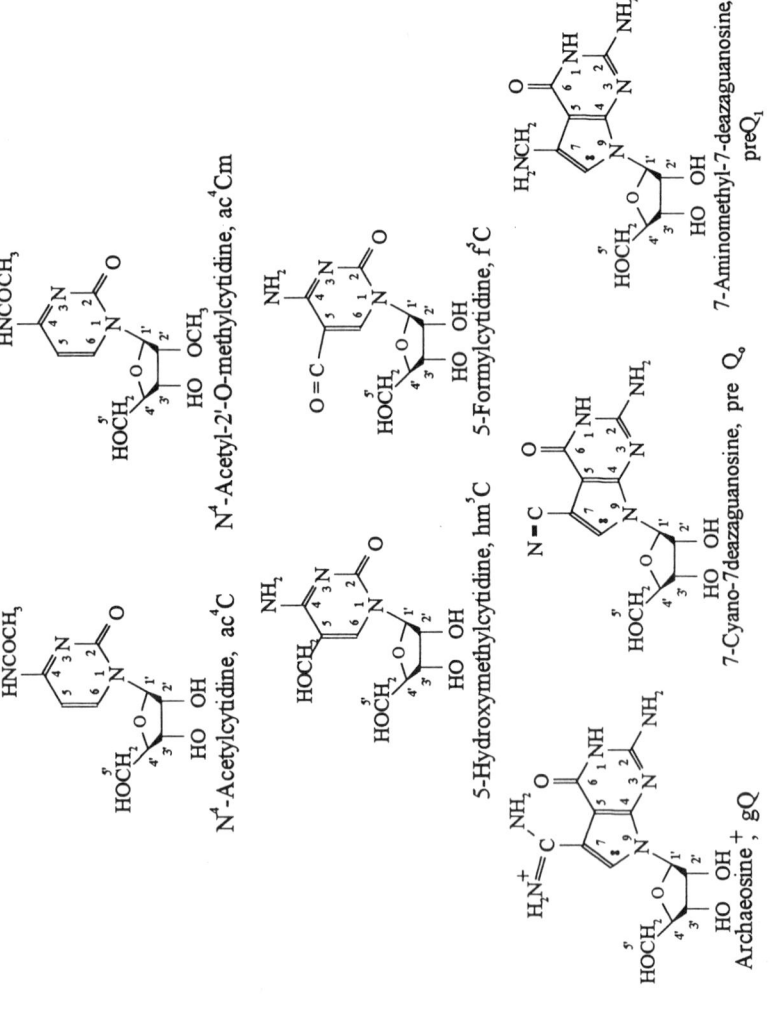

TABLE IIE
Modified Nucleosides That Increase Hydrophobicity

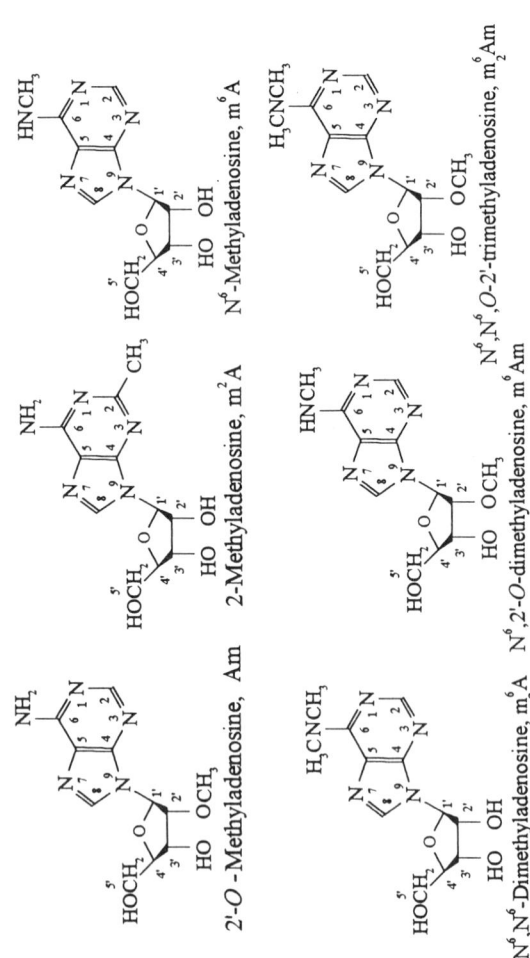

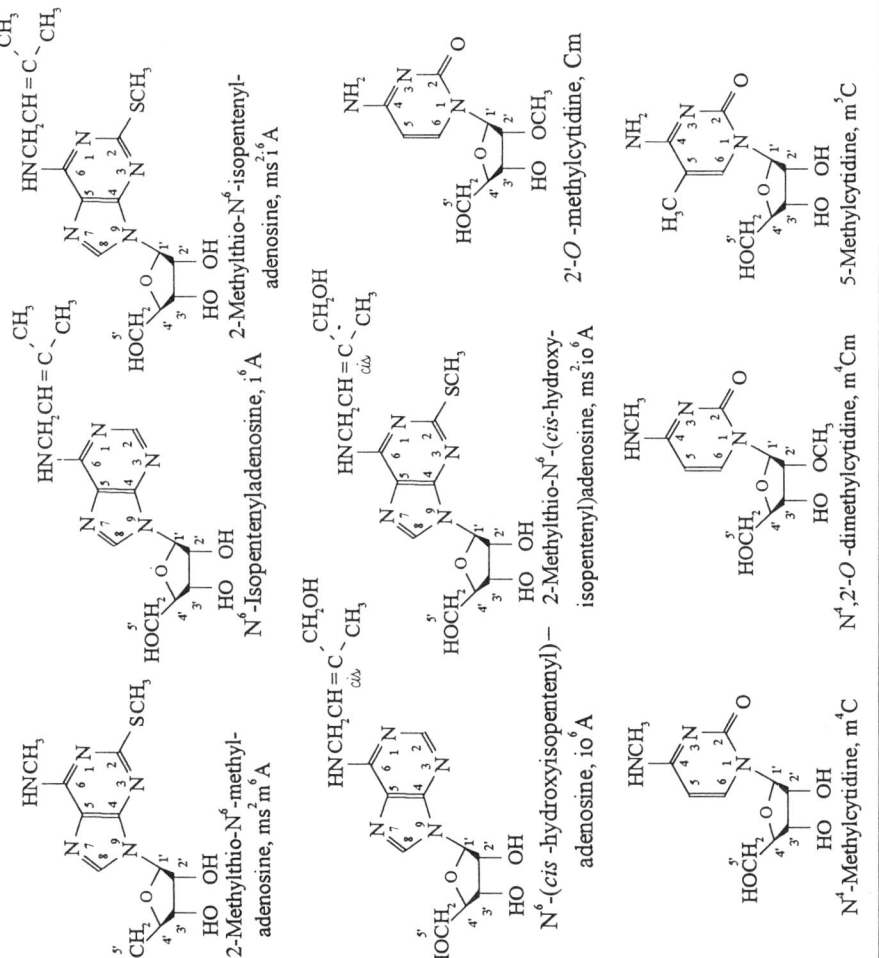

TABLE IIF
Modified Nucleosides That Increase Hydrophobicity

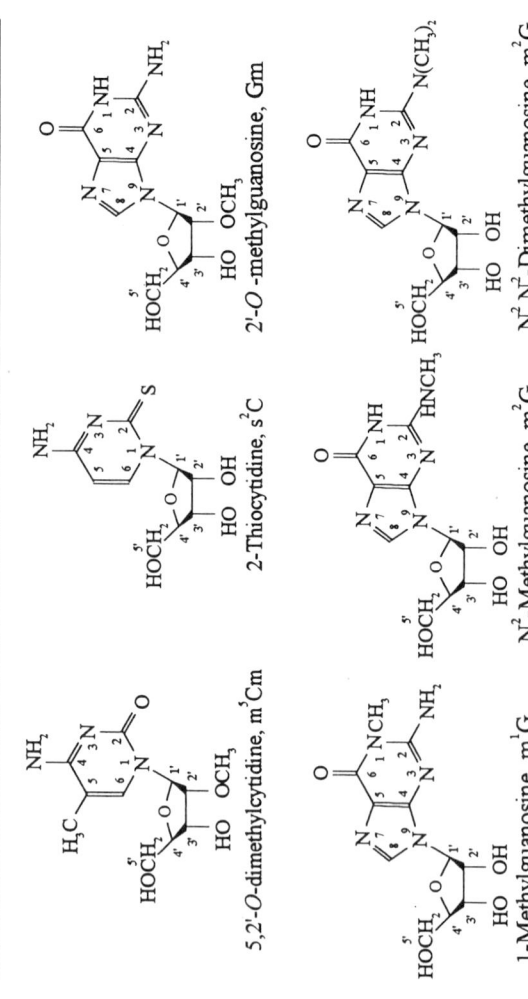

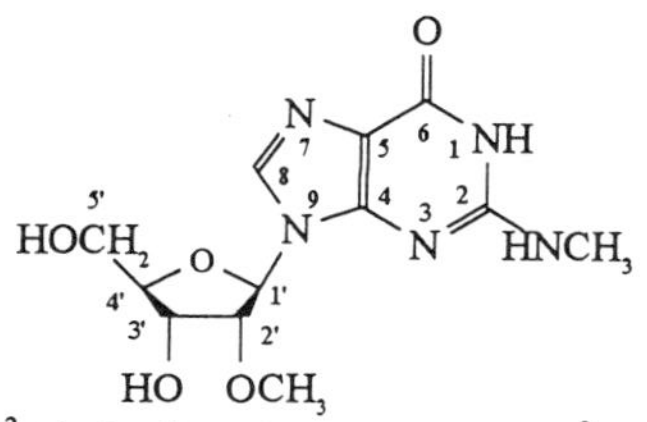

$N^2,2'$-O-dimethylguanosine, m^{2}Gm

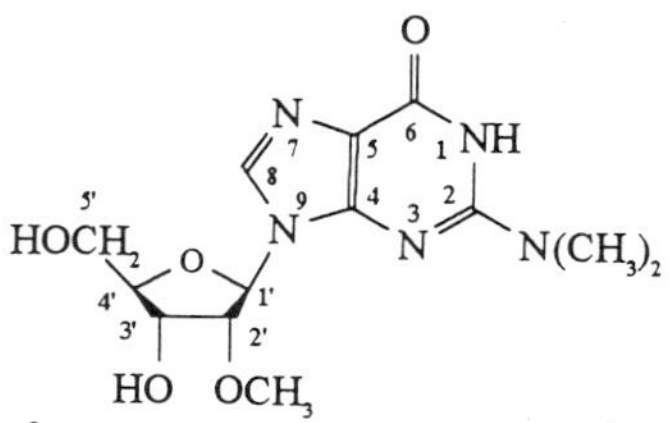

$N^2,N^2,2'$-O-trimethylguanosine, m$^2_2$Gm

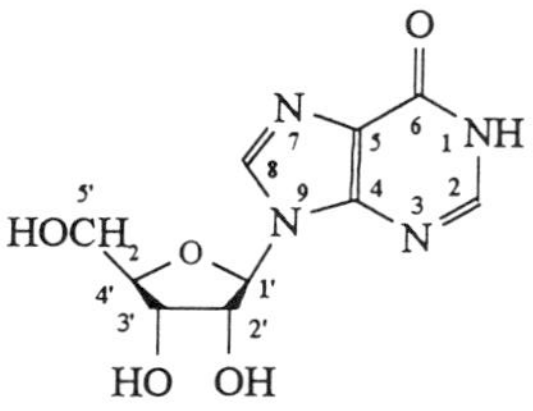

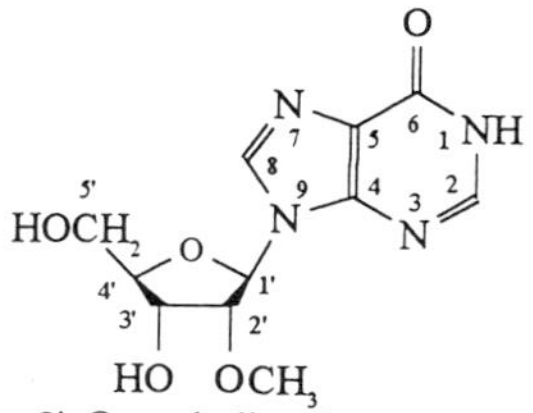

$2'$-O-methylinosine, Im

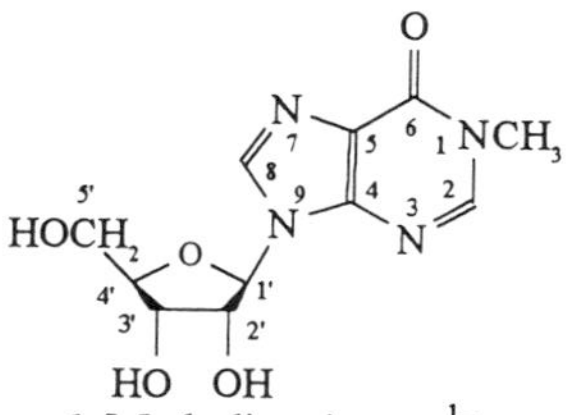

1-Methylinosine, m^{1}I

Inosine, I

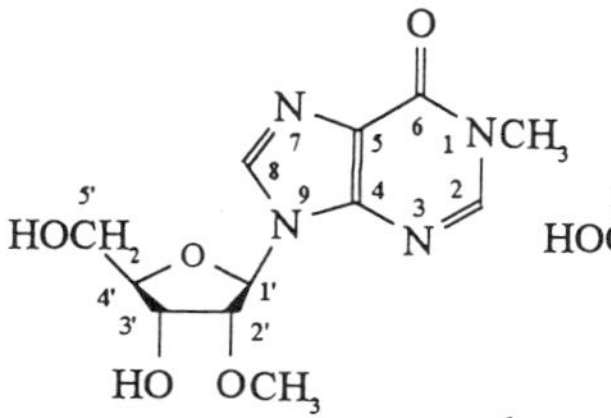

1,2'-O-dimethylinosine, m^{1}Im

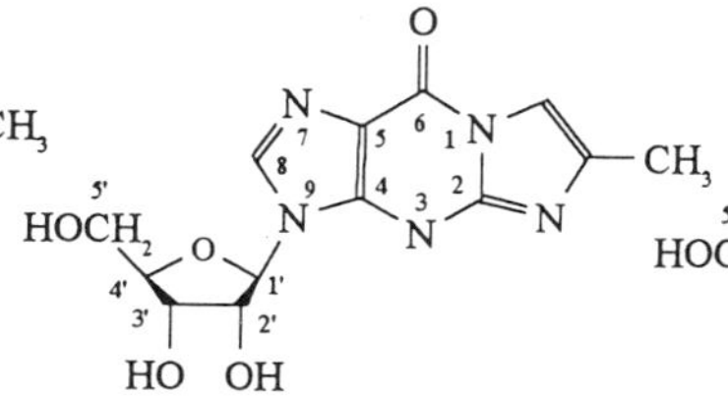

Wyosine, imG

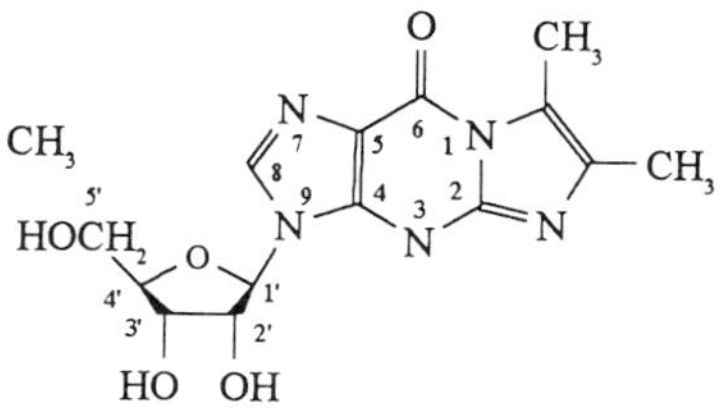

Methylwyosine, mimG

TABLE IIG
MODIFIED NUCLEOSIDES THAT INCREASE HYDROPHOBICITY

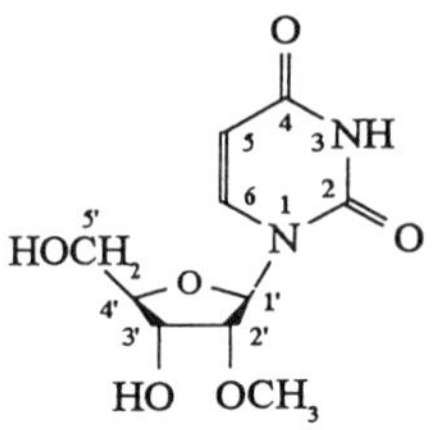

2'-O-methyluridine, Um

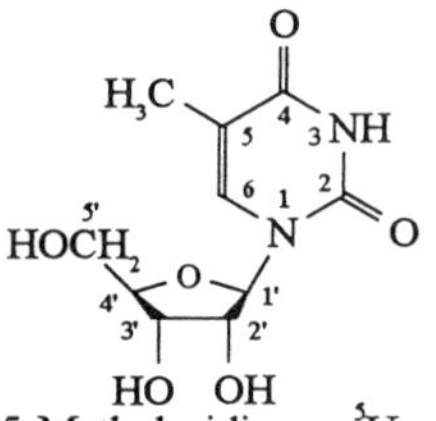

5-Methyluridine, m⁵U
Ribosylthymine, rT

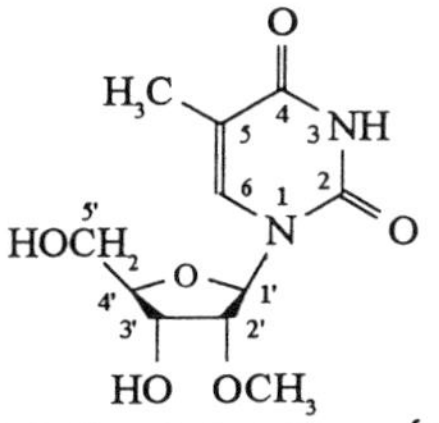

5,2'-O-dimethyluridine, m⁵Um
2'-O-ribosylthymine, rTm

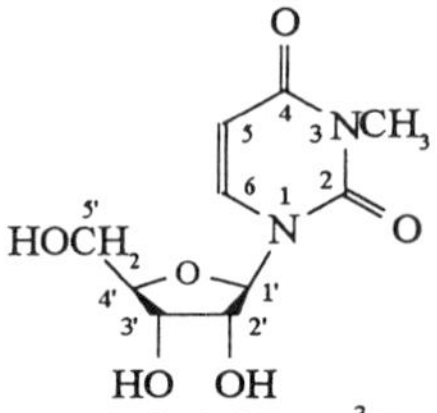

3-Methyluridine, m³U

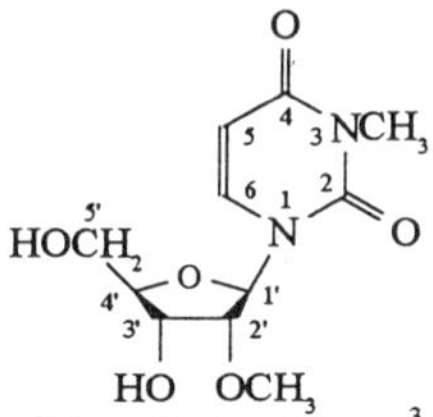

3,2'-O-dimethyluridine, m³Um

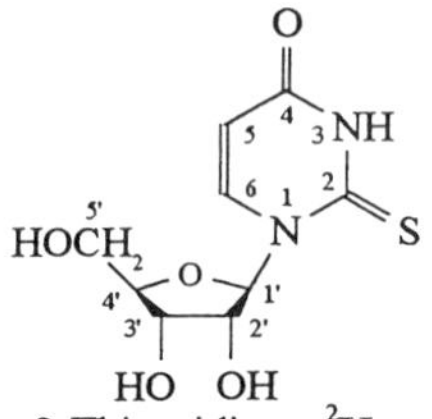

2-Thiouridine, s²U

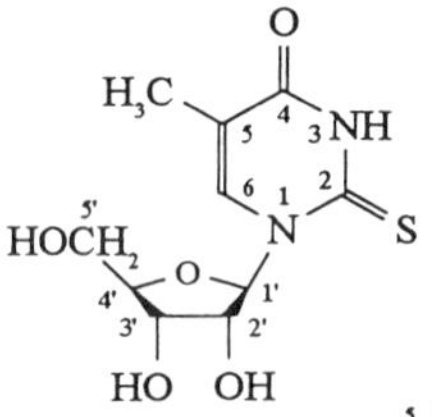

5-Methyl-2-thiouridine, m⁵s²U

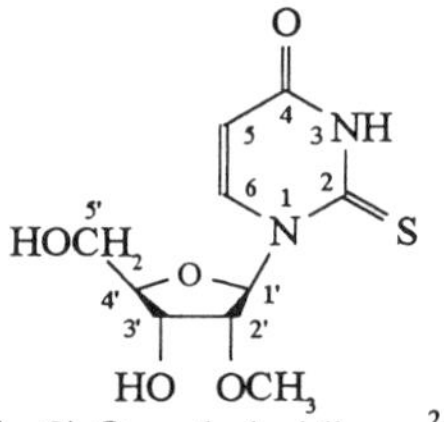

2-Thio-2'-O-methyluridine, s²Um

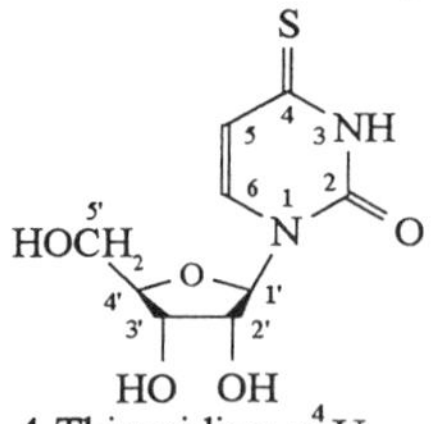

4-Thiouridine, s⁴U

tion of the 3-position of C has the analogous property of introducing a transient positive charge. All of the di- and trimethylated m⁷Gs are zwitterions at pH 7, but have the potential of being positively charged in an RNA, if protonated.

In contrast to m¹A⁺, m⁷G⁺, and m³C⁺, some ribonucleoside modifications would *introduce stable local charges to the RNA*, the second of the 10 physical chemical contributions (Table I). Amino-acid-derived hypermodifications can result in nucleoside zwitterions: k²C±, OHyW±, acp³U± (79), and m¹acp³ψ± (79). Local pH fluctuations and H-bonding within the RNA structure could produce localized charge. Fourteen modifications have the potential of introducing site-selective negative charges. Six amino-acid derivatizations yield negatively charged nucleosides: g^6A^-, t^6A^-, $ms^2t^6A^-$, $m^6t^6A^-$, hn^6A^-, and $ms^2hn^6A^-$.

The 5-position of uridine is a frequent target for both charged and polar modifications. Six 5-position-modified uridines are negatively charged: $cmnm^5U^-$, $cmnm^5Um^-$, $cmnm^5s^2U^-$, cm^5U^-, cmo^5U^-, and $chmo^5U^-$. In the absence of 2-thiolation, the lowest free-energy-state conformation of uridines modified with long side-chains at C-5 has a syn glycosyl bond angle (12, 80). Uridines in solution known to exhibit majority fractional populations with the syn conformation are designated in Table II with an arrow about the glycosyl bond.

Modifications that create a new equilibrium between alternative mononucleoside conformations represent the third class of physical contributions, that of altering and restricting nucleoside conformation (Table I). Other negatively charged nucleosides include $Ar[p]^-$ and $Gr[p]^-$, charged because of the phosphate associated with the 2'-O-ribosyl modification. An elaborate modification of guanosine yields nucleoside Q (queuosine), which in crystal structure is positively charged but in solution is uncharged (81). The plane of this nucleoside's 3,4-*trans*-4,5-*cis*-dihydroxycyclopent-1-ene modification is interestingly perpendicular to that of the purine ring, and the protons associated with the C-7 methylene are nonequivalent (81).

Polar, but uncharged modified nucleosides consist of amino, ester, and hydroxyl derivatizations (Tables IIC and IID). Representives of simple secondary and tertiary amine modifications include mnm^5U; the 2-thio derivatives, nm^5s^2U and mnm^5s^2U; and the 2-seleno derivative, mnm^5se^2U. Surprisingly, even the longest of 5-position modifications cannot overcome the large stereochemical energy barrier imposed by the 2-thio modification on the glycosyl bond transition from anti to syn (80, 82, 83). The amine and ester derivaties at C-5 of U are often found in wobble position 34 of tRNAs.

The polar grouping of ribonucleosides include four amides: ncm^5U and its 2'-O-methyl derivative (ncm^5Um), and ac^4C and its 2'-O-methyl derivative (ac^4Cm). Pseudouridine, ψ, the C-5 glycosyl isomer of uridine, is slightly more polar than uridine. However, the 1- and 2'-O-methylated ψ, $m^1\psi$ and ψm, would be less polar than ψ. The 7-deazaguanosines result in cyano and amine derivatives of G: $preQ_0$, $preQ_1$, and archaeosine (gQ).

Ester modifications at C-5 of U include mcm^5U and its 2'-O-methyl (mcm^5Um) and 2-thio- (mcm^5s^2U) derivatives, and -OH ($mchm^5U$) and oxy-($mcmo^5U$) derivatives. The wybutosines, yW, o_2yW, and OHyW, which are found at position 37, 3' adjacent to the tRNA anticodon, are more complex methyl esters of amino-acid side-chains of wyosine, a tricyclic derivative of G. Wyosine without further modification is more hydrophobic than G (70). Of interest is the absence from tRNAs of those methylated wyosines that would result in zwitterion or positively charged nucleosides at position 37, such as the m^7 derivative of wyosine (70).

Nucleoside polarity is increased with additions of the simplest of hydroxyls or with the complexity of a sugar-modified cyclopentenol substituent. Examples of -OH additions include ho^5U, hm^5C, and $mchm^5U$. The methoxy and formyl derivatives of C, mo^5C and f^5C, are also found. The latter occurs at wobble position 34 in bovine mitochondrial tRNA^{Met} and, as the mononucleoside, is conformationally restricted to the 3'-endo sugar pucker (84). More complex nucleoside modifications that add two or more -OH groups include the queuosines, 7-deazaguanosines, and their galactose and mannose derivatives: oQ, galQ, and mQ. Addition of hydrogen across the C-5–C-6' double bond of U destroys this base's aromatic character. Dihydrouridine, D, and its 5-methyl derivative, m^5D, are less hydrophobic than U and m^5U, and are unable to base-stack.

Nonpolar, hydrophobic derivatizations of the four major ribonucleosides are shown in Tables IIE–IIG. The simplest of these are the 2'-O-methylations, Am, Cm, Gm, and Um. Addition of methyl to the 2'-O-position, once thought to be a thermophile's protective mechanism from any enzymatic or metal-catalyzed hydrolysis of the phosphodiester bond that passed through the 2',3'-cyclic intermediate (85), stabilizes nucleoside conformation, in general, and raises the RNA melting temperature (86). Thus, *2'-O-methylations, and the more restrictive 2-thiolations of pyrimidines discussed below are additional examples of modifications altering and restricting nucleoside conformation* (Table I). Thermal stability of tRNA of hyperthermophilic archaea grown at 100°C may be dependent on the tRNA's unique content of both ac^4Cm and m_2^2Gm, and of m^5s^2U found at the position of m^5U_{54} (87).

Hydrogen-bonding of bases to ribose is inhibited by methylation of the 2'-OH. Therefore, 2'-O-methylations are examples of *modifications that reduce noncanonical base pairing*, the fourth physicochemical contribution of modifications (Table I). A number of single, double, and triple methylations and methylthiolations of adenosine are included among the nonpolar derivatizations: m^2A, m^6A, m_2^6A, m^6Am, m_2^6Am, and ms^2m^6A. The N^6-isopentenylation of A results in a more hydrophobic nucleoside: i^6A and ms^2i^6A. Hydroxylated isomers of the N^6-isopentenyladenosines, i^6A, are more polar than the parent modification: io^6A, ms^2io^6A. A set of mono-, di-, and trimethylated guanosines are similar to the exocyclic amine and ring methylations of A: m^1G, m^2G, m_2^2G, m^2Gm, m_2^2Gm. Inosine, I, and its methylated derivatives, m^1I, m^1Im, and Im, and wyosine, imG, and its methylated derivative, mimG, are less polar and thus more hydrophobic than G. Singly and doubly methylated Cs and Us include: m^4C, m^4Cm,

m^5C, m^5Cm, m^5U (rT), m^5UM (rTm), m^3U, and m^3Um. All are more hydrophobic than the parent pyrimidine.

Thiolations of the 2-position of C, s^2C, and of U, s^2U, m^5s^2U, s^2Um, etc., reduce the polarity of the nucleosides and result in distinct stereochemical restrictions to the glycosyl bond (anti) and sugar pucker (3′-endo, gauche$^+$) (12, 80, 82, 83). For instance, the energy barrier for the 3′-to-2′ sugar-pucker transition created by the 2-thiolation of U is considerably higher than that produced by 2′-O-methylation; 2-thiolation produces a nearly 100% population of the 3′-endo conformer (12, 80, 82, 83) versus 62% for 2′-O-methylation (86). The s^2Us are found at anticodon wobble position 34, whereas s^4U is common to position 8 of bacterial tRNAs. The evolution of the 93 or more modified ribonucleosides may be connected just as much with their site-selected positioning as with their unique altering of chemistry and structure.

III. Site-selective Positioning of Modifications

The complex three-dimensional structures of RNAs include secondary structure motifs: stem-and-loop "hairpins" (the most common element), internal loops (bubbles), bulges, single-stranded regions, and pseudoknots. Modified nucleosides are found in many different structures. Transfer RNA modifications exemplify the site-selective nature of all RNA modifications. Of all RNAs, tRNAs have the largest number and greatest variety of post-transcriptional modifications, 79 at the time of this writing (5, 29). Of the 79 modifications found in prokaryotic and eukaryotic tRNAs, 59 are unique to tRNA. Eukaryotic nuclear-encoded tRNAs tend to be more highly modified than their prokaryotic or mitochondrial-encoded counterparts; some 25% of the nucleosides of a eukaryotic tRNA could be modified in comparison to only 4% of a bacterial tRNA. Most tRNA modifications are found in many different amino-acid-accepting species.

In general, modifications are found at site-specific residues, and are not particular to a specific amino-acid-isoaccepting species. Modification-enzyme determinants can be as uncomplicated as sequences contiguous with, or local to, the target nucleoside. The tRNA-m^5U_{54} methyltransferase and the tRNA-guanine transglycosylase recognizes sequences near the target nucleoside, U_{54} or G_{34}, respectively (88, 89). Other modifying enzymes require distant sequence information, such as that for production of $m_2^2G_{26}$ in yeast tRNAs (90, 90a). Still others, such as the m^4G_{37} methyltransferase, require the complexity of the tertiary structure interactions (91, 91a).

The site-selected positioning of modified nucleosides in tRNA is functionally significant. The tRNA secondary structure consists of the aminoacyl stem (AA stem), dihydrouridine-containing (D) stem-and-loop, anticodon (AC) stem-and-loop, "extra arm" or variable (V) loop, and the m^{5}U-containing (T) stem-and-loop (Fig. 1). Nucleosides within loops of tRNAs are more often modified than are those in the double-helical stems. A review of the many positions in which modified nucleosides are found in the stem-and-loop cloverleaf secondary structure of tRNA clearly demonstrates that the loops are most often the sites of modifications (Fig. 1). RNA loops and bulges, whereby bases are more exposed to solvent, are good gargets for protein binding (92) and interactions with other RNAs. During translation, tRNA interacts with aminoacyl-tRNA synthetases, initiation and elongation factors, ribosomal proteins and rRNA, and mRNA. Among all tRNA sequences determined by either RNA or gene sequencing, the T stem-and-loop domain, composed of eight essentially invariant nucleoside positions, including two site-specific modifications, T_{54} and ψ_{55}, is the most highly conserved of the secondary structure's three stem-and-loop domains. Thus, functional interactions associated with the T domain of the molecule must be common to almost all tRNAs.

The highest frequency and greatest variety of modifications occur in the AC loop, adding to the uniqueness of tRNA species (Figs. 2 and 3). Although

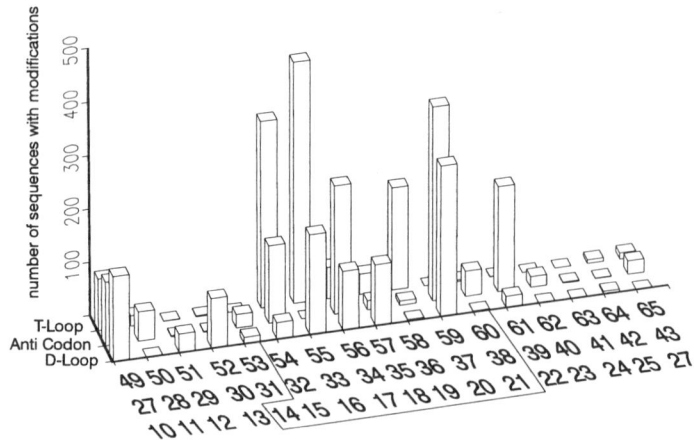

FIG. 2. Frequency of modification within the D, anticodon, and T stem-and-loop domains. The figure results from a compilation of 507 tRNA sequences. Nucleoside positions, numbered on the x axis, for each domain are in the standardized form for the 76-nucleotide tRNA molecule. The number of sequences having a modification at each residue in plotted on the y axis. Thus, position 55 of the T domain is frequently modified. The frequency of modification is generally high in the loop regions (boxed numbered area) for all three domains.

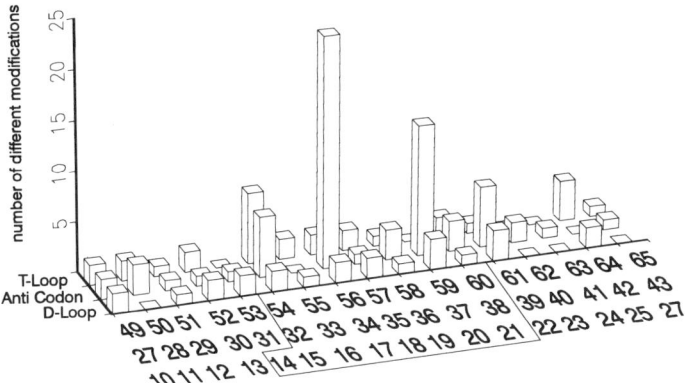

FIG. 3. Variety of modifications within the D, anticodon, and T stem-and-loop domains. Data are taken from 507 tRNA sequences. As in Fig. 2, the nucleotide position numbers are plotted on the x axis. The number of different modifications occurring at each position plotted on the y axis. Thus, very few types of modifications occur at position 55 of the T domain, but the position is frequently modified (Fig. 2). Residue 55 is commonly modified to ψ_{55}. Anticodon wobble position 34 and position 37, 3'-adjacent to the anticodon, are the positions in tRNA with the greatest variety of modified nucleosides.

modifications occur frequently at specific positions in the D and T loops of tRNA (Fig. 2), the variety of modifications occurring in these loops is far less than in the anticodon domain (Fig. 3). Approximately 50 of the 79 different types of tRNA-modified nucleosides are found in the AC stem and loop. Many are found exclusively in the AC domain and are located only at positions 34 or 37 (Fig. 3). Elaborately modified uridines and hypermodified purines are concentrated at wobble position 34 and position 37, 3' adjacent to the anticodon, respectively. Anticodon-domain modified nucleosides participate in aminoacyl-tRNA synthetase (aaRS) recognition of cognate tRNAs, aaRs discrimination of noncognate tRNAs, and ribosome-mediated codon binding.

Some modifications are found consistently at a single position: 5-position derivatives of s^2U at position 34, hypermodified purines at position 37, m^7G^+ at position 46, and m^5U at position 54. Others occur at a small number of positions in a single physical area of the molecule: m^2G and m^2_2G at positions 6, 7, 9, 10, and 26; D at residues 16, 17, 20, and 21. Still others are found in comparable loop or stem positions: m^1A_{14} in the D loop and $m^1A^+_{58}$ in the T loop; m^5C_{40} in the anticodon stem and m^5C_{49} in the T stem. Of all the modifications, by far ψ is found at the most widely different positions in all four stems and all four loops. *In vivo* site-selective positioning of these post-transcriptional modifications requires multiple enzymes for the complex modifications at a single site, and multiple enzymes for the simpler modifications at multiple sites. The ψ_{54} common to almost all tRNAs is produced by an enzyme different from that responsible for ψ production in the AC stem and loop domain.

The tRNA tertiary structure (Fig. 4) illustrates that some modified nucleosides are within the structure, though the modification may be accessible within a groove, whereas other modified nucleosides are fully accessible to solvent, proteins, and RNAs that interact with the molecule. (See Refs. 93–97 for structures of tRNA and tRNA/aaRS complexes derived from X-ray crystallography.) Examples of the former include the methyls of m^{2}G$_{10}$ and m$_2^2$G$_{26}$ and m^{1}A$_{58}^+$. *Alkylations of the A, C, and G exocyclic amines, the N-1 of A and G, and the N-3 of C exclude participation of the modified nucleoside in canonical Watson–Crick base pairing, thus contributing to conformational dynamics*, the fifth physical chemical alteration that modifications contribute to nucleic acid structure (Table I). Examples of modified nucleosides displayed to the outside of the tRNA molecule include the Ds of the D loop and the many different modifications at wobble position 34.

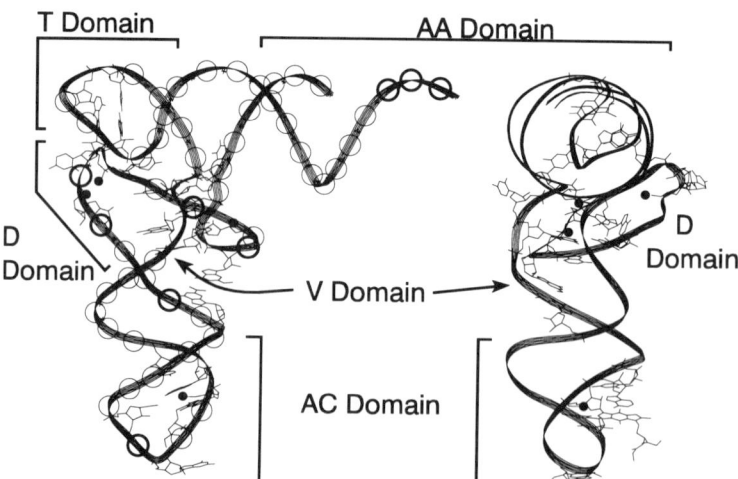

FIG. 4. Three-dimensional structure of tRNA. The backbone of the yeast tRNA^{Phe} 3-D structure is drawn as a ribbon from X-ray crystallographic data (93). A view from the side of the molecule is shown on the left and a view through the contiguous aminoacyl and T helices is shown on the right. Only modified nucleosides are drawn in detail; all other nucleosides are denoted by circles on the left-hand figure. Heavier circles denote the positions of the invariant and semiconserved nucleosides. The small solid circles represent the locations of the four tightly bound Mg^{2+} ions. Some modifications (D$_{16,17}$ and Gm$_{34}$) are exposed to solvent; others, such as yW$_{37}$, m^{5}C$_{40}$, T$_{54}$, and m^{1}A$_{58}$, are within the structure.

IV. The Study of Modified-nucleoside Contributions to Structure and Function

A. Syntheses of Site-selectively Modified RNAs

Site-specific positioning of modified nucleosides is important in the study of their contributions to RNA structure and function. There are four approaches to the synthesis of RNAs with site-selectively positioned modified nucleosides, one *in vivo* and three *in vitro*. The *in vitro* methods include manual solution or column-supported syntheses, enzymatic and automated chemical syntheses combined, and automated chemical synthesis alone. Physicochemical studies of modified nucleoside-containing RNAs requires milligram quantities of highly purified nucleic acid. The advantages and disadvantages of the four methods are compared in Table III. *In vivo* meth-

TABLE III
APPROACHES TO INCORPORATION OF MODIFIED NUCLEOSIDES INTO RNA FOR INVESTIGATIONS OF STRUCTURE AND FUNCTION

Approach	Advantages	Disadvantages
In vivo	Full-length RNA Mature RNA Native conformation	Low yields Labor intensive Waste of stable isotopes
Manual synthesis	Site-selected introduction of modified and/or stable isotope-labeled nucleosides that are sensitive to automated chemical synthesis Large quentities	Chain length is short Labor intensive
T7 polymerase and ligation	Longer length than chemical synthesis High yield Sparing of isotopes	Expensive in total materials Labor intensive Difficult to produce a single RNA with a variety of site-selected modifications and isotopes Chimeric RNA/DNA product difficult
Automated chemical synthesis	Commonly practiced chemistry Less labor intensive Site-select modified nucleosides and isotopes Ease in producing one RNA with a variety of modifications and isotopes Chimeric RNA/DNA product easy	Length limit Expense of phosphoramidites Less sparing of stable isotopes

ods, in which RNA species are purified from cells grown in culture, have the advantage of producing mature RNAs. This method is labor-intensive and can be quite expensive if it includes the enrichment of stable isotopes (^{13}C or ^{15}N) for NMR (98). In vivo methods do not easily permit investigation of the contributions of a single or small numbers of modified nucleosides. A few bacterial and yeast mutants lacking specific modifying enzymes are available (13).

Manual syntheses in solution or on a column support (83, 100–101) are also labor-intensive. These methods suffer from the significant disadvantage of the short lengths (dimers to heptamers) that can be made in quantities for NMR. Their advantage is solely in the ability to incorporate into RNAs modified nucleosides labile to phosphoramidite synthesis and the chemical cycling of an automated nucleic acid synthesizer. A number of dimers, trimers, and pentamers containing modified nucleosides labile to the chemistries of automated oligonucleotide syntheses have been produced in sufficient quantities for biophysical studies (12, 83, 99).

Large quantities of unmodified RNA are produced by T7 transcription of cloned genes or chemically synthesized template. Transcription with modified nucleoside triphosphates can be accomplished with those nucleosides accepted by the polymerase. However, because the synthesis is template-directed, a modified adenosine would be incorporated at all adenosine sites called for by the template. The same is true for the incorporation of stable-isotope-labeled nucleoside triphosphates, i.e., a labeled NTP would be incorporated at each and every residue at which the complementary base exists in the template (102–104). Although generally, if not uniform, ^{13}C and ^{15}N labeling by T7 transcription is advantageous for complete structural determination, it is neither required for the study of local contributions of modified nucleosides, nor is it necessarily desirable for investigations of RNA–protein and RNA–RNA interactions (112).

In order to achieve the synthesis of RNA with a single or small number of modified mnucleosides, T7 transcription of unmodified RNA can be combined with either the in vivo or the automated chemical syntheses of modified nucleoside-containing RNA. The former requires enzymatic fragmentation of the mature tRNA, separation of the fragments, and ligation of a fragment, selected for a particular modified nucleoside and sequence, with the T7 transcript (17). The latter requires site-selected positioning of modified and/or stable-isotope-labeled nucleosides by automated chemical synthesis and ligation of the resulting RNA to a T7 transcript. Although milligram quantities of RNA are produced from T7 transcription, ligation of the transcript with the modified nucleoside-containing RNA is considerably less efficient and requires almost stoichiometric amounts of DNA ligase.

In contrast, completely automated chemical synthesis produces milligram (micromole) quantities of RNA, DNA, or RNA–DNA chimeras with site-selectively incorporated modified nucleosides and/or stable-isotope-enriched nucleosides (Table III) (105, 112). The chemical synthesis of a small amount of biologically active *E. coli* tRNA^{Ala} was accomplished with the 2'-*O*-*tert*-butyldimethylsilyl-3'-*O*-(2-cyanoethyl-*N*-ethyl-*N*-methyl)nucleoside phosphoramidites with a coupling yield of 98% (107), the equivalent of the commercially available 2'-*O*-*tert*-butyldimethylsilyl-3'-*O*-(2-cyanoethyl-*N*-diisopropyl)nucleoside phosphoramidites (95–98%). Pseudouridine (ψ) and two other modified uridines, dihydrouridine (D) and ribothymidine (rT), were incorporated into the tRNA. Pseudouridine (ψ) has been introduced into an undecamer RNA duplex by automated chemical synthesis using nucleosides with standard protecting and coupling groups (108). Improvements in the chemistry and methods are published frequently (109, 110).

Thus, routine site-selected modified and stable-isotope-labeled nucleoside incorporation into RNA can be achieved through conventional automated chemical synthesis using standard phosphoramidite chemistry. This approach has the following advantages: commonly practiced chemistry, less labor-intensive than T7 transcription and ligation of half-molecules, ability to place modified nucleosides site-selectively, ease of producing one sequence with a variety of modified nucleosides and labels, synthesis of mature native nucleic acid, and preparation of chimeric product composed of ribo- and deoxyribonucleosides. The only significant disadvantage compared to the other techniques is a limitation on the length of RNA that can be produced in milligram quantities, approximately 35 residues.

An important consideration in choosing the chemistries for site-selected, automated incorporation of modified and stable-isotope-enriched nucleosides into RNA for biochemical and physical (CD and NMR) studies is the ability to take advantage of commercially available nucleoside phosphoramidites and instrumentation (111). The phosphoramidite coupling chemistry, 5'-dimethoxytrityl (DMT) protection, and 2'-*tert*-butyldimethylsilyl (tBDMSi) protection are commercially available and widely used for the synthesis of unmodified RNAs. However, the unique chemical contributions of the nucleoside modifications must be protected from reactions and degradation during automated chemical synthesis of the RNA, yet easily removed afterward. Eight 5'-*O*-(4,4'-dimethoxytrityl)-2'-*O*-*tert*-butyldimethylsilyl-modified ribonucleoside-3'-*O*-(2-cyanoethyl-*N*-diisopropyl)phosphoramidites and two stable-isotope-enriched, protected phosphoramidites have been synthesized (105; A. Malkiewicz, personal communication). All but one, s^{2}U, was site-selected incorporated into tRNA domains by automated chemical synthesis (105, 112), using standard phosphoramidite chemistry (113).

Coupling efficiencies in automated chemical synthesis, estimated by monitoring the released trityl group, averaged better than 98% (105). There is some noticeable decrease in coupling at the time of incorporation of a few of the modified nucleosides. RNAs are removed from the controlled pore glass (CPG) support under alkaline conditions (105). Purification of the full-length product from failed sequences is accomplished in a two-step process (105): first, by anion-exchange HPLC (Nucleogen 60-7 DEAE; 125 × 10 mm), capable of resolving sequences one to two nucleosides different in length; and second, reverse-phase HPLC for desalting and concentrating the sample. Yields of deprotected heptadecamer RNA synthesized on CPG columns range from 0.6–0.8 mg, and those from columns for 10-μmol syntheses, 4–6 mg.

B. The Choice of RNA and of Size

Biophysical investigations at atomic resolution—NMR and X-ray crystallography—require milligram quantities of nucleic acids, and circular dichroism, requires microgram quantities. Therefore, the molecule of choice should be easily produced with and without individual and combinations of modified nucleosides. In addition, stable isotope enrichment is important for NMR studies. The molecule of choice should be of a size amenable to structural studies, i.e., less than 40 nucleotides for NMR, and, if at all possible, biologically active.

The tRNA molecule is composed of separate physical domains that are also biologically active. A physicochemical approach, analogous to the study of biologically active domains of proteins, has been developed for determining modified nucleoside contributions to tRNA (112). A small biochemically active, physically separate, and stable domain of tRNA is synthesized with site-selectively positioned modified nucleosides and stable-isotope-labeled nucleosides for NMR (105). The three-dimensional structure of tRNA is that of a molecule with complex folding, but with two distinct domains: the amino-acid-accepting stem (tRNA$_{AA}$), and the anticodon stem-and-loop (tRNA$_{AC}$) (Fig. 2). These domains, and the TΨC stem-and-loop (tRNA$_T$) are functionally recognized (39–43, 72, 88, 114, 115). The conformation of the whole tRNA molecule is more than the sum of these parts. However, loops are common elements of RNA secondary structure (116); they often contain modified nucleosides and are often the building blocks for the three-dimensional folded architecture of RNA (117, 118).

NMR and other studies of selected, short sequences of both RNA and DNA show them to be stable as hairpins that are in the traditional sense without loops. They consist of Watson–Crick base-paired stems, but noncanonical conformational features (e.g., syn, C-2′ endo sugar pucker), noncanonical nucleoside pairings (e.g., G·U, A·G, A·C), and phosphate–base or 2′-OH–base bonding interactions across what normally would be considered a loop, and turns composed of as little as two nucleosides (117–122). A 5′ or

3' base stack is sequence-dependent on the location of the turn in the loop (93). It is clear that both types of nucleic acids readjust to a surprising extent by adopting nonconventional structures in order to minimize the free energy (123, 124). Intrahelical positioning of bulged nucleosides, such as the 3'-stacked bulged A found in the yeast tRNA^{Phe} anticodon domain and its DNA analog, tDNA$^{Phe}_{AC}$ (119), could facilitate switches between alternate conformations (125), a potentially important functional aspect of the tRNA anticodon (12, 119, 126).

V. Examples of Biophysical Contributions to Function by Modified Nucleosides

Modifications of RNA nucleosides that alkylate ring nitrogens, exocyclic amines, and ring carbons alter nucleoside chemical properties in two ways. Alkylated imines and amines disrupt canonical base pairing, the fifth modified nucleoside contribution to RNA structure (Table I). Together with ring carbon alkylations, alkylated imines and amines also *enhance base stacking interactions by producing more hydrophobic bases* (Table I, number 6). Introduction of the hydrophobic hypermodifications such as yW$_{37}$ and the isopentenyl-adenosine (i^{6}A$_{37}$) 3'-adjacent to the anticodon increase base stacking. The increased stacking occurs most importantly with the third anticodon base at residue 36, but also with the base at 38. The third anticodon base hydrogen-bonds to the first base of the codon. Temperature-jump studies of interacting, complementary tRNA anticodons (128) as a model for tRNA–codon interaction, and other studies, have shown the importance of modifications in positions 34 and 37 to base stacking in yeast tRNA^{Phe} and other tRNAs and, thus, presumably to codon binding (129–131). Therefore, increased stacking inteactions facilitate anticodon-to-codon base-pairing by promoting the 3' A-RNA base-stacked structure in a single-stranded region of the molecule. However, what are the physicochemical functions of some of the simpler modifications, the methyl and thio groups, pseudouridine and dihydrouridine?

Yeast tRNA^{Phe} and *E. coli* tRNA^{Glu} and tRNA^{Lys} have been chosen as model systems for the synthesis of milligram quantities of RNA domains with site-selected positioning of modified and stable-isotope-labeled nucleosides: pyrimidine ribonucleosides, m^{5}C, Cm, D, ψ, mnm^{5}U, rT, ^{13}CH$_3$-m^{5}U, and ^{15}N$_{1,3}$-uridine; and purine ribonucleosides, m^{1}G, m^{2}A, and Gm (105, 112). Micromole quantities of modified and stable-isotope-labeled tRNA domains were purified for CD, NMR, and biochemical analyses (72, 105, 112, 114, 127). To date, 25 RNAs with site-selectively positioned and differently modified nucleosides have been synthesized. The RNAs are 17 and 18 residues long. The octadecamer D domain and the heptadecamer T domain of yeast tRNA^{Phe} (tRNA$_D^{Phe}$ and tRNA$_T^{Phe}$, respectively) have been synthesized with their naturally occurring modified nucleosides located at the appropriate positions. The tRNA$_D^{Phe}$ domain was synthesized with D located at positions 16 and/or 17, and the tRNA$_T^{Phe}$ domain, with m^{5}C$_{40}$, T$_{54}$, or ^{13}CH$_3$-m^{5}U$_{54}$ and ψ_{55} (Fig. 5) (105, 112).

The anticodon domains of *E. coli* tRNA^{Lys} and tRNA^{Glu} (tRNA$_{AC}^{Lys}$ and tRNA$_{AC}^{Glu}$) are similar, having the modified nucleoside s^{2}mnm^{5}U$_{34}$, but have significantly different modifications at position 37, t^{6}A$_{37}$ for the former and m^{2}A$_{37}$ for the latter (Fig. 6). tRNA$_{AC}^{Glu}$ was synthesized successfully with mnm^{5}U at the wobble position 34. An attempt to incorporate s^{2}U failed because the thio group was oxidized during the trivalent-to-pentavalent phosphorus oxidation step of each cycle of the synthesizer (105). However, s^{2}U-containing pentamers corresponding to the anticodon loop have been produced by manual syntheses.

Variously modified yeast tRNA^{Phe} anticodon domains (tRNA$_{AC}^{Phe}$) have been synthesized with Cm$_{32}$, Gm$_{34}$, m^{1}G$_{37}$, ψ_{39}, and m^{5}C$_{40}$ (Fig. 7). Because wyosine, yW, is unstable as the phosphoramidite, its precursor, m^{1}G, was incorporated at position 37, 3′-adjacent to the anticodon, where it appears naturally in a number of tRNAs (European Molecular Biology Laboratory, tRNA sequence data bank, Heidelberg). In one synthesis, m^{2}A was incorpo-

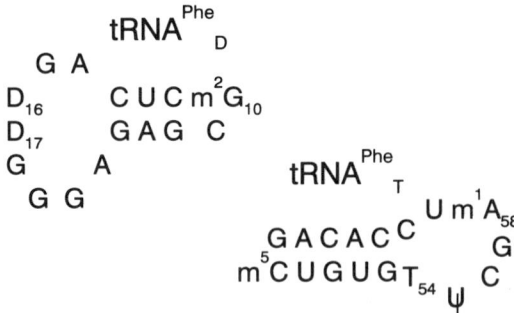

FIG. 5. Sequences and secondary structures of the yeast tRNA^{Phe} D and T domains. The fully modified hexadecamer D stem-and-loop domain and the heptadecamer T stem-and-loop domain sequences are shown in the hairpin conformation found in solution (98, 105, 112). tRNA$_D^{Phe}$ domains containing a single D at positions 16 or 17, and both Ds, have been synthesized with unmodified G at position 10. tRNA$_T^{Phe}$ domains have been synthesized with T$_{54}$, ^{13}CH$_3$–T$_{54}$, m^{5}C$_{40}$, and T$_{54}$, and with ψ_{55}. All sequences were produced with unmodified A at residue 58.

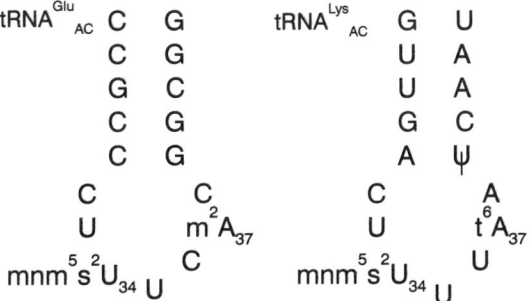

FIG. 6. *Escherichia coli* tRNA^{Glu} and tRNA^{Lys} anticodon domains. The tRNA^{Glu} anticodon domain was synthesized with mnm^{5}U$_{34}$ and with ^{15}N$_{1,3}$-uridine at positions 33 and 35. Four pentamers corresponding to the tRNA^{Lys} sequences at positions 33–37 were produced by manual synthesis with one or both of the modified nucleosides, mnm^{5}s^{2}U$_{34}$ and t^{6}A$_{37}$, and without modifications. A fifth pentamer was produced with s^{2}U$_{34}$ and t^{6}A$_{37}$.

rated at the position of A$_{38}$ for purposes of understanding, by NMR methods, the physical contributions of A$_{38}$ to the anticodon domain conformation. Investigations of modified nucleoside contributions to the structure of the yeast tRNA^{Phe} anticodon domain (tRNA$^{Phe}_{AC}$) and its DNA analog (tDNA$^{Phe}_{AC}$) have elucidated four physical chemical functions for three different methyla-

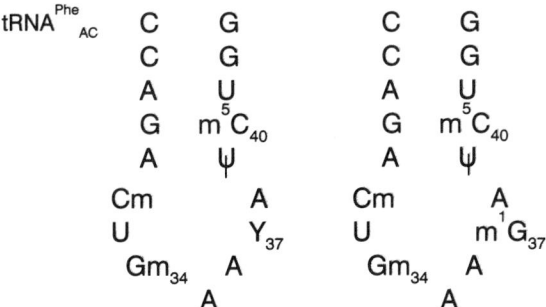

FIG. 7. Yeast tRNA^{Phe} anticodon domain. The fully modified anticodon domain sequence is shown. Sequences with one to four of the five modifications in various combinations have been synthesized by automated chemical synthesis. Because of instability of the yW phosphoramidite, yW$_{37}$ (shown as Y$_{37}$) had to be replaced by its biological precursor, m^{1}G. DNA analogs have been synthesized entirely with deoxyribose phosphoramidites, including dU, d(m^{1}G), and d(m^{5}C).

tions: *enhanced base stacking interaction, reordering of water, facilitated coordination of metal ions*, and *disruption of canonical base-pairing*.

A. Stabilization of Domain Structure and Induced Dynamics

1. MODIFIED NUCLEOSIDES REORDER WATER

The simple addition of methyl to the C-5 of pyrimidines stabilizes nucleic-acid structure through increased base-stacking. The thermal stability of RNA and DNA duplexes increases when T and dT are substituted for U and dU (*132*). Methylation of cytosine at position 5 (m^5C) also produces a less polar, more hydrophobic base (*133*). The nucleosides m^5C_{40} in $tRNA^{Phe}_{AC}$ and $d(m^5C_{40})$ in $tDNA^{Phe}_{AC}$ increase base-stacking on the 3' side of the anticodon stem (Fig. 7) (*133a*). The increased stacking stabilizes the stem structure. A second consequence of the increased stacking is potentially more important, *the stabilization of structure through the reordering of water* (Table I, number 7) (M. M. Basti, personal communication).

Results from X-ray crystallography of nucleic acids clearly indicate that water internal to and surrounding the molecules can be highly ordered (*134–136*). A difference-electron-density map between the complexes of *E. coli* glutaminyl-tRNA synthetase and cognate modified and unmodified $tRNA^{Gln}$ indicate that ψ reorders water by having H_2O specifically hydrogen-bonded between N-5 and their 5'-phosphates (*137*). Methylation of cytidine, by increasing the hydrophobicity of the nucleoside and its propensity for base stacking (*133*), may result in an ordering of water around a nucleic acid. In an X-ray crystallographic study of duplexes composed of self-complementary octamers, a significant increase in the ordering of water occurred when m^5C was substituted for C (*136*).

Although water that is ordered in nucleic-acid solutions is difficult to observe directly by NMR, differences in imino-base-paired ^1H-exchange rates with H_2O occur when Mg^{2+} is present in solutions of $tRNA^{Phe}_{AC}$–$d(m^5C_{40})$ (*116*) and its DNA analogs (*119, 126*). The differences in exchange rates could be indicative of conformers exchanging ordered and bulk water. NMR chemical-shift differences reported for exchangeable protons within tRNA structures and increased thermal stability correlated to the presence of modified nucleosides (*138*) also could be indicative of modified nucleoside-dependent reordering of water. A consequence of m^5C-dependent reordering of water may be the tight Mg^{2+} binding within the anticodon loop of yeast $tRNA^{Phe}$. Hydrated Mg^{2+} can be involved in a network of hydrogen bonds created between the nucleic acid and the immediate solvent layer (*139*).

2. Mg^{2+} Binds Modified tRNA and Affects Structure and Function

The literature is replete with data that, in general, correlate modification of RNA with the binding of Mg^{2+}, the result of which is a change in structure and/or biological activity. The m^5C_{40}-dependent binding of Mg^{2+} to the tRNA^{Phe} anticodon domain was the first documented example of how a single *modification facilitates the coordination of metal ions with RNA* (Table I, number 8). The importance of Mg^{2+} to the functioning of RNA was recognized early in studies of translation *in vitro* (*140*). The function of Mg^{2+} was attributed to the biochemistry of translation, as in the production of the aminoacyl-adenylate by the aminoacyl-tRNA synthetase (*141*). Although the potential of Mg^{2+} as a counterion to nucleotides and the RNA phosphodiester backbone was recognized (*142, 143*), only with the first RNA structure from X-ray crystallography, that of yeast tRNA^{Phe} (*93*), were inner-sphere Mg^{2+} ions found tightly bound within the RNA structure (*144*). At the time of these investigations, the relationship of modified nucleosides to tight Mg^{2+} binding sites and the resulting RNA conformation was neither obvious nor experimentally approachable.

The unmodified T7 transcript of yeast tRNA^{Phe} can be effectively aminoacylated. Therefore, modified nucleosides are not absolutely required for recognition by the cognate aminoacyl-tRNA synthetase (*145*). Yet, as the Mg^{2+} concentration is reduced to 5 mM, the unmodified tRNA molecule apparently adopts a nonnative conformation, accompanied by a decrease in recognition by the enzyme as compared to that of fully modified native tRNA. [Cytosolic free Mg^{2+} has been accurately determined to be between 0.6 and 2.1 mM in mammalian cells, *in vivo*, (*146*; R. E. London, personal communication).] Magnesium must play a significant role in anticodon conformation because a Mg^{2+}-induced change in the anticodon of tRNA^{Phe} has been detected by yW fluorescence and other methods (*147, 148*).

Effects of modifications on Mg^{2+} binding to tRNA are observed when unmodified transcripts are compared to native tRNA. Fully modified, native yeast tRNA^{Phe} has four tight Mg^{2+} binding sites (*144*) and approximately 50 weaker sites (*155*). The T7 transcript of yeast tRNA^{Phe}, devoid of modified nucleosides, requires higher than expected concentrations of Mg^{2+} (50 mM) in order for the ^{1}H NMR spectrum to approach that of the native molecule (*149*). Binding constants for Mg^{2+} are at least two orders of magnitude higher for the native *E. coli* tRNA^{Val} versus the unmodified transcript (*150*). In the absence of Mg^{2+} and at low Mg^{2+} concentrations, even fully modified *E. coli* and yeast tRNA^{Phe}s exist as mixtures of two or more conformations in slow exchange (*151–153*). The "melting" temperature of the unmodified tRNA^{Phe}

at concentrations of $MgCl_2 > 5$ mM is still 5°C lower than that of the native form (42). Rh(phen)$_2$phi^{3+}-catalyzed fragmentation of the unmodified transcript was enhanced in 10 mM MgCl$_2$, whereas increasing concentrations of Mg^{2+} actually decreased cleavage of the anticodon of fully modified tRNA^{Phe} (154). Although the general effects of modifications on Mg^{2+} binding and RNA structure are evident, the results do not clarify the individual and combined chemical contributions of the modified nucleosides.

The ability to introduce a single, site-selected modification showed that the tight Mg^{2+} binding to tRNA$^{Phe}_{AC}$ (127) and its DNA analog (119, 126) depends on the m^{5}C$_{40}$ modification (127). Only m^{5}C$_{40}$-containing tRNA$^{Phe}_{AC}$ bound Mg^{2+} within the anticodon loop and at an additional site in the upper part of the stem with a $K_d = 2.5 \times 10^{-9}$ M^2 (127; R. Guenther and P. F. Agris, unpublished). With the binding of Mg^{2+}, tRNA$^{Phe}_{AC}$–m^{5}C$_{40}$ and its DNA analog exhibit a conformational transition that was monitored by CD and NMR. What had been a seven-membered loop was closed by the addition of two canonical intraloop base-pairs, $C_{32} \cdot G_{37}$ and $U_{33} \cdot A_{36}$, resulting in a two-base turn (119, 126, 127). Anticodon domain stability increased by $\Delta G = -11.7$ kcal/mol, and NMR spectra indicated that two conformations are in slow exchange (H. Sierzputowska-Gracz and P. F. Agris, unpublished). Thus, in the absence of any other modifications, the m^{5}C-dependent, Mg^{2+}-induced structural transition actually restricts conformational freedoms of the anticodon domain by stabilizing the two intraloop base pairs.

3. Modifications Maintain an Open, Dynamic, and Functional Structure

Yeast phenylalanyl-tRNA synthetase (FRS) and its cognate tRNA represent a good system to test the nature of synthetase recognition of the tRNA's 3′ terminus and the distant highly modified anticodon domain. Three of five FRS identity elements are located in the tRNA^{Phe} anticodon (156) and the heptadecamer domain contains five modified nucleosides. The tRNA^{Phe} anticodon stem–loop domain (tRNA$^{Phe}_{AC}$) and its DNA analogs (tDNA$^{Phe}_{AC}$) have significant structural similarities to the native tRNA (133a). The conformationally restricted and closed-loop domain of tRNA$^{Phe}_{AC}$–m^{5}C$_{40}$ and DNA analogs inhibits FRS activity (114). In contrast, variously modified tRNA$^{Phe}_{AC}$ domains stimulate FRS. The fully modified anticodon domain of tRNA^{Val} stimulates the cognate VRS[4] activity (157). Three differentially modified tRNA^{Phe} anticodon domains with ψ_{39} alone, with m^{1}G$_{37}$ and m^{5}C$_{40}$, or ψ_{39}

---
[4] V, Valine; R, arginine; S, serine.

with m^1G_{37} and m^5C_{40} stimulated FRS activity. All three anticodon domains that stimulated FRS activity had open loop structures (114).

The fully modified anticodon stem and loop domains of tRNA will bind the 30-S ribosomal subunit with kinetics identical to that of the complete tRNA (158, 159). The anticodon domain binds the P-site (160). In contrast, the affinity of unmodified anticodon domain for the programmed 30-S subunit is decreased by two orders of magnitude (56) and the affinity of the conformationally restricted, closed loop $tRNA_{AC}^{Phe}-m^5C_{40}$ is reduced even further (72). The fully modified, heptadecamer yeast $tRNA_{AC}^{Phe}$ has five modified nucleosides. However, $tRNA_{AC}^{Phe}$ required only two modifications, m^5C_{40} and m^1G_{37} precursor to the hypermodification yW_{37}, for binding to programmed *E. coli* 30-S ribosomal subunits as effectively as native tRNA and thereby inhibiting the binding of native $tRNA^{Phe}$ (72). The m^5C_{40}-dependent, Mg^{2+}-induced structural transition stabilized the $tRNA_{AC}^{Phe}$ stem (127) and the m^1G_{37}-aided disruption of a $C_{32} \cdot G_{37}$ intraloop base-pair provided an open anticodon-loop conformation for ribosome binding. In fact, a survey of some 500 tRNA sequences proved that 95% of the 121 anticodon domains with the potential for two intraloop base-pairs have at least one modification blocking base-pair formation (72). The remaining 5% had the potential to form two A·U base-pairs.

Unmodified DNA analogs and chimeric RNA–DNA anticodon hairpins of the tRNA anticodon stem-and-loop domain do not bind the programmed ribosomal subunit (56, 72). The successful design and synthesis of a $tDNA_{AC}^{Phe}$ that bound the 30-S subunit and inhibited native tRNA from binding were accomplished only with the introduction of a m^5C_{40}-dependent, Mg^{2+}- induced structural stabilization of the stem and disruption of at least one of the two intraloop base pairs, $dC_{32} \cdot dG_{37}$ or $dU_{33} \cdot dA_{36}$ (72). The latter was accomplished either by the introduction of $d(m^1G)_{37}$ or by a base substitution, A for U_{33} or G for C_{32}. Biologically active $tDNA_{AC}^{Phe}$ has a dynamic anticodon loop with more than one conformation in fast exchange (133a).

Only those $tRNA_{AC}^{Phe}$ domains and their corresponding DNA analogs that have proper modifications and Mg^{2+}-stabilized structures with open, dynamic anticodon loops interact with the codon. However, in the absence of position-37 modifications, the tDNA anticodon–codon interaction can include a fourth, potentially frame-shifting base-pair. The results of experiments in which unmodified $tRNA_{AC}^{Phe}$ interacted with codon in solution, but $tRNA_{AC}^{Phe}-d(m^5C)_{40}$ did not, demonstrated that the open-loop structure of the tRNA anticodon domain is important for anticodon–codon interaction in the absence of ribosomes (P. F. Agris, V. Dao and M. M. Basti, unpublished). Although unmodified $tRNA_{AC}^{Phe}$ molecules interacted with the codon r(UUC) even in the absence of Mg^{2+}, the m^5C_{40}-dependent Mg^{2+} binding results in a significant stabilization of the tRNA anticodon conformation favored in ribosome-mediated codon recognition when coupled to a modification-dependent open anticodon loop (72, 127).

Tight Mg^{2+} binding within the anticodon loop neutralizes the negative electrical potential of the region, thus facilitating the association of tRNA with negatively charged mRNA (162). The anticodon domain of the X-ray crystallographic structure of yeast tRNA^{Phe} (93) and the NMR-derived solution structure of tDNA$^{Phe}_{AC}$ (133a) have bound Mg^{2+} within the loops, and both display a bend in the backbone at the location of the metal ion. Proteins with cationic surfaces induce bends in the DNA backbone by creating salt bridges to local phosphate negative charges, neutralizing the charge and producing an asymmetric charge distribution (163). The location of Mg^{2+} near the single-stranded RNA of the anticodon loop may produce an asymmetric charge distribution that results in a bend of the backbone.

B. Modification Influences Local Nucleoside Conformation and Dynamics

1. DIHYDROURIDINE

The modified uridines not only constitute a disproportionately large percentage of the known modifications (37%), but also exhibit a complex array of chemistries representative of the entire collection of modified nucleosides. Structures and the biosyntheses of uridine modifications have been studied for more than a quarter of a century. The conformations, and some dynamics, the ^1H and ^{13}C NMR chemical shifts, and J couplings (under physiological solution conditions) of 17 differently modified uridines (80, 82, 99) and 15 modified uridine di- and trinucleotides have been reported (83, 99). Modified uridines of particular interest include dihydrouridine (D), pseudouridine (ψ), and the 2-thiouridines (s^2U).

Dihydrouridine is found at various positions (16, 17, 20, and 21) in the D-domain and sometimes at position 47 of the variable (V) loop. It is one of the most frequent of all modifications in tRNA, and the only nonaromatic modification of any nucleoside. The mononucleoside, part of the D loop of tRNAs (79, 93–97), is in the 2-endo conformation. Modification of U to D significantly alters the nucleosidic sugar conformation. The 3′-endo pucker of A-form RNA is disfavored and the 2′-endo conformer predominates. Although the resulting 2′-endo sugar pucker is different from the 3′-endo conformation stabilized by 2-thiolation of U (12), as with s^2U, it is the stereochemistry of the D modification that is responsible for the altered conformation. The increased volume of the C-6 methylene of D produces the more energetically favored 2'-endo stereochemistry of the ribose in the crystal and solution structures (*164, 165, 165a*).

In tRNA crystal structures, nucleosides immediately 3' to D are also in the 2'-endo conformation. Another uridine modification, the hydrophilic acp^{3}U, occurs alone in the D-loops of some tRNAs and adjacent to D in the D-loops of tRNAs for Asn and Val. The acp^3 derivative of ψ occurs in ribosomal RNA (*79*). D, 5'-adjacent to acp^{3}U, in the D-loop model, Dp-acp^{3}UpA, influences the conformation and Mg^{2+} binding ability of the acp^3 side chain (*99, 165a*). CD and NMR spectra indicate that D-acp^{3}U-A and U-acp^{3}U-A bind Mg^{2+}, but that the former has two binding sites, whereas the latter has only one. The conformational restraints that D places on sequences 3'-adjacent may influence their chemistry and structure, producing the conformational and ion binding characteristics of tRNA D-loops. Therefore, D represents a modification that restricts phosphodiester bond conformation and/or nearest-neighbor nucleoside sugar pucker (Table I, number 9).

2. 2-THIOURIDINES

The 2-thiouridines (s^{2}U*) are found primarily at wobble position 34, and occur most often with modifications of carbon-5 of the pyrimidine ring. The 5-methyl derivative of s^{2}U (m^{5}s^{2}U or s^{2}T) is found at residue 54 in the T stem and loop, the position common for T. Another thiolated uridine (s^{4}U) is found at residue 8 of bacterial tRNAs, and occurs without further modification. As previously described, the 2-thio modification, no matter what the nature of the 5-position side-chain, restricts the nucleosidic conformation to 3'-endo, anti, gauche$^+$ (*80*), with a significant energy barrier to other conformations (*82*). Nonthiolated uridines with various 5-position modifications, such as amino-acid or methoxy additions, and s^{4}U are approximately 50% 3'-endo, gauche$^+$, and anti or syn, depending on the nature of the 5-derivatization. The highly restricted conformation of the 2-thiouridines is of biological significance because of their occurrence at the anticodon wobble position 34 of tRNAs. The local structure of the anticodon is affected by 2-thiolation at position 34. The torsion angle of the 3'-adjacent phosphodiester bond (i.e., toward the second anticodon position), but not that of the 5' invariant U$_{33}$, is affected by 2-thiolation of U$_{34}$ (*83*). Because tRNAs with s^{2}U$^*_{34}$ (Glu, Lys, and Ser) are restricted in anticodon nucleoside conformation, and because they almost exclusively recognize codons ending in A, i.e., lack the ability to wobble, a "modified wobble hypothesis" has been proposed (*12*). The s^{2}Us also exemplify modified nucleoside *restriction of phosphodiester bond conformation and/or nearest-neighbor nucleoside sugar pucker* (Table I, number 9).

The anticodon domains of the *E. coli* tRNAs for Lys and Glu are very similar in sequence, and include mnm^{5}s^{2}U$_{34}$, the notable difference being at position 37: t^6A_{37} in tRNA^{Lys} and m^2A_{37} in tRNA^{Glu} (Fig. 5). tRNA^{Lys} may have an anticodon conformation different from that of tRNA^{Glu} (*166*). Chemical and physical evidence from the study of small, chemically synthesized, oligoribonucleotide pentamers corresponding in sequence to the anticodon positions 33–37, U_{33}-mnm^{5}s^{2}U-U-U-t$^6A_{37}$, of *E. coli* tRNA^{Lys} and similar to that of tRNA^{Glu}, indicate that a unique interaction occurs between the two modifications in the anticodon loop, mnm^{5}s$^2U_{34}$ and t^6A_{37} (P. F. Agris and A. Malkiewicz, unpublished). For instance, H_2O_2 oxidation of the mnm^{5}s^{2}U of tRNA^{Lys} proceeds at a significantly lower rate than that of tRNA^{Glu} (*166*). The same was true for mnm^{5}s^{2}U within the doubly modified pentamer compared to the oxidation of mnm^{5}s^{2}U in a pentamer without the t^6 modification. NMR analysis of the doubly modified pentamer revealed that the RNA was a stable pseudocircular molecule, i.e., interaction of the two modifications completes the circle (P. F. Agris and A. Malkiewicz, unpublished). This is an example of the tenth physicochemical contribution of modified nucleosides: *interaction between modifications of modified nucleosides producing new conformations and chemistries* (Table I, number 10).

Studies of modified uridines and of t^6A indicate that the interaction of mnm^{5}s^{2}U with t^6A could be through hydrogen or ionic bonding, the latter perhaps including a salt bridge. In the crystal lattice of the related cmnm^{5}U, the amino-acid residue acts as a donor as well as an acceptor of protons in five separate hydrogen bonds (*167*). The C-5 modifications of U coordinate metal ions directly. Along with other charged and/or polar modifications, the C-5-modified Us represent modifications with the potential to directly participate in metal ion coordination to RNA (Table I, number 8). Seven uridines differently modified at position 5 coordinated strongly with Cu^{2+} and Ni^{2+}, but weakly with Zn^{2+} (*168*). N^6-Threonylcarbamoyladenosine coordinates well with some metals and not at all with others (*11*; A. Malkiewicz, personal communication).

By placing a distance constraint of no more than 5 Å between the amine of mnm^5 and the carbosyl of t^6 and by using molecular dynamics refinement, a model of the doubly modified pentamer was produced (P. F. Agris and A. Malkiewicz, unpublished). Characteristics of the energy-minimized model predict the pseudocircular structure composed of the anticodon loop backbone and the interaction of the mnm^5 modification with the t^6 modification. The model predicts an unusual positioning of the tRNA^{Lys} anticodon domain "U-turn" at mnm^{5}s$^2U_{34}$ instead of the invariant U_{33}. Also, the wobble base mnm^{5}s$^2U_{34}$, when facing t^6A, is neither able to interact with protein (KRS)[5] nor with the lysine-coding triplet. The mnm^{5}s$^2U_{34}$ is a positive determinant for ERS and KRS recognition of their cognate tRNAs (*17, 45*). Thus, an

---

[5] Defined in footnote 3.

unusual anticodon conformation of tRNA^{Lys} may be one way in which ERS and KRS discriminate between tRNA^{Lys} and tRNA^{Glu}. However, an anticodon domain conformational transition would be required for full base-pairing of the anticodon with the codon.

In this model system, the thio group served as one monitor of conformational change due to the interaction of mnm^{5}s^{2}U$_{34}$ and t^{6}A$_{37}$. However, it is very possible that C-5-modified uridines at position 34, as well as variously C-5-modified s^{2}U$_{34}$s, would interact with an amino-acid-modified A$_{37}$. Thus, the thio modification may be immaterial to the formation of the unusual anticodon loop.

In a survey of 507 tRNA sequences, in all tRNAs with t^{6}A$_{37}$ and a U at position 34, the U was C-5-modified with a potential proton acceptor and/or donor. The notable exceptions were unmodified U$_{34}$ of mitochondrial tRNAs. Unmodified wobble position Us in mitochondrial tRNAs afford the organelle with approximately 20 tRNAs, one-third the number of tRNA species in the cytoplasm, the ability to read all codons through wobble (*12*). Therefore, we postulate that t^{6}A$_{37}$ or its derivatives or g^{6}A would be capable of producing unusual anticodon loop conformations with many C-5-modified uridines, and that the resulting conformation may be used for synthetase determination and discrimination of cognate and noncognate tRNAs, respectively.

VI. Conclusions and Perspectives on the Importance of Being Modified

At the conclusion of Oscar Wilde's Victorian comedy, the audience recognizes that the meaning of "earnest" was beyond initial expectations of the word as used in the play's title. The importance of being modified exceeds the 10 contributions of modified nucleosides to nucleic acid structure listed in Table I. As with the play on words often used by Wilde, the contributions of modifications can be, and often are, in opposition. Modifications seem to have evolved to enhance stability of nucleic-acid structures, but the same or similar modifications, found at different locations within the RNA structure, enhance conformational dynamics. Both structure and dynamics are important for recognition, interaction, and function.

Nucleic-acid stability is enhanced through modifications that introduce local positive charges, neutralizing the anionic phosphate backbone. Charge produces structurally stabilizing salt bridges. In addition, charged and polar modifications stabilize structure through Mg^{2+} coordination. The divalent metal ion contributes to the neutralization of the backbone negative charge and to the reordering of water. Tight ion binding of hydrated Mg^{2+} by rRNA, modified tRNA, and its DNA analog result in structures that the binding of many other metals cannot duplicate (169–171). The biologically active structure of $tDNA_{AC}^{Phe}$ that results from the binding of hydrated Mg^{2+} is considerably altered by the substitution of Cd^{2+}, Co^{2+}, Cr^{2+}, Cu^{2+}, Ni^{2+}, Pb^{2+} VO^{2+}, or Zn^{2+}. Nanomolar concentrations of the transition metals are sufficient to denature the $tDNA_{AC}^{Phe}-d(m^5C)$ structure without catalyzing cleavage of the oligonucleotide. The reordering of water by hydrophobic modifications occurs through increased base-stacking, and thereby structure, particularly in RNA stem regions. Interaction of one modification with another has the potential of stabilizing novel nucleic-acid structures that, at first, may seem contrary to presently held ideas of structure–function relationships. At least one such interaction has become evident. The interaction of $mnm^5s^2U_{34}$ with t^6A_{37} stabilizes a $tRNA^{Lys}$ anticodon conformation that may determine KRS recognition, but is unable to bind the three bases of the lysine-coding triplet without a conformational transition.

In contrast to structural stabilization, modifications can, and do, give rise to conformational dynamics. Modifications of the N-1 of purines, N-3 of pyrimidines, and the exocyclic amines of A, C, and G block the formation of canonical base-pairs. Methylation of ribose inhibits noncanonical hydrogen bonding to the 2'-OH. Site-specific introduction of D interrupts base stacking. These are contributions not so much to structure, as to structural dynamics. Site-specific introduction of individual (and then various combinations of multiple) modifications to the $tRNA_{AC}^{Phe}$ and its DNA analogs has produced molecules with the same major nucleotide sequence, but exhibiting two types of dynamics (133a).

One form of dynamics is exhibited by conformationally restricted anticodon domains in slow exchange between two or three conformations separated by a significant energy barrier of ~10 kcal/mol. Intraloop hydrogen bonding contributes most significantly to the conformational restriction and to the anticodon's inability to bind the ribosome (72). Modifications at positions 32, 34, and 37 that inhibit canonical and noncanonical base pairing, in combination with a modification-dependent stabilization of the stem and upper loop region of the anticodon domain, produce the other form of dynamics, a fast exchange in the loop among a number of energetically accessible conformations. Anticodon domains exhibiting modification-dependent fast exchange are biologically active (72; R. Guenther and P. F. Agris, unpublished). The high frequency and large variety of post-transcriptional modifications at anticodon domain positions 34 and 37 (Figs. 2 and 3) contribute to the individuality of tRNA species.

The RNA World, defined as that period in evolution prior to the advent of the DNA genome, is conceived as having unmodified RNA genomes and enzymes. Certainly, the unmodified descendants, RNase P and the naturally occurring ribozymes, are evidential. However, some of the simpler modifications, methylations and thiolations, may have existed in the RNA World. Enzymatic activities of RNA are dependent on participation of metal ions in structure and chemistry. Modified ribonucleosides may have contributed to the coordination and bioinorganic chemistry required of RNA function in an RNA World.

Although some modern biological functions in which particular modifications play a role have been identified, the unique contributions of modified nucleosides to the functional chemistry and structure of RNA for the most part remain unknown. An understanding of the chemical–structure–function relationships of modified nucleosides may lead to a significant understanding of how they affect structure, stability, and dynamics through canonical and noncanonical hydrogen bonding and ion binding, and how they modulate RNA–RNA and RNA–protein interactions. With the derivation of modified nucleoside force-field parameters and an understanding of their chemical and physical contributions (112), molecular modeling of native and designed nucleic acids can be more accurately accomplished (172–174). The unique chemistry, structure, and dynamics that modified nucleosides contribute to RNAs provide investigators with exciting possibilities for engineering novel nucleic acid interactions with other nucleic acids and with proteins, expanding the genetic code, as well as generating basic knowledge of nucleic-acid evolution and function. Modified nucleosides have the potential to bring selectivity and specificity of targeting to therapeutics or diagnostics [SELEX-derived oligonlucleotides (175), triplex, antisense, and ribozyme technologies] and new chemistries to material science.

Appendix: Modified Nucleoside Symbols and Common Names

Symbol	Nucleoside Common Name
m^1A	1-methyladenosine
m^2A	2-methyladenosine
m^6A	N^6-methyladenosine
m^6_2A	N^6,N^6-dimethyladenosine
Am	2'-0-methyladenosine
m^6Am	$N^6,2'$-0-dimethyladenosine
m^6_2Am	$N^6,N^6,0$-2'-trimethyladenosine
ms^2m^6A	2-methylthio-N^6-methyladenosine
i^6A	N^6-isopentenyladenosine
ms^2i^6A	2-methylthio-N^6-isopentenyladenosine
io^6A	N^6-(cis-hydroxyisopentenyl) adenosine
ms^2io^6A	2-methylthio-N^6-(cishydroxyisopentenyl)adenosine
g^6A	N^6-glycinylcarbamoyladenosine
t^6A	N^6-threonylcarbamoyladenosine
ms^2t^6A	2-methylthio-N^6-threonylcarbamoyladenosine
m^6t^6A	N^6-methyl-N^6-threonylcarbamoyladenosine
hn^6A	N^6-hydroxynorvalylcarbamoyl adenosine
ms^2hn^6A	2-methylthio-N^6-hydroxy norvalylcarbamoyladenosine
Ar(p)	2'-0-ribosyladenosine (phosphate)
I	inosine
m^1I	1-methylinosine
Im	2'-0-methylinosine
m^1Im	1,2'-0-dimethylinosine
m^3C	3-methylcytidine
m^5C	5-methylcytidine
Cm	2'-0-methylcytidine
s^2C	2-thiocytidine
ac^4C	N^4-acetylcytidine
f^5C	5-formylcytidine
m^5Cm	5,2'-0-dimethylcytidine
ac^4Cm	N^4-acetyl-2'-0-methylcytidine
m^4C	N^4-methylcytidine
m^4Cm	$N^4,2'$-0-dimethylcytidine
hm^5C	5-hydroxymethylcytidine
k^2C	lysidine
m^1G	1-methylguanosine
m^2G	N^2-methylguanosine
m^7G	7-methylguanosine
$m^{2,7}G$	N^2,7-dimethylguanosine
Gm	2'-0-methylguanosine
m^2_2G	N^2,N^2-dimethylguanosine
$m^{2,2,7}G$	N^2,N^2,7-trimethylguanosine
m^2Gm	$N^2,2'$-0-dimethylguanosine
m^2_2Gm	$N^2,N^2,2'$-0-trimethylguanosine
Gr(p)	2'-0-ribosylguanosine (phosphate)
yW	wybutosine
o^2yW	peroxywybutosine
OHyW	hydroxywybutosine
OHyW*	undermodified hydroxywybutosine

Symbol	Nucleoside Common Name
imG	wyosine
mimG	7-methylwyosine
Q	queuosine
oQ	epoxyqueuosine
galQ	galactosyl-queuosine
manQ	mannosyl-queuosine
preQ$_0$	7-cyano-7-deazaguanosine
preQ$_1$	7-aminomethyl-7-deazaguanosine
gQ	archaeosine[6]
D	dihydrouridine
m^{5}D	dihydroribosylthymine
m^{5}U	ribosylthymine
Um	2'-O-methyluridine
m^{5}Um	5,2'-O-dimethyluridine
s^{2}U	2-thiouridine
s^{4}U	4-thiouridine
m^{5}s^{2}U	5-methyl-2-thiouridine
s^{2}Um	2-thio-2'-O-methyluridine
acp^{3}U	3-(3-amino-3-carboxypropyl)uridine
ho^{5}U	5-hydroxyuridine
cm^{5}U	5-carboxymethyluridine
mo^{5}U	5-methoxyuridine
cmo^{5}U	uridine 5-oxyacetic acid
mcmo^{5}U	uridine 5-oxyacetic acid methyl ester
chm^{5}U	5-(carboxyhydroxymethyl)uridine
mchm^{5}U	5-(carboxyhydroxymethyl)uridine methyl ester
mcm^{5}U	5-methoxycarbonylmethyl-uridine
mcm^{5}Um	5-methoxycarbonylmethyl-2'-O-methyluridine
mcm^{5}s^{2}U	5-methoxycarbonylmethyl-2-thiouridine
nm^{5}s^{2}U	5-aminomethyl-2-thiouridine
mnm^{5}U	5-methylaminomethyluridine
mnm^{5}s^{2}U	5-methylaminomethyl-2-thiouridine
mnm^{5}se^{2}U	5-methylaminomethyl-2-selenouridine
ncm^{5}U	5-carbamoylmethyluridine
ncm^{5}Um	5-carbamoylmethyl-2'-O-methyluridine
cmnm^{5}U	5-carboxymethylaminomethyl-uridine
cmnm^{5}Um	5-carboxymethylaminomethyl-2'-O-methyluridine
cmnm^{5}s^{2}U	5-carboxymethylaminomethyl-2-thiouridine
m^{3}U	3-methyluridine
m^{3}Um	3,2'-O-dimethyluridine
m^{1}acp^{3}ψ	1-methyl-3-(3-amino-3-carboxypropyl)pseudouridine
ψ	pseudouridine
m^{1}ψ	1-methylpseudouridine
ψm	2'-O-methylpseudouridine

[6] gQ for archaeosine was suggested by Waldo E. Cohn (g for guanidino).

Acknowledgments

The author thanks present and former members of his research group, his colleagues within the RNA Biology faculty at North Carolina State University, and members of the modified nucleoside research community worldwide for many years of constructive discussions on RNA chemistry, structure, and function. He thanks his family for their patience and support. This work was supported by grants from the National Institutes of Health (5-RO1-GM23037), the National Academy of Sciences/National Research Council Collaboration in Basic Sciences and Engineering, and the National Science Foundation (INT-9412828).

References

1. R. Hotchkiss, *JBC* **175**, 315 (1948).
2. R. Hall, "The Modified Nucleosides in Nucleic Acids." Columbia Univ. Press, New York, 1971.
3. P. F. Agris, "The Modified Nucleosides of Transfer RNA." Alan R. Liss, New York, 1980.
4. P. F. Agris and R. A. Kopper, eds., "The Modified Nucleosides of Transfer RNA, II." Alan R. Liss, New York, 1983.
5. C. W. Gehrke and K. C. T. Kuo, eds., "Chromotography and Modification of Nucleosides," Parts A, B, and C. Elsevier, Amsterdam, 1990.
6. P. F. Agris, and D. Söll, *in* "Nucleic Acid-Protein Recognition" (H. Vogel, ed.), pp. 321–344. Academic Press, New York, 1977.
6a. P. R. Srinivasan and E. Borek, *This Series* **5**, 157 (1966).
6b. P. R. Chambers, *This Series* **5**, 349 (1966).
6c. R. H. Hall, *This Series* **10**, 57 (1970).
6d. S. Nishimura, *This Series* **12**, 49 (1972).
7. S. Nishimura, *in* "Transfer RNA: Structure, Properties and Recognition" (P. R. Schimmel, D. Söll and J. N. Abelson, eds.), pp. 57–79. Cold Spring Harbor Laboratory, Cold Spring Harbor, New York, 1979.
8. S. Nishimura, *in* "Transfer RNA: Structure, Properties and Recognition" (P. R. Schimmel, D. Söll and J. N. Abelson, eds.), pp. 547–549. Cold Spring Harbor Laboratory, Cold Spring Harbor, New York, 1979.
8a. B. Singer and M. Krüger, *This Series* **23**, 151 (1979).
8b. S. Nishimura, *This Series* **28**, 50 (1983).
8c. R. P. Singhal, *This Series* **28**, 75 (1983).
9. H. Kersten, *This Series* **31**, 58 (1984).
10. R. W. Adamiak and Piotr Gornicki, *This Series* **32**, 27 (1985).
11. G. R. Björk, J. U. Ericson, C. E. D. Gustafsson, T. G. Hagervall, Y. H. Jonsson and P. M. Wilkstrom, *ARB* **56**, 263 (1987).
12. P. F. Agris, *Biochimie* **73**, 1345 (1991).
13. G. R. Björk, *in* "Transfer RNA in Protein Synthesis" (D. L. Hatfield, B. J. Lee and R. M. Pirtle, eds.), pp. 23–85. CRC Press, Boca Raton, Florida, 1992.
14. B. C. Persson, *Mol. Microbiol.* **8**, 1011 (1993).
14a. G. R. Björk, *This Series* **50**, 263 (1995).
14b. G. R. Björk, *in* "tRNA: Structure, Biosynthesis and Function" (D. Söll and U. Raj Bhandary, eds.), pp. 165–205. Am. Soc. Microbiol., Washington, D.C., 1995.
15. A. P. Bird, *Nature* **321**, 209 (1986).

16. T. Seno, P. F. Agris and D. Söll, *BBA* **349**, 328 (1974).
17. L. A. Sylvers, K. C. Rogers, M. Shimizu, E. Ohtsuka and D. Söll, *Bchem* **32**, 3836 (1993).
17a. K. C. Ragers, A. T. Crescenzo and D. Söll, *Biochimie* **77**, 66 (1995).
18. T. L. Helser, J. E. Davies and J. E. Dahlberg, *Nature (New Biol.)* **233**, 12 (1971).
19. K. Sirum-Connolly and T. L. Mason, *Science* **262**, 1886.
20. A. Bakin, B. G. Lane and J. Ofengand, *Bchem* **33**, 13475 (1994).
20a. B. G. Lane, J. Ofengand and M. W. Gray, *Biochimie* **77**, 7 (1995).
21. C. Isel, R. Marquet, G. Keith, C. Ehresmann and B. Ehresmann, *JBC* **268**, 25269 (1993).
22. Y.-L. Chen and R.-T. Wu, *Cancer Res.* **54**, 2192 (1994).
23. S. H. Kovacs, C. Rodi, V. K. Lin, B. S. Ortwerth and P. F. Agris, *NARes* **6**, 2275 (1979).
24. K.-K. Lin, T. D. Furr, S. H. Chang, J. Horwitz, P. F. Agris, and B. J. Ortwerth, *JBC* **255**, 6020 (1980).
25. V. K. Lin and P. F. Agris, *NARes* **8**, 3467 (1980).
26. V. K. Lin, W. R. Farkas and P. F. Agris, *NARes* **8**, 3481 (1980).
27. H. Kersten and W. Kersten, in "Chromatography and Modification of Nucleosides, B" (C. W. Gehrke and K. C. Kuo, eds.), pp. 69–107. Elsevier, Amsterdam, 1990.
29. P. A. Limbach, P. F. Crain and J. A. McCloskey, *NARes* **22**, 2183 (1994).
30. G. R. Björk and F. C. Neidhardt, *J. Bact.* **124**, 99 (1975).
31. A. K. Hopper, A. H. Furukawa, H. D. Pham and N. C. Martin, *Cell* **28**, 543 (1982).
32. H. Laten, J. Gorman and R. M. Bock, *NARes* **5**, 4329 (1978).
33. S. P. Eisenberg, M. Yarus and L. Söll, *JMB* **135**, 111 (1979).
34. F. Janner, G. Vogeli and R. Fluri, *JMB* **139**, 207 (1980).
35. L. R. Mandel and E. Borek, *BBRC* **4**, 14 (1961).
36. P. Munz, U. Leopold, P. F. Agris and J. Kohli, *Nature* **294**, 187 (1981).
37. H. D. Heyer, P. Thuriaux, J. Kohli, P. Ebert, H. Kersten, C. Gehrke, K. C. Kuo and P. F. Agris, *JBC* **259**, 2856 (1984).
38. A.-M. Grossenbacher, B. Stadelmann, W.-D. Heyer, P. Thuriaux, J. Kohli, C. Smith, P. F. Agris, K. C. Kuo and C. Gehrke, *JBC* **261**, 16351 (1986).
39. C. Francklyn, J.-P. Shi and P. Schimmel, *Science* **255**, 1121 (1992).
40. J.-P. Shi, S. A. Martinis and P. Schimmel, *Biochemistry* **31**, 4931 (1992).
41. C. Francklyn, K. Musier-Forsyth and P. Schimmel, *EJB* **206**, 315 (1992).
42. J. R. Sampson and O. C. Uhlenbeck, *PNAS* **85**, 1033 (1988).
43. Pak, Pallanck and L. H. Schulman, *Bchem* **31**, 3303 (1992).
44. M. Delarue, *Curr. Opin. Struct. Biol.* **5**, 48 (1995).
45. K. Tamura, H. Himeno, H. Asahara, T. Hasegawa and M. Shimizu, *NARes* **20**, 2335 (1992).
46. H. Bacha, M. Renaud, J.-F. Lefevre and P. Remy, *EJB* **127**, 87 (1982).
47. T. Muramatsu, K. Nishikawa, F. Nemoto, Y. Kuchino, S. Nishimura, T. Miyazawa and S. Yokoyama, *Nature* **336**, 179 (1988).
48. T. Niimi, O. Nureki, T. Yokogawa, N. Hayashhi, K. Nishikawa, K. Watanabe and S. Yokoyama, *Nucleosides Nucleotides* **13**, 1231 (1994).
49. V. Perret, A. Garcia, H. Grosjean, J. P. Ebel, C. Florentz and R. Giégé, *Nature* **344**, 787 (1990).
50. J. Putz, C. Florentz, F. Bensler and R. Giégé, *Nature Struct. Biol.* **1**, 580 (1994).
51. B. R. Baumstark, L. L. Spremulli, U. L. RajBhandary and G. M. Brown, *J. Bact.* **129**, 457 (1977).
52. C. E. Samuel and J. C. Rabinowitz, *JBC* **249**, 1198 (1974).
53. J. Desgres, G. Keith, K. C. Kuo and C. W. Gehrke, *NARes* **17**, 865 (1989).

54. S. Kiesewetter, G. Ott and M. Sprinzl, NARes 18, 4677 (1990).
55. C. Forster, K. Chakraburtty and M. Sprinzl, NARes 21, 5679 (1993).
56. O. V. Koval'chuke, A. P. Potapov, A. V. El'skaya, V. K. Potapov, N. F. Krinetskaya, N. G. Dolinnaya and Z. A. Shabarova, NARes 19, 4199 (1991).
57. J. F. Curran, NARes 23, 683 (1995).
58. T. G. Hagervall, J. Ericson, K. B. Esberg, L. Ji-nong and G. R. Björk, BBA 1959, 263 (1990).
59. R. K. Wilson and B. A. Roe, PNAS 86, 409 (1990).
60. C. Houssier and H. Grosjean, J. Biomol. Struct. Dyn. 3, 387 (1986).
61. V. I. Katunin, N. G. Soboleva, V. I. Makhno, E. A. Sedel'nikova, S. M. Zhenodarova and S. V. Kirillov, Mol. Biol. 28, 43 (1994).
62. G. R. Björk, P. M. Wikstrom and A. S. Byström, Science 244, 986 (1989).
63. D. Hatfield and S. Oroszian, Trends Biochem. Sci. 15, 186 (1990).
64. M. Yarus, Science 218, 646 (1982).
65. P. F. Agris, L. L. Spremulli and G. M. Brown, ABB 162, 38 (1974).
66. P. F. Agris, D. Söll and T. Seno, Bchem 12, 4331 (1973).
67. L. Johnson and D. Söll, PNAS 67, 943 (1970).
68. C. J. Green, H. O. Kammen and E. E. Penhoet, JBC 257, 3045 (1982).
69. B. Nawrot and A. Malkiewicz, Nucleosides Nucleotides 11, 1499 (1992).
70. H. Sierzputowska-Gracz, W. Folkman, R. H. Guenther, B. Golankiewicz and P. F. Agris, Magn. Reson. Chem. 29, 885 (1991).
71. H. Sierzputowska-Gracz, H. D. Gopal and P. F. Agris, NARes 14, 7783 (1986).
72. V. Dao, R. Guenther, A. Malkiewicz, B. Nawrot, E. Sochacka, A. Kraszewski, K. Everett and P. F. Agris, PNAS 91, 2125 (1994).
73. C. W. Gehrke, J. A. Desgres, K. O. Gerhardt, P. F. Agris, G. Keith, H. Sierzputowska-Gracz, M. S. Tempesta and K. C. Kuo, in "Chromatography and Modification of Nucleosides" (C. W. Gehrke and K. C. Kuo, eds.), pp. 159–223. Elsevier, Amsterdam, 1990.
74. J. A. McCloskey, Accts. Chem. Res. 24, 81 (1991).
75. M. Sochacki, Biol. Mass Spectrom. 23, 434 (1994).
76. J. A. Kowalak, S. C. Pomerantz, P. F. Crain and J. A. McCloskey, NARes 21, 4577 (1993).
77. D. P. Little, T. W. Thannhauser and F. W. McLafferty, PNAS 92, 2318 (1995).
78. P. F. Agris, H. Sierzputowska-Gracz and C. Smith, Bchem 25, 5126 (1986).
79. W. S. Smith, B. Nawrot, A. Malkiewicz and P. F. Agris, Nucleosides Nucleotides 11, 1683 (1992).
80. H. Sierzputowska-Gracz, E. Sochacka, A. Malkiewicz, K. Kuo, C. W. Gehrke and P. F. Agris, JACS 109, 7171 (1987).
81. H. Sierzputowska-Gracz, P. F. Agris and J. R. Katze, Magn. Reson. Chem. 26, 4 (1988).
82. P. F. Agris, H. Sierzputowska-Gracz, W. Smith, A. Malkiewicz, E. Sochacka and B. Nawrot, JACS 114, 2652 (1992).
83. W. S. Smith, H. Sierzputowska-Gracz, E. Sochacka, A. Malkiewicz and P. F. Agris, JACS 114, 7989 (1992).
84. G. Kawai, T. Yokogawa, K. Nishikawa, T. Ueda, T. Hashizume, J. A. McCloskey, S. Yokoyama and T. Watanabe, Nucleosides Nucleotides 13, 1189 (1994).
85. P. F. Agris, H. Koh and D. Söll, ABB 154, 277 (1972).
86. G. Kawai, Y. Yamamoto, T. Kamimura, T. Masegi, M. Sekine, T. Hata, T. Iimori, T. Watanabe, T. Miyazawa and S. Yokoyama, Bchem 31, 1040 (1992).
87. J. A. Kowalak, J. J. Dallüge, J. A. McCloskey and K. O. Stetter, Bchem 33, 7869 (1994).
88. X. Gu and D. V. Santi, Bchem 30, 2999 (1991).
89. S. Nakanishi, T. Ueda, H. Hori, N. Yamazaki, N. Okada and K. Watanabe, JBC 269, 32221 (1994).

90. J. Edqvist, K. Blomqvist and K. B. Stråby, *Bchem* **33**, 9546 (1994).
90a. J. Edqvist, K. B. Stråby and H. Grosjean, *Biochimie* **77**, 54 (1995).
91. W. M. Holmes, C. Andraos-Selim, I. Roberts and S. Z. Wahab, *JMB* **267**, 13440 (1992).
91a. W. H. Holmes, C. Andraos-Selim and M. Redlak, *Biochimie* **77**, 62 (1995).
92. K. Nagai, *Curr. Opin. Struct. Biol.* **2**, 131 (1992).
93. S. H. Kim, F. L. Suddath, G. J. Quigley, A. McPherson, J. L. Sussman, A. H. J. Wang, N. C. Seeman and A. Rich, *Science* **185**, 435 (1974).
94. M. A. Rould, J. J. Perona, D. Söll and T. A. Steitz, *Science* **246**, 1135 (1989).
95. M. Ruff, S. Krishnaswanmy, M. Boeglin, A. Poterszm, A. Mitschler, A. Podjarny, B. Rees, J. C. Thierry and D. Moras, *Science* **252**, 1682 (1991).
96. R. Basavappa and P. B. Sigler, *EMBO J.* **10**, 3105 (1991).
97. V. Bioiu, A. Yaremchuk, M. Tukalo and S. Cusack, *Science* **263**, 1404 (1994).
98. P. F. Agris, in "Encyclopedia of NMR" (D. M. Grant and R. K. Harris, eds.; S. Chan, section ed.). Wiley, New York, 1996. In press.
99. B. Nawrot, A. Malkiewicz, W. S. Smith, H. Sierzputowska-Gracz and P. F. Agris, *Nucleotides Nucleosides* **14**, 143 (1995).
100. A. Malkiewicz, E. Sochacka, A. F. Sayed Ahmed and S. Yassin, *Tetrahed. Lett.* **48**, 5395 (1983).
101. A. Malkiewicz and E. Sochacka, in "Biophosphates and their Analogues" (K. S. Bruzik and W. J. Stec, eds.), pp. 205–215. Elsevier, Amsterdam, 1987.
102. E. P. Nikonowicz, A. Sirr, P. Legault, F. M. Jucker, L. M. Baer and A. Pardi, *NARes* **20**, 4507 (1992).
103. B.-S. Choi, and A. G. Redfield, *Bchem* **31**, 12799 (1992).
104. R. T. Batey, M. Inada, E. Kujawinski, J. D. Puglisi and J. R. Williamson, *NARes* **20**, 4515 (1992).
105. P. F. Agris, A. Malkiewicz, S. Brown, A. Kraszewski, B. Nawrot, E. Sochacka, K. Everett and G. Guenther, *Biochimie* **77**, 125 (1995).
107. D. Gasparutto, T. Livache, H. Bazin, A.-M. Duplaa, A. Guy, A. Khorlin, D. Molko, A. Roget and R. Teoule, *NARes* **20**, 5159 (1992).
108. K. B. Hall and L. W. McLaughlin, *NARes* **20**, 1883 (1992).
109. I. Habus and S. Agrawal, *NARes* **22**, 4350 (1994).
110. N. N. Polushin, A. M. Morocho, B.-C. Chen and J. S. Cohen, *NARes* **22**, 639 (1994).
111. J. A. Grasby and M. J. Gait, *Biochimie* **76**, 1223 (1994).
112. P. F. Agris and S. C. Brown, in "Methods in Enzymology" (T. James, ed.), Vol. 261, pp. 277–299. Academic Press, Orlando, 1995.
113. N. Usman, K. K. Ogilvie, M.-Y. Jaing and R. J. Cedergren, *JACS* **109**, 7845 (1987).
114. R. H. Guenther, R. S. Bakal, B. Forrest, Y. Chen, R. Sengupta, B. Nawrot, E. Sochacka, J. Jankowska, A. Kraszewski, A. Malkiewicz and P. F. Agris, *Biochimie* **76**, 1143 (1994).
115. W.-D. Hardt, J. Schlegl, V. A. Erdmann and R. K. Hartmann, *Bchem* **32**, 13046 (1993).
116. G. Varani, C. Cheong and I. Tinoco, Jr., *Bchem* **30**, 3280 (1991).
117. G. Varani, and I. Tinoco, Jr., *Q. Rev. Biophys.* **24**, 479 (1991).
118. H. A. Heus and A. Pardi, *Science* **253**, 191 (1991).
119. R. H. Guenther, C. C. Hardin, H. Sierzputowska-Gracz and P. F. Agris, *Bchem* **31**, 11004 (1992).
120. J. R. Williamson and S. G. Boxer, *Bchem* **28**, 2819, 2831, and 2836 (1989).
121. M. J. J. Blommers, F. J. M. Van de Ven, G. A. Van der Marcel, J. H. Van Boom and C. W. Hilbers, *EJB* **201**, 33 (1991).
122. S. A. White, M. Nilges, A. Huang, A. T. Brunger and P. B. Moore, *Bchem* **31**, 1610 (1992).
123. J. SantaLucia, Jr., R. Kierzek and D. Turner, *Science* **256**, 217 (1992).

124. V. P. Antao and I. Tinoco, Jr., *NARes* **20**, 819 (1992).
125. S. A. White and D. C. Draper, *Bchem* **28**, 1892 (1989).
126. V. Dao, R. H. Guenther and P. F. Agris, *Bchem* **31**, 11012 (1992).
127. Y. Chen, H. Sierzputowska-Gracz, R. Guenther, K. Everett and P. F. Agris, *Bchem* **32**, 10249 (1993).
128. J. Weissenbach and H. Grosjean, *EJB* **116**, 207 (1981).
129. H. Grosjean and W. Fiers, *Gene* **18**, 199 (1982).
130. D. Labuda, G. Striker and D. Porschke, *JMB* **174**, 587 (1984).
131. V. I. Katunin, N. G. Soboleva, V. I. Makhno, E. A. Sedel'nikova, S. M. Zhenodarova and S. V. Kirillov, *Mol. Biol.* **28**, 43 (1994).
132. S. Wang and E. T. Kool, *Bchem* **34**, 4125 (1995).
133. L. C. Sowers, B. R. Shaw and W. D. Sedwick, *BBRC* **148**, 790 (1987).
133a. M. M. Basti, J. W. Stuart, A. T. Lam, R. Guenther and P. F. Agris, *Nature Struct. Biol.* **3**, 38 (1996).
134. S. R. Holbrook, J. L. Sussman, R. W. Warrant and S. H. Kim, *JMB* **128**, 631 (1978).
135. G. G. Prive, K. Yanaagi and R. E. Dickerson, *JMB* **217**, 177 (1991).
136. U. Heinemann and M. Hahn, *JBC* **267**, 7332 (1992).
137. J. G. Arnez and T. A. Steitz, *Bchem* **33**, 7560 (1994).
138. A. Kintanar, D. Yue and J. Horowitz, *Biochimie* **76**, 1192 (1994).
139. J. A. Cowan, *JACS* **113**, 675 (1991).
140. P. Lengyel and D. Söll, *Bacteriol. Rev.* **33**, 264.
141. A. Shearn and N. H. Horowitz, *Bchem* **8**, 295 (1969).
142. H. Sigel, *Chem. Soc. Rev.* **22**, 255 (1993).
143. J. Swiatek, *J. Coord. Chem.* **33**, 191 (1994).
144. G. J. Quigley, M. M. Teeter and A. Rich, *PNAS* **75**, 64 (1978).
145. J. R. Sampson, L. S. Behlen, A. B. DiRenzo and O. C. Uhlenbeck, *Bchem* **31**, 4161 (1992).
146. E. Murphy, C. Steenbergen, L. A. Levy, B. Raju and R. E. London, *JMB* **264**, 5622 (1989).
147. D. Labuda and D. Porschke, *Bchem* **21**, 49 (1982).
148. G. Striker, D. Labuda and M. C. Vega-Martin, *J. Biomol. Struct. Dyn.* **7**, 235 (1989).
149. K. B. Hall, J. R. Sampson, O. C. Uhlenbeck and A. G. Redfield, *Bchem* **28**, 5794 (1989).
150. D. Yue, A. Kintanar and J. Horowitz, *Bchem* **33**, 8905 (1994).
151. P. F. Agris, H. Sierzputowska-Gracz and C. Smith, *Bchem* **25**, 5126 (1986).
152. R. Kopper, P. G. Schmidt and P. F. Agris, *Bchem* **22**, 1396 (1983).
153. E. I. Hyde and B. R. Reid, *Bchem* **24**, 4315 (1985).
154. C. S. Chow, L. S. Behlen, O. C. Uhlenbeck and J. K. Barton, *Bchem* **31**, 972 (1992).
155. S. S. Reid and J. A. Cowan, *Bchem* **29**, 6025 (1990).
156. J. R. Sampson, A. B. DiRenzo, L. S. Behlen and O. C. Uhlenbeck, *Science* **243**, 1363 (1989).
157. M. Frugier, C. Florentz and R. Giégé, *PNAS* **89**, 3990 (1992).
158. C. Cantor, in "Transfer RNA: Structure, Properties and Recognition" (P. Schimmel, D. Soll and J. Abelson, eds.), pp. 363–392. Cold Spring Harbor Laboratory, Cold Spring Harbor, New York, 1979.
159. S. J. Rose, P. T. Lowary and O. C. Uhlenbeck, *JMB* **167**, 103 (1983).
160. D. V. Parfenov and E. M. Saminskii, *Mol. Biol.* **27**, 507 (1993).
162. K. A. Sharp, B. Honig and S. C. Harvey, *Bchem* **29**, 340 (1990).
163. J. K. Strauss and L. J. Maher III, *Science* **266**, 1829 (1994).
164. J. Emerson and M. Sundaralingam, *Acta Crystallogr. B.* **36**, 537 (1980).
165. J. Cadet, R. Ducolumb and F. E. Hruska, *BBA* **563**, 206 (1980).

165a. J. W. Stuart, M. M. Basti, W. S. Smith, B. Forrest, R. Guenther, H. Sierzputowska-Gracz, B. Nawrot, A. Malkiewicz and P. F. Agris, *Nucleosides Nucleotides* (1996). In press.
166. K. Watanabe, N. Hayashi, A. Oyama, K. Nishikawa, T. Ueda and K. Miura, *NARes* **22**, 79 (1994).
167. Z. Galdecki, B. Luciak, A. Malkiewicz and B. Nawrot, *Chem. Monthly* **122**, 487 (1991).
168. T. Kowalik-Jankowska, H. Kozlowski, I. Sovago, B. Nawrot, E. Sochacka and A. Malkiewicz, *J. Inorg. Biochem.* **53**, 49 (1994).
169. L. G. Laing, T. C. Gluick, and D. E. Draper, *JMB* **237**, 577 (1994).
170. M. Lu and D. E. Draper, *JMB* **244**, 572 (1994).
171. A. T. Lam, R. Guenther and P. F. Agris, *BioMetals* **8**, 290 (1995).
172. F. Major, M. Turcotte, D. Gautheret, G. Lapalme, E. Fillion and R. Cedergren, *Science* **253**, 1255 (1991).
173. D. Gautheret and R. Cedergren, *FASEB J.* **7**, 97 (1993).
174. H. Ogata, Y. Akiyama and M. Kanehisa, *NARes* **23**, 419 (1995).
175. L. Gold, *JMB* **270**, 13581 (1995).

Chemical and Computer Probing of RNA Structure

N. A. Kolchanov,*
I. I. Titov,* I. E. Vlassova*
and V. V. Vlassov†,[1]

*Institute of Cytology and Genetics
Siberian Division of Russian Academy
of Sciences
Novosibirsk 630090, Russia
†Institute of Bioorganic Chemistry
Siberian Division of Russian Academy
of Sciences
Novosibirsk 630090, Russia

I. Probing RNA Structure by Chemical and Enzymatic Approaches	133
A. General Principles of Chemical and Enzymatic Probing: Analysis of Modified RNA	133
B. Enzymatic Probes	140
C. Chemical Probes	142
D. Techniques for Probing RNA Structure	148
II. Computer Analysis of the Secondary Structure of RNA	164
A. Thermodynamic Parameters of RNA Secondary Structure	165
B. Thermodynamic Approach	166
C. Simulation of RNA Folding	172
D. Comparative Approach	174
E. Prediction of Pseudoknots	180
F. Statistical Analysis of RNA Secondary Structure	186
III. Concluding Remarks	190
References	191

Ribonucleic acids are one of the most important types of biopolymers. RNAs play key roles in storage and multiplication of genetic information. They are important in catalysis, RNA splicing, and the most important steps of translation. Studies in the past few years have demonstrated the possibility of developing RNA species (aptamers) that can recognize different biopolymers and synthetic organic molecules. Problems of investigation of RNA structure and functions, and recent exciting developments in the design of catalytic RNA molecules and specific RNA ligands, have been considered previously (1).

[1] To whom correspondence may be addressed.

The complicated natural functions of RNAs require specific interactions of these molecules with proteins and other nucleic acids. The specificity of these interactions of RNAs and their biological activities are determined by their three-dimensional structures. The three-dimensional (tertiary) structure of RNA is formed by hydrogen-bonding between functional groups of nucleosides in different regions of the molecule, by coordination of polyvalent cations, and by stacking between the double-stranded regions present in the RNA. Knowledge of the tertiary structure of RNAs and the possibility to predict RNA folding from nucleotide sequences are of key importance for understanding the principles of genetic information, for elucidation of relationships between the structure of RNA and its functions, for the design of functionally active polynucleotides, and for the selection of optimal oligonucleotide probes for the detection of specific RNAs and antisense oligonucleotides for modulating the functions of specific RNAs.

At present, the tertiary structures of only some small RNAs have been determined by high-resolution X-ray crystallographic analysis (2–6) and NMR analysis (7–8). For both of those physical methods, relatively large amounts of highly purified RNA are needed, and X-ray studies require high-quality RNA crystals, which are difficult to grow. These are serious and principal limitations, because the goal of researchers is investigation of the biologically active RNA structure, which is attained in solutions of definite composition, and specific protein factors are sometimes required for correct folding and functioning. Therefore, there is a need for methods allowing analysis of the folding of RNAs in solution and great attention has been paid to the development of approaches for prediction and investigations of RNA structure in complex biological systems.

The most widely used approach for investigation of RNA structure is chemical and enzymatic probing in combination with theoretical methods and phylogenetic studies allowing prediction of variants of RNA folding. Chemical and enzymatic probings allow determination of the reactivities of different functional groups of RNA that can be interpreted in structural terms. From such data it is possible to identify such structural features of RNA in solution as the occurrence of double-stranded regions and long-range interactions between nucleotides responsible for the three-dimensional RNA folding. Chemical methods allow detection of some specific spatial arrangements of nucleotides that bind metal ions and metal complexes. Cross-linking with chemical reagents allows the determination of intramolecular distances between certain nucleotides in the RNA tertiary structure. A key advantage of these methods is the ability to study RNAs too large for crystallographic and NMR studies and the possibility to investigate structures of nonpurified RNAs in complex systems containing various factors, or even RNA bound to specific proteins. Information from the probing experiments allows one to identify the real RNA structure among a number of potential structures that can be predicted from sequence data and free energy parameters, and from data of phylogenetic studies in which sequences of a particular RNA from diverse species are compared in order to infer the existence of base-paired regions.

In this essay we describe experimental methods for probing RNA structure and theoretical methods allowing prediction of thermodynamically favorable RNA folding. These methods are complementary, and together they provide a powerful approach to determine the structure of RNAs.

I. Probing RNA Structure by Chemical and Enzymatic Approaches

A. General Principles of Chemical and Enzymatic Probing: Analysis of Modified RNA

Most natural RNAs are globular molecules containing short single-stranded sequences and short double-stranded fragments formed by intramolecular interactions of complementary nucleotide sequences. The system of single-stranded and double-stranded regions formed by complementary nucleotide sequences of the molecule is called the secondary structure of RNA. The double-stranded helices of RNA assume the A form in physiological conditions, with 11 base-pairs per single turn. Long, regular, uninterrupted helices are rare in most RNAs. A typical element of local RNA structure is an 8- to 10-base-pair segment incorporating a bulge or mismatch (Fig. 1). Unpaired bases are often involved in interactions leading to the three-dimensional folding of the molecules. Due to these interactions between nucleotides and interactions with metal ions, the elements of the secondary structure of RNAs fold in a unique three-dimensional (tertiary) structure.

Clues for understanding the principles of the folding of RNA molecules come essentially from X-ray studies of tRNAs (2–6). tRNAs are 72–95 nucleotides long and are folded in a cloverleaf-like structure containing four stems and three loops (Fig. 2). The tertiary structure of tRNA is formed by interactions between nucleotides in the D and T loops. Besides the Watson–Crick base-pairing, a few other types of hydrogen-bonding occur in tRNAs. Thus a G15·C48 pair is formed by the bases in parallel RNA strands. m^{1}A58 and T54, also in parallel strands, form a reverse Hoogsteen pair.

The tertiary folding of tRNA also involves base triplets in which a third nucleotide forms hydrogen-bonds, with a Watson–Crick base-pair in the major groove of short helices. Thus, in yeast tRNA[Phe], the triplet

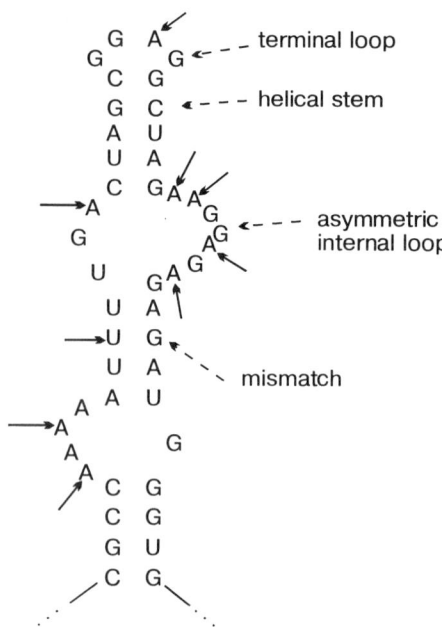

FIG. 1. A typical fragment of RNA secondary structure: a conserved RNA stem–loop structure in the packaging signal of the human immunodeficiency virus type 1. The RNA secondary structure was predicted theoretically and confirmed by probing with diethyl pyrocarbonate (reaction with adenosines in single-stranded regions of RNA) and S1 nuclease (cleavage of phosphodiester bonds in single-stranded regions of RNA). The positions attacked by the probes are indicated by arrows. It is seen that the modification and cleavage patterns are consistent with the RNA folding. Reprinted from T. Hayashi, Y. Ueno and T. Okamoto, *FEBS Lett.* **327**, 213 (1993), with kind permission from Elsevier Science—NL, Sara Burgerhart-Straat 25, 1055 KV Amsterdam, The Netherlands.

$(m^2G10 \cdot C25) \cdot G45$ is formed by the interaction of the third base with a Watson–Crick pair by one hydrogen bond. Triplets $(G22 \cdot C13) \cdot m^7G46$ and $(A23 \cdot U12) \cdot A9$ are formed by two hydrogen bonds of the third base with corresponding Watson–Crick base-pairs. Due to these interactions, the cloverleaf structure is folded into a three-dimensional L-shaped structure built of two helical domains formed by the stem regions of the molecule, because of stacking interactions (2–6) (Fig. 3).

RNAs are polyanionic molecules; they bind cations (metal ions and the organic polycations, spermine and spermidine). Specific active conformations are formed by RNAs only in the presence of certain concentrations of monovalent cations and magnesium ions. In the three-dimensional structure of tRNAs, there are some sites where particularly tight binding of cations occurs. These are sites where groups capable of interacting with the ions

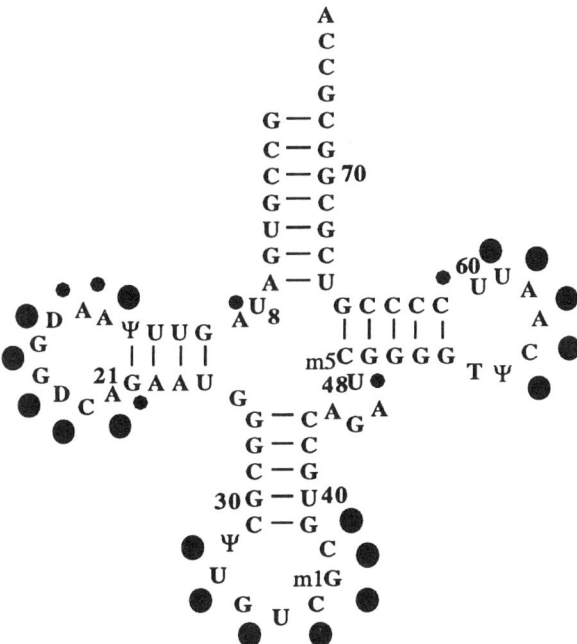

FIG. 2. Cloverleaf structure of yeast tRNA^Asp with imidazole-induced cleavage points. Phosphodiester linkages displaying enhanced susceptibility to hydrolysis by the imidazole buffer in conditions stabilizing the RNA structure are indicated by dots with diameters proportional to the intensity of the cuts (46).

(phosphates, nitrogens of heterocyclic bases, ribose oxygens) can be arranged optimally for simultaneous interaction with an ion (11).

RNAs are built of a large number of chemically similar monomers that possess a few chemical groups available for chemical modification or for attack by enzymes capable of hydrolyzing RNA. The microenvironment of these groups can be very different in the three-dimensional structure of RNA, and these differences can dramatically affect the reactivities of the groups toward chemical and enzymatic probes. When structural factors affecting the reactivities of specific groups to given probes are known, chemical modification data can be interpreted in structural terms.

The following main factors affect the reactivities of groups in RNAs.

1. A group can be partially or completely buried within the molecule, which interferes with reactions—in particular, with reactions with bulky probes. Stacking of heterocyclic bases decreases their reactivity toward reagents that attack the bases perpendicularly to their planes.

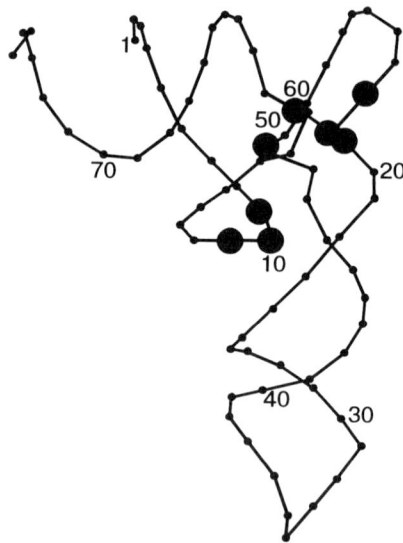

FIG. 3. Tertiary structure of yeast tRNA^{Phe}. Black circles indicate phosphates in the RNA structure protected from modification with a small probe, ethylnitrosourea. Reproduced from Ref. 10.

2. Participation of a group in hydrogen bonding or in coordination with a metal ion affects its nucleophilicity and results in structural shielding.
3. The electrostatic environment of a group affects its ionization. Because RNAs are polyanions they repel negatively charged species and attract positively charged reagents, which results in suppression or acceleration of the corresponding reactions.

The identification of factors suppressing the reactivity of a given base is possible by investigation of patterns of reactivities of nucleotides toward different chemical probes whose chemical specificity is known. Attempts have been made to combine steric and electrostatic factors in a form of a theoretical index [Accessible Surface Integrated Field Index (12)], which correlates with the reactivities of functional groups of RNA—e.g., reactivities of RNA phosphates toward ethylnitrosourea (10, 13).

Nucleotides within single-stranded and double-stranded regions of RNA can easily be distinguished using chemical probes reacting with functional groups of the bases participating in Watson–Crick interactions. In the double-stranded regions, the groups are shielded from reagents present in solution. Similarly, nucleotides in the double-stranded regions that participate in base-triplet formation are easy to identify because of shielding of the

N-7 atoms in the major groove of the double helix. Reduced reactivities of phosphates and nitrogen atoms of heterocyclic bases can reflect their involvement in coordination of metal ions.

When performing probing experiments, one should arrange experimental conditions in which the RNA under study will assume the desired structure and retain it throughout the experiment. RNAs acquire a biologically active structure only in a relatively narrow range of conditions ("physiological" conditions). In the course of isolation, an RNA structure is often denatured and it is necessary to transfer it to conditions allowing resumption of the biologically active structure. Therefore, before RNA is subjected to probing, it is essential to ensure that the population of molecules is homogeneous and to remove traces of denaturants used in the RNA isolation. It is recommended, when possible, to perform a heat treatment followed by a slow cooling down (renaturation) to allow the RNA to assume a thermodynamically favorable structure.

An important requirement of modification experiments aimed at probing the reactivities of different nucleotides consists in ensuring that the RNA is subjected to limited chemical modification or enzymatic hydrolysis providing statistically less than one cut or modification per RNA molecule. This guarantees that the molecule under study has not been changed in the course of investigation and allows obtaining quantitative data on the reactivities of specific residues. Reactions are performed in the presence of carrier RNA for controlling the reaction conditions. Incubation of RNA in experimental conditions without the reagent is performed as a control for detection of breakage caused by nonspecific factors potentially present.

Detection of cleavage sites and modification sites can be performed by two methods, the choice determined by the size of the RNA molecule and by the nature of the modification. One method uses end-labeled RNA molecules and allows detection of cleavages in the RNA by gel-sequencing. A limitation of this method is that it detects only scissions in RNA structure. This method can be used to study RNA with at least one homogeneous end, up to 300–400 nucleotides long, or terminal sequences of large RNAs.

Labeling of the 5' end of RNA can be performed enzymatically, using T4 polynucleotide kinase, transferring the γ-phosphate from [γ-^{32}P]ATP to the 5'-terminal ribose of RNA (14). If the RNA has a phosphate group at the 5' terminus, the phosphate can be removed by alkaline phosphatase prior to labeling. Alternatively, a T4 polynucleotide kinase-catalyzed exchange reaction between the γ-phosphate of [γ-^{32}P]ATP and the phosphate of RNA can be used to substitute the labeled phosphate for the cold one (15). Labeling of the 3' end of RNA can be performed by attaching [5'-^{32}P]pCp to the 3'-OH group of the RNA using T4 RNA ligase (16). tRNA can be 3' end-labeled by removing the terminal CCA sequence by phosphodiesterase and restoring the CCA end using tRNA nucleotidyl transferase and labeled CTP and ATP (17).

The principle behind the method is outlined in Fig. 4. Cleavage of RNA with an RNase or by a chemical probe in conditions allowing one hit per molecule generates pairs of fragments. The fragments are resolved by electrophoresis in polyacrylamide gel in denaturing conditions followed by autoradiography, which allows registration of the fragments originating from the labeled end of the RNA. To determine the size of the fragments, products of limited alkaline hydrolysis of the same RNA and fragments produced by some sequencing reactions are run on the same gel.

The second method for analysis of modified RNA uses reverse transcription (Fig. 5). The attacked position is identified by a stop in reverse transcription generated from a DNA primer. This method is most generally useful and an expedient approach to probe any RNA sequence, regardless of

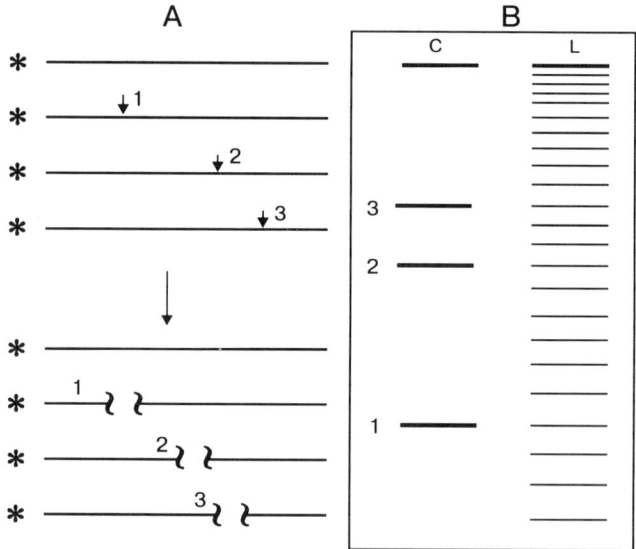

FIG. 4. Schematic representation of the method for detection of cuts in RNA structure. (A) A probe (chemical reagent or enzyme) attacks, under limiting conditions, three sites in a 5'-labeled RNA, which results in cleavage of the RNA. (B) Positions of the cleavages are mapped by electrophoresis on a denaturing polyacrylamide gel. The lengths of the produced labeled fragments are determined by comparison to the length standards, to locate precisely the reactive nucleotides. In the illustrating gel, the first line (C) might be, e.g., a partial T1 ribonuclease digest of the RNA containing three guanosine residues. The second line (L) is the ladder produced by limited alkaline hydrolysis of the RNA, providing statistical cleavages of all phosphodiester bonds.

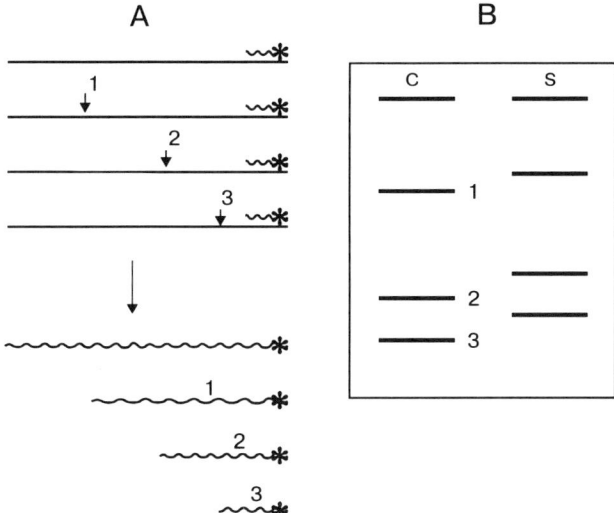

FIG. 5. The primer-extension procedure for detection of cuts and chemical modifications in RNA. (A) Arrows indicate positions of cuts or modified nucleotides. A radiolabeled oligodeoxyribonucleotide primer is annealed to the 3' end of the RNA. Reverse transcription of the modified molecules is terminated at the modified residues and yields shortened transcripts. (B) The length of the transcripts is determined by gel electrophoresis. The left lane (c) is the schematic presentation of the bands corresponding to the fragments, which might be, e.g., fragments transcribed from an RNA with three adenosine residues modified with diethyl pyrocarbonate. The second lane (s) shows the A-specific sequencing reaction. The relative shift of the bands in the lanes is explained by termination of the reverse transcription at a nucleotide preceding the modified residue.

size; it also allows one to detect chemical modifications that do not cleave RNA. Analysis of modified RNAs using the primer extension method is performed by annealing a complementary oligodeoxynucleotide primer to the 3' end of the RNA and synthesizing a cDNA copy of RNA using reverse transcriptase and dNTPs. Elongation proceeds from the 3' end of the primer and it is terminated prematurely when the enzyme meets scissions or chemically modified nucleotides and stops. When transcripts are resolved by gel electrophoresis, the stops in cDNA synthesis are detected as bands corresponding to a modified position in the RNA template.

Reverse transcription is effectively arrested by modifications of nucleotides at the groups involved in Watson–Crick base pairing. This was shown for G(N-1) and G(N-2), for modification with kethoxal; for A(N-1) and C(N-3) for modification with dimethyl sulfate; and for G(N-1) and U(N-3), for reaction with carbodiimides. Carbethoxylation of adenosine at N-7 with diethyl pyrocarbonate opens the imidazole ring, and the modified residue stops the reverse transcription. Methylation of guanosine at N-7 with Me_2SO_4 does not arrest transcription, but this modification can be transformed into a cleavage by treatment with aniline. The synthesized DNA fragments can be labeled by using 5' end-labeled primers or by the use of [γ-^{32}P]NTP. The produced fragments are identified by analyzing them in parallel with dideoxynucleotide sequencing reactions performed with the same unreacted RNA and DNA primer.

It should be mentioned that sequencing patterns obtained by using the primer extension method are shifted by one nucleotide relative to the modified residue, because the last nucleotide incorporated by the transcriptase is complementary to the one on the 3' side of the modified residue in the template RNA. If the RNA under study is long, it is necessary to perform reverse transcription with a few different primers to explore the whole RNA molecule.

Limitations of the method are related to the sensitivity of reverse transcriptase to different modifications. Some of them do not stop the enzyme. On the other hand, some naturally occurring modified residues (e.g., m^2G, m^6A) arrest reverse transcription, and some tightly folded regions of RNA slow down the transcription process. Therefore, a control reverse transcription reaction should be run on the unmodified RNA to detect pauses of natural origin caused by nonspecific breaks of RNA in the reaction conditions and by natural modifications of nucleotides and structural elements that may affect the transcription.

Excellent detailed experimental protocols for investigation of RNA structure with chemical and enzymatic probes and description of analytical techniques can be found elsewhere (18, 19).

B. Enzymatic Probes

Enzymes cleaving the ribose-phosphate backbone of RNA (ribonucleases) are the simplest and most widely used tools for probing RNA structure. Most of these enzymes attack single-stranded regions of the RNA structure showing different specificities to phosphodiester bonds adjacent to certain nucleosides. One enzyme, ribonuclease V1, cleaves RNA preferentially at double-stranded regions. Investigation of the susceptibility of different sequences within the RNA structure toward different ribonucleases allows identification of elements of the secondary structure of RNA.

Although enzymatic probes are popular because of easy handling and simplicity of detection of cleavage positions in RNA, these probes have some drawbacks. The mechanism of phosphodiester-bond cleavage is known; however, it is preceded by a step of enzyme–RNA recognition. Features of this step, involving noncovalent binding of enzyme probes with the surface of the

RNA, are not well understood. Binding of the enzyme to RNA can affect the polynucleotide structure. As a result, the produced cleavage pattern may characterize properties of a perturbed RNA structure rather than of its native structure. Thus, the small protein ribonuclease A shows a very high tendency to cleave Y–A linkages in single-stranded regions of RNA. However, the enzymes sometimes cuts at these sequences in double-stranded regions of RNA, apparently because binding of this cationic protein can unfold the substrate structure locally. Moreover, ribonuclease, noncovalently bound to RNA, can accomplish a few cuts in the same RNA molecule even under conditions of limited hydrolysis. These secondary cuts apparently do not reflect features of the native RNA structure. To detect such cuts, hydrolysis patterns of 5'- and 3'-end-labeled RNAs should be compared.

RNase U2 from *Ustilago sphaerogena* is used for probing adenines in single-stranded RNA sequences. The enzyme cleaves phosphodiester bonds adjacent to the 3' phosphate. The order of sensitivity of phosphodiester bonds to this enzyme is A > G ≫ C > U (20). The pH optimum of the reaction is 4.5; 7 M urea does not stop the hydrolysis.

RNase T1 from *Aspergillus oryzae* cleaves phosphodiester bonds after the 3' phosphate of unpaired guanosine residues. The reaction yields fragments with 3' phosphates and proceeds via the intermediate formation of guanosine 2':3'-cyclic phosphate (21). The presence of 7 M urea stimulates the enzyme activity, when the reaction is carried out at pH 4.5. The enzyme does not hydrolyze RNA after some naturally occurring modified guanosines (m^1G and m^7G).

RNase CL3 from chick liver is used as a probe, cleaving phosphodiester bonds after unpaired cytidines, and yields fragments with a 3' phosphate (22, 23). The enzyme activity is enhanced by spermine and magnesium ions; a pH effect depends on the nature of the buffer.

T2 RNase from *A. oryzae* cleaves RNA after unpaired adenosine residues yielding fragments with 3' phosphate, via formation of intermediates with 2':3'-cyclic phosphates (21). The enzyme has relatively low specificity and exhibits a strong activity to nucleotides at the apex of terminal loops. Internal loops are substantially less reactive. Although the pH optimum of the reaction is 4.5, the enzyme can be used at neutral pH (20). T2 RNase is inhibited by heavy metal ions.

S1 nuclease from *A. oryzae*, used as a probe, is capable of cleaving single-stranded regions in RNA and DNA (24). The enzyme yields fragments with a 5' phosphate. The pH optimum is at 4.5, although the enzyme is still active at neutral pH. The enzyme is stimulated by Zn^{2+}.

Neurospora crassa nuclease is used as a probe cleaving single-stranded regions in RNA and DNA. The hydrolysis generates fragments terminated by 5' phosphates. The pH optimum of the enzyme is 7.5–8 (25). Increasing pH and decreasing ionic strength result in decreasing sequence-specificity of the nuclease. Because the enzyme has Co^{2+} as prosthetic group, its activity is inhibited by EDTA.

RNase V1 from cobra venom cuts preferentially double-stranded and structured (stacked) regions of RNA, showing no apparent nucleotide specificity. The produced fragments contain 5' phosphate (26). The enzyme needs Mg^{2+} ions; it is active in the pH range of 4–9.

C. Chemical Probes

Several chemical reagents are available to probe reactivities of functional groups of heterocyclic bases and reactivities of phosphodiester bonds and ribose (Figs. 6 and 7). Detailed protocols for probing RNA with chemical reagents can be found in Refs. 18 and 19.

1. ALKYLATING REAGENTS

Dimethyl sulfate (Me_2SO_4) and derivatives of 2-chloroethylamine react with nucleophilic centers of heterocyclic bases. At neutral pH, the order of reactivities is G(N-7) > A(N-1), C(N-3) (27). The 7–8 double bond in an alkylated guanosine can easily be reduced by sodium borohydride. The resulting product provides a site for aniline-induced scission (28, 29). The reaction is used for detection of guanosines with N-7 atoms involved in hydrogen bonding or in coordination with metal ions. The reactivity of guanosines is affected, to some extent, by stacking. Alkylation at adenosines and cytidines is used to detect nucleotides not involved in Watson–Crick interactions. The modified residues can be detected by the primer extension method. For detection of modified cytidines, RNA can be cleaved at the modified residues by treatments with hydrazine and then with aniline. Hydrazine reaction results in some cleavage at uridines, but this reaction is structure-independent. The modifications can be detected by the primer extension method. Modification by Me_2SO_4 can be used for detection of hydrogen-bonding of adenosines in the *syn* conformation, which occurs in GA and Hoogsteen AU pairings.

Ethyl nitrosourea (ENU) is an alkylating reagent attacking both internucleotide phosphates and nucleophilic centers of heterocyclic bases in RNA with comparable efficiency (30, 31). The ethyl phosphotriesters formed are unstable; mild alkaline treatment results in breakage of the RNA chain at positions of the phosphotriesters. This reaction can be used to map phosphates not engaged in binding of metal ions or in hydrogen-bond formation (10, 13, 32).

2. CARBODIIMIDES

For modification of RNA, 1-cyclohexyl-3-(2-morpholinoethyl)carbodiimide metho-p-toluene sulfonate (CMCT) is usually used (33, 34). CMCT reacts with uridine (at N-3) and less efficiently with guanosine (at N-1). CMTC reacts also with some minor RNA components (thymidine, dihydrouridine, and pseudouridine). In pseudouridine, both nitrogen atoms are reactive, and both mono- and diadducts can be formed. Reaction with CMTC is stimulated by increasing pH; usually the reaction is performed at pH 8. At pH > 10, the adducts decompose, yielding nonaltered nucleosides. Because the reactive groups of nucleotides participate in Watson–Crick base-pairing, the reaction occurs only within single-stranded regions of RNA.

3. α-KETOALDEHYDES

Usually β-ethoxy-α-ketobutyraldehyde (kethoxal) is used for modification of RNA. The compound reacts with guanosine in single-stranded regions of RNA. The reaction yields a new ring involving N-1 and N-2 of the guanosine and both carboxyl groups of kethoxal (35). Because attack of the reagent occurs perpendicularly to the plane of the base, stacking inhibits the reaction.

Reaction is carried out at pH 7–7.5. The modification products are stable in slightly acidic medium; at basic pH, they decompose into the components. The adducts are stabilized by borate ion. The modification can be made irreversible by oxidizing the *cis*-diol group of the adducts with periodate.

4. DIETHYL PYROCARBONATE

This reagent carbethoxylates purines at N-7, favoring adenine over guanine in a reaction sensitive to the solvent exposure of the base (36–39). The reagent is used to provide information on the structural environment of purines by probing the involvement of N-7 of adenosine in tertiary interactions. DEPC is sensitive to stacking and poorly attacks purines in helical regions of RNA. The effective diameter of DEPC is 3.5 Å, which is similar to the width of the deep RNA major groove. The modification results in opening of the imidazole ring between atoms N-7 and C-8, and the RNA can be cleaved at the modified residues by aniline treatment. Minor modification of uridine (at N-3) in slightly basic medium can occur (38) and cytidine reacts in solutions with high concentration of salts.

5. BISULFITE

Bisulfite reacts with unpaired cytidines forming 5,6-dihydrocytidine 6-sulfonate. At pH 5–6, in the presence of high concentration of bisulfite,

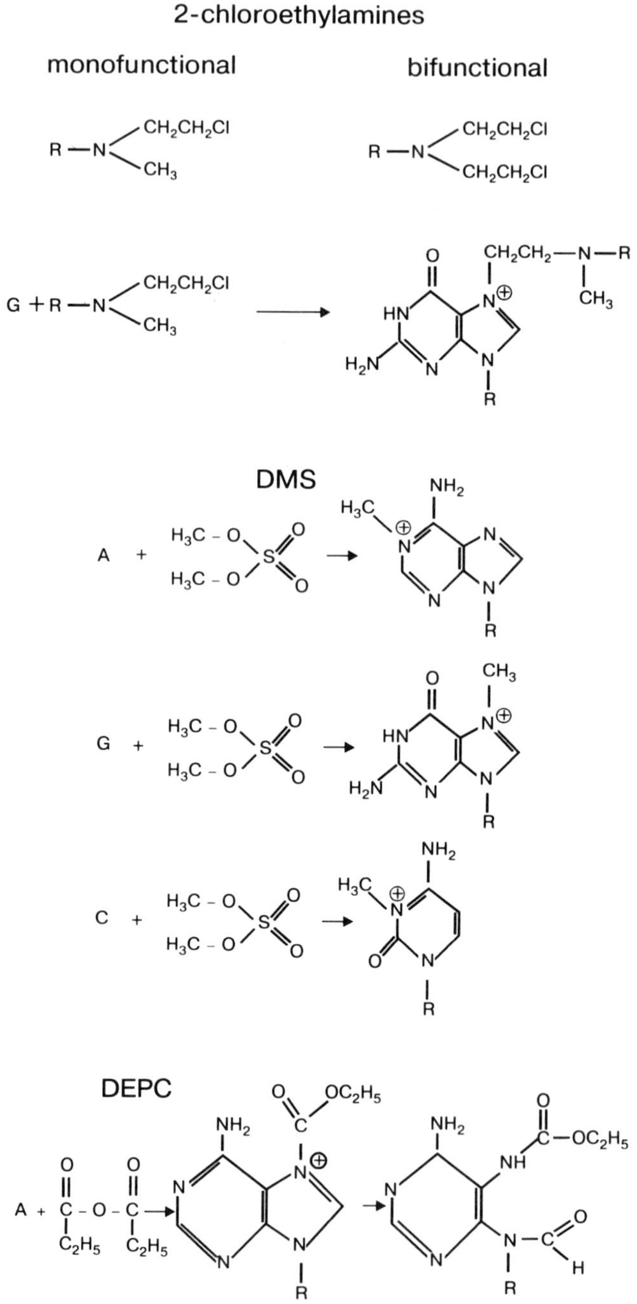

FIG. 6. Reactions of chemical probes with RNA nucleotides. DMS, Dimethyl sulfate; DEPC, diethyl pyrocarbonate; kethoxal, β-ethoxy-α-ketobutyraldehyde; CDI, carbodiimide; ENU, ethylnitrosourea.

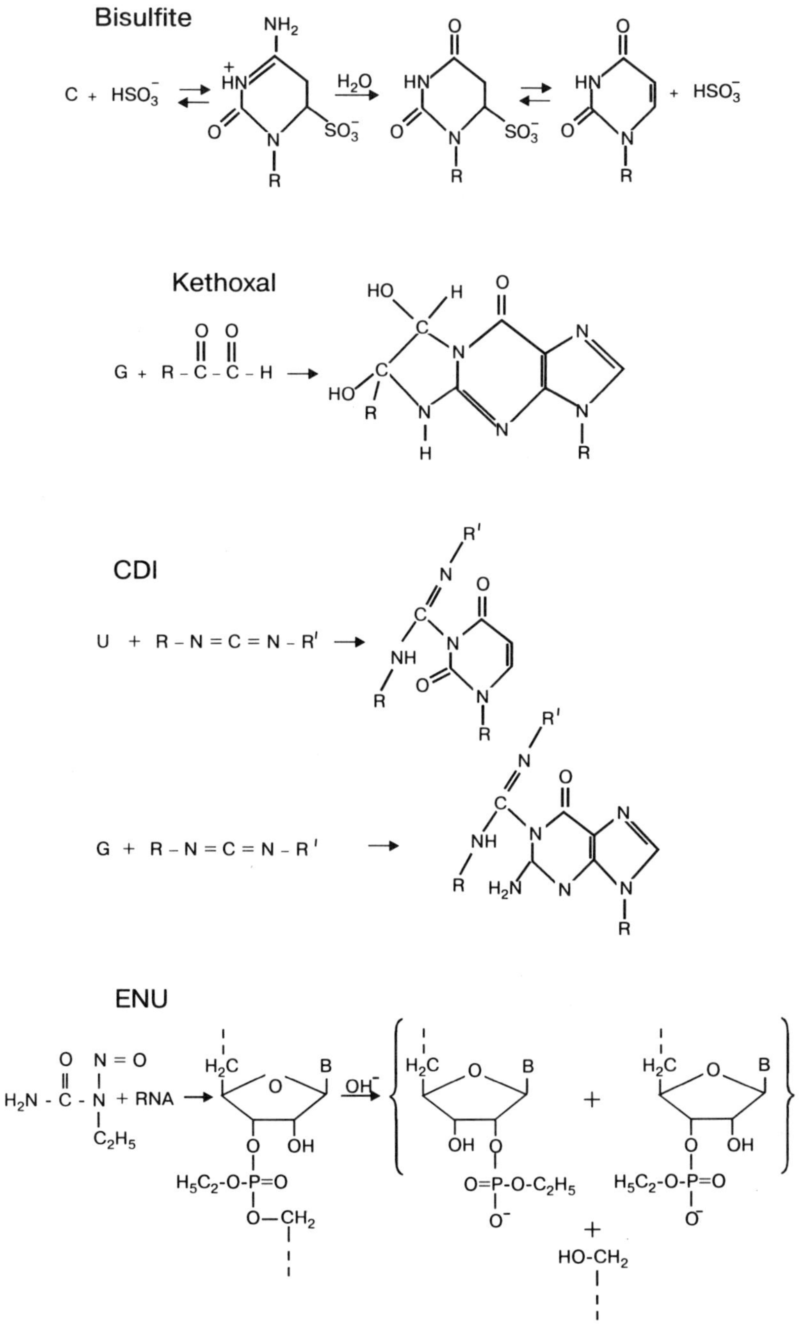

FIG. 6. *Continued*

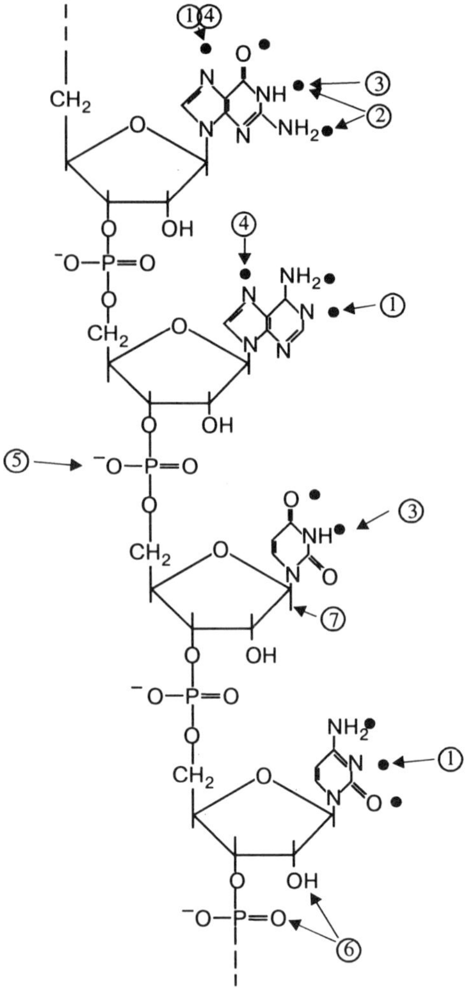

FIG. 7. Functional groups of nucleotides available for probing with chemical reagents. Black circles indicate groups of the bases participating in the Watson–Crick and Hoogsteen hydrogen bonding. Arrows indicate sites of reactions with chemical probes: 1, dimethyl sulfate; 2, kethoxal; 3, carbodiimides; 4, diethyl pyrocarbonate; 5, ethylnitrosourea; 6, imidazole; 7, OH radicals.

nucleophilic substitution at the exocyclic amino group of the cytidine derivative occurs. This results in formation of the corresponding derivative of uridine. Treatment of the compound with mild alkali removes the bisulfite moiety. Conversion of C to U can be detected by a U-specific sequencing reaction with hydrazine (38).

6. Fe(II)–EDTA

The negatively charged EDTA complex of iron(II) reacts with hydrogen peroxide in a Fenton reaction and generates hydroxyl radicals, which react with nucleic acids (39, 40).

$$\text{Fe(II)} + \text{H}_2\text{O}_2 \rightarrow \text{Fe(III)} + \text{OH}^- + \text{OH}\cdot$$

The complex of Fe(II) with EDTA is anionic and does not bind to RNA, so the hydroxyl radicals diffuse from the generation site to RNA. Although the highly reactive hydroxyl radicals attack both heterocyclic bases and ribose, only the latter modification results in strand breaks. The cleavage is initiated by abstraction of a hydrogen from ribose. The ribose radical produced decomposes, yielding as final products RNA fragments terminated by 5' and 3' phosphates. Apparently, the reaction is nonspecific with respect to the nature of the nucleotide. It can be used for identification of surface residues of RNA molecules (41–43).

7. METHIDIUM PROPYL–EDTA–Fe(II)

This conjugate contains a methidium moiety capable of intercalating in double-stranded regions of RNA and the OH-radical-generating Fe(II)–EDTA group (44). Due to the intercalating group, the reagent produces cleavages preferentially within the base-paired regions of RNAs.

8. IMIDAZOLE AND CONJUGATES BEARING IMIDAZOLE

Concentrated imidazole buffer catalyzes cleavage of phosphodiester bonds in RNA (45). The ionized and neutral components of the buffer catalyze the reaction similarly to imidazole residues in the active center of ribonuclease. Hydrolysis of phosphodiester bonds in the single-stranded regions of RNA occurs much more rapidly compared to phosphodiester bonds in the double-stranded regions of RNA (46), apparently due to higher rigidity of the latter, preventing conformational changes needed for the ribosephosphate to form a reactive intermediate. Therefore the reaction is useful for mapping single-stranded regions in RNA.

Conjugates of intercalating dyes with histamine, and spermine–histamine conjugates, in the presence of imidazole cleave RNA, with a specificity similar to that of ribonuclease A (46, 47). The most readily attacked are the Y–R sequences, in particular, C–A, in the single-stranded regions of RNA. The mechanism of the reaction is apparently the catalysis by imidazole residues brought into close contact to the RNA riboses and phosphodiester bonds.

D. Techniques for Probing RNA Structure

1. IDENTIFICATION OF NUCLEOTIDE INTERACTIONS IN RNA STRUCTURE AND ELUCIDATION OF RNA FOLDING

Nucleotides located in single-stranded and double-stranded regions in RNA structure can be distinguished by testing their reactivities toward chemical probes reacting with functional groups of nucleotides participating in Watson–Crick interactions. Because the ribosephosphate backbone within the single-stranded regions is less rigid, compared to the more structured regions of RNA, single-stranded regions are more susceptible to hydrolysis by imidazole buffer (46). Figure 8 shows the positions of those phosphodiester bonds sensitive to hydrolysis by imidazole in a small folded RNA molecule. It is seen that the cleavages occur within the single-stranded regions of the RNA. Spermine–imidazole conjugate in the presence of imidazole buffer cleaves preferentially Y–R sequences in the single-stranded regions of the molecule.

Direct correlations between chemical reactivities of nucleotides within RNA and conformation of the nucleic acids are well established by studies on tRNAs in which chemical reactivity patterns could be explained by crystallographic structures. The knowledge obtained on the reaction specificity allowed investigations of a great number of RNAs and RNA–protein complexes by chemical mapping procedures combined with modeling. In addition to identification of single- and double-stranded regions of RNA structure, chemical and enzymatic probes allow investigation of the stability of

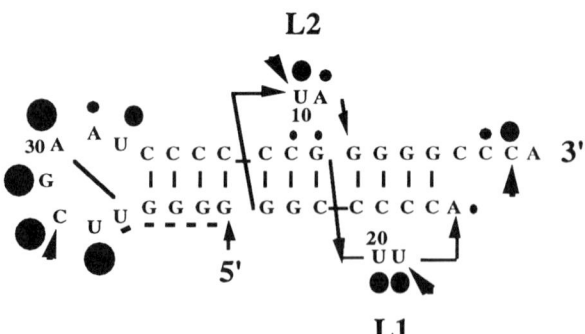

FIG. 8. Secondary structure of the RNA transcript derived from the TMV tRNA-like domain. Dots indicate nucleotides 5′ to the phosphodiester bonds susceptible to hydrolysis by imidazole buffer. The arrows indicate nucleotides 5′ to the phosphodiester bonds attacked by the binary chemical nuclease consisting of spermine–imidazole (2.5 mM) supplemented with 25 mM imidazole buffer, pH 7.0. L1 and L2 emphasize the two single strands of the pseudoknot crossing the deep and the shallow grooves, respectively. The dashed line indicates the nucleotides that could not be tested for methodological reasons (46).

RNA structures in different conditions and detection of interactions important for folding. Probing experiments can be performed under different conditions: under conditions providing the native structure, under various semidenaturing conditions (low salt conditions, variation of temperature), and under conditions wherein RNA is unstructured, at high temperature in the absence of salts. Tertiary interactions are destroyed in the semidenaturing conditions, whereas more stable elements of the secondary structure remain unchanged. Comparison of the data allows identification of elements of the tertiary folding of the molecule and provides information concerning stabilities of different elements of the secondary structure.

An example of a detailed investigation of the structure of an RNA molecule using chemical and enzymatic probes is the study of a 3'-terminal sequence of genomic brome-mosaic-virus RNAs (48). The terminal part of these RNAs can be specifically charged with tyrosine by tyrosyl-tRNA synthetase and it is recognized by other proteins interacting with tRNA. However, the proposed structural models of this RNA deviated considerably from the cloverleaf structure of canonical tRNAs. The 201-nucleotide RNA representing the 3' terminal part of the viral RNA was investigated (48) in solution using chemical and enzymatic probes (Figs. 9 and 10). Bases were probed with Me_2SO_4, CMCT, and DPC. Ribonucleases T1, U2, and V1 and nuclease S1 were used for detection of double-stranded and single-stranded regions of the molecule. Modifications and cleavages were detected by both the primer extension method and the direct gel-electrophoretic analysis of cleaved end-labeled RNA. In these experiments all base-paired nucleotides were identified in the double-stranded regions and long-range interactions between a number of bases in the single-stranded regions were identified. The results obtained on reactivities of various atomic positions toward chemical and enzymatic probes provided information needed for building a detailed structural model of the RNA.

In the model, a domain mimicking the shape and dimensions of tRNA were identified, which explains the ability of the RNA to interact with tRNA-related proteins. The model was built as follows. First, potential elements of the secondary structure were built using a computer program (49) capable of predicting helices, loops, and different folding motifs. The elements were then assembled in a global structure and atomic accessibilities for target sites for chemical reagents were calculated according to Richmond (50). Then the model was refined taking into account the results of the probing experiments. Figure 9 shows the data of the enzymatic mapping. It is seen that the nuclease-cleavage pattern is in good agreement with the shown secondary structure of the molecule.

The results presented in Fig. 9 support the existence of helices B1, B2, B3, and C, and D in the structure. The double-stranded regions are readily

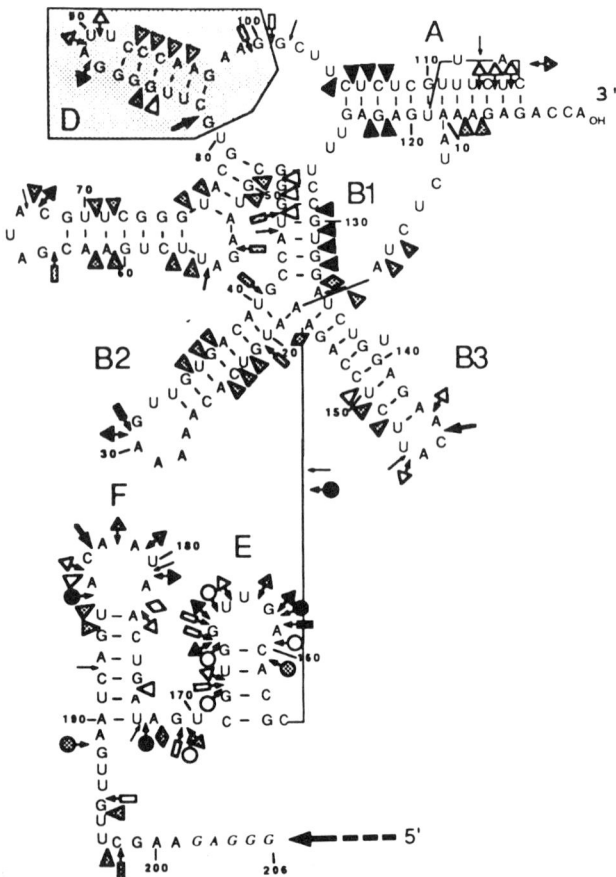

FIG. 9. Results of nuclease mapping of 3'-terminal sequence of the brome mosaic virus RNA (reproduced from Ref. 48). Indication of cuts: ◈, RNase U2; ▶, V1; ━▶, T1; ◀━, S1 at acidic pH; and ●▶, S1 at neutral pH. Open, stippled, and filled symbols correspond to weak, medium, and strong cuts, respectively. Arrows indicate fragile sites of RNA where spontaneous breakage was observed.

attacked by RNase V1. Nuclease S1 and ribonucleases T1 and U2 cut the RNA essentially within single-stranded regions. Some loops are cut less efficiently than others, apparently because of differences in accessibility to relatively bulky enzymes and differences in stability and compactness of the loops. Additional information about the mutual arrangement of the RNA domains was obtained in experiments with RNase V1 (Fig. 11). Continuous V1 cuts in a sequence U45–U51 suggested a stacking between the B1 and C arms of the structure. A symmetrical cleavage pattern in the two strands of

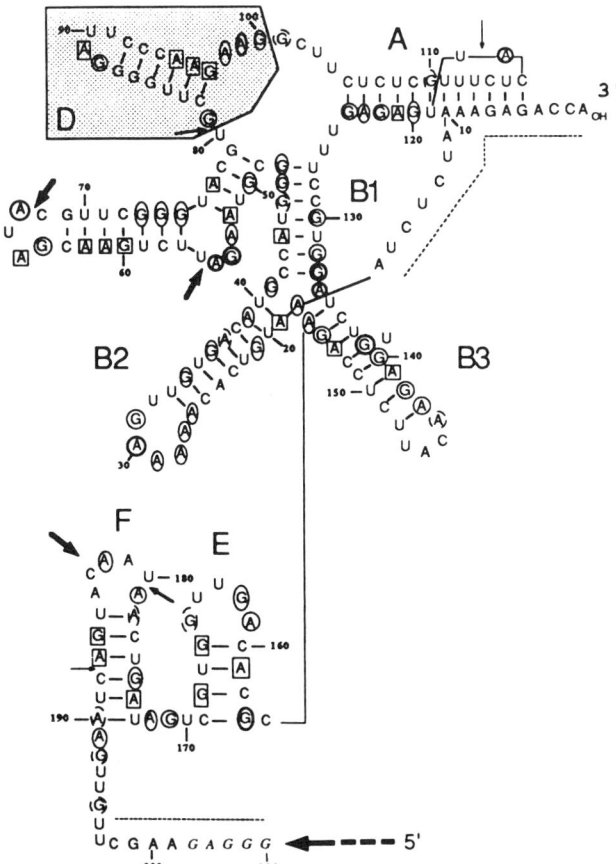

FIG. 10. Mapping of N-7 positions of purines in the 3' end of the brome mosaic virus RNA with dimethyl sulfate and diethyl pyrocarbonate (reproduced from Ref. 48). O, Reactive positions under native conditions; 0, nonreactive positions under native conditions, but reactive under semidenaturing conditions; □, nonreactive positions in both semidenaturing and native conditions. Bold, thin, and broken symbols correspond to strong, moderate, and marginal reactivities of the positions, respectively. Arrows indicate fragile sites in the RNA structure.

helices B and C indicated that both strands of the helices are accessible to the enzyme. In stem B3, only one strand was cut by RNase V1, which implies that another strand is not accessible for the enzyme. Figure 11 shows that the strongest V1 cuts are located at the most accessible external domains of the RNA model.

The results of probing the N-7 positions of the purines are shown in Fig. 10. All purines were reactive under semidenaturing and denaturing

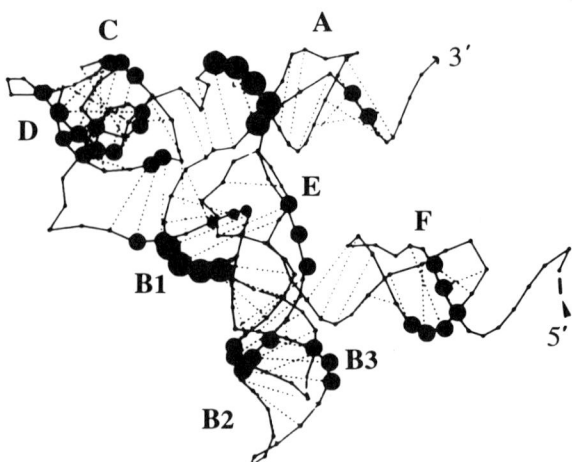

FIG. 11. Three-dimensional model of the 3' end of the brome mosaic virus RNA. The dots indicate sites of cuts by ribonuclease V1. Sizes of the dots correspond to the intensities of the cuts.

conditions. In conditions stabilizing the native structure, the purines presented the reactivity pattern that in general fits the proposed structure. Exceptions were G13, G132, and G133, which were reactive in native conditions, indicating that helix B1 has a distorted conformation. The presence of V1 cuts between nucleotides G195 and U196 and between U197 and C198, together with the data on protection of A181 and A182 in the hairpin loop and U194 and G195 in the 5' end of the RNA, provided evidence of the presence of a pseudoknot in which bases in the loop 181–184 bind to complementary sequence 194–197 in domain F, which was predicted from phylogenetic studies.

An example of tertiary long-range interactions identified in the molecule is a triple-helical region involving interactions (G41·A143)·A18 and (C42·G133)·A17, which include Hoogsteen binding of A17 and A18 with G133 and G41 in the major-groove side of the base-pairs C42·G133 and G41·A134. The data on the A18, G41, and A134 reactivities confirm the existence of these interactions.

2. Approaches for Investigation of Elements of Tertiary Structure of RNA

A straightforward approach to investigation of the tertiary structure of RNA is the identification of residues accessible at the surface of the molecule and residues buried within the structure. For such studies, reagents with broad specificity are needed that allow one to probe all types of nucleotides and that can potentially react with nucleotides irrespective of their involvement in the secondary interactions. A few reagents do allow probing the accessibility of universal constituents of RNA structure. Ethylnitrosourea reacts with phosphates in single-stranded and in double-stranded regions of RNA, if they are not buried in the structure (10, 13, 32). Figure 12 shows a good correlation between the experimentally determined reactivities of phosphates in tRNA^{Phe} and the theoretically calculated availabilities of the phosphates to small probes, taking into account the geometry of the molecule and electrostatic factors (12, 51). Positions of the most well-protected phosphates are shown in Fig. 3.

Another small probe that allows investigation of the accessibility of a universal constituent of the RNA structure, ribose, is the OH radical produced by the Fe(II)–EDTA complex. This probe has little specificity for RNA sequence or secondary structure, making it an attractive probe for the tertiary structure of RNA. It was tested on yeast tRNA^{Phe} (41) and it was shown that, in the native tRNA, riboses in the core of the molecule are protected from the modification. The reaction was used for investigation of structure of the self-splicing intron of *Tetrahymena thermophila* (41) and for monitoring the folding process of this catalytic RNA by probing the RNA structure in solutions containing different concentrations of metal ions (42, 43). From the modification data, an interior–exterior surface map of the folded RNA was constructed.

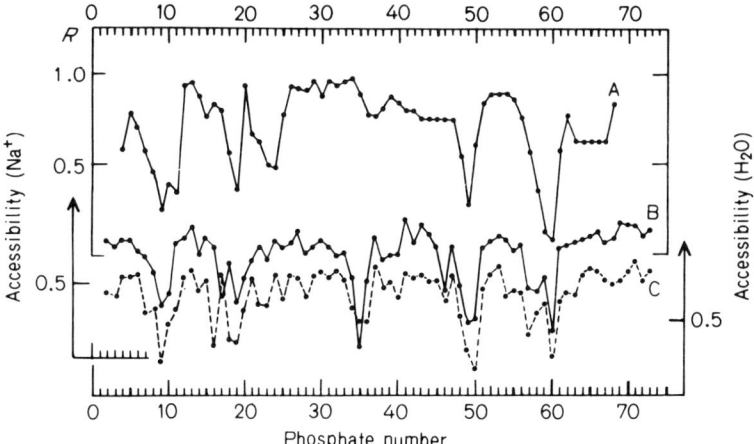

FIG. 12. Comparison of reactivities of phosphates in tRNA^{Phe} toward ethylnitrosourea (A) with calculated accessible areas of the anionic oxygens of the phosphates for Na$^+$ ions (B) and for water (C) in the tRNA structure. Reproduced from Ref. 32.

The determination of positions where the phosphodiester backbone of the RNA is on the inside or on the outside of the molecule provides constraints for modeling the three-dimensional structure of the ribozyme. It was found that the overall tertiary structure of this RNA forms cooperatively with the uptake of at least three magnesium ions, and that the high-order RNA foldings produced by Mg, Ca, and Sr ions are similar. Also, local folding transitions display different metal-ion dependencies, suggesting that the RNA tertiary structure assembles through a specific folding intermediate before the final structure is formed. The Fe(II)–EDTA cleavage was also used for probing the structures of mutated *Tetrahymena* ribozymes and to explore the role of individual structural elements in the tertiary folding of the RNAs. The results have allowed identification of different mutations that destabilize folding of the RNA and shift the optimal conditions of folding of the catalytic core to higher $MgCl_2$ concentrations (43).

Quantitative characterization of the reactivities of nucleotides toward a chemical agent when the chemistry is well-known provides high-resolution data on aspects of RNA structure such as the involvement of specific bases in stacking, local distortion of double helices, or even variations of the parameters of the RNA helices. Studies on RNA modification with DEPC have established a correlation between DEPC reactivity that reflects the accessibility of the major groove and the presence of either true or effective helix ends (38). In unstructured RNA, reaction with DEPC yields an even pattern of cleaving at each purine position, with adenosine being more reactive than guanosine. Purines in the uninterrupted helix are essentially unreactive toward DEPC.

Major groove inaccessibility is a result of the close approach of the phosphoribose backbone for helices of six or more base pairs. The minimum distance between phosphates across the major groove is approximately 10 Å for 7 bp, yielding a 4-Å groove width. This size of groove is not sufficient for the reagent to approach the reactive centers in a proper way. When the regularity of the helix is interrupted at the duplex termini, accessibility can be increased.

In the RNA duplex in the folded RNAs, susceptibility of each of the purines to the modification is strongly affected by position relative to helix termini, bulge, or internal loops. Asymmetric internal defects disrupt stacking regularity more than symmetric loops. Strongly coupled helices incorporate interhelix defects into a regular helix stacking geometry and are inaccessible to DEPC. A single-nucleotide bulge enhances accessibility in the major groove only modestly. Reactivity toward DEPC increases smoothly near helix termini and adjacent to the mentioned internal defects of the helix. Positions around large bulges are readily modified by DEPC. The accessibility extends one or two nucleotides further in the 5′ direction relative to a loop (equivalent to the 3' helix end) than in the 3' direction. However, the most reactive position is the purine 3' to the loop (equivalent to the 5' helix end), which is only half as accessible as the purines in the single-stranded RNA.

These differences in reactivities of purines can be explained by the geometry of the RNA helix. In the duplex terminus, the distance between phosphates across a single base-pair is 18 Å, which means a 12-Å groove width. Bases in an RNA duplex are tilted approximately 19° relative to the helix axis. Therefore, bases near the 3' end of the helix protrude from the major groove envelope, which enhances their accessibility (38). The observed stronger modification at the 5' base can be attributed to differences in stacking interactions, which makes the 5' base relatively more accessible.

3. Cross-linking and Intramolecular Modifications

An efficient experimental method for determining structural relationships between different parts of an RNA chain is chemical cross-linking. Sometimes cross-linking can be obtained by direct photoactivation of juxtaposed residues of RNA. In two cases, the exact chemical nature of the cross-links has been determined. Thus, in the case of bacterial tRNAs containing a s^4U residue in position 8, photoinduced cross-linking[2] between this minor nucleoside and cytidine C-13 occurs under irradiation with UV light (330 nm) (52, 53). The reaction is facilitated by a favorable relative orientation of the two residues in the tRNA structure. Another example is the UV-induced formation of the C48-U59 cyclobutane dimer in yeast tRNA[Phe] (54). In the folded structure of this tRNA, the two pyrimidine rings are adjacent to one another in the folded structure and their 5,6 double bonds are nearly parallel and juxtaposed. This allows efficient cross-linking to occur following irradiation with short-wavelength UV light. The efficiency and kinetics of these cross-linking reactions provide a simple approach for comparison of the state of structure around the reacting residues in different tRNAs and the effect of different conditions on tRNA structure. Experiments with different tRNAs show that the reaction is very sensitive to structural changes involving the nearby pyrimidines and therefore can be used for analysis of conformational state (53).

Cross-linking of proximal groups in RNA structure can be achieved by use of bifunctional chemical reagents (Fig. 13). Bifunctional reagents of variable sizes can be used as "molecular rulers" for the identification of groups at specific locations in an RNA. Bifunctional reagents that have been used for probing RNA structure include derivatives of psoralen (55, 56), N-acetyl-N'-(p-glyoxylylbenzene)cystamine (56), and bis-(2-chloroethyl)methylamine (57,

---

[2] See essay by E. I. Budowsky and G. G. Abdurashidava in Vol. 37 of this series [Eds.].

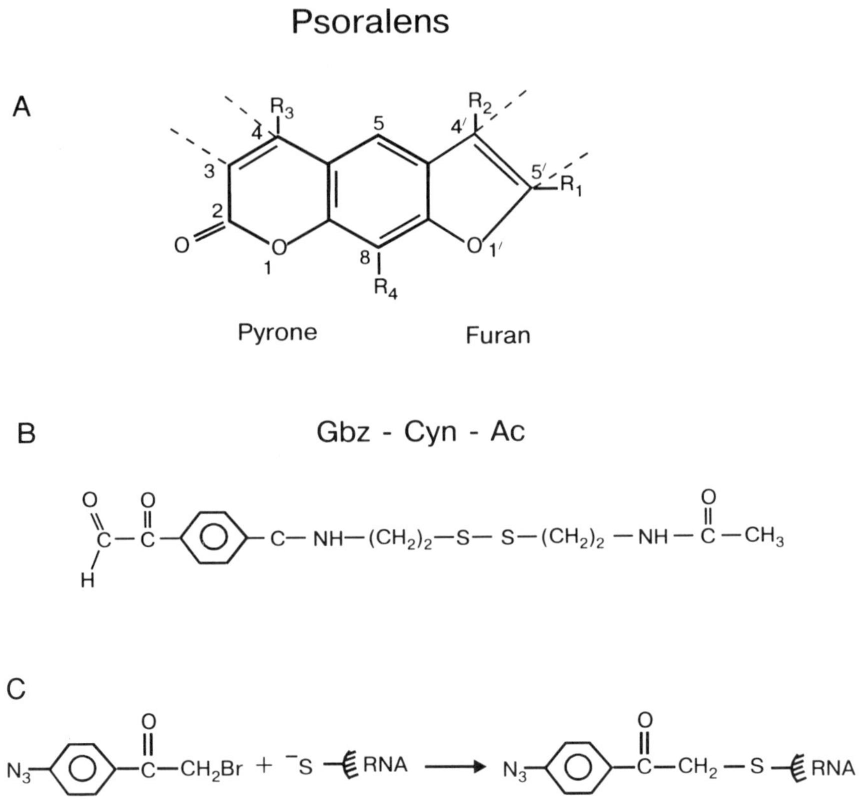

Psoralens

A

Pyrone — Furan

B Gbz - Cyn - Ac

C

FIG. 13. Bifunctional reagents for cross-linking RNA. (A) Psoralens: 4,5′,8-trimethylpsoralen, $R_1 = R_3 = R_4 = CH_3$, $R_2 = H$; 4′-(hydroxymethyl)-4,5′,8-trimethylpsoralen, $R_1 = R_3 = R_4 = CH_3$, $R_2 = CH_2OH$; 8-methoxypsoralen, $R_1 = R_2 = R_3 = H$, $R_4 = OCH_3$. (B) N-Acetyl-N′-(p-glyoxylylbenzoyl)cystamine (Gbz-Cyn-Ac). (C) Attachment of a photoreactive group to the phosphorothioate at the 5′ terminus of an RNA transcript (66).

58). This latter reagent can alkylate heterocyclic bases within the same polynucleotide chain or bases in two juxtaposed chains, when the distance between the reactive centers is less than 12–15 Å. The reagent shows little sequence specificity; it can cross-link residues located in single-stranded and double-stranded regions and form cross-links that are stable during analysis. An example of application of the reagent is the investigation of the structure of the small nuclear RNAs U1 and U2 (57, 58). In both RNAs, formation of two intramolecular cross-links was observed between residues for a part in the primary sequences of the molecules. The identification of the positions of the reactive residues was performed using the following procedure. After the reaction, individual forms of the cross-linked RNAs were isolated by electrophoresis. The RNAs were 3' end-labeled and the position of the modified residue closest to the 3' end was identified by comparing the products of partial enzymatic digests of the modified and the intact RNAs. Similar experiments with the 5' end-labeled RNAs allowed localization of the second modified residue, which is closer to the 5' end of the molecule. These studies have provided the information needed for reconstruction of the tertiary structures of the U1 and U2 RNAs.

Derivatives of psoralen are widely used for photocrosslinking of nucleic acids (59). These three-ring heterocyclic compounds (Fig. 13) intercalate into double-stranded regions of RNA; on irradiation with UV light (365 nm), they undergo a photochemical addition to heterocyclic bases of RNA located at a distance of about 8 Å. The intercalated psoralen attaches covalently to pyrimidine nucleosides, especially to uridine, by cyclobutane linkages to one nucleotide, producing a monoadduct, or to two nucleosides, producing cross-links. On activation with UV light, either the 3,4-pyrone double bond or the 4',5'-furan double bond of psoralen photoreacts with the 5,6 double bond of a pyrimidine to form a pyrimidine–psoralen monoadduct. Reaction of the 3,4-pyrone double bond destroys the coumarin nucleus and leads to monoadduct formation. If the reaction occurs with the 4',5'-furan double bond, the coumarin nucleus of the compound remains intact and can absorb light at 365 nm for its 3,4-pyrone double bond to react with the 5,6 double bond of another pyrimidine to form a cross-link.

Psoralen derivatives prefer to react with uracils near internal loops but not in loops or within perfect double helices (60). RNA molecules cross-linked by psoralen can be fractionated by gel electrophoresis in denaturing conditions. Although psoralen does not modify positions participating in Watson–Crick pairing, the cyclobutane adduct terminates reverse transcription because of the change in the uracil geometry. Therefore the sites of cross-links can be determined by primer-extensions with reverse transcriptase.

An example of cross-linking with psoralen for structural studies is the investigation of secondary structure of the SP6/mouse insulin precursor RNA (55). The RNA was treated with psoralen under conditions providing statistically less than one cross-link per RNA molecule, and individual fractions of the cross-linked RNA were isolated by gel electrophoresis in denaturing conditions. These RNAs were used as templates for reverse transcription to identify the cross-linking sites. A series of long-range contacts were detected within the 5'-half of the pre-mRNA that contains the intervening sequence. Because some of the interactions showed common sites, it was concluded that the RNA exists as a mixture of conformers. The pre-mRNAs with psoralen cross-links in different positions were used as substrates for *in vitro* splicing, and it was found that psoralen cross-linking of nucleotides in any of the double-stranded regions of RNA inhibited splicing, suggesting that a destabilization of secondary structure of the RNA precursor is required for splicing to occur *in vitro*. It was concluded that a melting of double-stranded regions within the pre-mRNA occurs before the endonucleolytic cleavage.

Cross-linking with psoralen has been used for investigation of the structure of 16-S RNA (56). Cross-linkage maps were generated for isolated 16-S RNA and for 16-S RNA within a 30-S ribosomal subunit. It was concluded that in both cases the RNA has equivalent regions of secondary structure. The data of cross-linking with different reagents were used for building a detailed model of 16-S RNA (61).

Cross-linkable groups can be introduced artificially in the single-stranded regions of RNA. This technique uses as a reagent N-acetyl-N'-(p-glyoxylylbenzoyl)cystamine (Gbz-Cyn-Ac; Fig. 13). This compound reacts with accessible guanosine residues in the single-stranded regions of tRNA by its glyoxal group, as shown in Fig. 6 for α-ketoaldehyde compounds. The adducts are stabilized by oxidation of their *cis*-diol groups to form N-acylguanosine derivatives; the disulfide bond of the derivatives is then reduced with sodium borohydride. After the reduction, each derivatized guanosine carries a free SH group (62). Treatment with hydrogen peroxide leads to formation of disulfide bonds between the modified guanosines, which are within a distance of approximately 17 Å. Cross-linking achieved by using this procedure allows identification of guanosines in single-stranded regions of RNA that are near one another in the folded molecule.

Identification of cross-linking sites is simplified considerably when a bifunctional reagent is attached in a specific position of an RNA. This approach is less general, but it allows a detailed study of the geometry of a specific part of an RNA. The technique was tested first on tRNAs. An aromatic 2-chloroethylamine residue, chlorambucil, was attached to the amino group of the amino-acid residue in aminoacylated yeast tRNA^{Val} (63). Intramolecular alkylation in this modified tRNA occurred within its acceptor stem: at the 5' phosphate and at residues of the CCA end in accordance with the solution structure of tRNA in which the CCA stem does not contact other parts of the molecule.

The approach was developed further in experiments with tRNA^{Phe}, to which a chlorambucil residue was conjugated via linkers of different lengths (64). A rigid linker in the constructs of the general formula chlorambucil-(prolyl)$_n$-[^{3}H]phenylalanyl-tRNA^{Phe} allowed variation of the probing group by changing the number of the prolyl residues. In the constructs with maximal length of the "molecular ruler" ($n = 15$, the distance between the 3' end of the molecule and the alkylating group is 62 Å), intramolecular alkylation of guanosine G20 in the D loop (located 60 Å from the 3' end of the molecule, according to the X-ray data) and rare nucleoside W (wyosine) in the anticodon was observed. The results are consistent with the known parameters of the tRNA structure.

Probing the tertiary structure of RNAs using autocleavage by the Fe(II)–EDTA group attached to a specific position has been tested in experiments with the yeast tRNA^{Phe} (65). The modified molecule was constructed by chemical incorporation of an EDTA-linked uridine into the 3' half-fragment of the tRNA at position 47. This modified 3' half of the tRNA was ligated enzymatically to the 5' half of the tRNA by T4 DNA ligase. The produced molecule was cleaved by lead ions (this cleaving probe is in Section I,D,4) similarly to the intact tRNA^{Phe}, which indicated that the uridine modification did not disturb folding of the molecule. Autocleavage of the molecule by the tethered group in the presence of Fe(II) and a reducing agent produced a set of fragments that were in general agreement with the three-dimensional structure derived from X-ray analysis. Because the cleavage is produced by diffusing species and because not every ribose at a fixed distance has identical reactivity toward the radicals, quantitative characterization of reactivities of individual riboses was not possible. However, for large RNAs the low-resolution information available from such experiments may be sufficient to discriminate between different structural models.

A method has been elaborated for the attachment of a photoactivatable cross-linking agent, the azidophenacyl (APA) group, to RNA molecules (66). To prepare the 5' end-labeled RNA, the RNA transcript was prepared using T7 RNA polymerase in the presence of guanosine monophosphorothioate. Inclusion of guanosine monophosphorothioate in a transcription reaction results in its incorporation only at the 5' end of the transcripts, because nucleoside monophosphates can initiate transcription but cannot be incorporated in the growing RNA chain. The phosphorothioate provides a unique site in the RNA for the conjugation of the APA residue (Fig. 13). This method was used first for conjugation of APA to the 5' terminus of tRNA, to prepare an affinity reagent for determination of sites in RNase P RNA that interact with tRNA substrate. Later it was shown that attachment of the APA group to specific sites in any part of RNA molecules can be achieved by tethering the group to the 5' terminus of circularly permuted RNA analogs (cpRNAs) (66, 67).

This methodology has been used for investigation of the three-dimensional structure of the catalytically active RNA component of ribonuclease P from *Escherichia coli* involved in tRNA maturation and for investigation of the tertiary structure of the RNase P RNA complexed to the tRNA^{Asp}.

The circularly permuted RNA analogs of RNase P were molecules represented by RNase P RNA with 5' and 3' ends connected with a nonnative

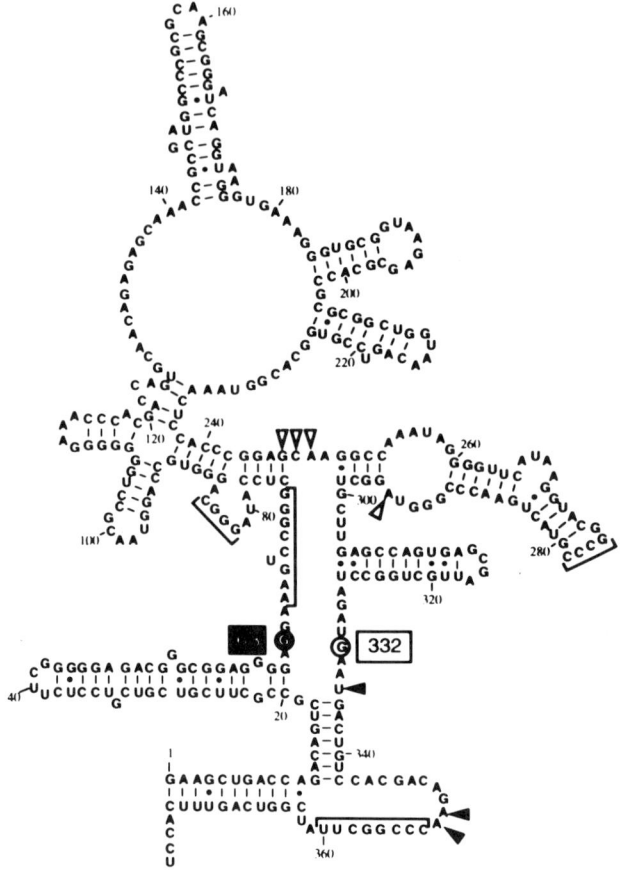

FIG. 14. Mapping of cross-links in circularly permuted RNase P RNA analogs. Reproduced from Ref. 68, M. E. Harris, J. M. Nolan, A. Malhotra, J. W. Brown, S. C. Harvey and N. R. Pace, *EMBO J.* **13**, 3953 (1994), by permission of Oxford University Press. Filled arrowheads indicate the positions of nucleotides attacked by the reactive group attached to the nucleotide shown by the filled circle. Open arrowheads indicate the residues modified by the group attached to the residue indicated by the open circle.

oligonucleotide linker (Fig. 14). The molecules contained discontinuities in the ribose phosphate backbone. Positions of the discontinuities were dictated by the DNA templates from which the RNAs were transcribed by T7 RNA polymerase. The end points were the specific photoreagent attachment sites for intramolecular cross-linking. Several cpRNase P RNAs with photoreactive groups located in different positions of the structure were prepared. The modified RNAs retained catalytic activity comparable to that of the natural RNase P RNA, proof that single interruptions introduced in the phosphoribose backbone and attachment of the APA did not alter significantly the RNA structure.

The modified RNAs were subjected to irradiation with UV light to convert the azido groups to nitrenes and produce cross-links. The cross-linking resulted in formation of "lariats," which were separated from the noncross-linked RNAs by gel electrophoresis. The particular nucleotides cross-linked to the 5' ends of the cpRNase P RNAs were determined by the primer-extension method. Investigation of the cross-linking has allowed determination of orientation and distance constraints between elements in the RNase P RNA and within the RNase P RNA–pretRNA complex. The cross-linking data together with the established secondary structure of RNase P RNA and the tertiary structure of tRNA were used with a molecular mechanism protocol to develop a model of the global structure of the core of the RNase P RNA–pretRNA complex (68).

Reactive groups can be introduced in any selected position of RNA by means of affinity modification with reactive derivatives of corresponding complementary oligonucleotides (69). RNA is alkylated with oligonucleotide derivatives bearing an aromatic 2-chloroethylamine at the terminal phosphate (Fig. 15). After the reaction, the modified RNA is incubated in mild acid conditions in which the phosphoramide bond between the reactive group and the oligonucleotide is hydrolyzed. The result is that a residue with aliphatic amino group is introduced in a specific position of the RNA structure. These groups can be reacted with the bifunctional reagent 2,4-dinitro-5-fluorophenylazide to attach the photoreactive azido group.

4. PROBES SENSITIVE TO SPECIFIC ELEMENTS OF RNA FOLDING

Some probes allow testing the state of the global structure of RNAs. One example is the photocrosslinking between pyrimidines in specific positions of some tRNAs (52–54), which occurs only in the native tRNA structure and can serve as a test for maintenance of the biologically active conformation. Another example of such probes is represented by cleavage of RNA with certain metal ions. Scission of RNA by coordinated metal ions is a simple and sensitive test for detection of the cation-binding regions and for probing the state of the RNA structure. Thus, highly specific hydrolysis of some tRNAs by Pb^{2+} occurs due to the presence of tight metal-binding sites in the RNA (70–73). The cleavage results in the formation of 2':3'-cyclic-phosphate and 5'-hydroxyl termini. Cleavage of $tRNA^{Phe}$ between nucleotides U17 and G18 was a sensitive way to identify and correctly position the two lead-coordinated pyrimidines. Nucleotide substitutions that disrupted the tertiary interactions of $tRNA^{Phe}$ reduced the rate of cleavages dramatically. This

FIG. 15. Attachment of reactive groups to arbitrary sites in RNA by means of affinity modification with derivatives of complementary oligonucleotides. (A) Schematics of the procedure. (B) Chemical reactions used for attachment of photoreactive groups to specific guanosine residues in RNA.

cleavage reaction has been exploited as a sensitive probe for the tertiary folding of RNA variants (71, 72).

An *in vitro* selection method has been developed to obtain RNA molecules that specifically undergo cleavage by Pb^{2+} ions (74). This selection method was applied for identification of different RNA motifs sensitive to cleavage by Pb^{2+} ions (75). The ability of tRNA^{Phe} to undergo a specific cleavage in the presence of Pb^{2+} ions was also used as a selective pressure for isolation of RNA molecules having a core part similar to that of natural tRNA^{Phe}. In these experiments, tRNA molecules with the anticodon hairpin replaced by some artificial sequences were constructed. From the rates of the site-specific cleavage by Pb^{2+} and the formation of specific UV-induced cross-links, it was concluded that certain tetranucleotide sequences can allow proper folding of the rest of the tRNA molecule (76).

The 5-S RNAs from a few bacterial species have been characterized by Pb(II)-induced hydrolysis. Investigation of the cleavages has allowed a refinement of the secondary structure model of 5-S RNA. The effect of binding ribosomal proteins L18 and L25 to the *E. coli* 5-S RNA on RNA cleavage was also investigated. Besides the shielding effect of the bound proteins, a highly enhanced cleavage in the RNA, between A108 and A109, was detected. This finding has supported the concept that the major L18-induced conformational change involves portions of helices A, B, and D of the RNA (77).

Cleavage of the RNase P RNA with different metal ions has been investigated in detail (78). A number of cations hydrolyze the RNA and five preferential cleavage sites have been characterized. Pb^{2+}-induced hydrolysis was suitable to sense different conformations of RNase P RNA (79). Good correlation of susceptibility to Pb^{2+} cleavage with catalytic activity was shown for the *T. Thermophilus* RNase P RNA under activity-assay conditions. This allows use of the test for studying conformation states of the RNA, to probe enzyme–substrate complexes, and to evaluate different salt and temperature conditions in reactions catalyzed by RNase P RNAs. The Pb^{2+}-cleavage assay was also applied to probe the tertiary structure of mutant RNase P RNAs. RNase P RNAs from three phylogenetically disparate organisms, *Chromatium vinosum, Bacillus subtilis*, and a few mutants from *E. coli* with deletions, were studied. Investigation of the patterns revealed some regions of identical structure that provide evidence for several ubiquitous metal-ion binding-sites in eubacterial RNase P subunits (79). Two cleavage sites occur at homologous positions in all the native RNAs regardless of sequence variations, suggesting common tertiary structural features. Such conservation in structure suggests that these regions are involved in some specific role of the RNA, for instance in substrate binding or catalysis. The cleavage sites in four deletion mutants of *E. coli* RNase P RNA differed from the native patterns, indicating alterations in the tertiary structures of the mutant RNAs.

Some complexes of transition metals can shape selective photoinduced cleavage of structured RNAs (80). Tris(1,10-phenanthroline)ruthenium(II) [Ru(phen)$_3^{2+}$], tris(3,4,7,8-tetramethylphenanthroline)ruthenium(II) [Ru(TMP)$_3^{2+}$], tris(4,7-diphenyl-1,10-phenanthroline)ruthenium(II) [Ru(TMP)$_3^{2+}$], tris(4,7-diphenyl-1,10-phenanthroline)rhodium(III) [Rh(DIP)$_3^{3+}$], and bis(phenanthroline)(9,10-phenanthrenequinonediimine)rhodium(III) [Rh(phen)$_2$phi^{3+}] are complexes that bind to RNA at some sites matching their shape; on photoactivation, they induce RNA strand scission, thereby marking sites of specific structural features.

Cleavage of a few complexes has been assayed on yeast tRNA[Phe] and a distinctive diversity in site-selective cleavage was shown. The RNA was irradiated with UV light in the presence of the complexes and then subjected to aniline treatment. Reactions with Ru(phen)$_3^{2+}$ and Ru(TMP)$_3^{2+}$ resulted in cutting preferentially at guanosine residues and formation of RNA fragments with terminal 5′ and 3′ phosphates. In these cases, the proposed mechanism of the reaction was attack on the nucleic acid base in a reaction mediated by singlet oxygen generated by photoexcitation of the ruthenium complex.

A different cleavage chemistry was observed for rhodium complexes. No preferred base composition for the attack was observed and aniline was not required for fragmentation. It was concluded that, in this case, the photoinduced cleavage occurs through a direct oxidation path and the target of the reaction is the RNA sugar. Different patterns of cleavage were observed for complexes with different ligands, which apparently reflect differences in their binding characteristics governed by their different molecular shapes. The rhodium complexes demonstrate a pronounced preference for some sites in the central part of the RNA. This structural preference was governed not by the cleavage chemistry, but rather by the presence in this part of appropriate binding sites fitting the shape of the compounds. $Rh(DIP)_3^{3+}$ induces strong cleavages at residues $\Psi 55$ and C70 with other weaker sites present at T54 and C56. The most interesting probe, $Rh(phen)_2phi^{3+}$, induces strong cleavages at residues G22, G45, U47, and U59. Under denaturing conditions, these sites were relatively unreactive, suggesting that the complex binds in the folded molecule to a unique region of RNA organized by parts of the D stem, T loop, and variable loop. The unusual reactivity pattern is consistent with recognition of a widened RNA A-like helix distorted by local formation of a base triplet that is open to permit intercalation by the bulky complex. $Rh(Phen)_2phi^{3+}$ was suggested to be a potential shape-selective probe targeting triple binding sites in RNA. This and other complexes may become useful for deducing tertiary structure features of RNA molecules.

II. Computer Analysis of the Secondary Structure of RNA

The problem of predicting secondary structure from nucleotide sequence dates back to the 1960s *(81)*. The question was how to determine a folding that provides the largest number of complementary bases. As the number of known RNA sequences increases the matter of prediction of secondary structure becomes of particular concern.

Up-to-date algorithms relevant to the subject fall into three basic groups: (1) those searching for the lowest energy secondary structures with the use of the thermodynamic parameters (thermodynamic approach); (2) those determining RNA secondary structure by simulation of folding (kinetic approach); (3) those searching for invariant secondary structures by comparing sets of homologous sequences (comparative approach).

The thermodynamic approach has been well-described (e.g., see Refs. *82–84*). Thus we will only outline it and mention its recent modifications. What we focus on are the kinetic and comparative approaches to the prediction of RNA secondary structure. The statistical aspects of organization and evolution of RNA secondary structure are discussed as well.

A. Thermodynamic Parameters of RNA Secondary Structure

Understanding the RNA folding code requires determination of its thermodynamic parameters. It is equally important to know both the parameters of the helices and the parameters of loops of any kind. The respective contributions of helices and loops to structure stability are opposed (82–84): the formation of a helix lowers free energy, and the closing of a loop usually increases free energy owing to entropy losses. The resultant stability depends on the balance between these two opposing factors.

Helix stability strongly depends on the ratio of A–U to C–C. However, it was shown as early as 1963 (85) that more precise estimates of helix stability require accounting for the sequence of bases in the RNA primary structure. This effect is due to the contribution of stacking interactions between the neighboring base-pairs within the helix that exceeds that of H bonds (86). That is why helix energy is evaluated under the nearest-neighbor model (87). According to the model, helix energy is postulated to be the sum of the energies of nearest complementary pairs within the helix.

Traditionally the thermodynamic parameters of neighbor pairs are estimated from melting experiments on duplicates of short synthetic oligonucleotides of varying base content (88). Spectroscopic and calorimetric methods are used (87). Thermodynamic parameters are inferred from the experimental data by using a two-state model. Under this model, only the completely melted or helical states are considered, the intermediate pairing states being ignored (87). Optical and calorimetric data are identical unless the model falls short of applicability. Today there is a great variety of sets of thermodynamic parameters of canonical complementary pairs in the RNA helix with different neighborhoods (82, 88–93). Table I presents the commonly used compilation of these parameters (88). Besides, thermodynamic parameters have been identified for the nearest-neighbor model of the non-canonical G·U pairs (82, 89), as well as the stabilization parameters of the nucleotides adjacent to the helix (82).

Despite unquestionable attainments, the errors of determination of the thermodynamic parameters (entropy and enthalpy) are still high. Free-energy assessments are much more accurate because the errors of entropy and the errors of enthalpy to some extent compensate for each other. The errors result from the limited applicability of the statements used in assessment of the parameters, and a rather broad melting profile of short helices. Stacking interactions in single-stranded oligonucleotides may also contribute to the error (94, 95).

TABLE I
FREE ENERGY INCREMENTS FOR RNA HELIX PROPOGATION[a]

Propagation sequence	AA→ UU←	AU→ UA←	UA→ AU←	CA→ GU←	CU→ GA←	GA→ CU←	GU→ CA←	CG→ GC←	GC→ CG←	GG→ CC←
ΔG°_{37} (kcal mol^{-1})	−0.9	−0.9	−1.1	−1.8	−1.7	−2.3	−2.1	−2.0	−3.4	−2.9

[a] In 1 M NaCl (82).
Reproduced, with permission, from the *Annual Review of Biophysics and Biochemistry*, Volume 17, © 1988, by Annual Reviews Inc.

The thermodynamic parameters of loops are also determined from calorimetric and spectroscopic measurements (87). Loop parameters are considerably less studied than those of helices. Studies on multiloops are few (96, 97). Little is known of the effect the nucleotide context has on loop energy. Calculations are based on averaged data not dependent on nucleotide context. The data on very stable loops, the so-called tetraloops (98, 99), are the only exception. Experimental estimates for loop free energy have been obtained only for those loops not longer than n_{max}, where n_{max} = 5, 6, and 9 for bulge, internal, and hairpin loops, respectively (92). For longer loops, the following approximation of free energy changes is used:

$$G(n) = G(n_{max}) + \alpha RT \ln(n/n_{max}),$$

where α is a parameter depending on which model of polymer chain is used [α = 1.5 for a phantom chain (100)].

Table II[3] presents a compilation of loop free energies (82). There is a continuous updating of thermodynamic parameters for helices and loops; to highlight details, a special review is required. The available thermodynamic parameters of helices and loops allow prediction of low-energy secondary structure, their statistical analysis, and the simulation of RNA folding.

B. Thermodynamic Approach

The thermodynamic approach is based on the assumption that the native secondary structure of RNA is the lowest energy form or one of the suboptimal forms of secondary structure of this molecule. In the programs implementing this approach (82, 83, 91, 92, 101–107), the free energy of RNA is calculated by use of the above-described thermodynamic parameters characterizing helix and loop formation. Two main directions of searching for low-energy secondary structure have been suggested: combinatorial and recursive (82).

[3] Table II is on page 188.

As has been noted (108), the number of possible RNA secondary structures increases with sequence length N as $1.85^N/N^{3/2}$. Hence, any algorithm for predicting low-energy secondary structure is faced with a problem: how to examine all the possible secondary structures, because the sequences of just about 100 nucleotides in length produce a very large number of potential secondary structures.

A quantitative prediction of RNA secondary structure by searching for the lowest energy secondary structure was pioneered by Tinoco et al. (109). Thermodynamic parameters were first used to predict the secondary structure of the short fragment of R17 RNA. The change of free energy was taken as -1.2 kcal/mol for the formation of A·U pairs, and -2.4 kcal/mol for G·C pairs. The change of free energy for loop formation was also taken into account. The lowest energy secondary structure was determined by trying all the complementary pairs obtained from complementarity matrix analysis. Further development of new methods for the prediction of RNA secondary structure and for the determination of the thermodynamic parameters of secondary structure was quite explosive, and resulted in a great variety of methods for the prediction of low-energy secondary structure (101–107, 110–112, 114–118).

An efficacious step was made when it was decided to take a helix, not a pair of complementary nucleotides, as an elementary object for analysis (101). The routine included (1) the research for the longest potential helices, (2) analysis for their compatibility by applying stereochemical constraints (Fig. 16), (3) reconstruction of the secondary structure from compatible helices, and (4) calculation of secondary structure energy in accordance with thermodynamic parameters. The first rule (Fig. 16A) forbids the concurrent presence of any RNA fragment at two helices; the other disallows pseudoknot formation (Fig. 16B). The time of calculation depends on the nucleotide

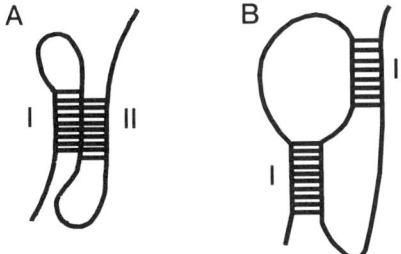

FIG. 16. Rules of stereochemical compatibility for helix pairs in the RNA secondary structure. (A) Ban on the concurrent presence of an RNA fragment in two helices; (B) ban on pseudoknots.

sequence length N as $N2^N$, which implies that the algorithm would not be applicable if sequences were longer than 70–80 nt. It was this algorithm that gave a start to the combinatorial approach to determination of the low-energy secondary structure of RNA by trying stereochemically compatible combinations of helices.

Use of the two rules of stereochemical compatibility accelerates examination of secondary structure remarkably, which is why they are now commonly exploited and form part of many later algorithms. However, they do not always reflect the features of real secondary structure. For example, if the termini of helices are partly open, two shorter nonoverlapping helices may exist (in violation of Rule 1).

This constraint was initially forestalled by regarding a "sliding" boundary between competing helices (*102*). Besides, examination of secondary structure was optimized by using routines of graph theory, which reduced time consumption to N^5 and allowed analysis of sequences of up to 150 nt in length.

Rule 2 (no pseudoknotting is allowed) is widely used in the most effective current means of secondary structure determination, the recursive algorithms (*103–107*). Pseudoknots are not allowed in these algorithms, which restricts their applicability. How to handle this restriction is described in Section II,E, where pseudoknot formation in RNA is discussed.

The key stage of secondary structure prediction by application of graph theory should be the construction of a graph of stereochemical compatibility, in which a helix is related to a vertex, two compatible helices being connected by an edge (*110, 111*). Finding low-energy secondary structure is equivalent to finding cliques. The method is good for prediction of secondary structure for RNA molecules up to 150 nt in length.

Another algorithm (*92, 112*) for prediction of low-energy secondary structures applying graph theory was based on the well-known method of branches and boundaries. This method was applicable to RNAs up to 200 nt in length.

One of the advantages of the combinatorial algorithms is that it is possible to determine not only the lowest energy secondary structure, but also a set of suboptimal secondary structures. What makes the combinatorial approach difficult is the necessity of examining an enormously large number of structures, which is rapidly rising with RNA nucleotide length. This was what encouraged the development of the currently most effective and rapid methods for RNA low-energy secondary-structure prediction by the recursive approach.

The recursive algorithms (*102–105, 107*) use the ideas of dynamic programming. A recursive approach was initially applied in 1966 (*113*) to determine the RNA secondary structure with the maximum number of complementary pairs. As there were no thermodynamic parameters of RNA available at that time, further development of the method was suspended. Later use of the recursive algorithm for maximization of number of complementary pairs was independently suggested and mathematically substantiated (106). Further development resulted in the first recursive algorithm for determination of the lowest energy secondary structure of RNA, using thermodynamic parameters (103).

The basic principle of the algorithm can easily be understood from the following example. Let us define a structure with the maximum number of pairs. Let the nucleotide sequence be presented as a circle (Fig. 17), with the possible complementary pairs as the arcs connecting the respective nucleotides. Now consider a circle section B_xB_y of length p and define the maximum number, $M(x, y)$, of pairs in it:

$$M(x, y) = \max \begin{cases} M(x, k-1) + M(k+1, y-1) + 1, \\ M(x, y-1) \quad x \le k < y = x + p. \end{cases} \quad (1)$$

Increasing x and y, we try all the sections of the entire sequence. Then the routine is iterated on the p values with an increment of 1. It is important that $M(x, y-1)$, $M(x, k-1)$, and $M(k+1, y-1)$ from Eq. (1) be evaluated at the previous step, thus providing easy calculation of the matrix element $M(x, y)$. Calculations terminate when the section B_xB_y corresponds to the entire sequence B_1B_n. By then, a matrix K has been filled. The element K_{ij} is the count number of the nucleotide that, when paired with the nucleotide B_j, provides the optimal folding of the sequence B_iB_j. The structure with the maximum number can readily be deduced from the matrix K.

The algorithm for searching for the structure with minimum energy is in

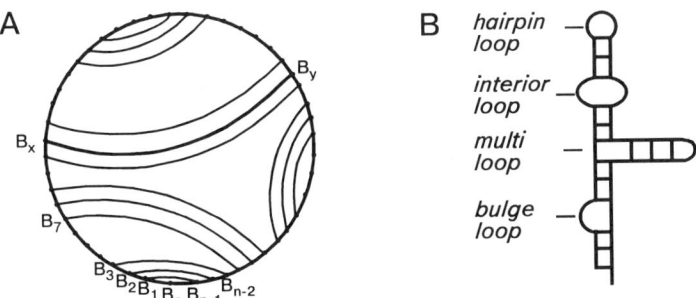

FIG. 17. Base-pairing in the planar secondary structure of an RNA. (A) Extended form; complementary base pairs are represented by arcs; nucleotides are designated by dots. (B) Condensed form of RNA secondary structure (103).

general the same, except that it is aimed at the lowest energy of an RNA section, not at the number of complementary pairs.

The same idea of recursive approach is exploited elsewhere. The suggested and widely used variants of the recursive algorithms (104) take account of the energy of helices as calculated by the nearest-neighbor method and the destabilizing energy of loops in accordance with thermodynamic parameters—those mentioned above and a range of others (82). On the whole, the works mentioned have given rise to a wide range of current recursive methods for determination of RNA secondary structure (105, 107, 114–117).

The main advantage of the recursive algorithms is that they are fast (82). Consumption time depends on sequence length as N^3; the secondary structure has been predicted for sequences up to 4217 nt in length (118). A remarkable advantage of the algorithms is that they easily make use of additional biological information, in particular the experimentally determined location of some nucleotides in single-stranded or double-stranded regions of the secondary structure (104). Thus, in this algorithm the thermodynamic parameters can be combined with various types of experimental data, which notably raises prediction accuracy. The common drawback of the recursive algorithms is that pseudoknots are not allowed in them, whereas they are a feature of real RNA secondary structure.

The characteristic feature of dynamic programming methods is that each step of analysis gives the only optimal structure. As a result, the initial versions of the above recursive algorithms produced only one optimal energy secondary structure. At the same time, the lowest energy secondary structure may not necessarily be the only functionally significant secondary structure for a given RNA molecule. There may also exist an ensemble of functionally important alternative secondary structures in dynamic equilibrium as, for example, the attenuators (119). It is also of importance that the RNA functioning in the "cellular context" interacts with other macromolecules (RNA, proteins, RNP particles, etc.), which can result in RNA refolding.

Finally, it is noteworthy that all the thermodynamic parameters used for determination of secondary structure are estimated with considerable errors. With loops, the error sometimes ranges between 15 and 50% (120).

It therefore seems reasonable to find a set of secondary structures within a certain energy window rather than the lowest energy secondary structure. Thus the probability of revealing the native secondary structure can be enhanced. Accordingly, the initial algorithm for prediction of secondary structure (104) was modified to target suboptimal secondary structures. The modified versions of the algorithm (105, 114) can bring up secondary structures within the window of energy defined by the user. Base-pairing probabilities can be determined with the energy weights of optimal and suboptimal structures (Fig. 18), the melting profile, and the RNA molecule specific heat (121). All of it can be effectively achieved by applying another dynamic programming approach (122), which calculates the secondary structure partition function and the probabilities of substructures. The partition function describes completely the equilibrium ensemble of secondary structure. Base-pairing probabilities, the melting profile of the RNA molecule, and other equilibrium parameters are also evaluated through the partition function (Fig. 19). The time consumed by the algorithm is estimated as N^3.

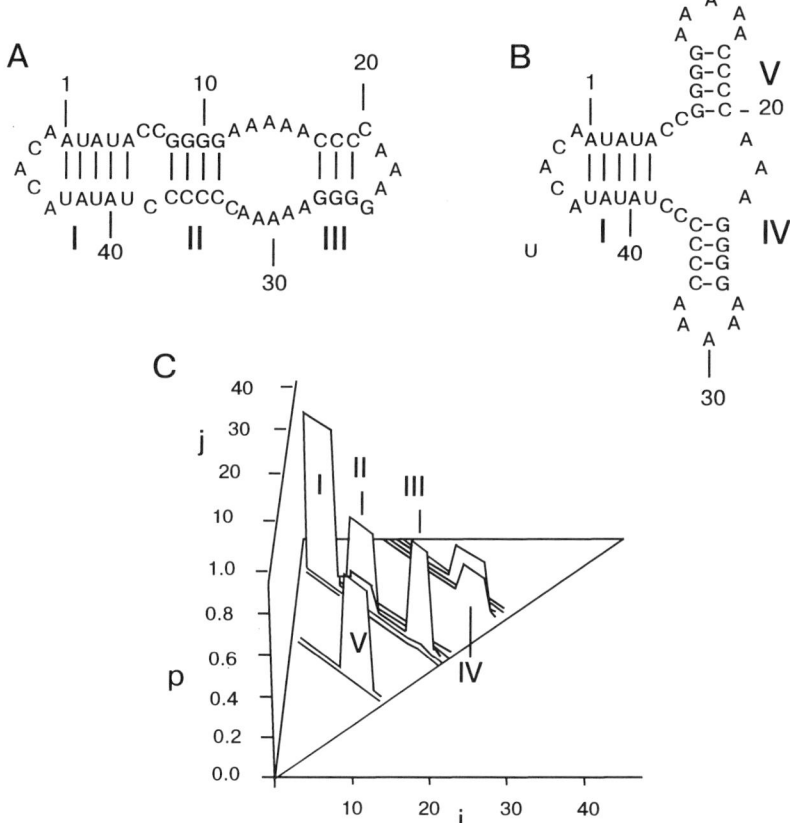

FIG. 18. Representation of the equilibrium distribution of secondary structures of a circular single-stranded RNA, 48 nucleotides long. Roman numbers I–V represent helices identical in each representation. (A) Optimal secondary structure at 35°C. (B) First suboptimal secondary structure. (C) Three-dimensional base-pairing plot at 35°C. Reproduced from Ref. 121, M. Schmitz and G. Steger, *CABIOS* **8**, 389 (1992), by permission of Oxford University Press.

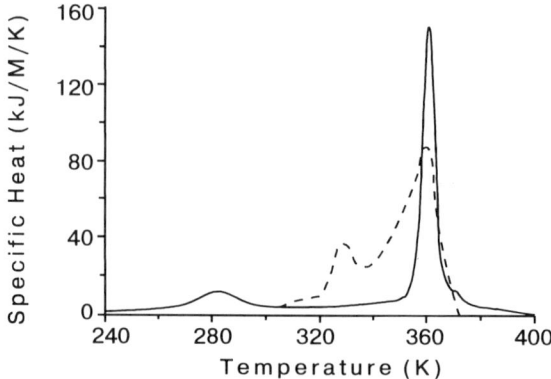

FIG. 19. The specific heat of *E. coli* 5-S RNA as a function of temperature [solid line represents the calculated curve (*122*), dashed line shows the experimental curve for the A form for 5-S RNA (*123*)].

C. Simulation of RNA Folding

The kinetic approach is based on the assumption that native secondary structure is the most kinetically attainable state of the RNA molecule and results from a multistage folding process (*124–126*). Levinthal was the first to deal with the problem of self-directed folding of a biopolymer into a unique spatial conformation (*127*). He noted that the total number of conformations of an N-monomer chain is g^N, where g is the number of stable conformations of the monomer. It is therefore clear that the native conformation of a long biopolymer cannot be achieved for any biologically reasonable time by randomly trying all the conformations. Initially, the idea was suggested for proteins (*128*). As first postulated (*129, 130*) and then demonstrated (*131, 132*), there must be a process underway such that the protein molecule passes a sequence of kinetically attainable states toward native spatial structure. This may also apply to RNA molecules. Presumably, during the formation of secondary structure the RNA molecule passes through kinetically attainable states with a gradual (in most cases) reduction in energy, certain elements of secondary structure being formed at each step (*133*). Generally, it is not yet clear if the secondary structure thus formed complies with the global minimum of energy.

The idea of secondary-structure prediction via folding simulation was initially realized in the simplest version (*134*). The suggested algorithm of predicting a secondary structure with the maximum number of complementary pairs was based on the step-by-step addition of new helices to the existing secondary structure. At each stage, there was added a helix with the maximum number of complementary pairs. A later algorithm (132) was based on the step-wise formation of secondary structure, the energy of which was calculated through thermodynamic parameters. At any step the structure would acquire such a helix from among all possible ones, whose formation would provide the maximum gain in free energy. Of a total of 80 tRNAs that had passed through the algorithm, 55% resulted in secondary structures of the "cloverleaf" type. A similar folding algorithm (133) used improved thermodynamical parameters. Its speed was N^2, which is much faster than the combinatorial and the recursive algorithms. Testing showed a good correspondence between the secondary structure predicted by this algorithm and the data on cleavage of RNA molecules by nucleases (133). Other algorithms have also been proposed for RNA secondary structure determination, on the basis of simulations of folding by step-wise addition of helices with the use of thermodynamic parameters (135–138).

Note that to add helices subsequently with the maximum gain in energy to secondary structure is in fact to select a way of folding characterized by maximum values of the equilibrium constants for the reactions of helix adding. Thus, under these algorithms, the travel of RNA through kinetically attainable states is substituted by a travel through local minima of energy.

In the mid-1980s an algorithm of RNA folding was suggested on the basis of the rates of helix formation and decay (124–126). Other algorithms of that kind have been implemented since (139–143). Under these algorithms, secondary structure formation is simulated by a Markovian random process. In the algorithm of Mironov and co-workers (124–126), the secondary structure transition from state to state leads either to the formation or decay of a helix. Which of the events will take place depends on the probabilities of the transitions, i.e., on the rates of the corresponding processes. It is assumed that the rate of decay of a helix of length N base pairs depends on its energy ΔG_h (144), which is calculated according to the nearest-neighbor model (87); the rate is given by Eq. (2):

$$k_d = Nk_0 \exp(-\Delta G_h/RT), \tag{2}$$

where k_0 ($= 10^{-6}$–10^{-8} sec^{-1}) is the rate of pairing of two bases adjacent to the helix (145). Note that the lifetime of a typical helix in the secondary structure is great. For example, Eq. (2) sets it at nearly one second for a helix of four G·C pairs at 37°C.

Helix formation is a kinetically controlled two-stage process. The rate-limiting stage is helix nucleus formation. First (after a spatial proximity has been achieved), a nucleus forms that is in fact one or several complementary pairs providing for sufficient stability of the intermediate structure. Second, helix formation proceeds rapidly by the "zipper" mechanism. That is why the rate of formation of a helix of length N is determined by losses of entropy

(ΔG_1) during the loop closure, which is calculated from tabulated values (*82, 89*):

$$k_f = Nk_0 \exp(-\Delta G_1/RT). \quad (3)$$

This way of determining the kinetic constants of helix formation or decay is the most reasonable because it satisfies the mass-action law.

Using the Monte Carlo method and Eqs. (2) and (3), secondary structure formation can be simulated for a time interval T. Multiple iterations would give the probability of certain secondary structures for the time $t < T$ (Fig. 20).

With this method, folding has been simulated for a number of tRNAs (*124, 126*), leader and intercistronic regions of mRNA transcribed from the *E. coli atp* operon (*124–126*), and the self-splicing YC4 intron of mitochondrial RNA of fungi (*141*). In this case (*141*), the experimentally predicted and estimated RNA secondary structures were in a good agreement. With the aid of the kinetic approach, a correlation was revealed (*124, 126*) between gene expression and RNA secondary structure. The correlation was found (*140*) between the refolding events of the RNA growing chain and the location of replication pause sites in MDV-1 RNA (*146*).

As is seen from Eqs. (2) and (3), the kinetic constants of helix formation or decay are strongly dependent on the length of helices and loops. Computer simulation of the folding of random RNA chains by the method described above demonstrated a power-law growth of the average time of conformational rearrangements (*142*). In fact, this implies that, in the course of folding, the secondary structure becomes more and more "frozen," which shows up in difficulties of large conformational rearrangements.

Accordingly, the kinetic algorithm (*139, 143*) was modified so as to account for a division of the kinetic ensemble of secondary structures into clusters. A cluster involves the conformations that are similar topologically and apt to undergo rapid transition into one another. From this viewpoint, the simulation of folding reduces to the description of kinetics of intercluster transitions. This simplified algorithm is more effective and allows calculation of the secondary structure of RNA sequences of hundreds of nucleotides in length, using personal computers (*139*).

D. Comparative Approach

Under the comparative approach, the matter of the native secondary structure of RNA correspondence to the global minimum of energy or to a kinetically attainable state is not considered. However, it is assumed that a functionally significant form of secondary structure must be evolutionarily conservative. This approach was applied to reconstruction of the secondary structure of practically all the classes of RNA shown experimentally to pos-

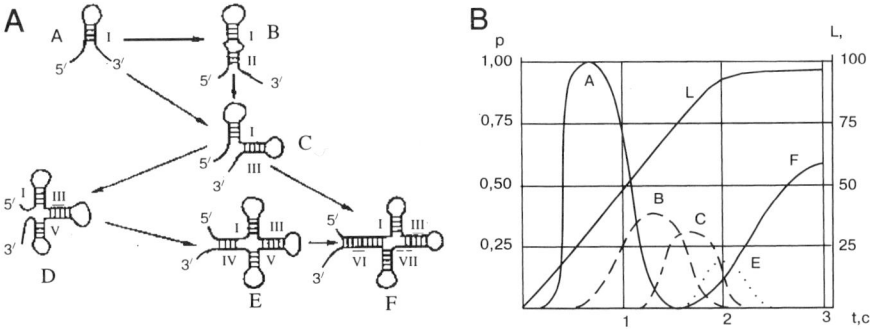

FIG. 20. The kinetic ensemble of RNA secondary structures of the pre-tRNA^Ala of *Bombyx mori*. (A) Secondary structure folding in the course of RNA chain elongation. (B) Probability distribution on the kinetic ensemble. The L curve designates the increase of RNA length with time during transcription *(124)*.

sess a secondary structure: tRNA *(147)*, 5-S RNA *(148)*, 16-S RNA, 23-S RNA *(149, 150)*, and many others.

The programs implementing the comparative approach are aimed at determination of evolutionarily invariant secondary structures in the families of isofunctional molecules of RNA. The known basic mechanism of secondary structure conservation in the course of RNA evolution is the fixation of so-called compensatory substitutions in the sequences, that is, nucleotide substitutions retaining helices.

With three types of complementary pairs (A·U, G·C, and G·U), 15 variants of compensatory substitutions are possible, i.e., substitutions of one complementary pair for another *(151)*. Of them, 11 are double substitutions (for example, A·U to G·C) and 4 are single (for example, A·U to G·U).

The probability of a double compensatory substitution in the helical part of RNA due to random combination of single nucleotide substitutions is 0.256; the probability of a single compensatory substitution under the same condition is 0.125. With these estimates, 28-S ribosomal RNAs of vertebrates were studied *(151)*. It was shown that the observed numbers of single and double compensatory substitutions in the helical parts of these RNAs were significantly higher than could be attributed to chance. Thus, fixation of compensatory substitutions in the helical parts of RNA is a mechanism that keeps RNA secondary structure evolutionarily conserved. Studies of 5-S and 5.8-S RNA molecules from different organisms *(152–154)* also provide evidence for the elevated number of compensatory substitutions within the helical parts of RNA compared with a random level.

Thus, the compensatory substitutions are the basic mechanism of maintaining the evolutionary conservatism of RNA secondary structure and to detect these substitutions is a common approach to predicting RNA secondary structure elements. The choice of algorithm for prediction of secondary structure essentially depends on the degree of homology between the sequences under comparison.

At a high level of homology, the method based on examining the aligned sequences is effective (148). All the possible locations of a pair of windows of size W are tried at a fixed distance R between them. For each location of windows, a helicity index is calculated. This index is defined as the number of the aligned sequences that have mutually complementary fragments corresponding to some helix of length not exceeding W in the given pair of windows. The maximum values of the helicity index indicates that there are invariant helices formed by complementary fragments about R bases apart. By interactively locally improving the alignment of the fragments comprising the helices, it is possible to increase the value of the helicity index. On the helicity plot (Fig. 21), it is signaled by a less fuzzy peak corresponding to the helix in question. By varying the distance R it is possible to reveal all the RNA regions that contain highly conserved helices. Figure 21 presents the superposition of the main peaks of the helicity index obtained by analyzing the 5-S RNA from *E. coli* and related organisms. Peaks I–IV correspond to four canonical helices of this RNA secondary structure that are invariant in most of the sequences studied. Besides, less clear-cut peaks (A–C) have been revealed. They correspond to the helices that are invariant within separate subgroups of the set of sequences in question.

There are also available methods for the automated secondary structure reconstruction based on alignment data. With these methods (155), a matrix of complementarity is constructed for the family of M aligned sequences of length N. One of the following symbols is assigned to an element $BP(i,j)$ of the matrix: **o** stands for nucleotides i and j that are complementary in all the sequences and these positions are absolutely conservative; * stands for those that are complementary in all the sequences, and compensatory substitutions are observed at these positions; **w** indicates those forming a G·U pair in most of the sequences; + is for those forming a complementary pair in most of the sequences (over some threshold), but not in all of them; and • indicates those with a considerable number of noncomplementary pairs (over some threshold).

Information on all the invariant helices for the family of sequences under consideration is coded in the matrix BP (i,j). To each invariant helix S in this matrix, there corresponds a continuous track of symbols **o**, *, **w**, or + normal to the diagonal. On calculating helix length h, loop length l, and defining the number of symbols * (A), **w** (B), and + (C), it is possible to estimate the "pseudopotential" F for the helix S. In general, the longer the helix and the higher the number of positions with compensatory substitu-

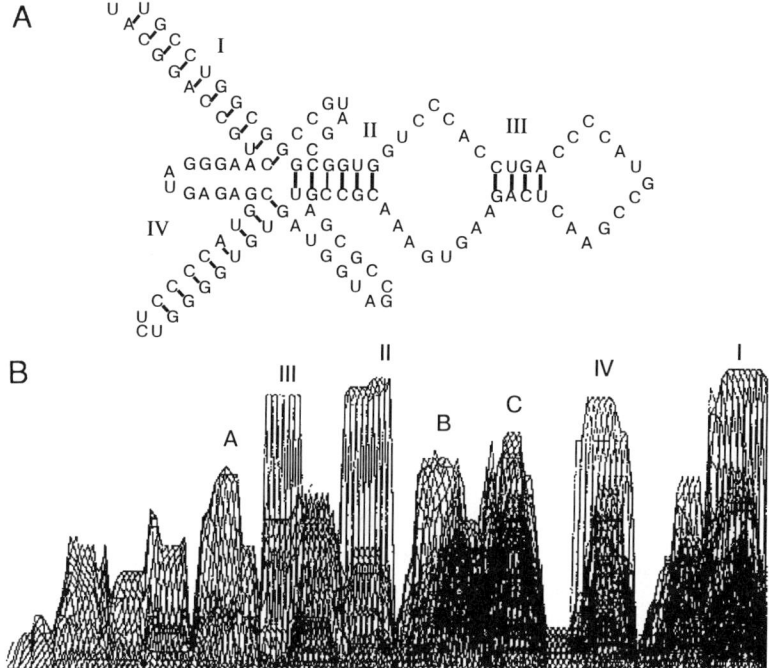

FIG. 21. Reconstruction of the secondary structure of 5-S RNA by comparison of the sequences from *E. coli* and related organisms (148). (A) The secondary structure of *E. coli* 5-S RNA. (B) The superposition of helicity peaks for the set of aligned sequences.

tions in it, the greater the value of pseudopotential $F(S)$. On the other hand, the higher the number of G·U pairs or noncomplementary pairs and the longer the loop formed, the less the pseudopotential $F(S)$. For example, $F(S) = h + A - 0.5B - 2C - 0.051$ proved good for revealing invariant helices (155).

Then an invariant secondary structure is constructed by step-wise selection of the helices with the maximum value of $F(S)$. In fact, this is a sort of simulation of RNA folding with only invariant helices allowed. The procedure follows the standard rules of stereochemical compatibility of helices (Fig. 16). A testing of the algorithm on the known secondary structures as tRNA, 5-S RNA, and 16-S RNA showed a good agreement between the predicted and canonical structures.

In fact, the method at issue (155) is the first practically effective comparative approach regarding storage capacity and time consumption. The time required for finding the invariant secondary structure for M sequences of length N is here $MN^2 + N^3$, and storage capacity is N^2 (155). Interestingly, the invariant RNA secondary structure reconstructed with this method (155) for the TAR fragment of 200 nt from type 1 human immunodeficiency virus (HIV-1) coincides with the lowest energy structure predicted for this fragment by the FOLD program (155). This result exemplifies that in some cases evolutionarily invariant secondary structure corresponds to the lowest energy form.

If homology of the considered RNA molecules is not high, analysis of multiple alignment is not applicable to determining the invariant secondary structure. If so, the following approach may be effective: (1) search for a set of suboptimal secondary structures for each RNA under study by one of the thermodynamic or kinetic methods; (2) compare the sets and the choice of secondary structure that is invariant for all sequences under comparison. Many methods have been proposed for comparison of secondary structures within the frame of this approach. All of them depend on the manner in which the secondary structure is coded and on further comparison of the resulting codes.

Under one of the versions, hairpin, internal, bulge, and multiple loops relate to tree vertices, and helices to edges (157–159). Under another version (160), helices relate to vertices, and loops to edges. Thus constructed, the trees can be compared by one of the methods.

For example, RNA secondary structure can be translated into a symbol string (158). In particular, the symbol of the corresponding local secondary structure can be assigned to each nucleotide. The resulting symbol strings can be aligned by one of the procedures of pairwise or multiple alignment. One of the advantages of this approach is that it is possible to use the routine of alignment. A significant drawback is that this approach provides a rough description of the compared secondary structures neglecting their individual features. This is not a serious problem when analyzing highly homologous sequences, but it surely is when the sequences are evolutionarily remote.

Another group of methods for RNA secondary-structure coding and comparison is based on tree-editing algorithms (158, 159, 161). These methods allow a detailed consideration of the RNA secondary structure. In particular, the method based on the construction of the so-called tree of cycles for RNA secondary structure provides its detailed description in linear or nonlinear code (162). With this method, types and the mutual disposition of loops, helices, and even the pairs of complementary nucleotides in helices can be described. To save computation time, at the next stage a condensed tree is constructed. In this tree a set of helices split by internal or bulge loops is regarded as a nonbranching helix, and a separate vertex is assigned to such a helix. With this approach, the reconstruction of an invariant secondary structure was performed for a highly variable domain D3 of the rRNA of large ribosomal subunit (162). The secondary structure of this domain has been reconstructed for each of the three major kingdoms: prokaryotes, eukaryotes, and archeaebacteria. What is essential is that, apart from having the common features (helices A, B, C, and D), these domains contain specific elements of secondary structure that are invariant only within a kingdom (Fig. 22).

The last result clearly demonstrates how the crucial limitation of any

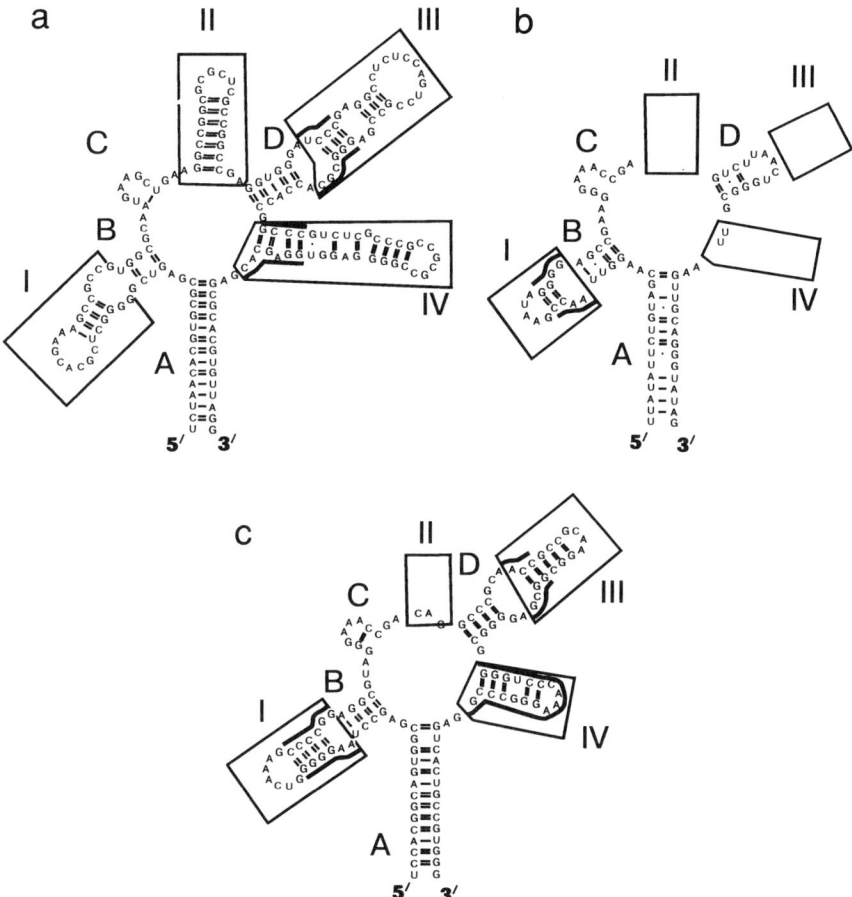

FIG. 22. Reconstruction of the invariant secondary structure of the D3 domain of large ribosomal RNA for three major kingdoms (162). (a) eukaryotes (*H. sapiens*); (b) eubacteria (*E. coli*); (c) archeaebacteria (*Desulfurococcus mobilis*). The helices invariant within all the three kingdoms are denoted by capital letters (A–D). Regions presenting structural variation among the three kingdoms are boxed and indicated by roman numerals. Secondary structure features conserved in each kingdom are shown by solid lines. Reproduced from Ref. 162, C. Chevalet and B. Michot, *CABIOS* 8, 215 (1992), by permission of Oxford University Press.

comparative approach works. This is the assumption that RNA secondary structure is invariant within any given group of sequences. Whether secondary structure is invariant should be checked in each individual case, because while the primary structure of RNA was evolving, so was the secondary structure. Nevertheless, this approach has a sound motivation, a pronounced hierarchy of the variability levels corresponding to the levels of RNA organization: primary structure is the most variable, secondary structure is less variable, and tertiary structure is highly conserved. This is well exemplified by the fragments of genomic RNA from some plant viruses considered in Section II,E. These fragments have a tRNA-like tertiary structure that displays a high structural similarity with the L-shaped tRNA (*163*). At the level of secondary structure, similarity is slight, and none of it can be detected at the level of primary structure.

Each of the above approaches relies on definite features of structure, function, and evolution of RNA secondary structure and, therefore, has certain advantages. Nonetheless, the most reliable results appear to be expected of a combination of experimental and computer methods. Such a combined approach was applied (*164*) to identify the packaging signal of HIV-1. Loops were localized using diethylpyrocarbonate (DEP) and RNase S1. The former is specific to the A sites of the single-stranded RNA, the latter to the A and U sites. Low-energy secondary structures were determined by the program FOLD of the software package GCG. The homologous sequences of HIV-1 strains were compared for the presence of evolutionarily invariant secondary structure. Indeed, the lowest energy secondary structure, namely, a stem of three helices separated by loops (Fig. 1), was revealed by DEP and S1-RNase cleavage of RNA.

The algorithms for prediction of RNA secondary structure by comparative analysis of sequences are additionally described in Section II,E, where pseudoknots are discussed.

E. Prediction of Pseudoknots

There is still a quite vague understanding of the factors accounting for tertiary structure formation. The conformations of single-stranded regions are one of them. The structure of these regions, which may be stabilized by stacking interactions, has been studied intensively since the 1960s (*87*), with synthetic oligonucleotides. The thermodynamics of such other features of tertiary structure as nucleoside triplets (*165*) and helix–helix interactions (*97*) have been less studied. Below we consider pseudoknots, one of the most important features of tertiary structure (*166, 167*).

A pseudoknot forms when the single-stranded region of a loop forms Watson–Crick pairs with a complementary fragment not within the loop (*166, 167*). Pseudoknots were discovered in studies of turnip yellow mosaic virus (TYMV) RNA (see *166*). As it turned out, this RNA resembles tRNA in several respects (Fig. 23). In particular, it is able to accept Val-tRNA at its 3′ end. However, further investigations (*168*) showed that the secondary structure of the 3′ end of this RNA, constituted by four helices (I–IV), is quite unlike a "cloverleaf." Meanwhile, it was possible to regard the 3′ end structure as a three-dimensional L-shaped tRNA by assuming that the pseudoknot forms owing to complementary interaction between the CCC fragment near helix II and the GGG fragment in the hairpin loop of helix I. Computer simulations showed that the tertiary structure of this fragment of RNA with a pseudoknot is similar to the L-shaped tertiary structure of tRNA (*163*). With the use of pseudoknot structure, spatial models were also suggested for the His- and Tyr-accepting domains at the 3′ ends of tobacco mosaic virus (TMV) RNA and brome mosaic virus (BMV) RNA (*170–172*).

There are four types of loops that may contain a single-stranded region involved in the formation of a pseudoknot, namely, hairpin, bulge, internal, and multiple. Note that the complementary region to be paired with may either be related to one of the loop types mentioned or be in the single-stranded region of RNA. Thus 14 different types of pseudoknots may exist, depending on the location of the complementary regions by which they are formed (*166*).

An H-type (hairpin) pseudoknot (Fig. 23A) that involves a hairpin loop is perhaps the best studied. It is formed by two helices, S1 and S2, and two

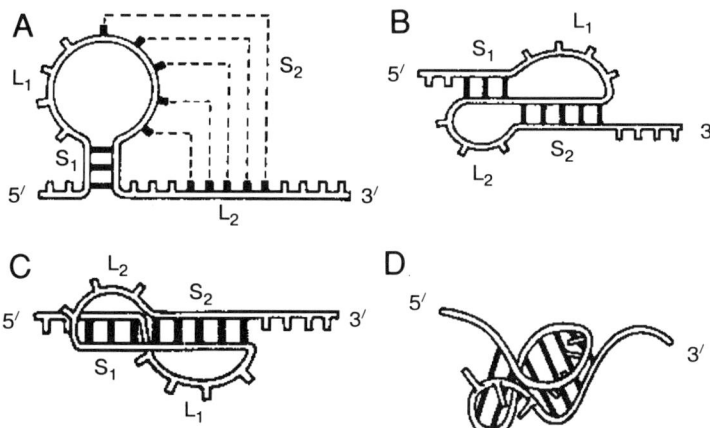

FIG. 23. Schematic representation of an H-type pseudoknot (*166*). (A) Basic elements of a pseudoknot: S1 and S2 are helices; L1 and L2 are connecting loops; (B) planar representation of a pseudoknot; (C) coaxial stacking of helices S1 and S2; (D) pseudoknot: three-dimensional schematic representation.

connecting loops, L1 and L2 (*166*). The helices are stacked coaxially in this structure to form a longer quasicontinuous helix S1 + S2 (*166*). While loop L1, which is the first from the 5' end, crosses the major groove of the quasicontinuous helix, loop L2 crosses the minor. It is supposed that stacking interactions between the termini of the two helices provide an additional gain in energy on the formation of pseudoknots.

As noted above, the existence of pseudoknots was initially demonstrated in plant viruses (*166*). Today there is experimental evidence for the existence of these structures in the RNA genomes of plant and some other viruses (see *166*, *167*).

Pseudoknots have been revealed in small subunit rRNAs (*169*, *173*, *174*). One of theme is H-type; another is formed by complementary pairing of the fragments of a hairpin loop and a multiple loop, and the last one by complementary pairing between a hairpin loop and a bulge loop of 16-S RNA.

Evidence has been gathered for an important role of pseudoknots in the autoregulation of prokaryotic mRNA expression (*166*). A pseudoknot has been found in the 5' noncoding leader region of the bacteriophage T4 gene-32 mRNA (*175*). Comparison of the operator sequences of phages T2, T4, and T6 containing this gene provides phylogenetic evidence for the conservation of this pseudoknot. The binding of the proteins encoded by the corresponding mRNA to the pseudoknots in the 5' leader nontranslated regions of phage mRNA can provide the autoregulation of translation of the corresponding mRNA (*166*).

Site-directed mutagenesis of *E. coli* operon mRNA (*176*) provides evidence for the existence of the pseudoknot, which is of importance for the binding of ribosomal protein S4, one of four protein products encoded by this operon.

Finally, there is evidence that pseudoknots play an important role in providing a translation frameshift for a range of eukaryotic mRNAs, in particular Rous sarcoma virus RNA (*177*) and avian coronavirus RNA (*178*). In the latter, the H-type pseudoknot is located six nucleotides away from a translation frameshift site.

It is also suggested that pseudoknots should be found in some ribozymes. In particular, type-I self-splicing introns carry the pseudoknots that provide the spatial proximity of the RNA regions involved in formation of the active center (*166*, *179*).

In general, the available data provide evidence for the existence of pseudoknots in various RNAs and for their functional significance.

The discovery of pseudoknots as new elements of RNA structure brought about a determination of their thermodynamic parameters. These parameters should be determined as dependent on the type of the pseudoknot, on the length of its helices and loops, and on their nucleotide sequences. Today, it is only being considered how to deal with the problem.

Pseudoknot formation was shown experimentally to depend on a number of factors, such as the content of the solution, the lengths of the involved loops and helices, the energies of the helices, the coaxial stacking energy of the helices, and the mutual arrangement of pseudoknot elements (*180–183*). The stability of such structures essentially depends on the conformation of the bases forming complementary pairs and on the steric constraints on the loops. It has been shown (*180, 181*) that the gain in enthalpy observed on pseudoknot formation is less than that calculated applying the nearest-neighbor model (*88*) to the helices involved. This effect can be due to unfavorable interactions between helices and loops of the pseudoknot or violation of conformation of complementary pairs from that typical for an A helix.

Studies of the pseudoknots formed by synthesized oligonucleotides demonstrated (*180, 181*) that some of them melt in a multistep way, whereas others can be described by the all-or-none model (cooperative melting). Such pseudoknots are just a little more stable (by 1.5–2 kcal/mol) than the most stable of the involved hairpins. Because there is no complete energy table for pseudoknots, simplified approximations are used in computer calculations. On the basis of experimentally confirmed pseudoknots, parameters were introduced (*137*) for the destabilizing energy of loops less than 15 nt in length for H-type pseudoknots. Constraints regarding the sizes of loops and helices and their mutual arrangement were also imposed on the pseudoknots. If the pseudoknot was allowed, the destabilizing energy of the loop was set at 4.2 kcal/mol. Another way to describe the energy of pseudoknots has also been suggested (*138*). The destabilizing energy of the resulting loops was set equal to the mean value of the energy of the loops involved in the pseudoknot (two loops if the helices were stacked coaxially, and three loops if the helices were slightly remote from each other). The energy of the helices within a pseudoknot was determined here by using standard parameters (*88*).

Allowance was made for the entropic interaction between the loops (*141, 184*). In this case closure of a loop facilitates formation of the next loop due to the cooperative effect. Cooperativity improved the simulation of kinetics of secondary structure formation for the self-splicing group-1 intron RNA (*141*).

Today there are suggested several approaches to pseudoknot detection in RNA secondary structure. One is based on comparison of homologs (*185*) and it was applied to pseudoknot detection in the V4 variable region of RNA of the small ribosomal subunit of eukaryotes. As in any comparative method the functionally significant pseudoknots are assumed to be evolutionarily conserved.

Analysis of 13 aligned sequences allowed the invariant helices to be listed. Then the compensatory substitutions were counted for each of them. Helices with the maximum number of compensatory substitutions were then selected into a subset. As a result, there was obtained a set of helices of maximal functional significance. These helices were used in developing a model of RNA secondary structure. Figure 24 exemplifies invariant pseudoknots revealed in the V4 region. That these pseudoknots are functionally significant is supported by the fact that the sequences belong to phylogenetically remote species (e.g., *Homo sapiens, Drosophila melanogaster*, and *Saccharomyces cerevisiae*).

Another empirical approach for predicting the location of H-type pseudoknots in RNA is also of interest (*186*). This one is based on assessment of the indices of statistical (St) and thermodynamic (Th) significance for a local region of RNA containing a potential pseudoknot. The index St shows to what extent the energy of a given short fragment containing a potential pseudoknot differs from the energy of random sequences of the same length. The index Th shows to what extent the energy of a given short fragment containing a potential pseudoknot differs from the energy of the other real fragments of the same RNA sequence. By surveying the whole sequence through a sliding window, one selects RNA fragments with the maximum value of the indices. Then all the potential pseudoknots for the selected regions are constructed. Further selection among them is performed by a range of empirical criteria. Let us consider a fixed pair of helix (S1) and loop

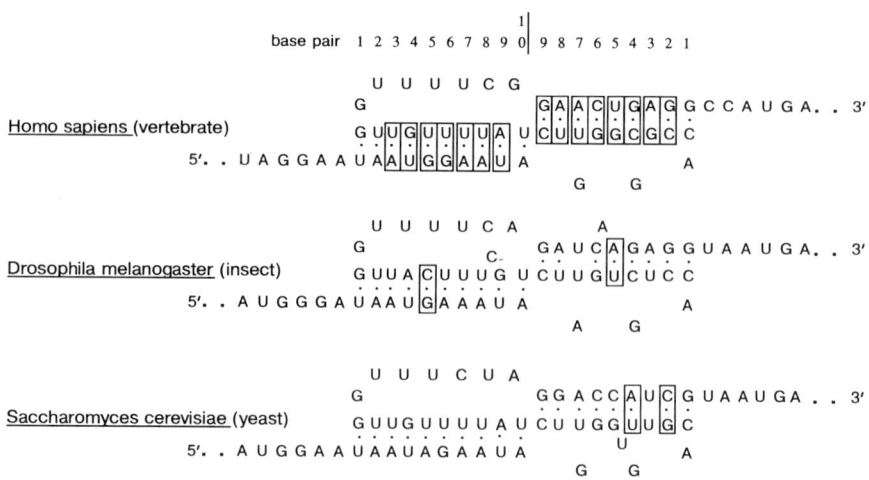

FIG. 24. Evolutionarily invariant pseudoknots in the variable region of eukaryotic small ribosomal subunit RNA. Reproduced from Ref. *185*, J.-M. Neefs and R. De Wachter, *NARes* **18**, 5695 (1990), by permission of Oxford University Press.

(L1). To simulate pseudoknot formation, it is necessary to indicate another helix (S2) and another loop (L2) within the preselected fragment (Fig. 23). Chosen out of all those available is the longest helix, S2, combined with loop L2 such that the minimum loss in free energy is provided. This method (*186*) was tested on RNA with pseudoknots in known locations. Its identification capacity proved to be quite high (*186*).

Most of the methods for predicting the RNA secondary structure described in Sections II,B–II,D forbid pseudoknot formation. This limitation was introduced in the first algorithms for prediction of RNA secondary structure. At that time there was no evidence of pseudoknots in RNA. It should be emphasized that this limitation is crucial for the most popular algorithms calculating low-energy secondary structures (*104*). Accordingly, it is not easy to consider pseudoknots in the framework of these algorithms. In particular, allowing pseudoknots makes combinatorial algorithms enormously complicated, and recursive algorithms impossible to be implemented.

This stimulated the development of algorithms that allow pseudoknots (*136–138*). They are based on a simulation of the step-wise folding of RNA. For example, RNA folding was simulated by step-wise selection of the helices compatible with those having been formed at the previous stages (*138*). Detected at each step were all the helices compatible with those previously formed and providing the energy gain higher than a threshold. Then a set of 100 random secondary structures was generated with these helices by the Monte Carlo method. The frequency of each helix involved in these structures was estimated. The product of the frequency and the free-energy gain at the helix formation was calculated. The helix with greatest product was included in the current secondary structure at the given simulation step. Energy parameters were taken from Freier *et al.* (*88*). The destabilizing energy of the pseudoknot was set equal to the average of the energies of the loops forming the pseudoknot.

This algorithm provided a correct prediction of 66% of phylogenetically conservative helices in the secondary structure of the small subunit rRNA from *E. coli* (*138*). Interestingly, allowing pseudoknots does not bring about any great number of them in the predicted secondary structures. As a rule each secondary structure involves three to five pseudoknots.

There is another similar approach to predicting RNA secondary structure with pseudoknots (*137*). Here, again, RNA folding is simulated via step-wise addition of a helix to the structure that has already been formed. The empirical values of the thermodynamic parameters of the pseudoknot are as described above (*137*). The predicting ability of this method was checked on a range of objects. With the LSU intron, 95 of 127 complementary pairs (75%) were identified correctly. As for the small subunit rRNA of *E. coli*, 26 of 65 (40%) phylogenetically conservatized helices were predicted correctly.

Using this method, the pseudoknot at the 3' end of TYMV RNA was predicted and good predictions were obtained for the TMV RNA 3' end secondary structure with five pseudoknots (137). It is noteworthy that this program is faster than those based on the recursive algorithms. Time consumption is about N^2 (where N is the sequence length), whereas the recursive algorithms (104) usually take about N^3.

One more program (136) for predicting the RNA secondary structure with pseudoknots was developed as a modification of the program RNAFOLD (135), initially intended for simulating step-wise RNA folding. Interestingly, although the algorithms for RNA folding simulations are not aimed at searching for the lowest energy structures, the resulting structures are often just like those corresponding to the global minimum of energy.

F. Statistical Analysis of RNA Secondary Structure

To understand the basic features of RNA secondary structure, it is of interest to analyze the dependence of fundamental characteristics (total energy, number of complementary base pairs, number of loops and helices, and some others) on the length and nucleotide content of the RNA molecule. The features of evolution of RNA secondary structure are of interest, too. This is of special importance because the comparative methods are highly effective means of RNA secondary structure prediction. An understanding of the matter may be achieved through analysis of secondary structure predicted for a variety of random RNA sequences (108, 187–190). By applying methods based on free energy minimization, it is possible to analyze the dependence of characteristics of secondary structure on RNA sequence length or on the mode of evolution. Such analyses usually cover sequences no longer than 150 nt (108, 187–190). With these lengths a correct prediction of low-energy RNA secondary structure is possible.

The totality of potential RNA sequences of length N forms an N-dimensional hypercube. The total number of M of RNA sequence variants depends on length L exponentially: $M = 4^N$, whereas the estimated number of different secondary structures of RNA is $S = N^{-3/2}(1.85)^N$ (108). This formula is applied to the planar secondary structures with hairpin loops more than 2 nt in length and helices not shorter than 2 bp. As length N approaches infinity, the M/S ratio also approaches infinity. This implies that at large N there is, on the average, a great many RNA sequences corresponding to any secondary structure. In other words, the RNA folding code is strongly degenerate (108, 189–191).

Prediction of low-energy secondary structure for random RNA sequences with homogeneous nucleotide content allowed the dependencies of the mean values of secondary structure characteristics on sequence length to be determined (187, 188). As it turned out, characteristics such as the total energy of secondary structure, the number of base pairs in helices, and the number of helices and loops depend linearly on sequence length (*187*). The dependence becomes linear for sequences with lengths exceeding 20 to 30 nt. On the average, a fragment of this length contains 1 to 3 helices and loops. At the same time, the mean length of loops and helix size tend to constant values as sequence length grows. The same is typical of the branching degree (the number of helices closed by a loop). The average helix length is 3 to 7 bp. The average loop length is not less than 3 to 5 nt, and the average branching index is 1.5 (*187*). What can be inferred from these results is that there must be some basic "module" of RNA secondary structure folding characterized by the parameters presented above.

The study of low-energy secondary structure formed by random RNA sequences characterized by different alphabets (*187, 188*) is also of concern. Considered were both the real alphabet $\{A,U,G,C\}$ and various model alphabets: $\{A,U\}$; $\{G,C\}$; $\{G,C,X,K\}$; $\{A,B,C,D,E,F\}$. In the last two, complementarity was set up as $X \cdot K$, $A \cdot B$, $C \cdot D$, $E \cdot F$ (in any, the energy was as in $G \cdot C$ pairs). Model alphabets help discriminate between effects of different nature and aid understanding the interference of the respective contributions to the natural RNA folding (alphabet volume, pairing rules, and base-pairing strength).

Any alphabet is characterized by two essential parameters: strength and "stickiness." Strength is defined by the energy of complementary base pairing. Thus $\{G,C\}$ is a stronger alphabet than $\{A,U\}$. Calculations show that a stronger alphabet favors the formation of more low-energy secondary structure. Besides, a stronger alphabet provides higher compactness of the resulting secondary structure (the less the number of loops and the higher the number of paired bases, the higher compactness). The reason here is that, *ceteris paribus*, the helices based on a strong alphabet are more stable than those on a weaker. High stability of helices provides possibilities of short-loop closure. As was indicated, loop formation entails energy loss (Table II). That is why loop closure is impossible unless there are highly stable helices involved. Besides, the stronger the alphabet, the higher the number of helices in RNA secondary structure, because even short helices become stable enough to compensate for the energy loss owing to the loop closure.

"Stickiness" characterizes the probability of two randomly selected bases forming a complementary pair in a homogeneous sequence. Thus the stickiness of the real alphabet $\{A,U,G,C\}$ with two complementary pairs ($A \cdot U$ and $G \cdot C$) is $1/4$, the same for the model alphabet $\{G,C,X,K\}$. Including $G \cdot U$ complementary pairs in the real alphabet gives stickiness as 0.375. Stickiness of the model alphabets $\{A,U\}$ and $\{G,C\}$ is 0.5.

At a fixed alphabet strength, the higher the stickiness, the higher the probability of complementary pairing for any nucleotide within the se-

TABLE II
FREE ENERGY INCREMENTS FOR LOOPS[a]

Loop size	Internal loop	Bulge loop	Hairpin loop
1	—	+3.3	—
2	+0.8	+5.2	—
3	+1.3	+6.0	+7.4
4	+1.7	+6.7	+5.9
5	+2.1	+7.4	+4.4
6	+2.5	+8.2	+4.3
7	+2.6	+9.1	+4.1
8	+2.8	+10.0	+4.1
9	+3.1	+10.5	+4.2
10	+3.6	+11.0	+4.3

[a] In units of kcal mol^{-1}, in 1 M NaCl, at 37°C (82).
Reproduced, with permission, from the *Annual Review of Biophysics and Biophysical Chemistry*, Volume 17, © 1988, by Annual Reviews Inc.

quence. Thus the mean helix length grows, the probability of new helices formation rises, and the loops shorten. That is, on average, as compactness of secondary structure increases, its energy decreases. The above estimates of the characteristics of the RNA secondary structure were obtained for random sequences, and it was interesting to compare them with the estimates for real DNA sequences.

The low-energy secondary structures of real RNA calculated by recursive algorithm (104) were compared (187) with the secondary structure of random sequences that had equal nucleotide frequencies. Mitochondrial, eubacterial 16-S rRNA and β-globin mRNA were analyzed. All the real sequences considered had approximately the same frequencies of every nucleotide. Five secondary structures with the lowest energies were considered for each of the RNA sequences.

With these structures the mean values of the following parameters were estimated: size of a loop, length of a helix, branching degree and helicity, i.e., the ratio of the number of complementary pairs, and sequence length (Table III). As is seen, the real and random RNA sequences have a similar mean branching degree, helicity, and helix size. The mean loop lengths of the real and random sequences differ by about 20%, whereas the respective mean values of the remaining parameters do not differ by more than 14%.

The results obtained suggest that the concept of "random edited biopolymer" proposed for natural aminoacid sequences (192) could be applied to RNA, too. According to the concept, globular proteins arose in the course of evolution from random aminoacid sequences due to the fixation of a limited number of aminoacid substitutions ("evolutionary editing"). This concept

TABLE III
Mean Values for Random and Natural Sequences[a]

Source	n_{BP}/N	n_{st}	n_{lp}	n_{BD}
Random sequences	0.29	4.57	5.42	1.82
β-Globin mRNAs	0.31	4.49	4.42	1.89
Mitochondrial rRNAs	0.26	4.44	6.53	1.74
Eubacterial rRNAs	0.33	4.59	4.62	1.92
Mitochondrial rRNAs	0.24	3.76	6.00	1.90
Eubacterial rRNAs	0.28	4.35	5.81	1.93

[a] n_{BP}, Mean number of base pairs; n_{st}, mean helix size; n_{lp}, mean loop size; n_{BD}, mean branching degree. Reproduced from Ref. *187*. Copyright © 1993. Reprinted by permission of John Wiley & Sons, Inc.

is supported by the similarity between mean values of different structural parameters of real globular proteins and the proteins formed from random sequences (*192*).

Again, the similarity of secondary structure parameters of random sequences and real RNA suggests that they can also be considered as random edited polymers. Computer experiments (*108, 189, 190*) provide a sound argument favoring evolutionary editing as a mechanism of incidence of the current RNA secondary structure. It was concluded that within a small neighborhood of any random point in the multidimensional space of RNA sequences there exists such a set of RNAs that can form all the possible low-energy secondary structures. In fact, it means that any random sequences can be transformed into a sequence with definite low-energy secondary structure via fixation of a limited number of certain mutations. The number of mutations to provide "editing" is 15–20 in a length of 100 nt (*108, 189, 190*). Thus, mutational editing was proved effective to obtain RNA sequences with given secondary structure from random sequences.

That secondary structures are vulnerable to mutations (*108, 189, 190*) is an important factor facilitating such editing. Experimental study of secondary structure of threonyl-tRNA synthetase mRNA from *E. coli* shows that this secondary structure can be drastically affected by a single nucleotide substitution (*193*). This is in a good agreement with the results of a computer simulation of mutation effects on secondary structure of random RNA sequences of 100 nt. It was shown (*108*) that even a few mutations (one to three) can result in significant alteration of the most low-energy secondary structure.

It is noteworthy that, if there were more than three random substitutions in the sequence, the probability of folding into the initial secondary structure was low. At 15–20 random mutations, the probability of sequence folding into the initial secondary structure was the same as the probability of two random sequences folding identically. It means that at this number of mutations, the sequence memory about the initial secondary structure is totally erased.

Meanwhile, analyses of isofunctional RNA molecules of different taxa show that a functionally significant secondary structure is, as a rule, highly conservative, whereas primary structures are quite variable. In some cases, there is the lowest similarity between primary structures (30–40%), yet the secondary structures of RNA have a similar pattern. As was pointed out in Section II,D, this must be due to a particular mode of RNA evolution, by which the helical regions of secondary structure were shown to fix the compensatory substitutions—that is, substitutions retaining complementarity. This feature of RNA evolution is taken into account for prediction of secondary structure using the methods described above.

Secondary structure conservation in the course of the evolution of an RNA sequence has been studied by many authors (*108, 190*), and was considered for various types of mutations (point substitutions, recombinations, deletions, and insertions) (*190*). As it turned out, deletions and insertions are the most effective in changing secondary structure; they thus provide the fastest possible evolutionary optimization of secondary structure. On the other hand, recombinations between strongly homologous RNAs causes only local changes in secondary structure. Synonymous substitutions cause fewer alterations in secondary structure than random point mutations.

It has been demonstrated (*108, 189*) that there are routes in the multidimensional space of sequences in which the initial low-energy secondary structure remains unaltered, no matter what the primary structure. Thus, in 22% of the cases, the route reaches an end point where the distance between the current and initial sequence is as great as possible, equal to the length of the sequence (zero homology). Such routes are implemented by the fixing of single mutations in nonhelical regions and compensatory mutations in helices. It implies that RNAs with similar secondary structures can be slightly homologous (*108, 189*). Approaches to revealing the invariant secondary structures for slightly homologous RNAs were discussed in Section II,D.

III. Concluding Remarks

Investigations of RNA structures with different enzymatic and chemical probes can provide detailed data allowing identification of double-stranded regions of the molecules and nucleotides involved in tertiary interactions. Combining results of probing experiments, and taking into account thermodynamic data, the data of phylogenetic studies, and the geometry of RNA units, it is possible to build models of RNA structures at the nucleotide level of resolution. Cross-linking approaches together with RNA shape-sensitive chemical probes and cleaving reagents that recognize specific structural features of RNA molecules provide easy means for monitoring conformational changes in RNA under different conditions or on the binding of various factors. Intrinsically, chemical probing is a high-resolution method because it allows investigation of the reactivities of individual functional groups of the RNA. It does not allow discovery of novel elements in an RNA. This can be done only by physical methods (X-ray analysis and NMR). However, when a structural element is discovered and characterized, it can be detected by probing techniques in novel RNA species and thus taken into account when building high-resolution models. Enzymatic and chemical probes are becoming increasingly more important for detecting structural variations, for monitoring conformational changes of RNA, for investigating effects of mutations on the RNA structure, and for investigating RNA–protein complexes. A great advantage of probing techniques is the possibility of investigating RNA structure in complex systems regardless of the presence of other biopolymers, when other methods are not applicable. Further development of the chemistry of probes and progress in computer modeling will provide researchers with new, simple, and reliable methods for investigation of RNAs at any level of complexity and for investigations of the dynamics of RNA–protein complexes in the cell.

Acknowledgments

The work of N.A.K. and I.I.T. was supported by the Russian Human Genome Project, the Russian Ministry of Science and Technique Politics, and by Russian Fundamental Research Foundation Grant No. 12757. I.E.V. and V.V.V. acknowledge the support of the Centre National de la Recherche Scientifique (CNRS), the Ministère de l'Education Nationale et de la Culture and Université Louis Pasteur (Strasbourg, France). We are grateful to V. Filonenko for translating Section II of this essay from Russian to English.

References

1. R. F. Gesteland and J. F. Atkins, eds., "The RNA world." Cold Spring Harbor Laboratory, Cold Spring Harbor, New York, 1993.
2. G. J. Quigley, A. H. J. Wang, N. C. Seeman, F. L. Suddath, A. Rich, J. L. Sussman and S. H. Kim, *PNAS* **72**, 4866 (1975).
3. R. Giege, D. Moras and J. L. Thierry, *JMB* **115**, 91 (1977).
4. A. Jack, J. E. Ladner and A. Klug, *JMB* **108**, 619 (1976).
5. J. L. Sussman, S. R. Holbrook, W. R. Wade, G. M. Church and S. H. Kim, *JMB* **123**, 607 (1978).
6. E. Westhof, P. Dumas and D. Moras, *JMB* **184**, 119 (1985).

7. P. B. Moore, *Curr. Opin. Struct. Biol.* **3**, 340 (1993).
8a. J. K. James and I. J. Tinoco, *NARes* **21**, 3287 (1993).
8b. H. W. Pley, K. M. Flaherty and D. B. Mckay, *Nature* **372**, 68 (1994).
9. T. Hayashi, Y. Ueno and T. Okamoto, *FEBS Lett.* **327**, 213 (1993).
10. V. V. Vlassov, R. Giege and J. P. Ebel, *FEBS Lett.* **120**, 12 (1980).
11. A. Jack, J. E. Ladner, D. Rhodes, R. S. Brown and A. Klug, *JMB* **111**, 315 (1977).
12. R. Lavery and A. Pullman, *Biophys. Chem.* **19**, 171 (1984).
13. P. Romby, D. Moras, H. Bergdoll, P. Dumas, V. V. Vlassov, E. Westhof, J. P. Ebel and R. Giege, *JMB* **184**, 455 (1985).
14. M. Silberklang, A. M. Gillam and U. L. RajBhandary, *NARes* **4**, 4091 (1977).
15. G. Chaconas, J. H. Van de Sande and R. B. Church, *BBRC* **66**, 962 (1975).
16. A. G. Bruce and O. C. Uhlenbeck, *NARes* **4**, 2427 (1978).
17. B. Rether, J. Bonnet and J. P. Ebel, *EJB* **50**, 281 (1974).
18. C. Ehresmann, F. Baudin, M. Mougel, P. Romby, J. P. Ebel and B. Ehresmann, *NARes* **15**, 9109 (1987).
19. A. Krol and P. Carbon, *in* "Methods in Enzymology," Vol. 180, p. 212. Academic Press, San Diego, 1989.
20. T. Uchida, T. Arima and F. Egami, *J. Biochem.* **67**, 91 (1970).
21. T. Uchida and F. Egami, *in* "Methods in Enzymology," Vol. XII, p. 228. Academic Press, New York, 1967.
22. M. S. Boguski, P. Hieter and C. C. Levy, *JBC* **255**, 2160 (1980).
23. C. Florentz, J. P. Briand, P. Romby, L. Hirth, J. P. Ebel and R. Giege, *EMBO J.* **1**, 269 (1982).
24. T. Ando, *BBA* **114**, 158 (1966).
25. S. Linn and T. R. Lehman, *JBC* **240**, 1287 (1965).
26. S. K. Vassilenko and V. C. Ryte, *Biokhimiya* **40**, 578 (1975).
27. P. D. Lawley and P. Brookes, *BJ* **89**, 117 (1963).
28. N. Wintermeyer and H. G. Zachau, *FEBS Lett.* **58**, 306 (1975).
29. V. S. Zueva, A. S. Mankin, A. A. Bogdanov and L. A. Baratova, *EJB* **146**, 679 (1985).
30. B. Singer, *Nature* **264**, 333 (1976).
31. B. Singer and H. Fraenkel-Conrat, *Bchem* **14**, 772 (1976).
32. V. V. Vlassov, R. Giege and J. P. Ebel, *EJB* **119**, 51 (1981).
33. B. J. Van Stolk and H. F. Noller, *JMB* **180**, 151 (1984).
34. R. Naylor, N. W. Y. Ho and P. T. Gihlam, *JACS* **87**, 4209 (1966).
35. R. Shapiro, B. I. Cohen, S. J. Shiuey and H. Maurer, *Bchem* **8**, 238 (1969).
36. D. A. Peatty, *PNAS* **76**, 1760 (1979).
37. D. A. Peatty and W. Gilbert, *PNAS* **77**, 4679 (1980).
38. K. M. Weeks and D. M. Crothers, *Science* **261**, 1574 (1993).
39. T. D. Tullis and B. A. Dombroski, *Science* **230**, 679 (1985).
40. T. D. Tullis and B. A. Dombroski, *PNAS* **83**, 5465 (1986).
41. J. A. Latham and T. R. Cech, *Science* **245**, 276 (1989).
42. D. W. Celander and T. R. Cech, *Science* **251**, 401 (1991).
43. B. Laggerbauer, F. L. Murphy and T. R. Cech, *EMBO J.* **13**, 2669 (1994).
44. R. P. Herztberg and P. B. Dervan, *Bchem* **23**, 3934 (1984).
45. R. Breslow, *Accts. Chem. Res.* **24**, 317 (1991).
46. V. V. Vlassov, G. Zuber, B. Felden, J.-P. Behr and R. Giege, *NARes* (1995). In press.
47. M. A. Podyminogin, V. V. Vlassov and R. Giege, *NARes* **21**, 5950 (1993).
48. B. Felden, C. Florentz, R. Giege and E. Westhof, *JMB* **235**, 508 (1994).
49. E. Westhof, P. Romby, C. Ehresmann and B. Ehresmann, *in* "Theoretical Biochemistry

and Molecular Biophysics" (D. Beveridge and R. Lavery, eds.), p. 399. Adenine Press, Guilderland, New York, 1990.
50. T. J. Richmond, *JMB* **173**, 63 (1984).
51. P. Thiyagarajan and P. K. Ponnuswamy, *Biopolymers* **18**, 2233 (1979).
52. D. E. Bergstrom and N. J. Leonard, *Bchem* **11**, 1 (1972).
53. A. Favre, R. Buckingham and G. Thomas, *NARes* **2**, 1421 (1975).
54. L. S. Behlen, J. R. Sampson and O. C. Uhlenbeck, *NARes* **20**, 4055 (1992).
55. P. L. Wollenzien, P. Goswami, J. Teare, J. Szeberenyi and C. J. Goldenberg, *NARes* **15**, 9279 (1987).
56. P. L. Wollenzien, R. F. Murphy and C. R. Cantor, *JMB* **184**, 67 (1985).
57. B. Datta and A. M. Weiner, *JBC* **267**, 4497 (1992).
58. B. Datta and A. M. Weiner, *JBC* **267**, 4503 (1992).
59. G. D. Cimino, H. B. Gamper, S. T. Isaaks and J. E. Hearst, *ARB* **54**, 1151 (1985).
60. J. Christiansen, *NARes* **16**, 7457 (1988).
61. R. Brimacombe, J. Atmadja, W. Stiege and D. Schuler, *JMB* **199**, 115 (1988).
62. A. Expert-Bezancon and D. Hayer, *EJB* **103**, 365 (1980).
63. M. A. Grachev and M. I. Rivkin, *NARes* **2**, 1237 (1975).
64. E. Wickstrom, L. S. Behlen, M. A. Renben and P. R. Ainpour, *PNAS* **78**, 2082 (1981).
65. H. Han and P. B. Dervan, *PNAS* **91**, 4955 (1994).
66. A. B. Burgin and N. R. Pace, *EMBO J.* **9**, 4111 (1990).
67. J. M. Nolan, D. H. Burke and N. R. Pace, *Science* **261**, 762 (1993).
68. M. E. Harris, J. M. Nolan, A. Malhotra, J. W. Brown, S. C. Harvey and N. R. Pace, *EMBO J.* **13**, 3953 (1994).
69. M. Zenkova, C. Ehresmann, J. Caillet, M. Springer, G, Karpova, B. Ehresmann and P. Romby, *EJB* (1995). In press.
70. R. S. Brown, J. C. Dewan and A. Klug, *Bchem* **24**, 4785 (1985).
71. W. J. Krzyzosiak, T. Marciniec, M. Wiewiorowski, P. Romby, J. P. Ebel and R. Giege, *Bchem* **27**, 5771 (1988).
72. L. S. Behlen, J. R. Sampson, A. B. DiRenzo and O. C. Uhlenbeck, *Bchem* **29**, 2515 (1990).
73. J. R. Rubin and M. Sundaralingam, *J. Biomol. Struct. Dyn.* **1**, 639 (1983).
74. T. Pan and O. C. Uhlenbeck, *Bchem* **31**, 3887 (1992).
75. T. Pan, B. Dichtl and O. C. Unlenbeck, *Bchem* **33**, 9561 (1994).
76. B. Dichtl, T. Pan, A. B. DiRenzo and O. C. Uhlenbeck, *NARes* **21**, 351 (1993).
77. J. Ciesiolka, S. Lorenz and V. A. Erdmann, *EJB* **204**, 575 (1992).
78. S. Kazakov and S. Altman, *PNAS* **88**, 9193 (1991).
79. J. Ciesiolka, W.-D. Hardt, J. Schlegl, V. A. Erdmann and R. K. Hartmann, *EJB* **219**, 49 (1994).
80. C. S. Chow and J. K. Barton, *JACS* **112**, 2839 (1990).
81. J. R. Fresco, B. M. Alberts and P. Doty, *Nature* **188**, 98 (1960).
82. D. H. Turner, N. Sugimoto and S. M. Freier, *Annu. Rev. Biophys. Biophys. Chem.* **17**, 167 (1988).
83. M. Gouy, *in* "Nucleuc Acid and Protein Sequence Analysis: A Practical Approach" (M. J. Bishop and C. J. Rawlings, eds.), p. 259. IRL Press, Oxford, 1987.
84. A. Wada and A. Suyama, *Prog. Biophys. Mol. Biol.* **47**, 113 (1986).
85. M. Chamberlin, R. L. Baldwin and P. Berg, *JMB* **7**, 334 (1963).
86. H. De Voe and I. Tinoco, Jr., *JMB* **4**, 500 (1962).
87. C. R. Cantor and P. R. Schimmel, "Biophysical Chemistry." W. H. Freeman, San Francisco, California, 1980.

88. S. M. Freier, R. Kierzek, J. A. Jaeger, N. Sugimoto, M. H. Caruthers, T. Nelson and D. H. Turner, *PNAS* **83**, 9373 (1986).
89. W. Salser, *CSHSQB* **62**, 985 (1977).
90. G. Steger, H. Hoffman, J. Fortsch, H. J. Gross, J. W. Randles and H. L. Saenger, *J. Biomol. Struct. Dyn.* **2**, 543 (1984).
91. J. Ninio, *Biochimie* **61**, 1133 (1979).
92. C. Papanicolaou, M. Gouy and J. Ninio, *NARes* **12**, 31 (1984).
93. J. A. Jaeger, D. H. Turner and M. Zuker, *PNAS* **86**, 7706 (1989).
94. V. V. Filimonov and P. L. Privalov, *JMB* **122**, 465 (1978).
95. D. W. Appleby and N. R. Kallenbach, *Biopolymers* **12**, 2093 (1973).
96. L. A. Marky, N. R. Kallenbach, K. A. McDonough, N. C. Seeman and K. J. Breslauer, *Biopolymers* **26**, 1621 (1987).
97. A. E. Walter and D. H. Turner, *Bchem* **33**, 12715 (1994).
98. H. Heus and A. Pardi, *Science* **253**, 191 (1991).
99. G. Varani, C. Cheong and I. Tinoco, Jr., *Bchem* **30**, 3280 (1991).
100. H. Jackobson and W. H. Stockmayer, *J. Chem. Phys.* **18**, 1600 (1950).
101. J. M. Paipas and J. A. McMagon, *PNAS* **72**, 2017 (1971).
102. G. M. Studnicka, J. M. Rahu, I. M. Cummings and W. A. Salser, *NARes* **5**, 3365 (1978).
103. R. Nussinov and A. B. Jackobson, *PNAS* **77**, 6309 (1980).
104. M. Zuker and P. Stiegler, *NARes* **9**, 133 (1981).
105. M. Zuker, in "Mathematical Methods for DNA Sequences" (M. S. Waterman, ed.). CRC Press, Boca Raton, Florida, 1987.
106. R. Nussinov, G. Pieczenik, J. R. Gribbs and D. J. Kleitman, *SIAM J. Appl. Math.* **35**, 68 (1978).
107. M. Zuker and D. Sankoff, *Bull. Math. Biol.* **46**, 591 (1984).
108. P. Schuster, W. Fontana, P. F. Stadler and I. L. Hofacker, *Proc. R. Soc. Lond. Ser. B* **255**, 279 (1994).
109. I. Tinoco, Jr., O. C. Uhlenbeck and M. D. Levine, *Nature* **230**, 362 (1971).
110. L. V. Omelyanchuk, Yu E. Bessonov and N. A. Kolchanov, in "Computer Systems" (N. G. Zagoruiko, ed.), p. 135. Institute of Mathematics, Novosibirsk, 88, 1981 (in Russian).
111. N. A. Kolchanov and L. V. Omelyanchuk, *Stud. Biophys.* **87**, 115 (1982).
112. J.-P. Dumas and V. P. Ninio, *NARes* **10**, 197 (1982).
113. V. G. Tumanyan, L. E. Sotnikova and A. E. Holopov, *Dokl. Akad. Sci. USSR* **166**, 1465 (1966) (in Russian).
114. A. L. Williams and I. Tinoco, Jr., *NARes* **14**, 299 (1986).
115. E. Comay, R. Nussinov and O. Comay, *NARes* **12**, 53 (1984).
116. P. Hogeweg and B. Hesper, *NARes* **12**, 67 (1984).
117. A. B. Jackobson, L. Good, J. Simonetti and M. Zuker, *NARes* **12**, 45 (1984).
118. A. B. Jackobson and M. Zuker, *JMB* **233**, 261 (1994).
119. R. Kolter and C. Yanofsky, *ARGen* **16**, 113 (1982).
120. I. Tinoco, R. N. Borer, B. Dengler, M. D. Levine, O. C. Uhlenbeck, D. N. Grothers and J. Gralla, *Nature* **245**, 40 (1973).
121. M. Schmitz and G. Steger, *CABIOS* **8**, 389 (1992).
122. J. S. McCaskill, *Biopolymers* **29**, 1105 (1990).
123. S. V. Matveev, V. V. Filimonov and P. L. Privalov, *Mol. Biol.* **16**, 990 (1982).
124. A. A. Mironov and A. E. Kister, *Mol. Biol.* **19**, 1350 (1985) (in Russian).
125. A. A. Mironov, L. P. Dyakonova and A. E. Kister, *J. Biomol. Struct. Dyn.* **2**, 953 (1985).
126. A. A. Mironov and A. E. Kister, *J. Biomol. Struct. Dyn.* **4**, 1 (1986).
127. C. Levinthal, *J. Chim. Phys. (Paris)* **65**, 44 (1968).

128. O. B. Ptitsyn, *Usp. Sovr. Biol.* **69**, 26 (1970) (in Russian).
129. O. B. Ptitsyn, *Izv. Acad. Nauk SSSR* **5**, 57 (1973) (in Russian).
130. D. B. Wetlaufer, *PNAS* **70**, 697 (1973).
131. T. E. Kreighton, *Prog. Biophys. Mol. Biol.* **53**, 231 (1978).
132. L. M. Gierasch and J. King, eds., "Protein Folding, Description of the Second Half of the Genetic Code." Am. Assoc. Adv. Sci., Washington, D.C., 1989.
133. V. B. Bokhonov and N. A. Kolchanov, *in* "Mathematical Models of Molecular Genetic Systems Controls" (V. A. Ratner, ed.), p. 124. Institute of Cytology and Genetics, Novosibirsk, 1979.
134. B. R. Jordan, *J. Theor. Biol.* **34**, 363 (1972).
135. H. M. Martinez, *NARes* **12**, 323 (1984).
136. H. M. Martinez, *in* "Methods in Enzymology," Vol. 183, p. 306. Academic Press, San Diego, 1990.
137. J. P. Abrahams, M. van den Berg, E. van Batenburg and C. W. A. Pleij, *NARes* **18**, 3035 (1990).
138. A. P. Gultyaev, *NARes* **19**, 2489 (1991).
139. A. A. Mironov and V. F. Lebedev, *Biosystems* **30**, 49 (1993).
140. A. Fernandez, *EJB* **182**, 161 (1989).
141. A. Fernandez, *Chem. Phys. Lett.* **183**, 499 (1991).
142. A. Fernandez, *Phys. Rev. Lett.* **64**, 2328 (1990).
143. A. Fernandez, *Phys. Rev. A* **43**, 1138 (1991).
144. V. V. Anshelevich, A. V. Vologodskii, A. V. Lukashin and M. D. Frank-Kamenetskii, *Biopolymers* **23**, 39 (1984).
145. D. Porschke, *Biophys. Chem.* **2**, 97 (1974).
146. D. R. Mills, C. Dobkin and F. R. Kramer, *Cell* **15**, 541 (1978).
147. M. Levitt, *Nature* **224**, 759 (1969).
148. M. S. Waterman, *in* "Methods in Enzymology," Vol. 164, p. 765. Academic Press, San Diego, 1989.
149. H. F. Noller and C. R. Woese, *Science* **212**, 403 (1984).
150. H. F. Noller, *ARB* **53**, 119 (1984).
151. M. T. Dixon and B. M. Hills, *Mol. Biol. Evol.* **10**, 256 (1993).
152. W. C. Wheeler and R. L. Honeycutt, *Mol. Biol. Evol.* **5**, 90 (1988).
153. F. Michel and B. Dujon, *EMBO J.* **2**, 33 (1983).
154. W. C. Curtiss and J. N. Vournakes, *J. Mol. Evol.* **20**, 351 (1984).
155. K. Han and H.-J. Kim, *NARes* **21**, 1251 (1993).
156. D. Sankoff, *SIAM J. Appl. Math.* **45**, 810 (1985).
157. D. Sankoff, J. B. Kruskal, S. Mainville and R. J. Cedergen, *in* "Time Warps, String Edits and Macromolecules: The Theory and Practice of Sequence Comparison" (D. Sankoff and J. B. Kruskal, eds.), p. 93. Addison-Wesley, Reading, Massachusetts, 1983.
158. B. A. Shapiro, *CABIOS* **4**, 387 (1988).
159. H. Margalit, B. A. Shapiro, A. B. Oppenheim and J. V. Maizel, Jr., *NARes* **17**, 4829 (1989).
160. S. Y. Le, J. Owens, R. Nussinov, J. H. Chen, B. A. Shapiro and J. V. Maizel, *CABIOS* **5**, 205 (1989).
161. A. S. Noetzel and S. M. Selkow, *in* "Time Warps, String Edits and Macromolecules: The Theory and Practice of Sequence Comparison" (D. Sankoff and J. B. Kruskal, eds.), p. 237. Addison-Wesley, Reading, Massachusetts, 1983.
162. C. Chevalet and B. Michot, *CABIOS* **8**, 215 (1992).
163. P. Dumas, D. Moras, R. Giege, P. Verlaan, A. van Belkum and C. W. A. Pleij, *J. Biomol. Struct. Dyn.* **4**, 707 (1987).

164. T. Hayashi, Y. Ueno and T. Okamoto, *FEBS Lett.* **327,** 231 (1993).
165. M. Chastain and I. Tinoco, Jr., *Bchem* **32,** 14220 (1993).
166. C. W. A. Pleij, *Trends Biochem. Sci.* **15,** 143 (1990).
167. R. M. W. Mans and C. W. A. Pleij, *in* "Nucleic Acids and Molecular Biology" (F. Eckstein and D. M. J. Lilley, eds.), p. 250. Springer-Verlag, Berlin and New York, 1993.
168. K. Rietveld, R. Van Poelgeest, C. W. A. Pleij, J. M. Van Boom and L. Bosh, *NARes* **10,** 1929 (1982).
169. C. W. A. Pleij, K. Rietveld and L. Bosh, *NARES* **13,** 1717 (1985).
170. K. Rietveld, K. Linschooten, C. W. A. Pleij and L. Bosh, *EMBO J.* **3,** 2613 (1984).
171. R. L. Joshi, S. Joshi, F. Chapeville and A. L. Haenni, *EMBO J.* **2,** 1123 (1983).
172. A. Van Belkum, J. P. Abrahams, C. W. A. Pleij and L. Bosh, *NARes* **13,** 7673 (1985).
173. S. Stern, B. Wieser and H. F. Noller, *JMB* **204,** 447 (1988).
174. C. R. Woese and R. R. Guttel, *PNAS* **86,** 3119–3122 (1989).
175. D. S. McPheeter, G. D. Stormo and L. Gold, *JMB* **201,** 517 (1988).
176. C. K. Tang and D. E. Drapper, *Cell* **57,** 531 (1989).
177. T. Jacks, H. D. Madhani, F. R. Masiarz and H. E. Varmus, *Cell* **55,** 447 (1988).
178. I. Brierley, P. Digard and S. C. Inglis, *Cell* **57,** 537 (1989).
179. R. W. Davies, R. B. Wawring, J. A. Ray, T. A. Brown and C. Scazzocchio, *Nature* **300,** 719 (1982).
180. J. D. Puglisi, J. R. Wyatt and I. Tinoco, Jr., *Nature* **331,** 283 (1988).
181. J. R. Wyatt, J. D. Puglisi and I. Tinoco, Jr., *JMB* **214,** 455 (1990).
182. J. R. Wyatt and I. Tinoco, Jr., *in* "RNA World" (R. F. Gesteland and J. F. Atkins, eds.), p. 465. Cold Spring Harbor Laboratory, Cold Spring Harbor, New York, 1993.
183. J. D. Puglisi, J. R. Wyatt and I. Tinoco, Jr., *JMB* **214,** 437 (1990).
184. H. S. Chan and K. A. Dill, *J. Chem. Phys.* **90,** 492 (1989).
185. J.-M. Neefs and R. De Wachter, *NARefs* **18,** 5695 (1990).
186. J.-H. Chen, S.-Y. Le and J. V. Maizel, *CABIOS* **8,** 243 (1992).
187. W. Fontana, D. A. M. Konnings, P. F. Stadler and P. Schuster, *Biopolymers* **33,** 1389 (1993).
188. W. Fontana, P. F. Stadler, P. Tarazona, E. D. Weinberger and P. Schuseter, *Phys. Rev. E* **47,** 2083 (1993).
189. P. Schuster, *Artificial Life* **1,** 39 (1994).
190. M. A. Huynen, D. A. M. Konnings and P. Hogeweg, *J. Theor. Biol.* **165,** 251 (1993).
191. A. M. Gutin, A. Yu. Grosberg and E. I. Shakhnovich, *J. Phys. A: Math. Gen.* **26,** 1037 (1993).
192. O. B. Ptitsyn, *Mol. Biol.* **18,** 574 (1984) (in Russian).
193. H. Moine, P. Romby, M. Springer, M. Grunberg-Manago, J.-P. Ebel, B. Ehresmann and C. Ehresmann, *JMB* **216,** 299 (1990).

Transcriptional Activation of Thymidine Kinase, a Marker for Cell Cycle Control[1]

QING-PING DOU[*,2] AND
ARTHUR B. PARDEE[†]

*Department of Pharmacology
University of Pittsburgh School of
  Medicine and Division of Basic
  Research
University of Pittsburgh Cancer Institute
Pittsburgh, Pennsylvania 15213
†Division of Cell Growth and Regulation
Dana-Farber Cancer Institute
  and Department of Biological
  Chemistry and Molecular
  Pharmacology
Harvard Medical School
Boston, Massachusetts 02115

I. Major Players in the G0 to S Transition	198
II. Transcriptional Regulation of Thymidine Kinase as a Model for Late G1 Events	202
III. Role of E2F Complexes in the Mouse Thymidine Kinase Transcription	204
IV. Role of New Complexes Yi1 and Yi2 in the Mouse Thymidine Kinase Transcription	208
References	214

Cell proliferation is tightly controlled in normal mammalian cells, but is defective in cancer cells (1, 2). The growth-factor-mediated signals that drive the cell-cycle progression and thereby cell proliferation have been linked

---

[1] Abbreviations: A31, BABL/c 3T3 clone A31 cells; bp, base pair(s); BPA31, benzo[a]pyrene-transformed A31 cells; cdk, cyclin-dependent kinase; CHX, cycloheximide; Cln, G1 cyclins found in budding yeast; DHFR, dihydrofolate reductase; E2F, transcription factor; EGF, epidermal growth factor; IGF-1, insulin-like growth factor; MT1(2,3), DNA sequence in promoter of mouse TK gene; PCNA, proliferating cell nuclear antigen; PDGF, platelet-derived growth factor; pRB, retinoblastoma protein; R point, restriction point; TK, thymidine kinase; TS, thymidylate synthase; Yi1, Yi2, protein complexes that are different from E2F and that bind to the DNA sequence MT3.

[2] To whom correspondence should be addressed.

recently to functions of cyclins, the regulatory subunits of cyclin-dependent kinases (cdks) (*3–6*). Evidence has now accumulated to show that the G1/S phase-specific cyclin/cdk kinases modulate one or more transcription factors that in turn up-regulate transcription of several S phase genes, such as thymidine kinase (TK), shortly before initiation of DNA synthesis (*3, 4, 6*). Understanding molecular mechanisms for regulation of these S phase genes should provide insights into the molecular basis of cell proliferation and cancer.

Conditions that permit or inhibit entry into S phase (DNA synthesis) also similarly modulate activation of synthesis of enzymes such as TK (*6a*). Thus a connection exists between these two kinds of events. Either regulatory mechanisms responsible for transcriptions of these enzymes and for initiation of DNA replication are related, or a transcriptional activation produces the limiting component for activity of the DNA replication complex, which may be an enzyme such as a helicase (*6b*).

The purpose of this essay is to summarize the current information about transcriptional regulation of the mouse TK gene and the roles in this process of cyclin/cdk kinases, growth suppressor proteins, and transcription factors.

I. Major Players in the G0 to S Transition

Cells become quiescent (G0) when deprived of serum, which contains growth factors and cytokines (*1, 7*). Addition of growth factors move the G0 cells back into the cell cycle, and this process includes at least two stimulations (*7, 8*). The first stimulation of fibroblasts requires platelet-derived growth factor (PDGF), but not some others, which makes the cells "competent" and emergent into the initial G1 phase. The second stimulation lets the cells progress through the G1 phase, which is divided into early and late portions. Epidermal growth factor (EGF) and insulin are required for progression of competent fibroblasts through the early G1 phase (*7–9*), whereas insulin-like growth factor-1 (IGF-1) is needed for progression through the late G1 phase (*9*) up to the restriction (R) point, which is located in late G1 about 2 hours prior to the onset of DNA synthesis (*10*). After the R point, cells are committed to synthesize DNA and no exogenous growth factors are required (*10*). The growth-factor-mediated signal transduction pathways controlling the G0 to S transition have been extensively studied, and the involved regulatory proteins have been investigated (*1–6*).

A. R-point Protein

Kinetic experiments from cell biology suggest that a key protein must accumulate up to the R point before a cell can enter S phase (*10–12*). This R protein was proposed to have the following three properties: (1) it is synthesized in G1, (2) it is unstable, with a short half-life in nontransformed cells, and (3) it is stabilized or overproduced in tumor cells (13).

The R-protein hypothesis was based on the following kinetic experiments (12). Nontransformed BALB/c 3T3 A31 cells (A31 cells) were synchronized by serum starvation and then restimulated by addition of fresh serum. Prior to the R point, the protein synthesis inhibitor cycloheximide (CHX) was added for a period of 4–6 hours and removed (pulse inhibition). The entry of these cells into S phase, as determined by labeling nuclei, was found to be delayed by a period of time that is longer than the length of the CHX pulse (= the pulse + an excess delay). The excess delay of reentry into S phase, observed in nontransformed A31 cells, was proposed to be caused by the degradation of the R protein while CHX was present; its half-life was calculated to be 2.5 hours (14). In contrast, benzo[a]pyrene-transformed A31 cells (BPA31 cells) had no such excess delay for the onset of DNA synthesis, suggesting that transformation had altered this protein in stability or amount. Recent results suggest that the R protein might be one or a combination of different G1/S cyclins. The molecular basis of the R-protein effect could be production and inactivation of these cyclins.

B. Cyclins and cdk Proteins

Cyclins were first identified in marine invertebrate embryos as proteins that accumulate during interphase but undergo an abrupt degradation at the end of each mitosis (or meiosis) (15). G1 cyclins, called Cln 1, 2, or 3, were first identified in budding yeast, *Saccharomyces cerevisiae* (16). These cyclins are required for the G1 to S transition, and "knockout" of all three, but not of any one, is lethal (17). Mammalian G1 cyclins (C, D1, D2, D3, and E) were identified from their yeast homologs.

1. CYCLIN-D-DEPENDENT KINASES

Mammalian cyclin D1 was identified as the protein able to rescue the G1 arrest of yeast caused by Cln deficiency (18, 19). This cyclin is induced in G1 by colony-stimulating-factor 1 to murine macrophages (20), and is a candidate oncogene involved in certain parathyroid adenomas (21). Subsequently, cyclins D2 and D3 were cloned based on sequence homology and apparent times of appearance in the cell cycle (22, 23).

Cyclin D1 associates with several cdk partners in different systems. It complexes with cdk4 in macrophages (20, 24), with cdk2, 4, 5, and 6 as well as proliferating cell nuclear antigen (PCNA) in WI38 human fibroblasts (25, 26), and with cdk6 predominantly in peripheral blood T cells (26). In reconstitution systems, each of the D-type cyclins can activate cdk4 and cdk6, whereas D2 and D3 but not D1 can also functionally interact with cdk2 (24,

26, 27). In an *in vitro* system, the cyclin D/cdk complexes preferentially phosphorylate the retinoblastoma protein (pRB) but not histone H1 (24, 26). Phosphorylation of pRB by cyclin D2/cdk kinase was also seen *in vivo* (27).

Overexpression of mouse D-type cyclins accelerates G1 phase in rodent fibroblasts (28). Inhibition of cyclin D1 by microinjecting its antisense vectors or antibody into a cell blocked entry of the cell into S; this inhibition was seen only when microinjection was performed during the G1 interval but not at or after the G1/S transition (29), consistent with cyclin D1 regulating G1 progression. The cyclin D1 gene was overexpressed and/or deregulated by clonal chromosome rearrangements or by amplification in B cell lymphoma (30), parathyroid adenoma (21, 31), and esophageal (32), breast (33–35), and other cancers. All these support cyclin D1 as a candidate for the R protein.

2. Cyclin-E-dependent Kinases

Another mammalian G1 cyclin, cyclin E, was discovered by its ability to rescue Cln-defective yeast (18, 36). It forms complexes almost exclusively with cdk2 (37, 38). Both cyclin E protein and its associated cdk2 protein kinase activity reach maximal levels in late G1, before the average cell in the population enters S. The cyclin E-associated kinase activity was correlated with the appearance of cyclin E–cdk2 complex (37–39), suggesting direct involvement at G1/S. Indeed, constitutive overexpression of cyclin E in human fibroblast cells shortened the duration of G1, decreased cell size, and diminished the serum requirement for the transition from G1 to S phase (40). Overexpression of cyclin E releases a pRB-induced G1 block (41). In addition, microinjection of a specific antibody to cdk2, which interacts with cyclin E as well as cyclin A (37, 38), results in G1 arrest (42). Cyclin E and its dependent histone H1 kinase activities satisfied all the properties of the R protein, which includes late G1 phase increase, an excess delay of appearance after inhibition of protein synthesis in nontransformed cells, and a faster recovery in transformed cells (39), suggesting it as a candidate for the R protein. Consistent with this, there is deranged expression of cyclin E protein in numerous tumor cell lines and tissue samples (43, 44).

3. Cyclin-A-dependent Kinases

Cyclin A functions in both S and G2/M phases (4–6). The human cyclin A gene was first identified as an E1A-binding protein (45), and as integration site of the hepatitis B virus in a hepatocellular carcinoma (46). In mammalian cells, the cyclin A–cdk2 kinase appears after the beginning of S phase (47,

48, 39). Overexpression of cyclin A also overcomes the pRB-mediated G1 block (41). Microinjection of either antisense oligonucleotides or antibodies against cyclin A blocks or delays DNA synthesis (47, 48). Cyclin A is also a possible candidate for the R protein because it possesses the three proposed properties (39).

C. Retinoblastoma Protein

The retinoblastoma protein (pRB), product of a tumor suppressor gene, serves as an inhibitor of cell proliferation in G1 phase when it is underphosphorylated (3, 4). As cells approach S phase pRB becomes hyperphosphorylated by cyclin-dependent kinases (3, 4) and it loses its suppressor activity (49–52). Introduction of the RB gene into a human osteogenic sarcoma cell line, which lacks full-length pRB, arrested these cells in G0/G1 phase in a metabolically active state; cotransfection of cyclins D2 (27), E, or A (41) prevents or overrides this pRB block, causing cell cycle progression. Consistent with these results, cdk2, which is activated by interactions of cyclins D, E, and A (3, 4, 27, 37, 38), phosphorylated pRB at G1/S (53). Overexpression of p107, a pRB-family protein, also inhibited proliferation in certain cell types, arresting sensitive cells in G1. However, growth inhibition by pRB and p107 did not occur through the same mechanism (54).

D. Transcription Factor E2F

E2F, originally found as a cellular transcription factor with a DNA binding activity (55), plays a role in the regulation of transcription of viral E1A and some cellular genes (56, 57). The "free" E2F binding activity, which is induced at G1/S (56, 57), is proposed to be responsible for activation of gene transcription. Most recently, E2F has been shown to be a family of proteins, and several E2F-like proteins have been cloned, including E2F-1, -2, -3, -4, and DP-1 (56–58).

The levels of E2F-1 mRNA were low in G0/G1 phase and increased at the G1/S boundary after serum stimulation (57), which may contribute partially to the S phase-specific "free" E2F binding activity. E2F activities may be also regulated by association with different cellular proteins in different stages of the cell cycle. E2F forms several different complexes during the cell cycle with underphosphorylated pRB (G1 and S phases) (59–61), p107 (G0 and G1 phases) (62, 63), p107/cyclin E/cdk2 (G1/S boundary) (64), or p107/cyclin A/cdk2 (S phase) (61–68). Overexpression of the E2F-1 gene induces quiescent cells to enter S phase (69), followed by p53-dependent apoptosis (70–72).

II. Transcriptional Regulation of Thymidine Kinase as a Model for Late G1 Events

At the onset of S phase, expression of several genes encoding products associated with DNA replication are induced, including histones, thymidine kinase (TK), thymidylate synthase (TS), and dihydrofolate reductase (DHFR) (1, 6, 7, 73). The R protein may control expression of these S phase genes as well as the onset of DNA synthesis, which was proposed based on their serum dependence as well as the following kinetic experiment (74). A several hour pause with CHX given to serum-stimulated A31 cells in G1 delays the increases of TK and *in vivo* TS activities as well as the onset of DNA synthesis by an interval several hours longer than the pause. Furthermore, the extra delay is not seen when transformed BPA31 cells are used (9, 14, 74). These observations suggest that expression of an S phase gene such as TK may be used as a molecular marker for G/S phase events. Studies of TK, which are technically far simpler to investigate at the molecular biological level than are investigations of DNA synthesis, should provide molecular clues for understanding the general G1–S cell-cycle-regulatory machinery.

TK is regulated at levels of transcription, post-transcription, translation (73, 75), and post-translation (76), thus ensuring a large increase in its activity when cells enter S phase. The transcriptional regulation of TK has been well studied. Comparison of the 5' sequences of TK genes from a number of species, including human, chicken, hamster, and mouse, indicates that TK promoter sequences have undergone extensive evolutionary divergence (77). The promoters of human, chicken, and hamster TK, although significantly divergent from each other, contain CCAAT and TATA elements (77). In striking contrast, the mouse TK gene promoter contains neither of these elements (78, 79).

The presence or absence of CCAAT and TATA elements in the promoter regions of the various TK genes may influence the sites of transcriptional initiation. Indeed, both human and hamster TK genes rely on one or two preferred sites of transcriptional initiation, whereas the mouse TK has over 20 different start sites that are distributed over a distance of about 200 bases upstream of the translational initiation site (80). In this regard, the mouse TK gene resembles other traditional "housekeeping" genes, such as those encoding TS and DHFR.

A 291-bp fragment from the 5' end of the mouse TK gene has promoter function in the context of a TK minigene (78). Further experiments both confirm and extend these findings (81). In addition, using a series of TK promoter–reporter gene fusion constructs, sequences located between −174 bp upstream and +159 bp downstream of the TK translation initiation site were found sufficient for S phase-specific expression (81).

To further define regulatory elements located in this region of the murine TK promoter by independent methods, we investigated DNA sequences within the TK promoter that are bound by murine nuclear proteins *in vitro*. Using DNase I footprint analysis with crude nuclear extracts prepared from exponential A31 cells, three protected regions were identified on both coding and noncoding strands of the upstream 170-bp fragment, named MT1 (−104 bp/−84 bp), MT2 (−83 bp/−67 bp), and MT3 (−43 bp/−28 bp), respectively (82) (Fig. 1). Subsequent studies demonstrated that these three MT elements bind transcription factors/complexes Sp1, E2F, and Yi1/TKE/Egr-1, respectively.

The MT1 region includes a perfect "GC" box (−96 CCCGCC −91), which is a consensus binding site for the transcription factor Sp1 (83). This is supported by the ability of a purified human Sp1 protein to bind and protect the MT1 region in footprint assays. The purified Sp1 protein did not protect MT2 and MT3 elements (82), consistent with the assertion that they are independent binding regions. Furthermore, a point mutation (changing a G residue to C at −93 bp), introduced into the MT1 region, completely eliminates Sp1 binding (82). The same point mutation on the MT1 site virtually blocks expression of a TK-reporter fusion gene in transfection assays (81), demon-

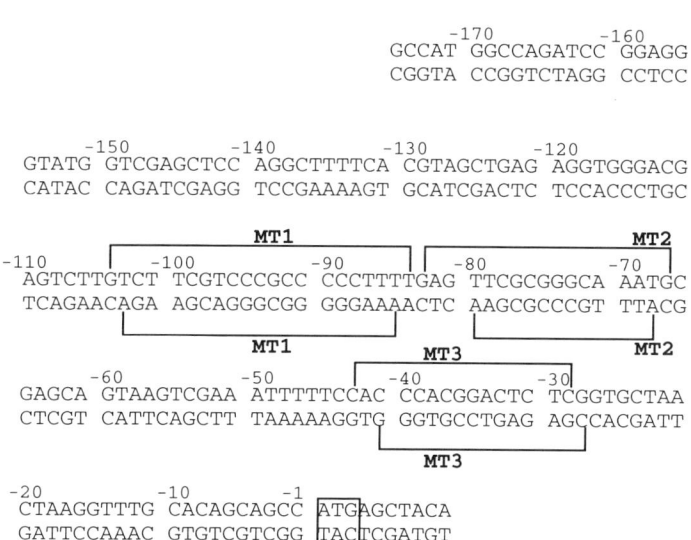

FIG. 1. Summary of DNase I footprint data. The sequence of the upstream promoter region of the murine TK gene is shown (78, 81, 82). The brackets indicate protected sequences, designated MT1, MT2, and MT3, respectively. The TK translational start site, ATG, is boxed. Reproduced from Ref. 82, with permission.

strating that Sp1 plays a critical role in determining expression from the TK promoter.

III. Role of E2F Complexes in the Mouse Thymidine Kinase Transcription

A. DNA Motif MT2 Binds E2F Complexes

MT2 includes an E2F-like binding site (GTTCGCGGGCAAA) (63). MT2 binds specifically to an affinity-purified fusion human E2F protein. Both MT2 and an authentic E2F site (TTTCGCGCGCTTT) bind specifically to similar or identical nucleoprotein complexes. Formation of both these DNA–protein complexes are cell-cycle-dependent: a G0/G1 phase-specific complex (E2F–G0/G1) was replaced by an S phase-specific complex(es) (E2F–S) while "free" E2F increased after the G1/S transition. Pulse inhibition of protein synthesis with CHX interchanged these complexes with similar kinetics. E2F–G0/G1, E2F–S, and "free" E2F complexes were eluted and analyzed by Western blot assay using a specific antiserum to human E2F-1; two forms of murine E2F (62 and 66 kDa) were observed. These MT2-bound complexes contain p107/cyclin E/cyclin A/cdk2 proteins (63). Also, MT2 is an E2F binding site involved in gene activation by polyoma virus large-T antigen (84).

1. G1/S-SPECIFIC BINDING OF NUCLEAR PROTEINS TO MT2

By performing a gel retardation assay using A31 cells synchronized by serum starvation, we investigated timing in the cell cycle of nuclear proteins binding to the MT2 site. Four cell-cycle-regulated bands were detected with either an MT2 or E2F sequence (63) (Fig. 2), named E2F–G0/G1, E2F–S, "free" E2F, and Yi2. The three E2F-containing bands, confirmed by several lines of experiments, were named according to the established E2F binding patterns and properties (66–68). The E2F–G0/G1 band was detected in G0 (0 hours) and G1 phase (6 hours) but disappeared before G1/S (12 hours), at which time cells began to synthesize DNA. The slower E2F–S band(s) (which may contain two overlapped complexes) appeared around G1/S (between 8 and 12 hours) (Fig. 2) and increased dramatically in S phase (17, 22, and 36 hours). The "free" E2F was low before 12 hours but increased dramatically after cells entered S phase with kinetics similar to those of the E2F-S complex. Similar cell-cycle-regulated E2F-containing complexes have also been found in other systems using the E2F consensus sequence as probe (66–68). Yi2 was undetectable in G0 and G1 phase, but increased as cells

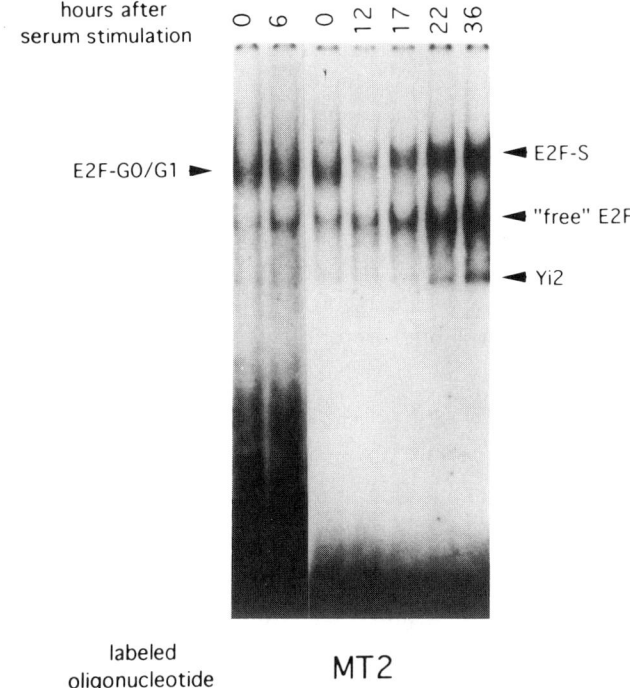

FIG. 2. Binding of mouse nuclear proteins to the MT2/E2F site is cell-cycle-regulated. Nuclear extracts (10 μg protein per reaction) were prepared from synchronized A31 cells harvested at the indicated times after serum stimulation. Labeled wild-type MT2 binding sequence was used in the bandshift assay. Complexed DNAs are indicated. Reproduced from Ref. 63, with permission.

entered S (see below). In contrast to these four cycle-regulated bands, binding of Sp1 to the MT1 site was constitutive.

To provide evidence for the "free" E2F, we used deoxycholate (DOC), which dissociates E2F from its associated proteins (66). DOC abolished both MT2-bound and G0/G1 and S phase-specific complexes, and simultaneously increased abundance of the "free" E2F (63), indicating release of free E2F protein from associated complexes by this treatment. Therefore, it is very likely that the slower mobilities of E2F–G0/G1 and E2F–S are caused by interactions of the "free" E2F with other proteins.

2. PULSE INHIBITION OF PROTEIN SYNTHESIS INTERCHANGED E2F-CONTAINING COMPLEXES

Changes of E2F complexes at G1/S (Fig. 2) could be caused by production of a labile G1/S phase-specific cyclin protein(s). If so, one would expect that pulse inhibition of protein synthesis affects MT2/E2F binding. CHX was added to a synchronized 12-hour BPA31 cell culture, which contained mainly E2F–S and "free" E2F, detected by using either an MT2 or E2F sequence (63). During the 2.5- or 5-hour CHX treatment, E2F–S was greatly reduced, whereas an E2F–G0/G1-like band(s) appeared simultaneously. After removing CHX from the cells, E2F–G0/G1 was first induced and then gradually decreased while E2F–S-like band(s) reappeared and remained as a main complex(es). The "free" E2F decreased during the CHX treatment but gradually recovered after removal of the inhibitor. In contrast, the Sp1–MT1 complexes were not changed during the pulse-chase process. These data support involvement of a labile protein(s) at G1/S in conversion of E2F–G0/G1 to E2F–S.

3. Presence of Murine E2F in All the MT2-bound Nuclear Protein Complexes

To provide direct evidence that murine E2F is involved in formation of the MT2-bound, G1/S-regulated protein complexes, E2F–G0/G1, E2F–S, and "free" E2F were excised from a wet gel and eluted. The eluted proteins were rerun on a sodium dodecyl sulfate–polyacrylamide gel, and detected in Western blot assay with a specific antiserum to human E2F-1. In all the three MT2-bound complexes, two major polypeptides of 62 and 66 kDa were identified (63), which may represent murine E2F proteins, because a similar doublet of human E2F has been reported (85).

4. Presence of p107/cdk2/Cyclin A/Cyclin E in E2F-MT2 Complexes

To investigate if members of pRB, cdk, and cyclin families are involved in MT2 bindings, as found in complexes binding to the consensus E2F site (59–68), we used antibodies to proteins of these families. We first tested different monoclonal culture supernatants to human p107 (SD6, SD9, and SD15) and to human pRB (XZ55, XZ91, and XZ104). One of the anti-p107 antibodies (SD9) reduced both E2F–G0/G1 and E2F–S. Another anti-p107 antibody (SD15) had no effect on the E2F–G0/G1 band, but slightly reduced E2F–S, and simultaneously generated a supershifted band. In contrast, none of the anti-pRB antibodies had any effect on either of these complexes.

We then used purified antibodies to the C terminus of cdk2 and cdc2 proteins. The anti-cdk2 antibody did not affect E2F–G0/G1 but reduced E2F–S and simultaneously gave a supershifted band. The anti-cdc2 had no effect on either of the complexes.

We also found that an antiserum to cyclin A inhibited, whereas an antiserum to cyclin E supershifted, the E2F–S complexes. Neither of these two antisera had effects on E2F–G0/G1. Because these two antibodies do not

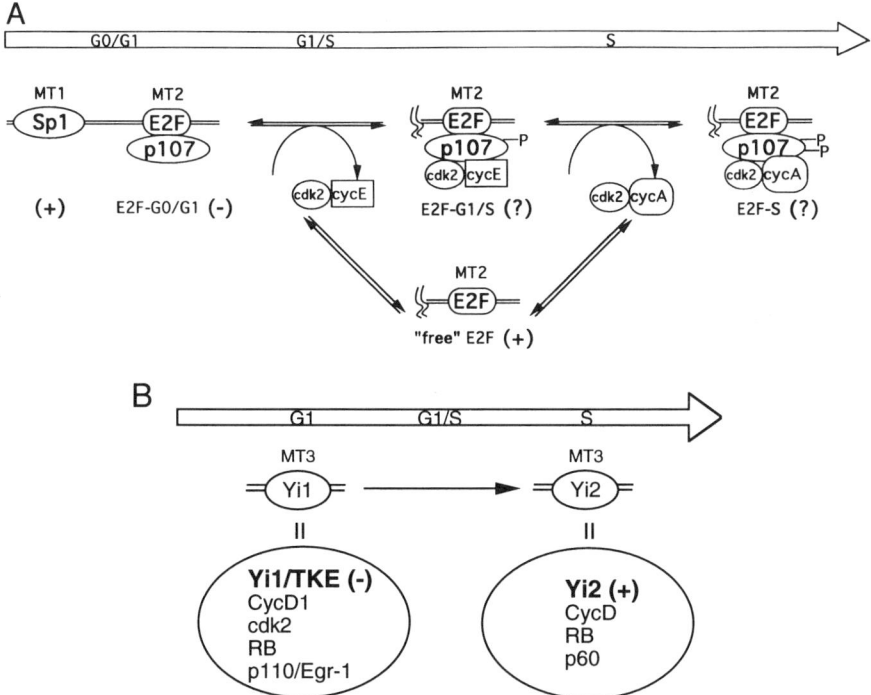

FIG. 3. Formation of E2F-containing (A) and Yi (B) complexes during the cell cycle. (+), A *trans*-activating function; (−), a repressing function; (?), an uncertain function. (A) Reproduced from Ref. 63, with permission.

cross-react with different cyclin proteins (37, 39), these data suggest that E2F–S may contain two overlapping bands. One contains cyclin A and another possesses cyclin E. These antibody studies suggest that E2F–G0/G1 contains p107 and E2F–S complex(es) contain p107, cdk2, cyclin A, and cyclin E (Fig. 3A).

B. Function of MT2-bound E2F Complexes

To study the function of the E2F complexes binding to MT2, the E2F-binding site within MT2 was removed from the promoter and replaced by a 9-bp substitution (86). This mutated TK promoter–reporter fusion gene was transfected into A31 cells, and stable transfectants were isolated. Cells containing different transfectants were arrested by serum deprivation, then restimulated to proliferate by the addition of fresh growth medium. Substitution of the E2F-binding site resulted in nearly constitutive expression of the reporter gene, despite a normal S phase induction of the endogenous TK message (86). The resultant gene expression in G0/G1 phase might be derived from Sp1 binding to the MT1 site (81, 82) (Fig. 3A). These data strongly suggest that MT2 bindings in G0/G1 phase, which were mainly in the E2F–G0/G1 form, act as a repressor.

Because E2F is a positive transcription factor for other genes (56, 57), the responsible repressor protein in this complex could be p107, the pRB-like protein. It is possible that in G0/G1 phase, p107 complexes with and inactivates the E2F protein; this E2F–p107 complex (E2F–G0/G1), when bound to MT2, actively represses the activity induced by the transcription factor Sp1 that binds to MT1, resulting in a block of gene transcription (Fig. 3A). At G1/S, the activated cyclin E/cdk2 kinase may interact with the E2F–p107 complex, producing an E2F/p107/cyclin E/cdk2 kinase complex and releasing the "free" E2F. Soon after cells enter S phase, E2F/p107/cyclin E/cdk2 may interact with the induced cyclin A/cdk2 kinase and form E2F/p107/cyclin A/cdk2 (62, 63). It has now been shown that cyclin A/cdk2 kinase phosphorylates E2F effectively, and this phosphorylation abolishes its ability to bind DNA and mediate transactivation (87).

IV. Role of New Complexes Yi1 and Yi2 in the Mouse Thymidine Kinase Transcription

A. DNA Sequence MT3 Binds Yi Complexes

1. DIFFERENT Yi COMPLEXES BIND TO MT3 DURING THE G0 TO S TRANSITION

MT3 (−43 bp to −28 bp) is a sequence different from MT1 or MT2 (82) (Fig. 1). When MT3-site oligonucleotides were used in gel retardation assays, a Yi1 complex (formerly TKE, which also contains an Egr-1-like protein) was found to peak in early G1; it gradually decreased afterward, and disappeared in late G1. In contrast, Yi2 binding activity dramatically increased after G1/S (82, 88) (Fig. 4). Treatment with alkaline phosphatase rapidly abolished Yi1 binding, and Yi2 binding was abolished more slowly. Phosphorylation activated the Yi2 complex, because incubation with ATP but not ATP-γ-S, an analog of ATP, increased Yi2 binding activity (88). These data indicate the involvement of a kinase in production of Yi2 at the G1/S boundary.

2. BINDING OF Yi1 TO MT3 IS SEQUENCE SPECIFIC

Yi1 binding was observed only when MT3, but not MT1 or MT2, was used, whereas Yi2 was bound to all three MTs (63, 82, 88). To find which base pairs are involved in Yi1 binding, we made different mutations in the

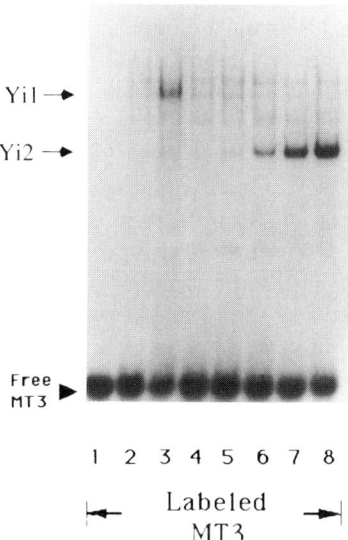

FIG. 4. G1/S phase-specific induction of Yi binding to an MT3 element. Labeled (~30,000 cpm/reaction) synthetic oligonucleotides containing the MT3 (−43 bp to −23 bp) sequence were mixed with 20 μg of crude nuclear extract prepared from A31 cells harvested at the indicated times following serum stimulation. Each binding reaction was performed in the presence of 0.63 μg of poly(dI–dC) as nonspecific competitor DNA. The free and complexed DNAs are indicated. Reproduced from Ref. 82, with permission.

MT3 region and used these mutated sequences in competition experiments (89). The most specific mutation in the MT3 region, a point mutation of C to A at −37, resulted in complete loss of competing ability (89), indicating that this is a critical binding base pair for Yi1. Mutations in MT3 had less effects on Yi2 binding, consistent with its lesser sequence specificity.

3. DNA-binding Proteins in Yi Complexes

To study the DNA-binding component of each complex, we performed a Southwestern blot analysis of nuclear extracts prepared from A31 cells in different cycle stages. Two major DNA-binding proteins of 60 (p60) and 110 (p110) kDa were detected with the MT3 oligonucleotide, but not by the AP1-binding site probe, indicating a relative sequence specificity. p60 binding was constitutive. In contrast, p110 activity was cycle-regulated, being very low in early G1 and peaking in mid-G1 phase (4 to 8 hours) (88).

Both p60 and p110 were found in heparin column fractions prepared from BPA31 nuclei. Column fractions with Yi2 activity contained p60, suggesting p60 as the DNA-binding protein in Yi2. p110 appears to be the DNA-binding component of Yi1, as supported by several lines of evidence: (1) the peaks of Yi1 and p110 from the heparin column cofractionated, (2) the peaks of both Yi1 and p110 were found in G1 phase, and (3) bindings of both Yi1 and p110 were abolished by alkaline phosphatase treatments.

4. Presence of RB/cdk2/Cyclin in Yi Complexes

Both Yi1 and Yi2 have molecular masses over 240 kDa, measured by Sephacryl S-300 gel filtration (unpublished data), suggesting that they are multiprotein complexes. To identify these proteins we incubated A31 nuclear extracts with different antibodies against pRB, cdk, or cyclins, and then assayed Yi by gel retardation (88, 89). Monoclonal antibody 245 to human pRB eliminated both Yi1 and Yi2 retardations; another human pRB monoclonal antibody 340 and goat antimouse IgG controls did not show these effects. With Western blot analysis of crude A31 nuclear extracts, antibody 245, but not 340, revealed two forms of murine pRB, p100 and p110. These data show specific interactions between antibody 245 and pRB in both Yi complexes.

In addition, anti-cdk2, but not anti-cdc2, selectively eliminated the Yi1 complex, and simultaneously generated two bands with faster mobility (indicated by F1 and F2, Fig. 5) (89). To determine if cyclins are involved, we used antibodies to different cyclins. An affinity-purified antibody to cyclin D1 at a very low dose (0.25 $\mu g/\mu l$) completely inhibited Yi1 and simultaneously generated a new band with a faster mobility, similar to that of band F1 (Fig. 5, lane 7 vs. lane 6). Antisera to cyclins A, E, and B had no effect on formation of Yi1. Because both cyclin D1 (23, 24) and Yi1 (82, 88) are early G1 phase-specific, these data strongly suggest that cyclin D1 is present in Yi1. Therefore, Yi1 may contain cyclin D1/cdk2 kinase.

None of the antibodies to cdc2 and cdk2 had any clear effects on Yi2 formation, and antibodies to cyclins A, B, and E had no effect on Yi2. However, the purified anticyclin D1 antibody, which did not affect Yi2 at a low dose (0.25 $\mu g/\mu l$, lane 6 vs. lane 7 in Fig. 5), reduced Yi2 formation at a high dose (0.5 $\mu g/\mu l$) (lane 6 vs. lane 8). The same amount of anti-cdc2 had no such effects (lane 6 vs. lane 3). These data suggest that Yi2 may contain the same or a different D-type cyclin with respect to Yi1 (see Fig. 3B).

B. Cloning of MT3-binding Proteins

To identify recombinant clones that encode proteins capable of binding to the mouse TK MT3 element, a λgt11-3T3 fibroblast-cDNA expression library was screened with a labeled MT3 oligonucleotide probe. Several recombinant clones that reproducibly bound the multimeric MT3 probe were isolated, including two that were named clone 17 (90) and clone 18 (unpublished data, and see Ref. 89).

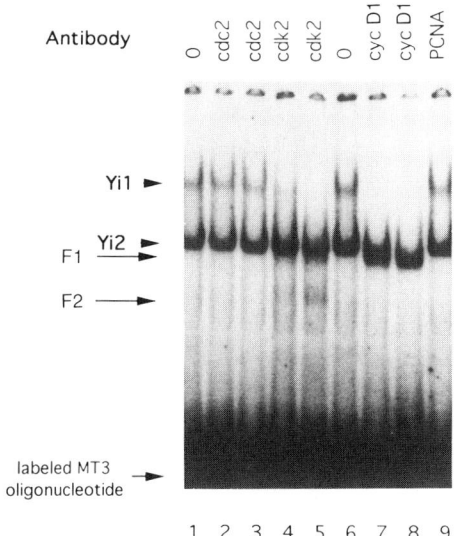

FIG. 5. Effects of antibodies on Yi1 and Yi2 complexes. A nuclear extract (5 μg per reaction), prepared from G1/S phase A31 cells, was first incubated for 3 hours with a purified antibody to the C terminus of cdc2 (lanes 2 and 3), the C terminus of cdk2 (lanes 4 and 5), to cyclin D1 (lanes 7 and 8), or to PCNA (lane 9) at a final concentration of 0.25 μg/μl (lanes 2, 4, 7, and 9) or 0.50 μg/μl (lanes 3, 5, and 8), and then assayed for MT3 binding. No antibody was added to lanes 1 and 6. Free MT3, Yi1, Yi2, and two other complexes (F1 and F2, formed after addition of antibodies) are indicated. Reproduced from Ref. 89, with permission.

1. CLONE 17/Egr-1

Clone 17 contains an insert of about 1.5 kb (90). The DNA sequence shows a perfect identity with the murine transcription factor Egr-1 between nucleotide 235 and 1552 of the Egr-1 nucleotide sequence, including the three zinc fingers of the DNA-binding domain (91).

To assess the sequence specificity of clone 17/Egr-1 binding to MT3, fusion proteins were produced from either the full-length clone 17 or a more soluble 243-aminoacid peptide corresponding to the DNA-binding domain of Egr-1. These fusion proteins bound to MT3, as demonstrated by Southwestern and gel-retardation assays (90). In addition, binding of clone 17/Egr-1 to MT3 was competed by a consensus Egr-1 binding sequence, but not by others, such as the MT2/E2F-binding site.

Binding of nuclear factors to the Egr-1 consensus sequence peaked in early G1 phase (1–2 hours), indistinguishable from that of Yi1/TKE to MT3 (82). The relationship of serum-induced Yi1/TKE activity to Egr-1 was studied using an antiserum that specifically recognizes Egr-1. Addition of this antiserum to the gel-shift reaction eliminated serum-stimulated Egr-1 binding to the EGR probe, and this antiserum also inhibited the serum-induced Yi1/TKE binding activity to the MT3 probe. Thus, the DNA-binding component protein in Yi1/TKE complex is probably Egr-1 protein (see Fig. 3B).

2. CLONE 18

Clone 18 encodes a novel protein. A 460-bp cDNA fragment, corresponding in sequence to 153 amino acids contained within this MT3-binding protein, was expressed in a prokaryotic system using the pGEX expression vector. The expressed polypeptide was gel purified and used for generating a polyclonal antiserum (anti-C12). Western blot analysis of a nuclear extract prepared from G1/S phase BPA31 cells showed that the C12 antibody bound strongly to two major cellular proteins, p110 and p60 (89). These results are consistent with the previous finding that two major DNA-binding proteins, p110 and p60, bound to the labeled MT3 probe in Southwestern blot assays (88). Therefore, the clone 18 product may also be related to the DNA-binding proteins in Yi complexes.

We expected that addition of C12 antiserum in a gel-retardation assay would affect formation of Yi complexes. Indeed, anti-C12, but not its preimmune serum, blocked Yi1 formation, and simultaneously increased the level of Yi2. This antiserum generated a faint supershifted band, indicating a specific reaction between the anti-C12 and a component protein (which may be p110) in Yi1, supporting the suggestion that Yi2 is derived from or related to Yi1. Even though anti-C12 interacted with a p60 under denatured conditions, it did not block formation of Yi2 in this bandshift assay. Two possible interpretations are that (1) the recognition site of p60 by anti-C12 may be hidden when p60 is complexed with other proteins such as pRB in Yi2 (88) and (2) the p60 recognized by anti-C12 under Western blotting conditions may be different from the one present in Yi2.

Because Yi1 contains at least the DNA-binding protein p110 (88) and cyclin D1/cdk2 kinase (89) (Fig. 3B), these three proteins might interact. If so, C12 antiserum would co-immunoprecipitate cyclin D1/cdk2 kinase, which can phosphorylate pRB or p107 *in vitro* (24). Indeed, anti-C12, but not its preimmune serum, precipitated a kinase(s), from G1 phase BPA31 cell nuclear extracts, which phosphorylated added GST–p107 fusion protein in the *in vitro* kinase assay. Anti-C12 also co-immunoprecipitated the same or similar kinase activity from S phase BPA31 cells (89). These results suggest that either transformed BPA 31 cells have altered kinetics of cyclin-D1-containing Yi1 complex formation or another D-type cyclin is present in the S phase-specific Yi2 complex in these cells.

C. Potential Functions of Yi Complexes

That Yi1 is different from E2F complexes is supported by the following evidence. First, Yi1 and E2F complexes have different consensus binding sequences. The necessary DNA sequence for Yi1 includes CCCAC (82, 88, 89), whereas the consensus binding sequence for E2F is TTCGCG (56, 57, 63) (Fig. 1). Second, Yi1 and E2F complexes contain different DNA-binding proteins. Yi1 contains the DNA-binding protein p110 (88), not E2F as does MT2 (63). In addition, the cDNAs that were cloned, as based on the MT3-binding activity of their gene products, encode proteins that have different amino-acid sequences from E2F (89, 90). In contrast, a purified human GST–E2F protein interacted only with E2F-binding sequences but not with MT3 DNA (63). Third, Yi1 and E2F complexes contain different cyclin components. Yi1 contains cyclin D1/cdk2 kinase (Fig. 5) (89), whereas E2F complexes contain cyclin E/cdk2 and cyclin A/cdk2 kinases (56, 57, 63) (Fig. 3).

Egr-1 activates the mouse TK promoter through the MT3 element. To investigate whether Egr-1 could activate transcription of a reporter gene driven by the MT3-containing TK promoter sequence, previously shown to be serum responsive (81), an Egr-1 expression plasmid was cotransfected with the TK–CAT reporter, which led to sixfold stimulation of CAT reporter gene activity (90). A control expression plasmid encoding a truncated version of Egr-1 did not increase the reporter gene activity. To demonstrate that activation of the TK promoter by Egr-1 depends on the MT3 binding site, a site-directed mutation was introduced into MT3, which prevented Egr-1-activated TK promoter activity. Because Egr-1-mediated activation of the TK reporter gene was detected in serum-starved cells and not in a population of serum-fed cycling cells (90), these data suggest that the role of Egr-1/TKE serves as a G0-to-G1 transition factor.

Yi is connected to proliferation and cancer. Yi1 is formed in early G1 (82, 88, 89), when cyclin D1 appears and has its effects (23, 24). In contrast, Yi2 appears in late G1 (82, 88), when cyclins E (3, 4, 39) and D3 (24) are active. However, the presence of cyclin D1/cdk2 kinase in Yi1 links an oncogenesis-related D cyclin (21) to a cell-cycle event, which is a specific DNA-binding activity that is probably involved in regulating later TK transcription and DNA synthesis. The cyclin D-regulated Yi complex might coordinate with cyclin E/cyclin A-regulated E2F complexes during the middle of G1 to S transit, in controlling transcription of S phase genes (such as TK) and thereby DNA synthesis (see Fig. 3). The target molecule of cyclin D1/cdk2 in the Yi1 complex should be the pRB protein (88). The functional significance of this association to form the Yi1 complex is not yet known; however, we propose a down-regulatory effect on MT3, which would work in cooperation with the E2F complex on the MT2 site. This double suppressor effect in early G1 phase could compensate the effect of the permanently serum-stimulated Sp1 site.

Formation of Yi2 might also be controlled by accumulation of the R protein inasmuch as its stability and cell cycle regulation are different in nontransformed A31 and transformed BPA31 cells (92). The differences in regulation were observed after serum withdrawal, when serum levels were returned and while protein synthesis was inhibited. After serum reduction, Yi2 was unstable in nontransformed A31 but was significantly more stable in transformed BPA31 cells. After serum concentration was returned to high levels, Yi2 activity patterns were also different between A31 and BPA31 cells. Although protein synthesis was inhibited, again Yi2 is much more stabilized in BPA31 than in A31 cells (92). Consistent with that finding, expression of Yi2 is at least fivefold higher in human breast tumor tissues than in adjacent normal tissues (unpublished data).

The information that we and others have derived from molecular studies on regulation of the mouse TK transcription clearly indicates involvement of cyclin/cdk kinases, pRB, E2F, and Yi complexes. The challenge for the future will be to investigate further how external and internal signals regulate these G1/S major players. A better understanding of these processes in normal cells should shed light on how these controls become deranged in tumor cells.

Acknowledgments

This research was supported by Public Health Service Grant GM24571, a start-up fund from Department of Pharmacology, University of Pittsburgh School of Medicine, and a grant from the Council for Tobacco Research.

References

1. A. B. Pardee, *Science* **246**, 603 (1989).
2. L. H. Hartwell and M. B. Kastan, *Science* **266**, 1821 (1994).
3. C. J. Sherr, *Cell* **79**, 551 (1994).
4. T. Hunter and J. Pines, *Cell* **79**, 573 (1994).
5. P. Nurse, *Cell* **79**, 547 (1994).
6. A. B. Pardee, K. Keyomarsi and Q.-P. Dou, *in* "Colony Stimulating Factors: Molecular and Cell Biology" (J. Garland, ed.), in press, 1995.
6a. D. L. Coppock and A. B. Pardee, *MCBiol* **7**, 2925 (1987).
6b. B. Stillman, *Cell* **78**, 725 (1994).

7. R. Baserga, "The Biology of Cell Reproduction." Harvard Univ. Press, Cambridge, Massachusetts, 1985.
8. C. D. Stiles, *Cell* **33**, 653 (1983).
9. H. C. Yang and A. B. Pardee, *J. Cell. Physiol.* **127**, 410 (1986).
10. A. B. Pardee, *PNAS* **71**, 1286 (1974).
11. P. W. Rossow, V. G. H. Riddle and A. B. Pardee, *PNAS* **76**, 4446 (1979).
12. J. Campisi, E. E. Medrano and A. B. Pardee, *PNAS* **79**, 436 (1982).
13. A. B. Pardee, *Cancer Res.* **47**, 1488 (1987).
14. A. B. Pardee, J. Campisi, and R. G. Croy, *Ann. N.Y. Acad. Sci.* **397**, 121 (1982).
15. T. Evans, E. T. Rosenthal, J. Youngblom, D. Distel and T. Hunt, *Cell* **33**, 389 (1983).
16. J. A. Hadwiger, C. Wittenberg, H. E. Richardson, M. de Barros Lopes and S. I. Reed, *PNAS* **86**, 6255 (1989).
17. H. E. Richardson, C. Wittenberg, F. Cross and S. I. Reed, *Cell* **59**, 1127 (1989).
18. D. J. Lew, V. Dulic and S. I. Reed, *Cell* **66**, 1197 (1991).
19. Y. Xiong, T. Connolly, B. Futcher and D. Beach, *Cell* **65**, 691 (1991).
20. H. Matsushime, M. F. Roussel, R. A. Ashmun and C. J. Sherr, *Cell* **65**, 701 (1991).
21. T. Motokura, T. Bloom, H. G. Kim, H. Juppner, J. V. Ruderman, H. M. Kronenberg and A. Arnold, *Nature* **350**, 512 (1991).
22. H. Kiyokawa, X. Busquets, C. T. Powell, L. Ngo, R. A. Rifkind and P. A. Marks, *PNAS* **89**, 2444 (1992).
23. T. Motokura, K. Keyomarsi, H. M. Kronenberg and A. Arnold, *JBC* **267**, 20412 (1992).
24. H. Matsushime, M. E. Ewen, D. K. Strom, J.-Y. Kato, S. K. Hanks, M. F. Roussel and C. J. Sherr, *Cell* **71**, 323 (1992).
25. Y. Xiong, H. Zhang and D. Beach, *Cell* **71**, 505 (1992).
26. M. Meyerson and E. Harlow, *MCBiol* **14**, 2077 (1994).
27. M. E. Ewen, H. K. Sluss, C. J. Sherr, H. Matsushime, J.-Y. Kato and D. M. Livingston, *Cell* **73**, 487 (1993).
28. D. E. Quelle, R. A. Ashmun, S. A. Shurtleff, J. Y. Kato, D. Bar-Sagi, M. F. Roussel and C. J. Sherr, *Genes Dev.* **7**, 1559 (1993).
29. V. Baldin, J. Lukas, M. J. Marcote, M. Pagano and G. Draetta, *Genes Dev.* **7**, 812 (1993).
30. C. L. Rosenberg, E. Wong, E. M. Petty, A. E. Bale, Y. Tsujimoto, N. L. Harris and A. Arnold, *PNAS* **88**, 9638 (1991).
31. C. L. Rosenberg, H. G. Kim, T. B. Shows, H. M. Kronenberg and A. Arnold, *Oncogene* **6**, 449 (1991).
32. W. Jiang, S. M. Kahn, N. Tomita, Y.-J. Zhang, S.-H. Lu and I. B. Weinstein, *Cancer Res.* **52**, 2980 (1992).
33. J. Bartkova, J. Lukas, H. Muller, M. Strauss, B. Gusterson and J. Bartek, *Cancer Res.* **55**, 949 (1995).
34. E. A. Musgrove, J. A. Hamilton, C. S. L. Lee, K. J. E. Sweeney, C. K. W. Watts and R. L. Sutherland, *MCBiol* **13**, 3577 (1993).
35. M. F. Buckley, K. J. E. Sweeney, J. A. Hamilton, R. L. Sini, D. L. Manning, R. I. Nicholson, A. deFazio, C. K. W. Watts, E. A. Musgrove and R. L. Sutherland, *Oncogene* **8**, 2127 (1993).
36. A. Koff, F. Cross, A. Fisher, J. Schumacher, K. Leguellec, M. Phillippe and J. M. Roberts, *Cell* **66**, 1217 (1991).
37. A. Koff, A. Giordano, D. Desai, K. Yamashita, J. W. Harper, S. Elledge, T. Nishimoto, D. O. Morgan, B. R. Franza and J. M. Roberts, *Science* **257**, 1689 (1992).
38. V. Dulic, E. Lees and S. I. Reed, *Science* **257**, 1958 (1992).
39. Q.-P. Dou, A. H. Levin, S. Zhao and A. B. Pardee, *Cancer Res.* **53**, 1493 (1993).
40. M. Ohtsubo and J. M. Roberts, *Science* **259**, 1908 (1993).

41. P. W. Hinds, S. Mittnacht, V. Dulic, A. Arnold, S. I. Reed and R. A. Weinberg, *Cell* **70**, 993 (1992).
42. M. Pagano, R. Pepperkok, J. Lukas, V. Baldin, W. Ansorge, J. Bartek and G. Draetta, *J. Cell Biol.* **121**, 101 (1993).
43. K. Keyomarsi, N. O'Leary, G. Molner, E. Lees, H. J. Fingert and A. B. Pardee, *Cancer Res.* **54**, 380 (1994).
44. J. Gong, B. Ardelt, F. Traganos and Z. Darzynkiewicz, *Cancer Res.* **54**, 4285 (1994).
45. A. Giordano, P. Whyte, E. Harlow, B. R. Franza, D. Beach and G. Draetta, *Cell* **58**, 981 (1989).
46. J. Wang, X. Chenivesse, B. Henglein and C. Brechot, *Nature* **343**, 555 (1990).
47. F. Girard, U. Strausfeld, A. Fernandez and N. J. C. Lamb, *Cell* **67**, 1 (1991).
48. F. Zindy, E. Lamas, X. Chenivesse, J. Sobczak, J. Wang, D. Fesquet, B. Henglein and C. Brechot, *BBRC* **182**, 1144 (1992).
49. K. Buchkovich, L. A. Duffy and E. Harlow, *Cell* **58**, 1097 (1989).
50. P.-L. Chen, P. Scully, J.-Y. Shew, J. Y. J. Wang and W.-H. Lee, *Cell* **58**, 1193 (1989).
51. J. A. DeCaprio, J. W. Ludlow, D. Lynch, Y. Furukawa, J. Griffin, H. Piwnica-Worms, C.-M. Huang and D. M. Livingston, *Cell* **58**, 1085 (1989).
52. K. Mihara, X.-R. Cao, A. Yen, S. Chandler, B. Driscoll, A. L. Murphree, A. T'Ang and Y.-K. T. Fung, *Science* **246**, 1300 (1989).
53. T. Akiyama, T. Ohuchi, S. Sumida, K. Matsumoto and K. Toyoshima, *PNAS* **89**, 7900 (1992).
54. L. Zhu, S. van den Heuvel, K. Helin, A. Fattaey, M. Ewen, D. Livingston, N. Dyson and E. Harlow, *Genes Dev.* **7**, 1111 (1993).
55. A. S. Yee, P. Raychaudhuri, L. Jakoi and J. R. Nevins, *MCBiol* **9**, 578 (1989).
56. J. R. Nevins, *Science* **258**, 424 (1992).
57. P. J. Farnham, J. E. Slansky and R. Kollmar, *BBA* **1155**, 125 (1993).
58. D. Ginsberg, G. Vairo, T. Chittenden, Z.-X. Xiao, G. Xu, K. L. Wydner, J. A. DeCaprio, J. B. Lawrence and D. M. Livingston, *Genes Dev.* **8**, 2665 (1994).
59. S. P. Chellappan, S. Hiebert, M. Mudryj, J. M. Horowitz and J. R. Nevins, *Cell* **65**, 1053 (1991).
60. L. R. Bandara, J. P. Adamczewski, T. Hunt and N. B. La Thangue, *Nature* **352**, 249 (1991).
61. L. Cao, B. Faha, M. Dembski, L.-H. Tsai, E. Harlow and N. Dyson, *Nature* **355**, 176 (1992).
62. J. K. Schwarz, S. H. Devoto, E. J. Smith, S. P. Chellappan, L. Jakoi and J. R. Nevins, *EMBO J.* **12**, 1013 (1993).
63. Q.-P. Dou, S. Zhao, A. H. Levin, J. Wang, H. Helin and A. B. Pardee, *JBC* **269**, 1306 (1994).
64. E. Lees, B. Faha, V. Dulic, S. I. Reed and E. Harlow, *Genes Dev.* **6**, 1874 (1992).
65. S. H. Devoto, M. Mudryj, J. Pines, T. Hunter and J. R. Nevins, *Cell* **68**, 167 (1992).
66. M. Mudryj, S. H. Devoto, S. W. Hiebert, T. Hunter, J. Pines and J. R. Nevins, *Cell* **65**, 1243 (1991).
67. M. Pagano, G. Draetta and P. Jansen-Durr, *Science* **255**, 1144 (1992).
68. S. Shirodkar, M. Ewen, J. A. DeCaprio, J. Morgan, D. M. Livingston and T. Chittenden, *Cell* **68**, 157 (1992).
69. D. G. Johnson, J. K. Schwartz, W. D. Cress and J. R. Nevins, *Nature* **365**, 349 (1993).
70. X. Wu and A. J. Levine, *PNAS* **91**, 3602 (1994).
71. X.-Q. Qin, D. M. Livingston, W. G. Kaelin and P. D. Adams, *PNAS* **91**, 10918 (1994).
72. B. Shan and W.-H. Lee, *MCBiol* **14**, 8166 (1994).
73. L. F. Johnson, *Curr. Opin. Cell Biol.* **4**, 149 (1992).
74. D. L. Coppock and A. B. Pardee, *J. Cell. Physiol.* **124**, 269 (1985).
75. G. B. Knight, J. M. Gudas and A. B. Pardee, *Jpn. J. Cancer Res.* **80**, 493 (1989).

76. Z.-F. Chang, D.-Y. Huang and N.-C. Hsue, *JBC* **269,** 21249 (1994).
77. S. S. Arcot, E. K. Flemington and P. L. Deininger, *JBC* **264,** 2343 (1989).
78. H. B. Lieberman, P.-F. Lin, D.-B. Yeh and F. H. Ruddle, *MCBiol* **8,** 5280 (1988).
79. C. Seiser, M. Knofler, I. Rudelstorfer, R. Haas and E. Wintersberger, *NARes* **17,** 185 (1989).
80. J. M. Gudas, J. L. Fridovich-Keil, M. W. Datta, J. Bryan and A. P. Pardee, *Gene* **118,** 205 (1992).
81. J. L. Fridovich-Keil, J. M. Gudas, Q.-P. Dou, I. Bouvard and A. B. Pardee, *Cell Growth Diff.* **2,** 67 (1991).
82. Q.-P. Dou, J. L. Fridovich-Keil and A. B. Pardee, *PNAS* **88,** 1157 (1991).
83. J. T. Kadonaga, K. A. Jones and R. Tjian, *Trends Biochem. Sci.* **11,** 20 (1986).
84. E. Ogris, H. Rotheneder, I. Mudrak, A. Pichler and E. Wintersberger, *J. Virol.* **67,** 1765 (1993).
85. W. G. Kaelin, W. Krek, W. R. Sellers, J. A. DeCaprio, F. Ajchenbaum, C. S. Fuches, T. Chittenden, Y. Li, P. J. Farnham, M. A. Blaner, D. M. Livingston and E. K. Flemington, *Cell* **70,** 351 (1992).
86. J. L. Fridovich-Keil, P. J. Markell, J. M. Gudas and A. B. Pardee, *Cell Growth Diff.* **4,** 679 (1993).
87. B. D. Dynlacht, O. Flores, J. A. Lees and E. Harlow, *Genes Dev.* **8,** 1772 (1994).
88. Q.-P. Dou, P. J. Markell and A. B. Pardee, *PNAS* **89,** 3256 (1992).
89. Q.-P. Dou, G. Molnar and A. B. Pardee, *BBRC* **205,** 1859 (1994).
90. G. Molnar, A. Crozat and A. B. Pardee, *MCBiol* **14,** 5242 (1994).
91. V. P. Sukhatme, X. Cao, L. C. Chang, C.-H. Tsai-Morris, D. Stamenkovich, P. C. P. Ferreira, D. R. Cohen, S. A. Edwards, T. B. Shows, T. Curren, M. M. Le Beau and E. D. Adamson, *Cell* **53,** 37 (1988).
92. D. W. Bradley, Q.-P. Dou, J. L. Fridovich-Keil and A. B. Pardee, *PNAS* **87,** 9310 (1990).

Eukaryotic Gene Expression: Metabolite Control by Amino Acids[1]

Roney O. Laine,[2]
Richard G. Hutson
and Michael S. Kilberg

*Department of Biochemistry and
  Molecular Biology
University of Florida College of Medicine
Gainesville, Florida 32610*

I. Amino-acid Control in Bacteria	219
II. Amino-acid Control in Yeast	220
III. Adaptive Regulation of Eukaryotic Plasma Membrane Amino-acid Transport	226
IV. Aminoacid-dependent Regulation of Asparagine Synthetase Expression	232
V. Identification and Characterization of Aminoacid-regulated Ribosomal-protein Genes	238
VI. Summary	245
References	245

I. Amino-acid Control in Bacteria

Depletion of a single amino acid from the growth medium of most unicellular organisms leads to increased transcription of genes coding for the enzymes in the cognate pathway. The consequences of amino-acid starvation are reduction of both the intracellular pool and the corresponding charged

---

[1] Abbreviations: tRNA, transfer RNA; GCN, general control, nondepressable; GCD, general control, derepressed; DAI, double-stranded RNA-activated inhibitor; HCR, heme-controlled repressor; UTR, untranslated region; AIB, 2-aminoisobutyric acid; cDNA, complementary DNA; eIF, eukaryotic initiation factor; MDCK, Madin–Darby canine kidney cell line; AS, asparagine synthetase; BHK, hamster kidney cell line; CAT, chloramphenicol acetyltransferase; AARE, amino-acid regulatory element; KRB, Krebs–Ringer bicarbonate buffer; MEM, minimum essential medium; RP, ribosomal protein; AA+, culture medium containing amino acids; AA-, culture medium lacking amino acids; cpc-1, cross-pathway-control gene; FIRE, *fos* intragenic regulatory element; Rev, regulator of virion protein expression; *Fau*, Finkel–Biskis–Reilly murine sarcoma virus-associated ubiquitously expressed gene; tum, tumor-specific transplantation antigens that arise on mouse tumor cells treated with mutagens.

[2] To whom correspondence may be addressed

tRNA. These effects, along with ribosome stalling, are the basis for transcription termination control (attenuation) in bacteria (1). For example, the attenuation process of the tryptophan operon depends on the simultaneous translation and transcription of the newly synthesized RNA (1, 2). Within the leader sequence are two tryptophan codons (UGG); when the cellular content of this amino acid is sufficient, the ribosome translates through these codons, triggering an mRNA secondary structure that acts as a termination signal for further transcription. However, if the cellular tryptophan content is too low, translation stalls at the tandem UGG codons, resulting in a unique RNA stem–loop structure that permits transcription beyond the attenuator site. Transcriptional initiation also is regulated in an aminoacid-dependent manner through a tryptophan-activated repressor protein that specifically binds to the tryptophan operon promoter (1).

In contrast to unicellular organisms, additional levels of control exist in mammals. *In vivo*, a change in metabolite concentration can cause corresponding changes in enzyme activity through complex hormonal or neural processes rather than through direct transcriptional or translational control by substrate signaling. Amino-acid or protein deprivation of the entire animal produces significant changes in circulating hormones that, in turn, activate or inhibit numerous pathways. Therefore, dissection of the cellular pathways for direct nutrient-dependent regulation of gene expression must be accomplished by *in vitro* studies. In mammalian cells, for a heterogeneous nuclear RNA (hnRNA) molecule to be converted to a functionally mature mRNA, it first must be capped, polyadenylylated, spliced within the nucleus, and then exported out of the nucleus (3–6). Splicing steps are targets for metabolite control (7), and proposed control of nuclear RNA export involving the nuclear matrix, nucleolus, and nuclear pore complex increases the number of potential nuclear regulatory steps (8). Of course, once in the cytoplasm, nutrient-dependent mRNA turnover and translational control provide additional regulatory steps for expression.

II. Amino-acid Control in Yeast

One of the most studied mechanisms of aminoacid-dependent gene expression is that of the yeast, *Saccharomyces cerevisiae*. In addition to "specific control" of genes involved in the synthesis of individual amino acids, this yeast employs a "general control" process for nitrogen metabolism (GCN, general *c*ontrol *n*ondepressable) whereby over 40 different genes in more than 10 different biosynthetic pathways are regulated by starvation of the cell for a single amino acid (9, 10). Although transcription of the affected genes is ultimately altered (10), the evidence suggests that the initial signal for the general control response increases translation of the mRNA for the transcription factor GCN4 (9). Thus, during amino-acid starvation, there is a decrease in general protein synthesis due to a limited supply of substrate, yet the synthesis of GCN4, and other selected proteins, is increased.

Depletion of the intracellular pool for a single amino acid is sufficient to increase the expression of GCN4 and the genes under its control, although amino-acid depletion is not an absolute requirement. For example, there is induction of the GCN-associated genes in the yeast *ils1* mutant, which contains a temperature-sensitive isoleucyl-tRNA synthetase and expresses low levels of the active enzyme at the nonpermissive temperature (11). Although the intracellular isoleucine pools remain near normal, the cells are deficient in isoleucyl-tRNA, and enzymes in at least four amino-acid biosynthetic pathways are induced under these conditions (11). The increased gene expression in the *ils1* mutant is blocked by a mutation in the positive regulator protein GCN1. This mutation also prevents the starvation-induced control, suggesting that a reduction in acylated-tRNA content and amino-acid depletion affect GCN4 control through a common pathway (11).

Similar results are observed with the yeast *mes1* mutant, which produces a temperature-sensitive methionyl-tRNA synthetase (12). Starvation for purines also stimulates translation of GCN4 by the same mechanism as starvation for amino acids, suggesting that the GCN4 response may be a global sensor of cellular nitrogen (13).

Although the regulatory mechanism and many of the interactions of the gene products involved in the general amino-acid control pathway in yeast have been elucidated biochemically, genetic approaches have played a major part in the deciphering of this mechanism. A common method used by us in identifying genes involved in this type of control is to impose amino-acid starvation by culturing wild-type cells in the presence of inhibitors of amino-acid biosynthesis. For example, 3-aminotriazole inhibits the activity of the HIS3 gene product, imidazole glycerol phosphate dehydratase, causing histidine starvation. Another approach is to culture a strain that is deficient for an amino-acid biosynthetic enzyme in a medium deficient in the required amino acid. The genes involved in this regulatory system were identified from yeast mutants that cannot derepress GCN4 translation, or mutants that are constitutively derepressed (GCD, general *c*ontrol *d*erepressed). This methodology has led to the identification of numerous mutants in both the GCN and GCD classes. The interactions between the protein products of the various GCN and GCD genes has been reviewed (9, 14, 15). Many of the known GCD and GCN genes encode subunits of the initiation factor eIF-2B and the eIF-2B-specific protein kinase GCN2.

Based on the interactions between the GCN and GCD mutations, the GCN4 protein has been shown to affect the derepression of the amino acid

Fed Condition

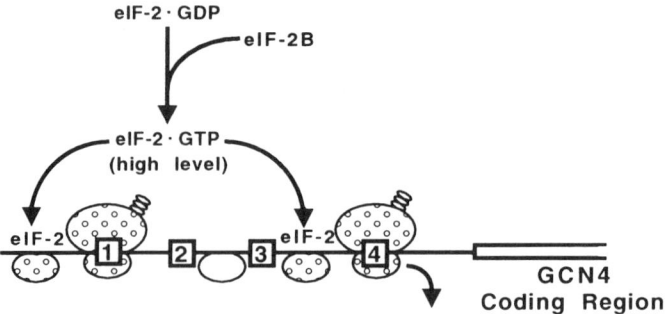

Starved Condition

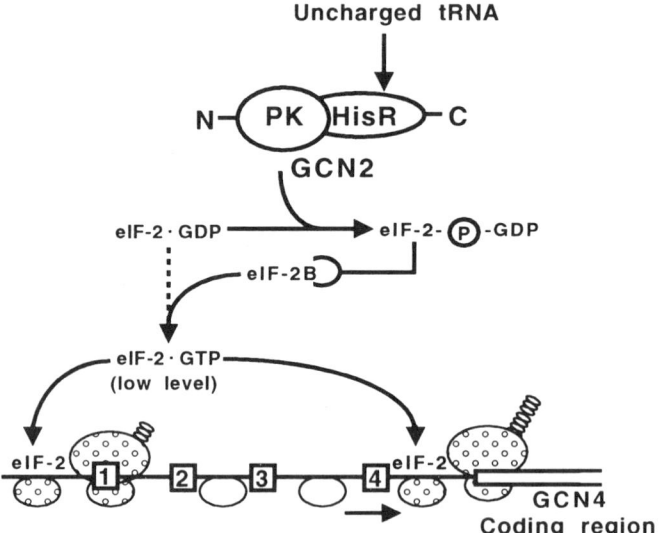

FIG. 1. Model for the aminoacid-dependent translational control of GCN4 synthesis in yeast. The GCN4 mRNA is shown with uORFs 1–4 and the GCN coding region (indicated as boxes). Ribosomal subunits (40 S) associated with eIF-2 and thus competent to initiate translation contain dots. Those subunits not associated with eIF-2 and not competent to initiate translation do not contain dots. As discussed in the text, under aminoacid-fed conditions, eIF-2–GDP is converted to eIF-2–GTP, which associates with the 40-S subunit, forming a ternary complex of eIF-2, GTP, and Met-tRNA. After translation of uORF1, the 40-S subunit continues scanning the RNA until the necessary factors reassociate with the subunit once again, making it competent to reinitiate translation. On translation of uORF4, translation is terminated, and the GCN4 coding region is not translated. Under amino-acid starvation conditions, the level of eIF-2–GTP is reduced due to GCN2 phosphorylation of eIF-2–GDP, and the

biosynthetic genes more directly than the other identified GCN and GCD genes (16). Additionally, the GCN4 protein increases in response to amino-acid starvation, and this increase is the result of increased translation of the GCN4 mRNA (17). The GCN4 protein is a member of the bZIP family of transcriptional activators that bind to DNA as homodimers (18). The GCN4 protein binds to the sequence TGACTC found in the promoter of all genes under its control (10, 19). Serial deletions of the GCN4 protein have identified separate regions responsible for DNA binding and transcriptional activation located along the linear amino-acid sequence (20). The DNA binding and dimerization domain of GCN4 is located in the extreme C-terminal 56 amino acids of the protein (20–22), and is homologous to the mammalian oncogene *jun* (23). A transcriptional activation domain is believed to be located in the center of the protein in a region rich in acidic amino acids (20). The deletion of the N-terminal 95 amino acids of GCN4, leaving the central activation domain intact, reduces the efficiency of transcriptional activation and the dependence on a TATA element (24). A recent study employed serial deletions of the GCN4 protein expressed under control of its native promoter and translational control elements (25). These deletion experiments indicate that GCN4 contains two activation domains, each of similar strength, that can function independently to promote essentially wild-type levels of HIS3 and HIS4 target-gene transcription (25). Each of these domains contains two or more subdomains that are dependent on hydrophobic amino acids for their function and that may mediate interactions between GCN4 and other proteins in the transcription initiation complex (25).

Aminoacid-dependent synthesis of the GCN4 transcription factor depends on translational control of the GCN4 mRNA. A model put forth by Hinnebusch and colleagues is summarized in Fig. 1. Translation initiation is a multistep biochemical pathway directed toward positioning the ribosome and associated factors at an AUG initiation codon at the beginning of an mRNA coding sequence. Two pathways, cap-dependent and internal, are known. Internal initiation appears to be associated primarily with a number of viral mRNAs, in an attempt to evade cellular defense mechanisms. The major pathway of translation initiation is cap-dependent, commonly referred to as the scanning model. This process involves the attachment of initiator Met-tRNA to the 40-S ribosomal subunit, binding of this complex at or near the m^{7}GpppG cap structure at the 5' end of the mRNA, migration of the ribosome along the mRNA to search for AUG codons (scanning), and recognition of the initiator codon and its flanking sequences. The aminoacid-dependent

---

reduced availability of eIF-2B. Ribosomal subunits competent to reinitiate translation are not formed until after the 40-S subunit has scanned beyond uORF4; thus, translation is reinitiated at the GCN4 coding region (model modified from Ref. 12, copyright 1992 Cell Press).

control of GCN4 mRNA translation relies on the proposed scanning mechanism of translational initiation; the data generated in the elucidation of the translational control of GCN4 support the scanning mechanism.

As shown in Fig. 1, the 591-nucleotide leader sequence of the GCN4 mRNA contains four short upstream open reading frames (uORF), which precede the actual GCN4 coding sequence (12). Generally, uORFs occur rarely in most eukaryotic transcripts, and their occurrence inhibits translation of the downstream product (26–28). This inhibitory effect has been explained by proposing that ribosomes must begin at the 5' end of the transcript and scan the mRNA leader until they come to the first AUG codon where translation begins. When an uORF is present, initiation occurs at that site and precludes the downstream coding sequence, apparently because of the time required for reinitiation (27, 28). The uORFs in the 5' untranslated region (UTR) of the GCN4 transcript act as a translational control element, allowing ribosomal translation of the GCN4 coding sequence only during aminoacid-starvation conditions. Removal of the four uORFs from the GCN4 mRNA transcript results in constitutive derepression (9, 14, 15). Upstream ORFs 2 and 3 do not appear to be necessary to maintain translational regulation of GCN4, because a combination of uORF1 and 4 is sufficient for nearly wild-type regulation of GCN4 expression (14).

The regulatory functions of uORF1 and uORF4 can be mimicked by various uORFs found in heterologous promoters, although less efficiently, suggesting that no requirement exists for certain nucleotide sequences, secondary structure, or the peptide end-products of these uORFs (15). Reversing the 5'–3' order of uORFs 1 and 4 completely abolishes the regulation of GCN4 expression, showing that uORF1 and uORF4 are not functionally interchangeable. uORF4 is sufficient to repress GCN4 translation under nonstarvation conditions because it acts as a translational barrier and allows GCN4 expression to only a few percent of that seen in the absence of uORFs (14). Transfer of the last codon plus the 10 following nucleotides of uORF4 to the uORF1 is sufficient to convert uORF1 into a strong translational barrier, suggesting that the characteristics of translation termination at uORF4 are responsible for its inhibition of GCN4 translation (29). Grant and Hinnebusch separately randomized the sequence of the third codon and the termination region of uORF4, showing that many different A–U-rich triplets present at the third codon can overcome the inhibitory effect of this translation termination region (30).

Efficient reinitiation is not associated with codons specifying a certain amino acid or isoacceptor tRNA, suggesting that reinitiation is impaired by stable base-pairing between nucleotides surrounding the uORF4 stop codon and the tRNA that pairs with the third codon, the rRNA, or sequences located in the GCN4 mRNA (30). These interactions delay the scanning ribosome following peptide chain termination, and thus lead to ribosome dissociation from the mRNA (30).

Insertions of sequences around uORF4 with the potential to form secondary structure abolish aminoacid-dependent derepression, indicating that ribosomes reach the GCN4 coding region by traversing uORF4 rather than by binding internally to the GCN4 start site (31). Showing that wild-type regulation occurred when the reading frame of uORF4 was extended beyond the GCN4 initiator codon provided evidence that the ribosomes that translate GCN4 do so by ignoring the uORF4 initiator codon (31). Overall, the function of the uORF4 is to delay the scanning ribosome following peptide chain termination, and thus lead to ribosome dissociation from the mRNA (31).

In contrast to uORF4, uORF1 is a "leaky" translational barrier, reducing GCN4 expression to 50% of that seen in the absence of uORFs (15). uORF1 is required to overcome the inhibitory effect of uORF4 and permit stimulation of GCN4 translation in aminoacid-starved cells. Single amino-acid substitutions in uORF1 have no effect on GCN4 expression, whereas point mutations that lengthen or shorten uORF1 by a single codon significantly lower GCN4 expression under starvation conditions. Increasing the distance between uORF1 and uORF4 to equal that distance between uORF1 and GCN4 impairs the ability of uORF1 to induce GCN4 translation. These results suggest that under aminoacid-rich conditions the ribosomes translate uORF1, reinitiate at uORF4, terminate translation of uORF4, and dissociate from the mRNA without translating the GCN4 sequence (Fig. 1). Under aminoacid-starvation conditions, ribosomes translate uORF1 and then fail to reassemble the necessary factors needed to reinitiate at uORF4, continue scanning along the mRNA, and become translationally competent at the GCN4 coding sequence (31).

The product of the yeast *GCN2* gene is essential for translational control of GCN4 (9, 12). A portion of the GCN2 protein sequence is homologous to histidinyl-tRNA synthetases from *S. cerevisiae*, humans, and *Escherichia coli*, and this sequence is juxtaposed to a protein kinase domain (11, 32). The role of the GCN2 protein kinase was elucidated by comparison with the mammalian protein kinases, i.e., double-stranded RNA-activated inhibitor (DAI) and heme-controlled repressor (HCR), both of which are involved in the regulation of translation in mammalian cells (11). During translation initiation, eukaryotic initiation factor 2 (eIF-2) forms a ternary complex with GTP and methionyl-tRNA^{Met} and functions to promote binding of methionyl-tRNA^{Met} to the 40-S ribosomal subunit (33–35). Phosphorylation of the eIF-2 α subunit by the protein kinases DAI or HCR inhibits mammalian translation under conditions in which these kinases are active (11). GCN2 phosphorylates the yeast eIF-2 α subunit in an aminoacid-starvation-dependent manner on the same serine residue that is modified by the mammalian kinases (12).

The GCN2 protein is associated with ribosomal subunits and polysomes (36). Under conditions in which polysomes are dissociated, the fraction of GCN2 protein cosedimenting with polysomes is reduced, and when the association of 40-S and 60-S subunits is prevented by omission of Mg^{2+}, almost all of the GCN2 protein comigrates with the 60-S subunit. GCN2 can be dissociated from the 60-S subunit with 0.5 M KCl, suggesting that it is loosely associated with ribosomes rather than being an integral ribosomal protein (36). The extreme carboxyl-terminal portion of GCN2 is necessary for its interaction with ribosomes, and its ability to up-regulate GCN4 translation (36). These observations, along with the preferential association of GCN2 with free ribosomal subunits during exponential cell growth, suggest that GCN2 interacts with ribosomes during translation initiation, and is probably involved in the initiation of GCN4 translation.

Thus, it is proposed that, in yeast, the GCN2 protein lacks specificity for tRNA^{His} and recognizes an increased number of uncharged tRNAs, thereby activating the GCN2-associated protein kinase to phosphorylate eIF-2 (32). When eIF-2 is phosphorylated, it binds more tightly to the guanine exchange factor eIF-2B and does not exchange GDP for GTP, a necessary part of translation initiation (33, 34). The reduced level of eIF-2–GTP leads to a decreased rate of reinitiation allowing the 40-S subunit to scan beyond uORF4 before eIF-2–GTP associates with the subunit (Fig. 1). As a result, the ribosome reinitiates at the GCN4 coding sequence. These changes in reinitiation result in corresponding changes in GCN4 translation as outlined above, and therefore represent an important component of the starvation-response signaling pathway.

A similar mechanism of gene regulation has been proposed for other fungi. In *Neurospora crassa*, expression of the cross-pathway-control gene (*cpc-1*) is required for the cross-pathway-mediated regulation of amino-acid biosynthetic genes (36a). In the 720-bp 5′ UTR of *cpc-1* are two open reading frames preceding the *cpc-1* coding region, suggesting that expression of *cpc-1* may be regulated translationally, in addition to transcriptionally. The deduced amino-acid sequence of the protein encoded by *cpc-1* shows regions of high homology to the DNA-binding domain and transcriptional activation domain of GCN4 (36a). The structural and functional similarities between *cpc-1* and GCN4 indicate that *cpc-1* may act in an analogous manner in *Neurospora*.

III. Adaptive Regulation of Eukaryotic Plasma Membrane Amino-acid Transport

Translocation of amino acids across the plasma membrane of mammalian cells is mediated by a large number of independent transport systems (see

Refs. 37–39 for reviews). The individual systems have been defined by determining ion dependence, pH sensitivity, substrate specificity, regulation, and localization to a specific domain of the plasma membrane. System A, a sodium-dependent transporter for neutral amino acids, is present in nearly all mammalian cells and its activity is regulated by the concentration of extracellular substrate amino acids. Gazzola et al. (40) and Riggs and Pan (41) first reported that incubation of cells or tissue in an aminoacid-free medium results in an up-regulation of System A-mediated transport. Since those original descriptions, enhancement of this transport activity by substrate deprivation has been demonstrated in a wide variety of cell types (42–44). There are two components of transporter control in response to amino-acid deprivation. Following removal of amino acids from the medium, there is an immediate increase of System A transport activity of twofold or less that results from depletion of the intracellular substrates. The transporter is thought to be "trans-inhibited" as a result of amino-acid and/or sodium binding to the carrier in the intracellular orientation. Because the sodium electrochemical gradient strongly favors translocation in the inward direction, the transporter becomes locked in this "cytoplasmic" orientation. Transfer to aminoacid-free medium results in partial depletion of the intracellular pools, release from trans-inhibition, and increased inward transport of extracellular amino acid (45).

The second phase of increased transport occurs 1–2 hours after aminoacid removal and continues for 12–18 hours (44). This latter phase is termed "adaptive up-regulation" or "derepression," because it is thought to result from increased expression of a System A-associated gene. Unfortunately, this hypothesis cannot be confirmed because neither the System A protein nor the corresponding cDNA have been isolated. In the absence of molecular tools to investigate directly transcriptional and translational control, inhibitors of macromolecular synthesis have been used to provide indirect evidence (Fig. 2). Unlike release from trans-inhibition, adaptive up-regulation is completely blocked by inhibitors of general protein synthesis (44–46), glycoprotein biosynthesis (47), total RNA synthesis (44–46), and poly(A) polymerase (43). The inhibition by tunicamycin (46) and retention of the stimulated System A transport activity within plasma membrane vesicles (45) suggest that the necessary protein is a cell-surface glycoprotein. Support for the substrate-dependent synthesis of a plasma membrane glycoprotein was reported by Fong et al. (45), who demonstrated full retention of the increased transport activity after detergent solubilization from isolated plasma membrane vesicles and then reconstitution into artificial proteoliposomes.

Amino-acid refeeding of cells previously incubated in aminoacid-depleted medium results in a down-regulation of the enhanced System A transport activity (43–53). The substrate-dependent down-regulation can be separated

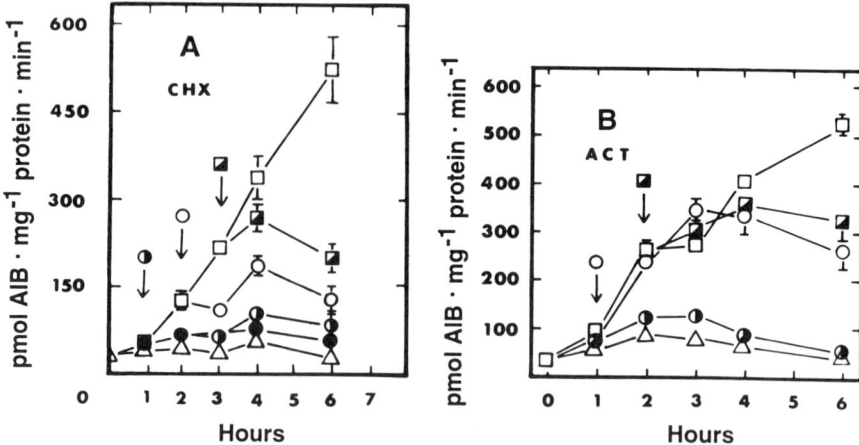

FIG. 2. Effect of cycloheximide or actinomycin D on the adaptive regulation of System A. H4 hepatoma cells were transferred to aminoacid-rich medium (△) or aminoacid-free medium (□). In some cases, 0.1 mM cycloheximide (A) was added after 0 (●), 1 (◐), 2 (○), or 3 (■) hours of incubation in the AAF medium, or 4 μM actinomycin D (B) was added after 0 (◐), 1 (○), or 2 (□) hours. The sodium-dependent uptake of 50 μM [³H]AIB, a selective substrate for System A, was measured for 1 minute at 37°C. Where not shown, the standard deviation bars are contained within the symbol (from Ref. 46).

into two components. The first is a rapid, protein synthesis-independent effect caused by trans-inhibition resulting from increased intracellular amino-acid pools (45, 51). As mentioned above, trans-inhibition is a result of trapping the transporter in its cytoplasmic orientation.

The second component of the down-regulatory response is referred to as "repression" and can be distinguished from trans-inhibition because it is prevented by inhibition of RNA or protein synthesis (43–46). Addition of the repressor amino acids for 60 minutes prior to including actinomycin D results in a decay of the starvation-induced transport at a rate similar to that seen in the absence of the inhibitor (43, 50). In contrast, inhibition of protein synthesis at specific intervals after initiating repression by substrates causes the transport activity to be "frozen" along the decay curve, essentially at the level existent when the inhibitor is added. Collectively, these observations indicate that amino acids induce the synthesis of an mRNA within an hour of refeeding substrate-deprived cells, and that this mRNA codes for a relatively short-lived protein.

Interestingly, in contrast to the starvation-dependent increase in transport activity, the substrate-induced decay is not prevented by inhibitors of poly(A) polymerase, such as araA or cordycepin (Fig. 3) (43). This observation suggests that the aminoacid-induced RNA necessary for the down-

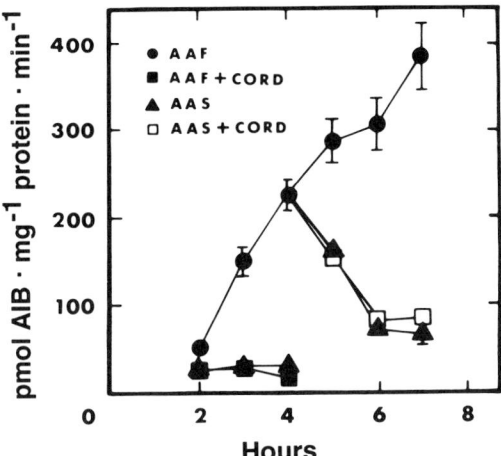

FIG. 3. Effect of cordycepin on adaptive regulation in cultured rat hepatocytes. Cells were transferred to aminoacid-free (AAF) medium (●), AAF medium containing 100 μM cordycepin (■), aminoacid-supplemented (AAS) medium (▲), or AAS medium containing 100 μM cordycepin (□). The System A activity was assayed as the sodium-dependent uptake of 50 μM [^{3}H]AIB for 1 minute at 37°C. Similar results were obtained regardless of whether the aminoacid-supplemented medium was Krebs–Ringer bicarbonate buffer containing 10 mM asparagine or Waymouth's culture medium. The data are the averages ±SD of four determinations. Where not shown, the standard deviation bars are contained within the symbol (reproduced from Ref. 43, with permission).

regulation of the transport activity may lack the typical poly(A) tail, a trait typically associated only with the histone mRNAs. The exact function of the corresponding protein is unknown, but it may serve as a repressor of transcription for the System A transporter, or a necessary component.

Additional evidence for a protein repressor of System A gene expression comes from research in Englesberg's laboratory [reviewed by Englesberg and Moffett (49)]. Generation of Chinese hamster ovary (CHO) mutant cell lines has permitted these investigators to gain insight into the aminoacid-dependent regulation of System A transport and led to the proposal that a transporter regulatory protein exists. One of the mutant cell lines, Ala^{r}4, expresses a constitutively elevated level of System A transport activity of fivefold or more compared to the parental CHO cells. In contrast to the parental CHO-K1 cells, substrate starvation of the Ala^{r}4 CHO mutants causes no further increase in System-A-mediated uptake. Furthermore, the enhanced transport activity in the mutant cell line cannot be repressed by incubation in an aminoacid-rich medium. Similarly a transformed variant of MDCK cells, termed MDCK-T$_1$, also exhibits a constitutively elevated level of System A activity that is not inducible by substrate starvation and is not repressible by amino acids (54). Based on genetic analysis of a large number of CHO cell mutants, it was concluded (49) that the constitutive increase in transport and lack of substrate control were the result of a single mutation in a regulatory gene (termed *R1*) that codes for an apo-repressor-inactivator.

Several independent models suggest that adaptive up-regulation results from increased gene expression of either the System A transporter protein or a transporter-associated regulatory component (44, 48, 49). The model originally proposed (44) hypothesized that amino-acid binding by the System A transporter may initiate a signal to enhance transcription of a gene encoding a transport "inactivator protein" that inactivates the transporter protein in a manner requiring substrate binding. Simultaneously, transporter occupation by substrates would initiate a decrease in the transcription of the System A transporter gene itself.

Englesberg and Moffett (49) (Fig. 4) also propose the existence of a "regulatory gene." Their model suggests that the regulatory protein serves as an aminoacid-binding apo-repressor. Like the Gazzola et al. (44) model, this model proposes that the aminoacid/repressor complex serves to down-regulate expression of the System A transporter gene and to act as an inactivator of the transporter protein.

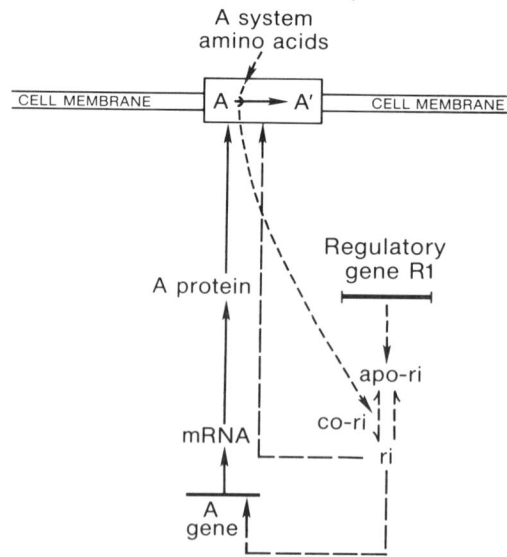

FIG. 4. A model for aminoacid-dependent regulation of the System A transporter. The general features of the model are supported by data from numerous laboratories, although there remain discrepancies regarding the details. The diagram was kindly provided by Ellis Englesberg.

However, the two models differ with regard to the role of the transport step. The Englesberg and Moffett (49) proposal does not require binding or transport of the substrate per se, but rather recognition of amino acid within the cell. This feature of their model is based on their observations (52, 53), as well as work in other laboratories, that in some cells repression of System A activity is mediated by amino acids that serve neither as competitive inhibitors nor as substrates. The System A transporter exhibits the greatest affinity for neutral amino acids with unbranched sidechains.

In general, the rate of transport of a particular amino acid is related to its ability to function as a co-repressor and cause down-regulation of System A transport activity. However, there is no absolute correlation, and a number of studies show that amino acids that do not serve as substrates for the transporter can mediate down-regulation and vice versa (52–54).

Our investigations of rat hepatocytes in primary culture reveal that those amino acids transported most effectively by the System A carrier (asparagine, alanine, serine, glycine, proline, and threonine) are also the best repressors (48, 50). Because histidine and glutamine are transported into rat hepatocytes by a sodium-dependent carrier termed System N (55), little or no transport is mediated by System A, yet both amino acids serve as repressors of System A activity (56). Likewise, amino acids with net charge, anionic or cationic, are effective nonsubstrate repressors in MDCK cells (54). Furthermore, the converse is also true. For example, in CHO cells 2,4-diaminobutyric acid is an inhibitor of System A transport activity, but does not serve as a repressor (49, 56, 57).

Another form of aminoacid-dependent regulation associated with System A transport activity is the induction following incubation of cells in a hypertonic medium (57–60). Interestingly, the stimulation of transport by hypertonic treatment requires prior incubation in aminoacid-free medium. Thus, the stimulation by tonicity is an additional level of control superimposed on adaptive up-regulation. Like the starvation-dependent component, the hypertonic response is prevented by cycloheximide, but not by the glycoprotein synthesis inhibitor tunicamycin (60). Furthermore, the constitutively derepressed regulatory mutant CHO cell line Ala^{r}4 exhibits further increases in transport in response to hypertonic medium (60). These results suggest that shrinkage of nutrient-deprived cells results in the synthesis of a regulatory protein that may activate existent transporters.

In contrast to System A, complete amino-acid deprivation of mammalian cells produces little or no effect on System L transport activity. However, substrate-dependent regulation of sodium-independent System L transport has been documented in a mutant CHO cell line (ts025C1) containing a temperature-sensitive leucyl-tRNA synthetase (61, 62). Maintenance at the nonpermissive temperature effectively results in starving the cells for leucine with regard to protein synthesis and causes a two- to threefold increase in System L transport activity. Incubation of parental CHO-K1 cells in a medium containing reduced levels of individual System L substrates also results in elevated transport activity, but to a lesser extent than that seen in the tRNA synthetase mutant. The increase in System L transport is blocked by the presence of protein synthesis inhibitors, but not by actinomycin D, suggesting that the aminoacid-dependent regulation occurs at the level of translation rather than transcription (62).

From the original temperature-sensitive (ts025C1) mutant, Collarini *et al.* (63) isolated a regulatory mutant (C11B6) that constitutively expresses elevated System L activity. Although the C11B6 cell line retains the leucinyl-tRNA mutation, growth of the cell line is insensitive to temperature because the increased transport provides sufficient leucine to support tRNA charging. Fusion of the ts025C1 cells with the C11B6 mutant results in a cell line that exhibits temperature-sensitive growth and regulated System L transport (63). The authors conclude that the CHO cell System L transporter is regulated by a dominant-acting factor that is defective in the C11B6 cell mutant.

A number of mutant CHO cell lines containing different temperature-sensitive aminoacid-tRNA synthetases exhibit increased transport for the corresponding amino acid following incubation at the restrictive temperature (62). For example, the cell line containing a defective asparaginyl-tRNA synthetase exhibits elevated asparagine transport when grown at the nonpermissive temperature. As discussed in Section IV, regulation of asparagine synthetase expression may also result from a cellular process that senses the degree of tRNA acylation (64).

These observations suggest that the level of tRNA charging within the cell is an important step in the amino-acid signaling pathway for "specific control" mechanisms triggered by deprivation of a single amino acid. This process contrasts with the regulation of System A in which starvation of a single amino acid does not induce transport, and repression of the induced activity responds to a broad spectrum of individual amino acids.

IV. Aminoacid-dependent Regulation of Asparagine Synthetase Expression

Most mammalian cells express the enzyme asparagine synthetase (AS), which is responsible for the biosynthesis of asparagine from aspartate and glutamine. Asparagine synthetase varies in its abundance depending on the tissue, but the protein content is highest in the rat pancreas, testes, brain, and spleen (65). These results are consistent with early studies in which the level of AS activity was examined in various mammalian and avian species as well as in different tissues (66).

Many cancer cells have an increased need for nutrients to sustain rapid growth and cell division, and therefore up-regulate enzymes involved in the corresponding metabolic pathways. Clinical treatment of certain cancers have been devised to take advantage of this need. One example is the treatment of certain types of leukemia with the enzyme asparaginase (67). Some leukemias have little or no expression of asparagine synthetase, and are thus dependent on the plasma level of L-asparagine, which is tightly controlled by the liver (67, 68). Asparaginase therapy of leukemia is based on the concept that the enzyme degrades the circulating asparagine which, in turn, deprives those cancer cells lacking sufficient asparagine synthetase for growth. The basal activity of AS and the ability to up-regulate this enzyme have been correlated with lymphoma sensitivity to asparaginase treatment (69).

Even before the gene for AS had been isolated, it was known that AS activity was affected by the availability of substrate. After transfer of CHO cells to medium lacking asparagine, the level of aminoacylation of tRNA^{Asn} decreases and the level of AS activity increases (70, 64). Refeeding asparagine to these cells causes an increase in the level of asparaginyl-tRNA^{Asn} and a concurrent decrease in AS activity. Consistent with those results, when a mutant cell line containing a temperature-sensitive asparaginyl-tRNA synthetase is transferred to the nonpermissive temperature, asparaginyl-tRNA^{Asn} declines while AS activity increases (64). CHO mutants with temperature-sensitive leucyl-, methionyl-, and lysyl-tRNA synthetases also have increased AS activity when grown at the nonpermissive temperature, even though the level of asparaginyl-tRNA^{Asn} remains unchanged (64). The increased AS activity in these aminoacyl-tRNA mutants is not due to a decrease in the rate of protein synthesis, because the addition of translation inhibitors, such as cycloheximide or emetine, causes a decrease in protein synthesis without affecting the level of AS activity (64). Overall, not only do these results suggest that the level of tRNA charging is important for the sensing of asparagine starvation, but they also suggest that a signaling mechanism exists in mammalian cells that is similar to the general control response of yeast, such that starvation of any one of a number of amino acids results in the up-regulation of enzymes not necessarily involved in their biosynthesis (9).

Asparagine synthetase has been cloned from a number of species and shows a high degree of homology between species (71–76). Ray et al. cloned the AS cDNA from CHO cells that overexpressed the AS mRNA after selection by resistance to the amino-acid analog albiizzin (76). Andrulis et al. cloned the human AS cDNA by screening a human fibroblast cDNA library with the above CHO cDNA (73). Complementation of the asparagine-requiring Jensen rat sarcoma cell was used to confirm that the isolated human cDNAs contained the complete AS coding region (73). The Jensen rat sarcoma cell is auxotrophic for the normally nonessential amino acid asparagine because it does not express AS mRNA, presumably due to hypermethylation of the AS promoter (73, 77). The human gene has been mapped to chromosome 7q21.3 by fluorescence *in situ* hybridization (78). The genomic promoter region has features common to many so-called housekeeping genes, such as a high (G + C) content, absence of a TATA box, and multiple putative RNA start sites.

The cellular AS mRNA content is regulated during the cell cycle. Basilico and co-workers (71) identified a human cDNA that could complement mutant hamster BHK *ts11* cells, which is specifically blocked in progression through the G1 phase of the cell cycle when grown at the nonpermissive temperature. By sequence homology and the ability of exogenous asparagine to bypass the *ts11* mutation, the human cDNA was identified as AS. At the nonpermissive temperature the BHK *ts11* cells produce an inactive AS protein (72), resulting in a depletion of cellular asparagine and a corresponding increase in AS mRNA content. The addition of exogenous asparagine to the medium leads to cell growth and a corresponding decrease in AS mRNA levels (71). The AS mRNA is induced at approximately mid-G_1 phase in human, mouse, and hamster cells, and the addition of asparagine to the culture medium prevented the G1 induction in serum-stimulated cells (71, 82). AS activity is induced during lymphocyte activation by phytohemagglutinin, and the increase in activity coincides with the rate of DNA synthesis (81). TSH stimulation of quiescent rat thyroid cells causes entry of the cells into S phase, and AS mRNA content is increased, which is additional evidence that the AS gene is regulated in a cell-cycle-dependent manner (80).

We cloned the rat AS cDNA using oligonucleotide primers corresponding to the human sequence and the reverse transcriptase polymerase chain reaction (74). Rat, hamster, and human cells each express a predominant AS mRNA species of approximately 2.0 kb, and hamster and rat cells express a larger AS mRNA of approximately 2.5 kb (71–74). In the rat, three AS mRNA species of 2.0, 2.5, and 4.0 kb are observed, and all three of these mRNAs are coordinately induced by amino-acid starvation (74). In addition to a full-length rat cDNA clone, we have isolated a number of cDNA clones that extend beyond the apparent poly(A) tail associated with the 2.0-kb mRNA. Northern analysis with the 3'-extended sequence of the longer rat AS cDNA revealed hybridization to the 2.5- and 4.0-kb species only, suggesting either alternative polyadenylylation or splicing of the larger AS mRNAs (74). Given that the 2.0-kb AS cDNA does not contain this additional 3' untranslated sequence and yet is regulated by amino-acid availability to the same degree as the 2.5- and 4.0-kb species, this sequence does not appear to be involved in nutrient regulation of these mRNAs (74).

The steady-state level of AS mRNA, relative to other cellular mRNAs, has been shown to be induced by amino-acid starvation, as well as by other hormonal or environmental stimuli, in both mammals and plants (74, 75, 79–81). Complete amino-acid starvation of rat hepatoma cells results in a time-dependent increase in AS mRNA content over a 12-hour period (Fig. 5). Interestingly, starvation of either HeLa cells or wild-type BHK-21 cells for individual amino acids such as leucine, isoleucine, or glutamine causes an increase in the level of AS mRNA as well (79). Likewise, depletion of many other nonessential amino acids results in an increase in the level of AS mRNA in a rat hepatoma cell line, but the degree of induction of AS mRNA varies with the particular amino acid depleted from the media and none of the single amino-acid depletions are as effective as complete amino-acid starvation (Fig. 6) (74).

Of those amino-acid depletions examined, the enhancement in AS mRNA was greatest following complete amino-acid starvation (10.1- ± 0.8-fold increase), depletion of histidine (5.6- ± 2.0-fold increase), threonine (5.4- ± 1.9-fold increase), or tryptophan (5.3- ± 1.9-fold increase). Cell culture in the absence of leucine and phenylalanine gave intermediate results, whereas depletion of the amino acids isoleucine and glutamine from an otherwise complete culture medium did not dramatically affect the level of AS mRNA in the rat hepatoma cells (74), in contrast to other results (79). These data suggest that the nutrient-dependent control of AS mRNA expres-

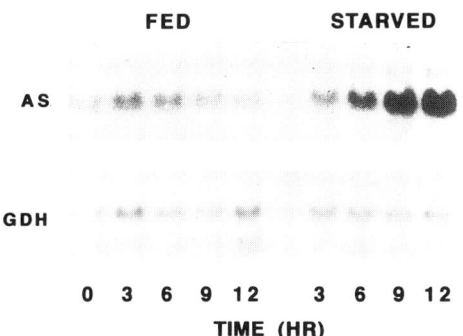

FIG. 5. Time course of the induction of steady-state AS mRNA level during amino-acid starvation. Rat Fao hepatoma cells were transferred to either aminoacid-rich (MEM) or aminoacid-free (KRB) medium and total RNA was isolated after 3, 6, 9, and 12 hours. The level of AS and glutamate dehydrogenase (control) mRNA was determined by hybridization with a ^{32}P-labeled cDNA probe for the corresponding mRNA (modified from Ref. 74, with permission).

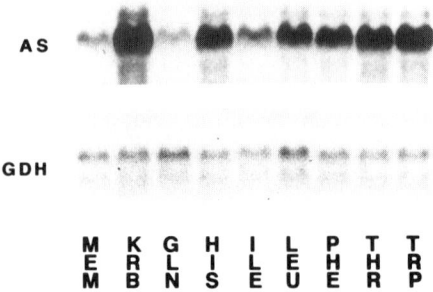

FIG. 6. Induction of AS mRNA by depletion of individual amino acids from complete MEM medium. Rat Fao hepatoma cells were transferred for 12 hours to aminoacid-rich medium (MEM), aminoacid-free medium (KRB), or MEM deficient in a single amino acid. Total RNA was isolated and the level of AS or GDH mRNA was determined by hybridization with a ^{32}P-labeled cDNA probe for the corresponding mRNA (modified from Ref. 74, with permission).

sion is not specific for asparagine, but rather the cellular control mechanism senses the availability of several other amino acids as well. Conversely, aminoacid-free media supplemented with even a single amino acid can maintain partial repression of the level of AS mRNA (74). Of the amino acids tested, glutamine showed the greatest ability to suppress the AS mRNA level, but alanine, asparagine, leucine, proline, serine, threonine, and aspartate also were effective, supporting the hypothesis that a general control mechanism is responsible. Interestingly, glutamine, a substrate of the AS reaction, could substantially down-regulate the level of AS mRNA at a concentration one-tenth that of its plasma level (74).

Regulation of AS gene expression involves cis-acting elements contained within the mRNA sequence as well as in the genomic promoter (79, 83). Initial studies showed that both RNA and protein synthesis were necessary for the aminoacid-dependent down-regulation of AS mRNA in the BHK hamster cell, suggesting that post-transcriptional mechanisms are involved in the regulation of AS mRNA expression (79). To verify this hypothesis, plasmid DNA constructs were created in which the hamster AS cDNA was placed under control of the SV40 early promoter. Hamster *ts11* cells were transfected with these constructs and the cells were transferred to the nonpermissive temperature in the absence of asparagine. In contrast to the chloramphenicol acetyltransferase (CAT) control, the level of AS mRNA increased at the nonpermissive temperature in the absence of asparagine. Given that the SV40 promoter is presumed to be unaffected by amino-acid starvation and the CAT mRNA content remained constant, the results suggest that the AS mRNA is regulated post-transcriptionally by amino-acid starvation (79). Insertion of a nonsense codon into the first third of the AS coding region eliminated the starvation-dependent stabilization of the AS mRNA, indicating that translation of the full AS coding region may be necessary (79). One possible mechanism for this regulation could be similar to one proposed in which a protein associated with polysomes protects a region of the c-*myc* mRNA from endonuclease attack (84). By analogy, inhibiting ribosomes from translating the complete AS mRNA may leave a nuclease-sensitive region unprotected.

In addition to post-transcriptional control mechanisms, the level of AS mRNA is regulated at the level of transcription (79, 83). Cells transfected with cDNA constructs in which the CAT reporter gene was placed under control of a 3.4-kb fragment of human AS genomic DNA that contained a putative promoter region and the first two exon and intron sequences showed an increase in the level of CAT mRNA in response to asparagine starvation. The level of CAT mRNA under the control of the thymidine kinase (TK) promoter decreased under the same conditions (79, 83). Deletion analysis of the AS genomic promoter region connected to the CAT reporter gene showed that the minimum AS 5' sequence that retains full inducibility by amino-acid starvation spans from nucleotide −164 to +44 (33).

DNA sequence analysis of the AS gene revealed an element located just downstream of the major transcription start site within the first untranslated exon as having a high degree of homology to the FIRE element of the *fos* gene (83). Interestingly, this sequence is located in the first untranslated exon of three other genes, c-*myc*, c-*jun*, and the gene for ornithine decarboxylase, which also appear to be regulated by amino-acid starvation (83, 85). Gel-shift experiments using HeLa cell extracts revealed that two DNA single-strand binding proteins bind specifically to this sequence, but the binding of these proteins does not change with amino-acid starvation (83). Furthermore, deletion of this element results in increased basal expression, but aminoacid-dependent regulation is still maintained, arguing against this sequence as an amino-acid regulatory element (83).

Scanning the −150 to +1 region of the AS promoter by site-directed mutagenesis revealed several additional elements involved in AS promoter regulation. Mutagenesis of the sequence between nucleotides −70 and −64 destroyed the aminoacid-dependent regulation of the AS promoter and this sequence, CATGATG, has been termed the aminoacid-regulatory element (AARE) (83). A protein that binds specifically to this element was detected by gel-retardation experiments with nuclear extracts from HeLa cells (83). Mutations in the AARE sequence that abolish the aminoacid-dependent regulation of the AS promoter prevent binding of this protein, suggesting that it is involved in the regulation of the AS gene. However, the amount of protein bound as a result of amino-acid starvation does not change (83).

V. Identification and Characterization of Aminoacid-regulated Ribosomal-protein Genes

We have successfully identified a number of proteins for which synthesis is increased following amino-acid deprivation of rat hepatocytes in primary culture and rat Fao hepatoma cells. Use of the cultured cell line Fao simplifies most experiments and we have observed no differences in the response between the two cell types. We have utilized two basic approaches to identify aminoacid-regulated genes and mRNAs. The first approach used pulse-chase labeling of cells with [^{3}H]leucine following incubation of normal rat hepatocytes in culture medium containing (AA+) or lacking (AA−) amino acids. The aminoacid-free medium was Krebs–Ringer bicarbonate (KRB) buffer and the aminoacid-containing medium was either a tissue culture medium [Waymouth's or Eagle's minimal essential medium (MEM)] or KRB supplemented with amino acid(s) known to down-regulate the activity/ mRNA being tested. Following a protein-labeling procedure, a crude membrane fraction (10,000 g pellet) was prepared and analyzed by two-dimensional PAGE and fluorography. Several proteins were identified as up-regulated more than threefold after amino-acid starvation (86). One of these proteins actually separates on two-dimensional PAGE as a family of three "spots" having the same molecular mass of 66 kDa (MP66), but differing in pI values. V8-protease digestion of the first and third spots cut from multiple gels resulted in identical peptide fragment patterns.

A second protein identified in this manner was a polypeptide of approximately 73 kDa (MP73). Antibodies have been prepared against this polypeptide by excising these spots from several nitrocellulose blots and subcutaneously implanting them into rabbits (86). This novel approach permits the production of monospecific polyclonal antibodies from complex mixtures without protein purification and is useful for proteins of low abundance. Antibody screening yielded a partial (2-kb) cDNA clone that identifies a single mRNA species of approximately 4 kb that is highly regulated by amino-acid availability in normal rat liver or rat hepatoma cells (87). Following 6 hours of amino-acid starvation, the level of MP73 mRNA is induced by more than 20-fold (87). MP73 is an integral membrane protein located in the inner mitochondrial membrane, and the limited amount of coding sequence obtained thus far shows no significant homology to sequences currently in the protein or DNA data bases.

The second approach used to identify aminoacid-regulated mRNAs was by "plus–minus" screening of a subtracted cDNA library (88). First-strand cDNA, corresponding to poly(A)$^+$ mRNA prepared from cells incubated in AA− medium for 6 hours, was hybridized to a 10-fold excess of poly(A)$^+$ mRNA prepared from cells incubated in AA+ medium. After separating the

RNA·DNA hybrids, the single-stranded cDNA fraction was used to prepare a λgt11 phage library that contained 200,000 independent clones. Duplicate lifts were probed by plus–minus screening using labeled cDNA from amino acid-fed or amino acid-starved cells and one of the resulting positive clones was plaque purified (88). A full-length cDNA of 612 bp was obtained that hybridized to a single mRNA in all rat tissues and hepatoma cells in culture (89). Following amino-acid starvation of hepatoma cells, the mRNA increased from three- to sixfold and returned toward basal levels following refeeding (88).

Following sequencing, homology to bacterial ribosomal protein (RP) L22 was recognized, but the mammalian counterpart had not been identified previously (89). To confirm the ribosomal localization, antipeptide antibodies were generated and shown to recognize a polypeptide of approximately 22–23 kDa associated specifically with the 60-S ribosomal subunit (89). Ira Wool (personal communication) confirmed that we had cloned the cDNA for mammalian RP L17; he obtained the same sequence by independent means (90).

Of 12 other randomly chosen RP cDNAs screened for their response to amino-acid starvation, 11 showed little or no change. In contrast, like L17, the mRNA for RP S25 was induced following 9 hours of amino-acid starvation (Fig. 7) (91, 92). The S25 cDNA was cloned by others using subtractive hybridization to identify genes that were overexpressed in an adriamycin-resistant human cell line (93). S25 mRNA was elevated in these drug-resistant cells, but there was no concurrent increase in the abundance of the S25 protein (94). This lack of translation is consistent with our observation of nuclear retention of S25 mRNA following amino-acid starvation. The rat S25 sequence is identical to the original human clone (95). Interestingly, mammalian S25 shows a strong sequence homology to the yeast ribosomal protein S31, which has been suggested to play an important role in assembly of the small ribosomal subunit and, therefore, is essential to ribosomal maturation (96). Whether L17 plays a key role in 60-S ribosome assembly is not known, but such a function for these two proteins may explain their uniqueness with regard to regulation by amino-acid availability.

Previous work has suggested that, in general, genes for ribosomal proteins are constitutively transcribed, but increased mRNA content has been observed for selected ribosomal protein genes in growing cells, as compared to resting cells (97–100). Inhibition of protein synthesis prevents the starvation-dependent induction of both L17 and S25 mRNAs, demonstrating a requirement for *de novo* synthesis of a regulatory or signaling protein(s) in the starvation response (Fig. 7). Cycloheximide, in the absence or presence of amino acids, did not change mRNA levels from control values, indicating that the putative regulatory protein(s) only serves to enhance S25 or L17 mRNA content, and that the L17 and S25 mRNAs are not degraded at a

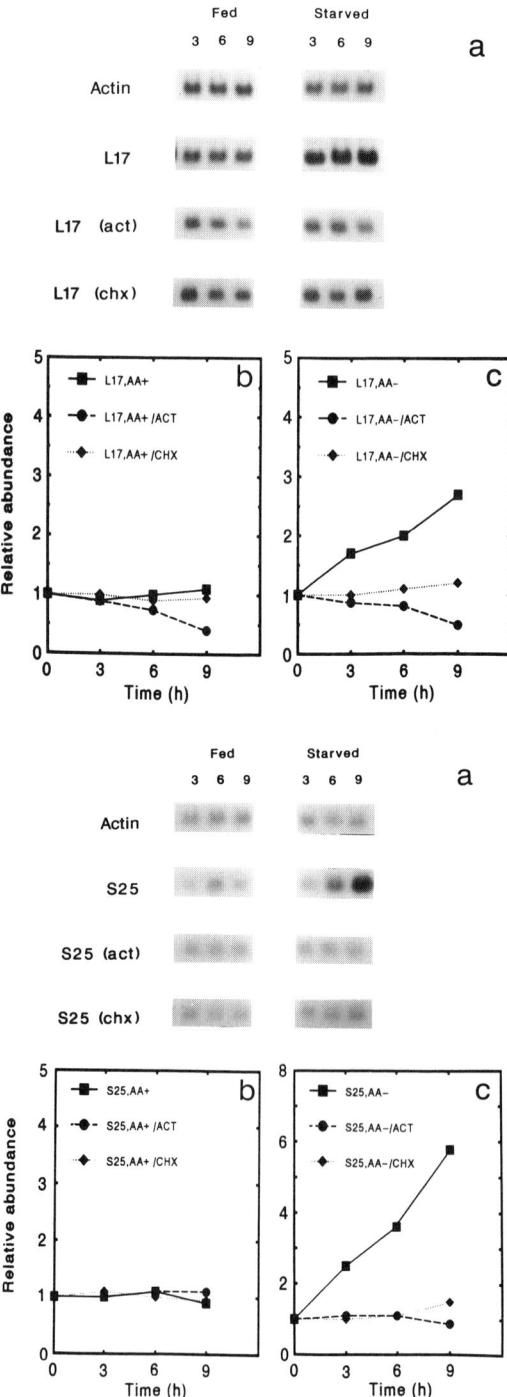

faster rate in the absence of protein synthesis or in the presence of repressor amino acids. In this context, the function of this regulator could be analogous to the action of the yeast trans-acting factor, GCN4, discussed in Section II. Over the 9-hour period tested, the basal level (cells in AA+ medium) of S25 mRNA was not changed by the presence of actinomycin D, although the starvation-dependent increase was completely prevented by the inhibitor (Fig. 7). The presence or absence of amino acids had no effect on the mRNA content in the presence of actinomycin D, suggesting that substrate availability does not alter the turnover of the S25 mRNA. In contrast, actinomycin D not only prevented the increase in L17 mRNA content, but also reduced the basal L17 mRNA level, reflecting a half-life of less than 10 hours (Fig. 7).

These data demonstrate the need for continued mRNA synthesis to maintain the basal cellular level of L17 mRNA, and show that the half-life of L17 mRNA is measurably shorter than that for S25. Like S25, amino acids had no significant effect on the turnover of the L17 mRNA in the presence of actinomycin D, suggesting that substrate-dependent mRNA stabilization is not a major regulatory factor (Fig. 7). Nuclear run-off assays demonstrated that L17 and S25 exhibit transcriptional control in response to nutrient deprivation (91). These results, in conjunction with our previous ones (90), confirm the transcriptional control of ribosomal protein L17 and S25 genes.

The specificity of aminoacid-dependent repression for L17 and S25 is similar to, but distinct from, that for the System A transporter (50) and asparagine synthetase (83). Glutamine, aspartate, and 2-aminoisobutyric acid (AIB), a nonmetabolizable alanine analog, are the most effective repressors of L17 induction. Asparagine, ammonia, and leucine are equal in their potency to a complete culture medium. These results suggest that there is a common sensing mechanism for amino-acid availability, and it is the substrate specificity of that process that is revealed by amino-acid refeeding experiments, regardless of how amino-acid deprivation is assayed (i.e., System A activity, AS mRNA, or L17/S25 mRNA). As in yeast, that sensing

---

FIG. 7. Induction of L17 (top) and S25 (bottom) mRNA during amino-acid starvation and effect of metabolic inhibitors on cellular content of L17 and S25 mRNA. (a) Total RNA (7 µg) from rat Fao cells cultured in either AA+ (MEM) or AA− (KRB) medium for 3, 6, and 9 hours in the presence or absence of actinomycin D (25 µM) or cycloheximide (100 µM) was used for Northern blot analysis with either RP S25 or L17 cDNA as the probe. The ethidium-bromide-stained 18-S rRNA, quantified by densitometry, serves as a reference standard for comparison of RNA loading. Rat Fao hepatoma cells were incubated for the indicated time in (b) aminoacid-containing (AA+, MEM) or (c) aminoacid-free (AA−, KRB) medium in the presence of either actinomycin D (25 µM) or cycloheximide (100 µM). Northern analysis was performed with either S25 or L17 cDNA as the probe and the data were quantified by laser scanning densitometry. Each time point is representative of at least three independent determinations (reproduced from Ref. 91, with permission).

mechanism may be synthesis of a specific signal protein for which translation is altered in response to tRNA charging. Consistent with this hypothesis are the observations of N. Shay and M. S. Kilberg (unpublished results) using L-azetidine-2-carboxylic acid and L-histidinol, amino-acid analogs of proline and histidine, respectively. These analogs mimic starvation for amino acids with respect to protein synthesis, because they compete for tRNA charging and, therefore, do not permit elongation (*101, 102*). Including the analogs in the amino acid-free condition produced no further enhancement in L17 mRNA content; however, when an excess (10 mM) of either of these analogs is added to tissue culture medium containing a complete amino-acid mixture, the level of L17 mRNA is induced. Similar results are observed for asparagine synthetase (*74*).

These observations, plus the more recent data that *de novo* protein synthesis is required for the starvation-dependent increase (*91*), suggest that the function of the putative mammalian protein regulator may be analogous to the yeast trans-acting factor, GCN4 (*9*). Furthermore, as in yeast, starvation of mammalian cells may initiate the starvation signaling pathway by enhancing only the translation for the proposed regulator(s) (*92*).

To determine if starvation-induced S25 or L17 mRNAs were available for translation, Fao cells, cultured for 6 hours in either AA+ or AA− medium, were fractionated to isolate RNA specifically associated with the cytoplasm, polysomes, or nucleus (*91*). Northern analysis of these samples, using L17 and S25 cDNAs as probes, as well as three ribosomal protein cDNAs (S13, L7a, L3), showed little or no change in total mRNA content in response to amino-acid deprivation. The change in mRNA content for the polysomal fractions was twofold or less for each of the five ribosomal proteins tested, and the nonpolysomal, cytoplasmic mRNA content was reduced by starvation in each case. These results are consistent with preliminary observations by immunoblotting that the L17 protein content is not increased significantly following amino-acid deprivation (unpublished results).

Measurement of mRNA in the nuclear fraction demonstrated the novel observation that the nuclear content for RP L17 and S25 paralleled the increase detected in total cellular RNA (Fig. 8). The specificity of this nuclear retention for RP L17/S25 is documented by three lines of evidence: (1) ribosomal mRNAs not enhanced by amino-acid starvation did not show a change in nuclear mRNA content; (2) probing these same blots with AS cDNA revealed that the increase in AS mRNA is associated with the polysomal fraction, not the nucleus (R. G. Hutson and M. S. Kilberg, unpublished results); and (3) starvation-induced System A transport activity requires translation (*46, 45*); obviously the responsible mRNA leaves the nucleus.

The Fao cells can be serum-deprived for 12–18 hours prior to initiating the changes in amino-acid availability to eliminate growth arrest as a variable,

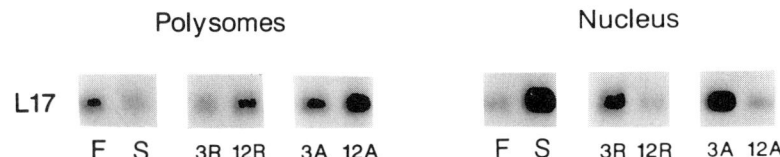

FIG. 8. L17 mRNA content in the polysomes or nuclei during amino-acid starvation and refeeding. Rat Fao cells, incubated in either MEM (F) or KRB (S) conditions for 6 hours, were used to obtain fractions for isolation of RNA associated with the polysomes or nuclei. Cells previously starved for 6 hours were then transferred to MEM medium in the presence (A) or absence (R) of actinomycin D (10 μM) and RNA samples were isolated after an additional 3 hours (3R or 3A) or 12 hours (12R or 12A) of refeeding. Northern analysis was performed (8 μg/lane) using RP L17 cDNA as the probe under conditions published previously (reproduced from Ref. 91, with permission).

and the starvation response is unchanged. Furthermore, the starvation-dependent increase in L17 mRNA, System A activity, and MP73 mRNA (the only examples tested thus far) is observed in primary cultures of rat hepatocytes that do not actively divide. Collectively, the results argue that the adaptive response to amino-acid deprivation is not the result of a block in cell growth. A few of the RP mRNAs accumulate during periods of cell-growth arrest (103, 104), but there are no large concurrent changes in synthesis or turnover of the corresponding ribosomal proteins.

It has been proposed that the translation of these mRNAs is somehow selectively inhibited. The data suggest that the translational efficiency of RP mRNAs is controlled by the immediate cellular need for ribosomes, whereas transcription of the corresponding genes is more anticipatory (105). Thus, it is noteworthy that in response to amino-acid starvation, nearly all of the increased L17 and S25 mRNA (not shown) is retained within the nuclear compartment (Fig. 8). Apparently, amino-acid starvation affects RP L17 and S25 RNA synthesis, processing, and/or transport to the cytoplasm. Aminoacid-dependent change in RNA splicing does not appear to be involved because a single mRNA species, unchanged in size, is detected in cells cultured in the presence or absence of amino acids (91). It is plausible that mRNAs coding for ribosomal proteins are synthesized and stored in a nontranslatable form during periods of decreased amino-acid availability, thereby allowing a rapidly elevated and selective rate of ribosomal protein synthesis to be initiated immediately on replenishment. In support of this interpretation, a preliminary experiment showed that the accumulated nuclear L17 (Fig. 8) and S25 (not shown) mRNAs were shifted to the polysomes after refeeding (91). The results were the same in the presence of actinomycin D during the refeeding, eliminating the possibility that the increase seen in the polysomal fraction was due to new mRNA synthesis and export.

Other data that have indicated sequestration of transcripts within the nucleus involve the Rev-responsive element of human immunodeficiency virus type 1 (HIV-1) (106). This process appears to differ from that for L17 and S25 in that inefficient splicing signals were deemed to cause the retention, and only late in the infection do the incompletely spliced RNAs accumulate within the cytoplasm. HIV-1 splice donor and acceptor sites are not required for nuclear retention of envelope-derived mRNAs and there is a cis-acting sequence responsible for the nuclear retention of HIV-1 envelope mRNAs that may partially coincide with the element responsible for Rev binding and function (106). This is intriguing in that there is now a precedent in proposing a signaling mechanism for our nutrient-controlled nuclear retention.

Limited information is available about the coordinated synthesis of vertebrate ribosomal proteins, and even less about the involvement of ribosomal protein gene expression in tumorigenesis. Preliminary data indicate that ribosomal proteins L17 and S25 are unique in their regulation by amino-acid availability. In the same manner, a limited number of mammalian ribosomal protein genes have been observed to be overexpressed in tumor cells, whether as an apparent secondary change or as a causal event. Interestingly, one of the two nutrition-regulated ribosomal protein genes, S25, is overexpressed in adriamycin-resistant human leukemia cells (94). Insight into the adaptive regulation of this ribosomal protein gene and other aminoacid-unregulated ribosomal protein genes would allow one to study the potential role that this process may play in controlling cellular growth and metabolism, and at the same time gain insight into the impact of ribosomal gene expression in tumorigenesis.

Although it would seem reasonable that the increased level of ribosomal protein transcripts in tumor tissue is not a causal event in carcinogenesis, two examples of ribosomal fusion proteins isolated from transformed cells appear to play an active role in transformation. First, the *trk*-2h oncogene from NIH 3T3 cells transformed with DNA from a human breast carcinoma cell line was isolated and found to code for 41 amino acids of ribosomal protein L7 fused to the receptor kinase domain of the *trk*-protooncogene (107), and second, the transforming retrovirus, Finkel–Biskis–Reilly murine sarcoma virus, increases its transforming capacity by acquiring the genomic *Fau* gene. This gene encodes the ribosomal protein S30 fused to a ubiquitin-like protein (108). In contrast, specific ribosomal proteins moderate tumorigenesis or stimulate the immune response. For example, rat RP L13a shows 98% amino-acid sequence identity to the mouse tum transplantation antigen P198. This protein is expressed in tum-variants of the mouse mastocytoma tumor cell line P815 in which tum variants do not form tumors in syngeneic mice. Therefore, P198 generates a new epitope on the cell surface that is recognized by syngeneic T cells (109). Also, a 12-kDa protein isolated from the facultative bacterium *Brucella abortus* has been shown to induce strong lymphocyte proliferation of peripheral blood lymphocytes in cattle. This specific protein shares 67% amino-acid homology with *E. coli* RP L7/L12 and has potential as a subunit of a genetically engineered candidate vaccine (*110*).

VI. Summary

Our understanding of the metabolite control in mammalian cells lags far behind that in prokaryotes. This is particularly true for aminoacid-dependent gene expression. Few proteins have been identified for which synthesis is selectively regulated by amino-acid availability, and the mechanisms for control of transcription and translation in response to changes in amino-acid availability have not yet been elucidated. The intimate relationship between amino-acid supply and the fundamental cellular process of protein synthesis makes aminoacid-dependent control of gene expression particularly important. Future studies should provide important insight into amino-acid and other nutrient signaling pathways, and their impact on cellular growth and metabolism.

REFERENCES

1. R. Kolter and C. Yanofsky, *ARGen* **16**, 113 (1982).
2. I. P. Crawford and G. V. Stauffer, *ARB* **49**, 163 (1980).
3. J. Hamm and I. W. Mattaj, *Cell* **63**, 109 (1990).
4. M. P. Terns and S. T. Jacob, *MCBiol* **9**, 1435 (1989).
5. W. Gilbert, *Nature* **271**, 501 (1978).
6. M. Robash and B. Seraphin, *Trends Biochem. Sci.* **16**, 187 (1991).
7. L. A. Burmeister and C. N. Mariash, *JBC* **266**, 22905 (1991).
8. K. C. Carter, D. Bowman, W. Carrington, K. Fogarty, J. A. McNeil, F. S. Fay and J. B. Lawrence, *Science* **259**, 1330 (1993).
9. A. G. Hinnebusch, *Microbiol. Rev.* **52**, 248 (1988).
10. I. A. Hope and K. Struhl, *Cell* **43**, 177 (1985).
11. C. E. Samuel, *JBC* **268**, 7603 (1993).
12. T. E. Dever, L. Feng, R. C. Wek, A. M. Cigan, T. F. Donahue and A. G. Hinnebusch, *Cell* **68**, 585 (1992).
13. R. J. Rolfes and A. G. Hinnebusch, *MCBiol* **13**, 5099 (1993).
14. A. G. Hinnebusch, *Trends Biochem. Sci.* **19**, 409 (1994).
15. A. G. Hinnebusch, *Trends Biochem. Sci.* **148** (1990).
16. A. G. Hinnebusch and G. R. Fink, *PNAS* **80**, 5374 (1983).
17. A. G. Hinnebusch, *PNAS* **81**, 6442 (1984).
18. I. A. Hope and K. Struhl, *EMBO J.* **6**, 2781 (1987).
19. A. G. Hinnebusch and G. R. Fink, *JBC* **258**, 5238 (1983).
20. I. A. Hope and K. Struhl, *Cell* **46**, 885 (1986).

21. T. E. Ellenberger, C. J. Brandl, K. Struhl and S. C. Harrison, *Cell* **71**, 1223 (1992).
22. E. K. O'Shea, J. D. Klemm, P. S. Kim and T. Alber, *Science* **254**, 539 (1991).
23. P. K. Vogt, T. J. Bos and R. F. Doolittle, *PNAS* **84**, 3316 (1987).
24. D. Pellman, M. E. McLaughlin and G. R. Fink, *Nature* **348**, 82 (1990).
25. C. M. Drysdale, E. Duenas, B. M. Jackson, U. Reusser, G. H. Braus and A. G. Hinnebusch, *MCBiol* **15**, 1220 (1995).
26. M. Kozak, *NARes* **15**, 8125 (1987).
27. M. Kozak, *NARes* **12**, 3873 (1984).
28. C. C. Oliveira and J. E. G. McCarthy, *JBC* **270b**, 8936 (1995).
29. P. F. Miller and A. G. Hinnebusch, *Genes Dev.* **3**, 1217 (1989).
30. C. M. Grant and A. G. Hinnebusch, *MCBiol* **14**, 606 (1994).
31. J. P. Abasado, P. Miller, B. M. Jackson and A. G. Hinnebusch, *MCBiol* **11**, 486 (1991).
32. R. C. Wek, B. M. Jackson and A. G. Hinnebusch, *PNAS* **86**, 4579 (1989).
33. W. C. Merrick, *Microbiol. Rev.* **56**, 291 (1992).
34. J. W. B. Hershey, *ARB* **60**, 717 (1991).
35. R. E. Rhoads, *JBC* **268**, 3017 (1995).
36. M. Ramirez, R. C. Wel and A. G. Hinnebusch, *MCBiol* **11**, 3027 (1991).
36a. J. L. Paluh, M. J. Orbach, T. M. Legerton and C. Yanofsky, *PNAS* **85**, 3728 (1988).
37. C. Cheeseman, *Prog. Biophys. Molec. Biol.* **55**, 71 (1991).
38. H. N. Christensen, *Physiol. Rev.* **70**, 43 (1990).
39. M. S. Kilberg, B. R. Stevens and D. Novak, *Annu. Rev. Nutr.* **13**, 137 (1993).
40. G. C. Gazzola, R. Franchi, V. Saibene, P. Ronchi and G. G. Guidotti, *BBA* **266**, 407 (1972).
41. T. R. Riggs and M. W. Pan, *BJ* **128**, 19 (1972).
42. M. A. Shotwell, M. S. Kilberg and D. L. Oxender, *BBA* **737**, 267 (1983).
43. M. S. Kilberg, *Trends Biochem. Sci.* **11**, 183 (1986).
44. G. C. Gazzola, V. Dall'Asta and G. G. Guidotti, *JBC* **256**, 3191 (1981).
45. A. D. Fong, M. E. Handlogten and M. S. Kilberg, *BBA* **1022**, 325 (1990).
46. M. S. Kilberg, H.-P. Han, E. F. Barker and T. C. Chiles, *J. Cell. Physiol.* **122**, 290 (1985).
47. E. F. Barber, M. E. Handlogten and M. S. Kilberg, *JBC* **258**, 11851 (1983).
48. M. S. Kilberg, D. S. Bracy and M. E. Handlogten, *FP* **45**, 2438 (1985).
49. E. Englesberg and J. Moffett, *J. Membrane Biol.* **91**, 199 (1986).
50. D. S. Bracy, M. E. Handlogten, E. F. Barker, H.-P. Han and M. S. Kilberg, *JBC* **261**, 1514 (1986).
51. M. F. White and H. N. Christensen, *JBC* **258**, 8028 (1983).
52. J. Moffett and E. Englesberg, *J. Cell. Physiol.* **126**, 421 (1986).
53. J. Moffett and E. Englesberg, *MCBiol* **4**, 799 (1984).
54. P. Boerner and M. H. Saier, Jr., *J. Cell. Physiol.* **122**, 308 (1985).
55. M. S. Kilberg, M. E. Handlogten and H. N. Christensen, *JBC* **255**, 4011 (1980).
56. M. E. Handlogten, M. S. Kilberg and H. N. Christensen, *JBC* **257**, 345 (1982).
57. G. C. Gazzola, V. Dall'Asta, F. A. Nucci, P. A. Rossi, O. Bussolati, E. K. Hoffmann and G. G. Guidotti, *Cell. Physiol. Biochem.* **1**, 131 (1991).
58. Y. Atsushi, A. Miyai, K. Yokoyama, T. Itoh, T. Kamada, N. Ueda and Y. Fujiwara, *Am. J. Physiol.* **267**, C1493 (1994).
59. C. Soler, A. Felipe, F. J. Casado, J. D. McGivan and M. Pastor-Anglada, *BJ* **289**, 653 (1993).
60. B. Ruiz-Montasell, M. Gomez-Angelats, F. J. Casado, A. Felipe, J. D. McGivan and M. Pastor-Anglada, *PNAS* **91**, 9569 (1994).
61. P. A. Moore, D. W. Jayme and D. L. Oxender, *JBC* **252**, 7427 (1977).

62. M. A. Shotwell, P. M. Mattes, D. W. Jayme and D. L. Oxender, *JBC* **257**, 2974 (1982).
63. R. J. Collarini, G. S. Campbell and D. L. Oxender, *J. Cell. Biochem.* **56**, 544 (1994).
64. I. L. Andrulis, W. Hatfield and S. M. Arfin, *JBC* **254**, 10629 (1979).
65. S. Hongo, M. Fujimori, S. Shioda, Y. Nakai, M. Takeda and T. Sato, *ABB* **295**, 120 (1992).
66. H. A. Milman and D. A. Cooney, *BJ* **142**, 27 (1974).
67. S. Amadori, M. Tribalta, L. Pacilli, C. Delaurentis, G. Papa and F. Mandelli, *Cancer Treat. Rep.* **64**, 939 (1980).
68. J. S. Woods and R. E. Handschumacher, *Am. J. Physiol.* **224**, 740 (1973).
69. M. D. Prager and N. Bachynsky, *ABB* **127**, 645 (1968).
70. S. A. Arfin, D. R. Simpson, C. S. Chiang, I. L. Andrulis and G. Wesley, *PNAS* **74**, 2367 (1977).
71. A. Greco, S. S. Gong, M. Ittman and C. Basilico, *MCBiol* **9**, 2350 (1989).
72. S. S. Gong and C. Basilico, *NARes* **18**, 3509 (1990).
73. I. L. Andrulis, J. Chen and P. N. Ray, *MCBiol* **7**, 2435 (1987).
74. R. G. Hutson and M. S. Kilberg, *BJ* **304**, 745 (1993).
75. K. M. Davies and G. A. King, *Plant Physiol.* **102**, 1337 (1993).
76. P. N. Ray, L. Siminovitch and I. L. Andrulis, *Gene* **30**, 1 (1984).
77. R. H. Sugiyama, S. M. Arfin and M. Harris, *MCBiol* **3**, 1937 (1983).
78. H. H. Q. Heng, X. M. Shi, S. W. Scherer, I. L. Andrulis and L. C. Tsui, *Cytogenet. Cell. Genet.* **66**, 135 (1994).
79. S. S. Gong, L. Guerrini and C. Basilico, *MCBiol* **11**, 6059 (1991).
80. G. Colletta and A. M. Cirafici, *BBRC* **183**, 265 (1992).
81. S. Hongo, M. Takeda and T. Sato, *Biochem. Int.* **18**, 661 (1989).
82. A. Greco, M. Ittmann and C. Basilico, *PNAS* **84**, 1565 (1987).
83. L. Guerrini, S. Gong, K. Mangasarian and C. Basilico, *MCBiol* **13**, 3202 (1993).
84. P. Berstein, D. J. Herrick, R. D. Prokipcak and J. Ross, *Genes Dev.* **6**, 642 (1992).
85. P. Pohjanpelto and E. Holtta, *Cell. Biol.* **10**, 5814 (1990).
86. T. C. Chiles, T. W. O'Brien and M. S. Kilberg, *Anal. Biochem.* **163**, 136 (1987).
87. T. C. Chiles, R. O. Laine, N. F. Shay, H. S. Nick, M. E. Handlogten and M. S. Kilberg, *BBRC* **193**, 1068 (1993).
88. N. F. Shay, H. S. Nick and M. S. Kilberg, *JBC* **265**, 17844 (1990).
89. R. O. Laine, P. J. Laipis, N. F. Shay and M. S. Kilberg, *JBC* **266**, 16969 (1991).
90. K. Suzuki and I. G. Wool, *BBRC* **178**, 322 (1991).
91. R. O. Laine, N. F. Shay and M. S. Kilberg, *JBC* **269**, 9693 (1994).
92. M. S. Kilberg, R. G. Hutson and R. O. Laine, *FASEB J.* **8**, 13 (1994).
93. M. Li, C. Latout and M. S. Center, *Gene* **107**, 329 (1991).
94. M. Li and M. S. Center, *FEBS Lett.* **298**, 142 (1992).
95. Y. Chan and I. G. Wool, *BBRC* **186**, 1688 (1992).
96. R. T. M. Nieuwint, C. M. T. Molenaar, J. H. van Bommel, M. M. C. van Raamsdonk-Duin, W. H. Mager and R. J. Planta, *Curr. Genet.* **10**, 1 (1985).
97. M. G. Denis, C. Chadeneuck and M. Lecabellee, *Int. J. Cancer* **55**, 275 (1993).
98. K. Pogue-Geile, *MCBiol* **11**, 3842 (1991).
99. M. Kitagawa, S. Takasawa, N. Kikuchi, T. Itoh, H. Teraoka, H. Yamamoto and H. Okamoto, *FEBS Lett.* **283**, 210 (1991).
100. O. Meyuhas, in "Recombinant DNA and Cell Proliferation" (G. S. Stein and J. L. Stein, eds.), p. 243. Academic Press, Orlando, Florida, 1984.
101. B. S. Hanson, M. H. Vaughan and L. Wang, *JBC* **247**, 3854 (1972).
102. O. A. Scornik, M. L. S. Ledbetter and J. S. Malter, *JBC* **255**, 6322 (1980).
103. D. Faliks and O. Meyuhas, *NARes* 789 (1982).

104. O. Meyuhas, E. A. Thompson and R. P. Perry, *MCBiol* **7**, 2691 (1987).
105. L. M. Angerer, Y. Qing, J. Liesveld, P. D. Kingsley and R. C. Angerer, *Dev. Biol.* **149**, 27 (1992).
106. D. W. Brighty and M. Rosenberg, *PNAS* **91**, 8314 (1994).
107. A. Ziemiecki, R. G. Muller, F. X. Chang, N. E. Hynes and S. Kozma, *EMBO J.* **9**, 191 (1990).
108. K. Kas, L. Michiels and J. Merrigaert, *BBRC* **187**, 927 (1992).
109. Y. Chan, J. Olvera, A. Gluck and I. G. Wool, *JBC* **269**, 5589 (1994).
110. S. C. Oliveira and G. A. Splitter, *FASEB J.* **8**, A197 (1994).

Molecular Recognition in the Assembly of the Segmented Reovirus Genome[1]

WOLFGANG K. JOKLIK[2] AND
MICHAEL R. RONER

Department of Microbiology
Duke University Medical Center
Durham, North Carolina 27710

I.	Basic Features of the Reovirus Multiplication Cycle	252
II.	The Nature of Reovirus Genome Segments	255
III.	The Nature of Reovirus Defective Interfering Particles	259
IV.	The Nature of Reovirus Genome Segment Reassortment and of the Reassortants That Are Generated	260
V.	The Functions of Reovirus Proteins	261
VI.	Reovirus Genome Segment Assortment Studied with Monoclonal Antibodies against Reovirus Proteins	269
VII.	The Infectious Reovirus RNA System	270
VIII.	The Interplay of Signals Required for the Insertion of Genetic Information into the Reovirus Genome	274
	References	277

Reoviruses were first isolated in the early 1950s when, fueled by efforts to develop a vaccine against poliomyelitis virus, the new technology of tissue culture permissive for virus growth was developing rapidly (1–4). Initially, cytopathic agents isolated from the alimentary tracts of ill as well as healthy subjects were lumped together as enteric cytopathogenic human orphan viruses, or Echoviruses. But it soon become apparent that some Echoviruses were larger than others and produced different cytopathic effects; these

---

[1] Abbreviations used: ATI, A-type inclusion; CMV, cytomegalovirus; CPV, cowpox virus; CSK, cytoskeleton; DI, defective interfering; dsRCC, double-stranded RNA-containing complex; ISVP, intermediate subviral particle; MAB, monoclonal antibody; MEM, minimal essential medium; NS, nonstructural; PFU, plaque-forming unit; PKR, dsRNA-dependent p68 kinase; PRRL, primed rabbit reticulocyte lysate; RRL, rabbit reticulocyte lysate; ssRCC, single-stranded RNA-containing complex; ST1, 2, and 3, serotypes 1, 2, and 3; SVP, subviral particle; L, M, S, large, medium, and small sizes of ds segments; l, m, s, large, medium, and small sizes of ss (plus-strand RNA) segments.

[2] To whom correspondence may be addressed.

Sabin (5) placed in a separate group designated respiratory enteric orphan viruses (reoviruses), because they could be isolated from the alimentary and respiratory tracts of healthy as well as ill subjects. At the same time it was found that the same reoviruses isolated from humans could also be isolated from virtually all species of mammals (6–9); closely related viruses have also been isolated from reptiles, bats, and frogs.

Reoviruses attracted attention when it was shown that their genome consisted of dsRNA (10–13). Shortly thereafter, it was shown (14, 15) that the small RNA fragments extractable from highly purified reovirus particles are not random breakdown fragments, but distinct and unique RNA molecules. Shortly thereafter, it was shown (16) that there are 10 of them. Subsequently, many other viruses, including insect, plant, and fish viruses, were also shown to possess segmented genomes comprising 10, 11, or 12 segments of dsRNA and all such viruses were therefore placed into a family designated the Reoviridae, grouped into seven genera, three of which—the orthoreoviruses (generally referred to as the reoviruses), the rotaviruses, and the orbiviruses—share hosts. All are capable of infecting and multiplying in mammals, including humans.

Little is known concerning the assembly of segmented viral genomes from their component segments. Viruses whose genome segments are protein covered, such as the helical nucleocapsids of influenza virus, and therefore are incapable of recognizing each other appear to solve the problem by encapsidating "head-full" amounts of random genome segments. For example, the influenza virus genome comprises eight species of genome segments; but influenza virus particles appear to contain random collections of 11 or 12 of these, so that about 1 in 25 virus particles contains one of each of the eight species of genome segments and is therefore infectious (17, 18).

The situation is different for reoviruses, which must use a specific assortment mechanism because in their case the ratio of the number of total to infectious particles is close to 1 (19, 20). This assortment mechanism exhibits two remarkable features. First, it is highly accurate because virtually each virus particle possesses one of each of the 10 genome segments. Thus the mechanism not only recognizes individual RNA species, but also identifies them within the context of each of the others; that is, it "counts." We review below the information available concerning the nature of the RNA–RNA, RNA–protein, and protein–protein interactions that might be involved in this mechanism.

Its second remarkable feature is that this mechanism can recognize and accept heterologous or "foreign" genome segments. The first indication of this was when it was found that the progeny of infections with pairs of mutants with temperature-sensitive (ts) lesions in different genome segments include wild-type (wt) virus (21). This formed the basis of the phenomenon of reovirus genome segment reassortment.[3] It was soon found that genome segment reassortment occurs not only among genome segments that are very similar, like those of ts mutants, but also among genome segments that are very dissimilar (but genome segments of viruses that are in different genera, like those of rotaviruses and orbiviruses, are not accepted).

When cells are infected with two reoviruses that belong to different serotypes (see Section IV), reassortants that contain n genome segments of one parent and $10 - n$ of the other are formed at frequencies close to those expected on the basis of random reassortment; that is, there is little linkage between any pair or combination of genome segments (22). This means that cognate genome segments that have diverged almost to randomness are accepted as efficiently as genome segments that differ in sequence by no more than 5% (see Section IV). Thus genome segments must possess assortment signals.

It has also been found that the introduction of heterologous, or modified homologous, genome segments into the reovirus genome requires the presence of what appear to be acceptance signals on certain genome segments of recipient genomes (see Section VIII). Identification of these signals is of vital importance: it permits identification of the functional domains of reovirus proteins (catalytic, protein–protein, or protein–RNA binding, etc.), assessment of the effects of modifying protein function on viral phenotype, construction of reovirus strains with any desired characteristic, and the development of reoviruses—viruses with very little if any pathogenicity for humans—as expression vectors that could have far-reaching clinical applications.

In order to lay the groundwork for examining molecular recognition interactions in reovirus genome segment assortment and reassortment, we review (1) the basic features of the reovirus multiplication cycle, (2) the nature of the genome segment sets of the three reovirus serotypes, (3) the nature of reovirus defective interfering (DI) particles that—reassortants derived from parents of different serotypes being genome segment *sequence* reassortants—are genome segment *size* reassortants, (4) the nature of reovirus genome segment reassortment and of the reassortants that are generated, (5) the functions of the various reovirus proteins, (6) the nature of the association of reovirus mRNA molecules with reovirus proteins to form the precursors of assortment complexes and progeny genomes, (7) the nature of the infectious reovirus RNA system, and (8) the nature of the signals essential for the generation of reassortants that have been identified as a result of analyses using the infectious reovirus RNA system.

---

[3] The assembly of homologous and heterologous genome segment sets is referred to as assortment and reassortment, respectively.

I. Basic Features of the Reovirus Multiplication Cycle

The basic features of the reovirus multiplication cycle are depicted in Fig. 1. Reovirus particles (Fig. 2) (23) adsorb, via their cell attachment protein σ1 (24, 25), to glycoprotein receptors that are rather nonspecific [which explains their very wide host range (26–31)]. They then enter cells in phagocytic vesicles that fuse with lysosomes (32), from which they are liberated into the cytoplasm (33) in a form in which the capsomers of the outer capsid shell are extensively degraded proteolytically. Protein σ3 is removed completely, and μ1C is cleaved once (34) and its C-terminal portion is also lost, leaving its (major) N-terminal portion, known as protein δ, still attached to the core shell (35–40). This conversion of virus particles to what are known as intermediate subviral particles (ISVP) causes a profound change in the ability of the RNA polymerase/transcriptase (see Section V) to transcribe the 10 dsRNA genome segments within them (38, 41–43). In virus particles, the transcriptase can only effect abortive transcription (44, 45); only up to

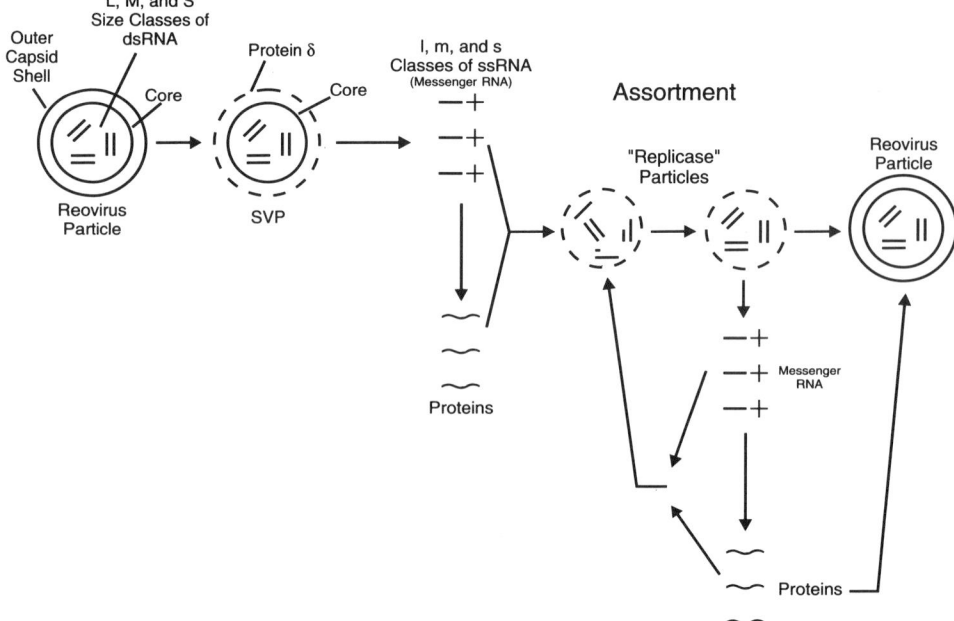

FIG. 1. The reovirus multiplication cycle.

four 5'-terminal residues [which are the same in all 10 genome segments (46)] of minus strands are transcribed, which leads to the formation of a series of small oligonucleotides, GC, GCU, GCUA, GCUA(U)$_{1-4}$, and GCUA(A)$_{1-4}$, as well as (A)$_{2-20}$ (47–50). Clearly, movement of template relative to the transcriptase is precluded. The disruption of the outer capsid shell, however, removes this constraint, so that entire genome segments can now be transcribed [although most transcription initiation events are still abortive (44)].

The products of the transcription [which is conservative (51)] are perfect plus strands (52) that serve both as mRNAs and as templates for the synthesis of minus strands, which starts at about 3 hours after infection. The newly synthesized minus strands then remain associated with their plus strand templates, thereby generating the progeny dsRNA genome segment (53–56). The major question is, at what stage are the 10 genome segments assorted into unique sets of 10? Because single strands can presumably recognize each other much more efficiently than double strands, it is generally assumed that assortment occurs with ssRNA genome segments. However, no particles containing unique sets of 10 ssRNA molecules have ever been isolated or demonstrated, and there has been only partial success (57–59) with attempts to discover the nature of the complexes within which genome segments are assorted by physically isolating (either by density gradient centrifugation or by gel electrophoresis under nondenaturing conditions) complexes or particles with the ability to transcribe endogenous plus strands into minus strands, that is, "replicase particles."

"Replicase particles" are about 40 nm in diameter; their sedimentation coefficients range from 180 to 550 S and their plus-stranded RNA templates are sensitive to RNase digestion. Interestingly, the particles with the lowest sedimentation coefficients been reported to synthesize predominantly genome segments of the S size class; those with intermediate S values synthesize increased amounts of M size class genome segments, and those with the largest S values synthesize predominantly L size class genome segments (59). There is no synthesis of plus stands in these particles, nor is there initiation, only completion, of minus strand synthesis. Once synthesis of minus strands is complete, all particles sediment with about 550 S, and presumably contain complete progeny genomes.

Morphogenesis is then thought to proceed with the formation of the core shell and attachment of the projections/spikes and of the outer capsid shell capsomers (see Fig. 1), which shuts down transcription of the dsRNA genome segments (38, 43). However, it is not known how the "replicase" particles, which contain neither of the two core shell components $\lambda 1$ and $\sigma 2$, are converted to cores.

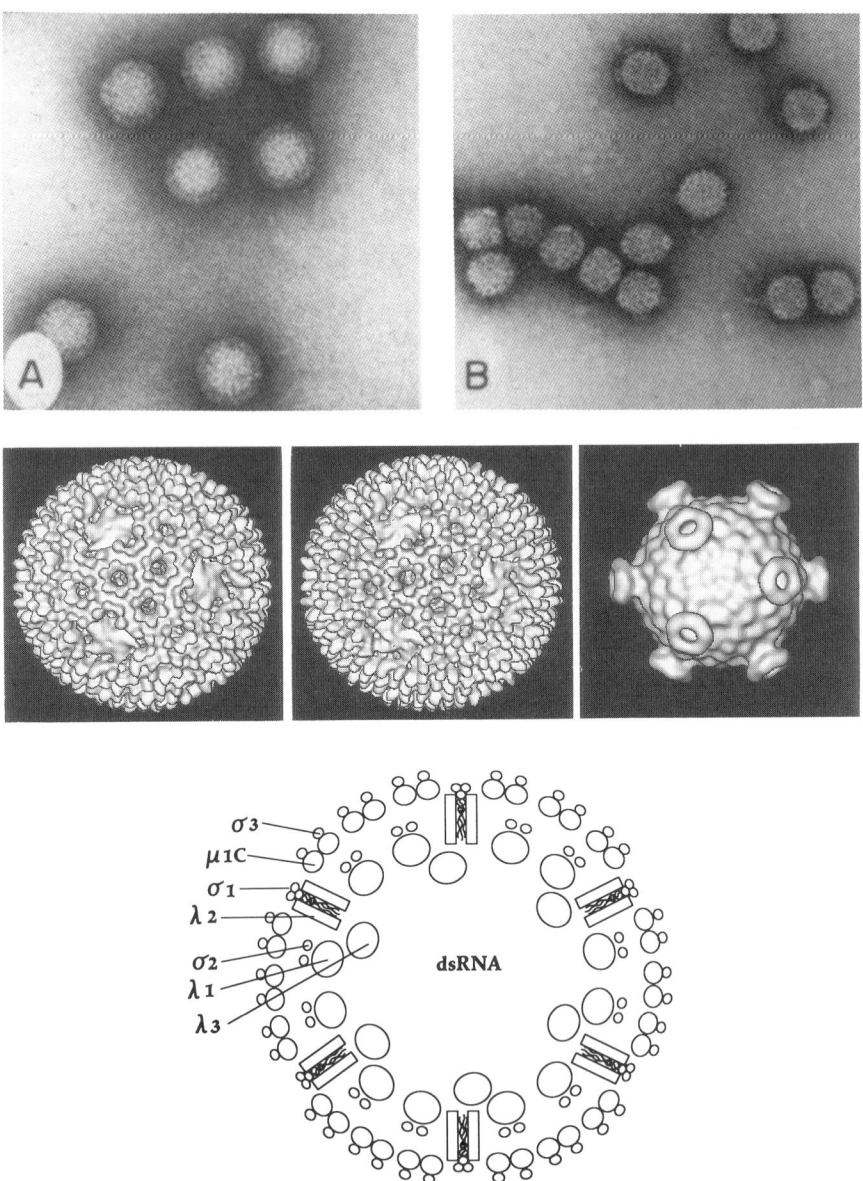

Fig. 2. The orthoreovirus particle. Reovirus consist of cores made up of an icosahedral shell composed of proteins λ1 and σ2 (molar ratio 1:2), at the vertices of which projections/spikes composed of pentamers of protein λ2 (which is also the mRNA capping enzyme) are located. Into the projections/spikes are inserted trimers of protein σ1, the cell attachment protein and the protein on which the epitopes that elicit the formation of type-specific neutraliz-

II. The Nature of Reovirus Genome Segments

All 10 genome segments of ST3 have been sequenced; for ST1 and ST2, all except the L2 and M3, and the L2, M1, and M3 genome segments, respectively, have been sequenced. This has permitted an interesting analysis of the evolutionary history of reoviruses (Table I). The following salient features have emerged.

1. All genome segments of ST1 and ST3, except the one that specifies serotype, are extremely closely related (97% or more). For discussions of sequence relationships, as here, they are essentially indistinguishable. They are substantially less closely related to the genome segments of ST2. Thus the genomes of ST1 and ST3 on the one hand, and that of ST2 on the other, represent homologous genome segment sets.
2. The lone exception is genome segment S1, the ST1 and ST2 versions of which are the most closely related (60). Here an exchange occurred, at some time in the past, between the S1 genome segment of ST2 and that of either ST1 or ST3. The following scenario proposes how this exchange might have occurred. Reassortment among ST1, ST2, and ST3 genome segments occurs readily in the laboratory (see Section IV), but reassortants usually cannot, under natural conditions, compete successfully with homologous genome segment sets, probably because they generate less well-fitting and therefore less efficiently assembled capsids (so that their yield would be lower), or less stable capsids. Occasionally, however, variants arise for which the fit between structural proteins is better than for the then-dominant homologous genome segment set, and these variants then become the dominant population components via selective sweeps and provide the origins for new phylogenetic trees (60, 61). This happens quite fre-

---

ing antibodies are located. The outer capsid shell, through about one-half of which the projections/spikes penetrate, is made up of capsomers composed of S—S-bonded dimers of protein μ1C complexed with two molecules of protein σ3. Cores contain not only the 10 genome segments composed of dsRNA, but also the RNA polymerase, the catalytic site of which is located on protein λ3 and which is thought to be located at the base of the projections/spikes. Cores also contain protein μ2, which, like λ3, is present to the extent of about 12 molecules. Neither its location nor its function (there is suggestive evidence that it may be part of the polymerase) is known. *Top:* (A) Reovirus particles; (B) reovirus cores [both courtesy of Drs. R. B. Luftig and W. K. Joklik; Zinsser Microbiology, 20th Ed. (W. K. Joklik *et al.*, eds.), p. 757; Appleton and Lange, Norwalk, Connecticut, 1992]. *Middle:* Surface representations of icosahedral reconstructions of reovirus ST2, strain D5/Jones (left); reovirus ST3, strain Dearing (middle); and a reovirus ST3 core (right). From Ref. 23, P. Metcalf, M. Cyrklaff and M. Adrian, *EMBO J.* **10**, 3129 (1991), by permission of Oxford University Press. *Bottom:* Schematic representation of the reovirus particle.

TABLE I
EVOLUTIONARY DIVERGENCE PATTERNS OF REOVIRUS GENOME SEGMENTS

	Serotype pair[a]																											
	1:3										1:2										2:3							
	L1[b]	L2[c]	L3[d]	M1[e]	M2[f]	M3[g]	S1[h]	S2[i]	S3[j]	S4[k]	L1[b]	L2[g]	L3[d]	M1[l]	M2[e]	M3[g]	S1[h]	S2[i]	S3[j]	S4[k]	L1[b]	L2[g]	L3[d]	M1[l]	M2[e]	M3[g]	S1[h]	S2[i]	S3[j]	S4[k]
First-base codon position	2	2	6	2	4		85	6	5	4	17		31	11	11		55	13	25	12	17		33		11		85	14	25	13
Second-base codon position	1	2	3	0.5	2		68	0	2	1	5		31		2		40	3	8	5	5		28		2		68	4	8	5
Third-base codon position	13	29	6	7	53		96	54	48	23	77		87	82			79	75	79	73	77		87		83		91	73	74	77

[a] Values are percent divergence; the numbers are percent mismatches multiplied by 1.333 to provide an estimate of percent divergence toward randomness. [b] From Ref. 62.
[c] Based on first 87 codons (66, 62). [d] Based on first 94 codons (67, 62).
[e] From Refs. 68 and 69. [f] From Refs. 34 and 70.
[g] ST1 and ST2 sequences not yet available. [h] From Refs. 60, 65, and 71–73.
[i] From Refs. 74–76 and 166. [j] From Refs. 77 and 169.
[k] From Refs. 78 and 80. [l] ST2 sequence not yet available.

quently for ST1 and ST3 reassortants, but never for ST1/ST3 and ST2 reassortants (see Section IV), with one exception: i.e., when the ST2 S1 genome segment exchanged with that of ST1 or ST3, probably because the resulting new fits of the respective σ1s in their heterologous capsids resulted in improved survival.

3. Extent of divergence toward randomness for third-base codon positions provides an evolutionary clock. Because these numbers are very high for all genome segments for serotype pairs 1:2 and 2:3, it is clear that (a) ST2 separated from the ST 1:3 precursor a very long time ago, (b) the genome segment S1 serotype 1/3:2 exchange also occurred a long time ago, and (c) ST 1/3 genome segments do not become stably associated with the ST2 genome; that is, reassortants between S1/3 and ST2 do not survive for any length of time under natural conditions.

4. For the ST 1:3 pair, the third-base codon divergence percentages vary tremendously, from 6 to 96%. This suggests that divergence started at different points in time for each genome segment (62). The most likely mechanism for this occurring is that described above; that is, the precursor in each case was a reassortant in which an ST1 or 3 genome segment became, at some point in time, stably associated with the heterologous genome segment set, which resulted in a selective sweep. As a result, contemporary ST1 and 3 genome segments are the descendants of genome segments that became associated with heterologous genome segment sets a fairly long time ago for M2, S2, and S3, and only fairly recently for L3 and M1.

Another expression of this type is the stable exchange of ST1 and ST3 genome segments; this came to light recently in studies in which sequence diversity among S2 (63) and S4 (64) genome segments of random isolates was examined. In both cases phylogenetic trees based on RNA sequence could be constructed for the dozen or so isolates examined; because most mismatches occur in third-base codon positions, these trees established yet again the fact that these isolates are the descendants of stable ST1/ST3 genome segment exchanges that occurred at different times. Further, not surprisingly, all σ2 and σ3 proteins of the isolates were equally related to each other, as well as to those of the Lang and Dearing strains; they were indistinguishable as far as sequence relationship was concerned (see above). It would be interesting to extend this sort of analysis to ST2 in order to test experimentally the postulate that the genomes of ST1/ST3 and ST2 are two homologous genome segment sets.

5. The proteins that have diverged the most, by far, are the three σ1 proteins. Of the several factors that control evolutionary divergence

rates—such as the error proneness of polymerases, the ability of proteins to accept sequence changes, and selective pressures/opportunities—the most important during evolution is the ability of proteins to accept sequence changes without losing their ability to function. By contrast, in short-term situations such as mutations in oncogenes/antioncogenes, error proneness of polymerases is of primary importance. For the evolution of the σ1 protein, both the second and the third factor listed above are of critical importance.

As far as we know, protein σ1 is subject to few structural constraints. One is that the N-terminal region must maintain heptad repeats so that it can assume the coiled-coil helix configuration necessary for it to be inserted into the λ2 projection/spike channels (Fig. 2) (65). However, this is only a weak constraint, because any hydrophobic amino acid can function in any of the hydrophobic aminoacid-specific locations, and a wide assortment of amino acids is tolerated elsewhere. Another is that the functional domains of epitopes, including the one that elicits the formation of type-specific neutralizing antibodies, and of the receptor-binding site, must be preserved. But these are short regions, well separated, and probably not restricted to specific locations. A third restriction is that the function, whatever it is, of protein σ1S (see below) must be maintained. In spite of these constraints, σ1 is an exceptionally "plastic" protein, that is, it is capable of accepting numerous amino-acid exchanges without major loss of function. Added to the above is the fact that as the bearer of the epitopes that elicit the formation of neutralizing antibodies, it is subject to immunologic selection pressures, both in established as well as in newly acquired hosts. This accounts for the very extensive divergence of the three allelic forms of protein σ1.

As for the other nine proteins, the one that has diverged the most, as judged by the extent of first-base codon-position divergence, is the major core shell component λ1, which is surprising for a structural protein [but the estimate is based on only the 5'-terminal 94 (out of 1267) codons], and it is followed by σNS, a nonstructural ssRNA-binding protein. All the rest are highly resistant to sequence change.

The relevance of this discussion of the nature and evolutionary history of reovirus genome segments to the problem of molecular recognition during the assembly of the reovirus genome is that the sequences of some cognate genome segments are very similar among the three serotypes, and those of others are very different. Yet the latter are accepted into heterologous genome segment sets as efficiently as the former, generating completely functional reassortants. Hence acceptance into heterologous genome segment sets must depend not on overall sequence or shape, but on possession of *reassortment signals*. As long as the nature of these signals is not known, novel genetic information cannot be introduced experimentally into the reovirus genome.

III. The Nature of Reovirus Defective Interfering Particles

When reovirus ST3 is passaged repeatedly at high multiplicity, defective interfering (DI) particles begin to accumulate after six to eight passages (81–86). These are particles that lack an intact L1 genome segment; instead, they contain a truncated L1 genome segment that is only about 1800 bp long, compared with the 3854 bp of the wt L1 genome segment (62). Particles that contain this truncated nonfunctional genome segment can multiply readily in the presence of wt virus, which acts as helper, supplying the λ3 protein that the DI particle cannot provide. Because the truncated genome segment is smaller than the wt genome segment, DI particles outgrow wt particles, so that they account for more than 98% of virus yields. There is a limit to this ratio because DI particles can multiply only in cells infected with at least one wt particle. Interestingly, if high-multiplicity serial passaging is continued, particles arise that contain not only L1, but also L3 truncated genome segments (85, 87).

When reovirus ST1 is passaged similarly, DI particles accumulate that lack an intact genome segment M1, but contain a truncated M1 genome segment (88), and certain ST1 × ST3 reassortants also delete first the M1, and then also the L2 genome segment (89). In the smallest truncated M1 genome segment, 1960 bp are deleted from the center of the wt genome segment (2304 bp) (90). By contrast, ST2 virus can be passaged indefinitely under conditions of high multiplicity without truncated genome segments being generated (88).

Reassortants that contain genome segments belonging to different serotypes are *sequence* reassortants; DI reovirus particles are *size* reassortants. Because genomes containing truncated genome segments are produced in large numbers, variation in size is clearly no impediment to assortment. What is of interest is that most truncated genome segments in DI virus particles are almost exactly the same size, and that the deletions always occur in almost exactly the same location. This indicates either that sequences are present at the 5'- and 3'-terminal boundaries of deletions that cause the RNA polymerase to detach and reattach [and which appear to contain short regions of homology (90)], or that assortment and reassortment signals essential for the introduction of novel genetic information into the viral genome are located there (90). It would be logical to begin searching for such signals in these regions.

IV. The Nature of Reovirus Genome Segment Reassortment and of the Reassortants That Are Generated

Reovirus genome segment reassortment was first shown for pairs of reovirus ts mutants by Fields and Joklik (21), who found that, in cells infected with pairs of mutants with ts lesions in different genome segments, ts$^+$ virus is generated. The phenomenon was analyzed in a comprehensive series of studies involving two-factor and three-factor crosses of ts mutants, which demonstrated that reassortment is essentially random (91–93).

The finding that established genome segment reassortment analysis as a very powerful technique was the demonstration that the dsRNA genome segments and proteins of the three reovirus serotypes possess readily distinguishable electrophoretic migration rates (22). This permitted identification of the parental origin of each genome segment and protein of reassortants generated in mixed intertypic infections (crosses of pairs of serotype prototype strains). In such crosses, up to 15% of the yield is reassortants that contain genome segments of the two parents in an approximately Poisson distribution, there being only limited linkage between genome segments (89). The reason for the upper limit of about 15% reassortants in yields was not evident; if the reassortant yield was lower, no significance was attached to this phenomenon. Ability to identify the genome segments and proteins of each genome segment set also provided the tools necessary for assigning ts mutations to specific genome segments (94–96), which resulted in a map of the reovirus genome (97). The results obtained agreed completely with the biochemical approach (98) that translated individual denatured dsRNA genome segments *in vitro*.

Fields and co-workers then developed genome segment reassortant analysis into a powerful technique for assigning *functions* to the proteins encoded by the various genome segments. The technique consists of selecting some parameter in which the members of two serotypes differ, examining a collection of random reassortants as to which parent they resemble, and then identifying the genome segment that accompanies that parameter. Properties ranging from resistance of viruses to chemical reagents to elimination of virus from infected hosts have been tied in this way to specific genome segments (99–111).

Moody and Joklik (112) realized some time ago that if sets of reassortants containing nine genome segments of one serotype and the tenth of another could be isolated, they would provide powerful reagents for genome segment reassortant analysis. An attempt was made to isolate such sets by injecting cells infected with one serotype with individual genome segments of another. Although the injected genome segments were functional, they were not introduced into progeny genomes. Interestingly, however, when cores of two serotypes were lipofected into cells (cores, not possessing protein σ1, are not infectious; but when introduced into cells by lipofection, possess the same specific infectivity as virions), 10–15% of the progeny were reassortants, and they were almost exclusively monoreassortants.

Thus there is a major difference between these three modes of genome interaction: formation of multireassortants in cells infected with *virions*, formation of monoreassortants in cells lipofected with *cores*, and failure of *injected genome segments* to be accepted into the reovirus genome. The last of these observations could conceivably be explained on topologic grounds, but the difference in outcome between infections initiated by virions and cores suggests that the outer capsid shell, either in the form of protein δ or of degraded parental σ3 and/or μ1C molecules, plays some role in controlling reassortment (though not assortment: the yield of infectious virus is essentially the same in cells infected with virions or with cores).

V. The Functions of Reovirus Proteins

As discussed in Section VI, reovirus mRNA species, soon after they are transcribed, associate with three viral proteins to form complexes that are assembled into progeny genomes under the influence of several other viral proteins. The functions of these proteins are therefore directly relevant to the assembly of reovirus genomes (Table II).

A. Protein λ1

Protein λ1 is a 1233-aminoacid protein with a predicted α-helix and β-sheet content of 23.6 and 28.3%, respectively (67). It binds σ2, with which it forms the capsomers of the core shell (see Fig. 2), as well as λ2 and λ3 (113). In fact, in cells infected with recombinant vaccinia viruses expressing λ1 and σ2, particles are formed that resemble reovirus cores lacking the 12 projections/spikes, and in cells infected with vaccinia viruses that express λ1, σ2, and λ2, particles are formed that resemble reovirus cores (114). Protein λ1 possesses a nucleotide-binding site, TKGKSGG, starting at residue 8; dsRNA-binding activity (as demonstrated by Northwestern filter-binding assays) mediated by its 187-aminoacid N-terminal region (115); and a zinc-finger motif centered around residue 194 (67), which is not involved in

TABLE II
INFORMATION CONTENT OF THE REOVIRUS ST3 GENOME

Genome segment	Protein	No. of codons	No. of molecules/virus particle	Function
L1	λ3	1267	12[a]	Polymerase/transcriptase
L2	λ2	1289	60	Component of projections/spikes; mRNA guanylyl transferase
L3	λ1	1233	120[a]	Major core shell component
M1	μ2	736	12[a]	Unknown; core component
M2	μ1	708	600[a]	Major outer capsid shell component[b]
M3	μNS	719	—	Nonstructural; binds ssRNA
S1	σ1	455	36	Cell attachment protein; elicits formation of type-specific neuralizing antibodies
	α1S	120		Unknown; nonstructural
S2	σ2	418	240[a]	Core shell component
S3	σNS	366		Nonstructural; binds ssRNA
S4	σ3	365	600[a]	Outer capsid shell component; binds dsRNA and ds regions in ssRNA

[a] Best estimate.
[b] In the form of cleavage product μ1C.

binding nucleic acids (115). The dsRNA-binding ability of λ1 is both expected, because it is a major component of the core shell, and surprising, because in order to be transcribed, the dsRNA segments must move relative to it.

B. Protein λ2

Pentamers of λ2 form the reovirus core projections/spikes (116). In cells infected with recombinant vaccinia virus expressing λ2, it is formed as a monomer (117), but in cells infected also with the recombinant vaccinia viruses expressing λ1 and σ2, particles are formed, as pointed out above, that resemble cores; that is, λ2 pentamerizes in the presence of core shells (114). Protein λ2 possesses two motifs near residues 900 and 1000 that match similarly spaced dual-consensus sequence elements that are conserved in GTP-binding proteins (66); it binds GMP via a phosphoamide bond to the Lys at position 226 (118) and it is a key component of the reovirus mRNA capping complex (119), possessing guanylyl-transferase activity both as a monomer and as a pentamer (117). However, it possesses no methyl-transferase activity as a monomer (117), although it possesses a motif reminiscent of methyl-transferase motifs (66); nor does it possess the nucleoside-triphosphate phosphohydrolase activity that converts 5'-PPP to 5'-PP groups prior to guanylylation (117) and that is present in reovirus particles (120). Perhaps it possesses these two activities when it is in the form of pentamers or complexed with λ1, λ3, and σ2. Interestingly, it is a component of the complexes in which plus-strand RNA genome segments are transcribed to form complexes that contain all 10 dsRNA-containing genome segments in equimolar amounts (121) (see Section VI).

C. Proteins λ3 and μ2

Protein λ3 is a key protein for reovirus RNA transcription and genome generation (see Section VI). It accumulates in infected cells in perinuclear inclusion bodies (122) and is a stable, structural component of reovirus cores that lose dsRNA when treated with 3 M urea, but not λ3 (122). It has been purified from cells infected with the recombinant vaccinia virus that expresses genome segment L1 (113); it is formed as a monomer that transcribes poly(C) into poly(G) with absolute precision, but fails to transcribe either ds or ss reovirus RNA.

The fact that reovirus encodes a poly(C)-dependent poly(G) polymerase is reminiscent of the situation with the ssRNA-containing bacteriophages Qβ, MS2, and R17. These phages encode a protein that contains the polymerase catalytic site, but possess no enzymatic activity; in combination with 30-S ribosomal protein S1, it is a poly(C)-dependent poly(G) polymerase, and only in combination with the two protein synthesis elongation factors, EFTs and EFTu, does it catalyze the specific transcription of viral RNA (123).

The problem thus arises as to the nature of the factor or cofactor that enables λ3 to transcribe first ds reovirus RNA and then plus-stranded reovirus RNA. Any role for a host protein would appear to be ruled out, at least for transcribing dsRNA, because such transcription occurs within cores. That leaves four virus-encoded proteins as possible candidates for conferring affinity and specificity for dsRNA (and perhaps also for ssRNA): the three major core components, λ1, λ2, and σ2, and the minor core component, μ2. So far all efforts to demonstrate the conversion of λ3 from a polymerase that transcribes only poly(C) to one that transcribes ds or ss reovirus RNA have been negative. With respect to the core components λ1, λ2, and both together, all form complexes with λ3; but they possess undiminished poly(C)-dependent poly(G) polymerase activity, and no RNA polymerase activity. This situation is unchanged in the added presence of σ2, which becomes part of the complex by virtue of its affinity for λ1 (113).

Extensive efforts have been made to determine whether protein μ2 might provide the missing function. Its α-helix content (α-helix/β-sheet, 1.2) is much higher than that of typical viral structural proteins (α-helix/β-sheet, 0.5–0.9), but lower than that of typical nonstructural proteins such as μNS and σNS (α-helix/β-sheet, 2.5). It possesses no motifs indicative of function.

Like other reovirus genome segments, the genome segment that encodes it (M1) has been inserted into the TK gene of vaccinia virus under the control of the CPV ATI protein gene promoter (see Refs. 117 and 124), and μ2 is expressed in cells infected with this recombinant vaccinia virus to the extent of 0.1 to 0.5% of total cell protein (88).

There are two reasons why μ2 is the leading candidate as a λ3 cofactor. First, it is present in reovirus particles in seemingly equimolar amounts with λ3 [about 12 molecules per virus particle, consistent with a position for both proteins at the base of each projection/spike, which would enable templates to move past the enzyme and the product to exit through the projections/spikes with mRNA capping capability (see above)]. Second, it may be significant that the two most common DI particles are particles in which either genome segment L1, which encodes protein λ3, is truncated (ST3 virus), or genome segment M1, which encodes protein μ2 (ST1 virus) (see Section III). However, all attempts to detect a direct association of λ3 and μ2, either in virus particles or in infected cells, have failed. Does that mean that λ3 and μ2 do not interact? Also, why does λ3 appear to have no affinity for RNA, when it has sufficient affinity for poly(C) to be a very active poly(C) transcriptase? Does it have such excess affinity for poly(C) and GTP that it cannot transcribe RNA unless it is modified by combination with some other protein? Or is the affinity of λ3 for μ2, on the one hand, and for RNA, on the other, not an issue because their concentrations in the complexes in which transcription occurs are extremely high: for example, the concentration of λ3, μ2, and dsRNA molecules in cores, assuming that there are 10–12 molecules of each, is almost molar!

D. Protein μ1

Protein μ1 is synthesized in large amounts in myristoylated form (125) in reovirus-infected cells and combines, soon after it is formed, with protein σ3 (126). This association causes it to be cleaved to μ1C (127), the major component of reovirus particles as well as of their outer capsid shell, and an N-myristoylated 42-aminoacid peptide that apparently remains associated with the μ1C–σ3 complexes because it is present in reovirus particles (128, 129), where its function may be to effect the release of ISVPs from endoplasmic vesicles/lysosomes (see Section I) (129). Although the affinity between μ1 and σ3 is sufficient for them to bind prior to their incorporation into the outer capsid shell (in which they exist in the form of σ3/μ1C–S–S–μ1C/σ3 complexes (128, 34), possibly in the form of trimers), about 10% of intracellular μ1 is free (130).

Cleavage of μ1 to μ1C requires both σ3 and myristoylation of its N terminus (131). The proteolytic site is apparently not on σ3, which does indeed possess a motif similar to the picornavirus protease motif, but which is not functional here, because truncated forms of σ3 that lack it still combine with μ1 and cause its cleavage (132). Rather, the cleavage appears to be autocatalytic, activated by combination with σ3. Unfortunately definitive experiments involving the use of ST3 μ1 purified to homogeneity are not possible because μ1 is the only reovirus protein not formed in amounts sufficient for isolation or purification in cells infected with the recombinant vaccinia virus that contains the genome segment that encodes it (88). Numerous vaccinia virus constructs have been tested; a recombinant Sindbis virus that expresses it has been constructed; and the effect of coinfection with vaccinia viruses expressing other reovirus proteins has been examined; but in all cases only very small amounts of μ1 are formed. The reason for this may be its cytotoxicity. Protein μ1 causes chromium release from cell membranes (133); it interferes with virus multiplication in special situations (134); and it is implicated in blocking transmembrane transport (111). Interestingly, ST1 μ1 is considerable less cytotoxic than ST3 μ1 (113, 134, 111); it, or ST2 μ1, which has not yet been examined, may be expressed by expression vectors in much larger amounts than the ST3 protein.

E. Protein μNS

Protein μNS could well play a very key role in genome segment assortment. It is a 719-aminoacid phosphoprotein (68) that exists in cells not only as μNS, but also as μNSC (130), a protein originally thought to be a cleavage product, but actually the protein formed when translation starts at the *second* initiation codon, at residue 42, rather than at the first (68).

Protein μNS has several interesting properties. First, it is an ssRNA-binding protein (121). It binds to reovirus mRNAs within minutes after they are transcribed (see Section VI); in fact, μNS is the first of three such proteins that bind. It is a component of all ssRNA-containing complexes, and it is a prominent component of the complexes within which ssRNA is transcribed into dsRNA, as well as of the resulting dsRNA-containing complexes in which all 10 genome-segment species are present in equimolar amounts (121).

Second, μNS possesses affinity for proteins of the cytoskeleton (CSK) (135). In fact, a monoclonal antibody against μNS (130) stains the CSK, indicating either that the epitope against which it is directed contained both μNS and CSK elements, or that the μNS epitope against which it is directed is identical to one present on a CSK component. The full significance of the association of μNS with the CSK is not clear; for example, does it indicate that assortment and reassortment are highly compartmentalized on elements of the CSK?

Third, μNS has a very high α-helix content (50%). Dot-matrix analysis reveals that its C-terminal region, like myosins, exhibits periodic sequence similarity elements indicative of helical structure (68); but tandem heptapeptide repeats with hydrophobic amino-acid residues in positions 1 and 4, indicative of a coiled-coil helical structure, such as are present in protein σ1, are not present in μNS. Rather μNS contains numerous short (maximum length, 25 amino acids) helices distributed throughout the molecule, with a preponderance in the C-terminal region. Interestingly, it possesses a helix–loop–helix motif of the type present in DNA-binding domains.

Finally, nothing is known concerning the role of μNSC. About one-half of μNS is generally present in the form of this slightly shorter protein that lacks the 42 N-terminal amino acids, the composition of which is unremarkable (almost one-half are A, S, or T; nine are hydrophobic and seven are charged [(five basic, two acidic) (68)].

F. Protein σ1

Protein σ1 (ST1, ST2, and ST3; 470, 462, and 455 amino acids, respectively) (60, 65, 71–73) is an extremely important protein: it is the cell attachment protein (24, 25), that is, it controls all interactions between reovirus and cells, including tissue tropism (136, 137) and pathogenicity (138). Protein σ1 also possesses the epitopes that are responsible for most of the interactions of reovirus with the immune system (139–143), including those that elicit the formation of type-specific neutralizing antibodies (99). Protein σ1 spontaneously trimerizes (124, 144) because it possesses, in its N-terminal portion, known as the tail, a heptad repeat region with hydrophobic amino acids in positions 1 and 4 that causes it to assume a coiled-coil helical configuration (65) and that is inserted into the 12 icosahedrally located projections/spikes. The C-terminal portion of σ1 is a globular structure (145, 146), the head, which contains, well separated from each other, the cell attachment site recognized by receptors, and the epitopes. In many cases, the tail is not completely buried in the projection/spike, but extends outward from the virus particle so that the head is projected 40 nm or more away from the virus particle surface (147), which gives reovirus an appearance similar to that of adenovirus. This indicates that the tail is larger than predicted by the coiled-coil sequence region and suggests that additional sequences in σ1 contribute to it.

Protein σ1 is a very well-studied protein. Its structure has been worked out in detail by genetic, chemical, and biophysical methods so that a detailed picture of its structural/functional nature has emerged (124, 145, 146, 148–157). Interestingly, when reovirus is electrophoresed in agarose gels, 13 well-separated and distinct bands are obtained. These bands are caused by the fact that not all reovirus particles contain 12 trimers of σ1; rather reovirus particle populations comprise the complete range of particles that contain from 0 to 12 σ1 trimers (158), the median number of trimers per virus particle being 7. The specific infectivity of all particles that contain two or more trimers is the same; those that contain only one trimer are about one-third as infectious (still remarkably high!); and particles that contain no σ1 trimers are, of course, not infectious. The reason for the low number of σ1 trimers in reovirus particles is probably the fact that only small amounts of it are made in infected cells (*159*), s1 being a poorly translated mRNA [about 10% as efficiently as s4, the most efficient (*160*)]. Because reovirus particles with only one σ1 trimer are still quite infectious, there is clearly no selective pressure for the S1 genome segment, which already provides seven times the amount of σ1 necessary for infectivity, to evolve into a form from which σ1 is translated more efficiently.

G. Protein σ1S

Genome segment S1 possesses a second open-reading frame, in a different reading frame from that for σ1, that encodes a 120-aminoacid (for ST3; 119 and 125 amino acids for ST1 and ST2, respectively) basic nonstructural protein variously designated σ1S, σNS, and p14 (*60, 161–164*). Its function is not known; but it is known to be localized to the nucleus, more specifically, to the nucleolus (*165*).

H. Protein σ2

Protein σ2 is one of the two major components of the reovirus core shell. It possesses strong affinity for λ1; in cells infected with the recombinant vaccinia viruses expressing λ1 and σ2, particles resembling core shells are formed (*114*). Protein σ2 is iodinated to a much lesser extent than λ1 (*167*), which suggests that it is located primarily on the inner, rather than on the outer surface of the core shell (the low degree of labeling not being caused by a low Tyr content). Like protein λ1, it binds dsRNA in filter-blotting assays (*76*). The sequence between amino-acid residues 354 and 374 is strikingly similar to the sequence between residues 1008 and 1031 in the β subunit of *E. coli* DNA-dependent RNA polymerase, which may exist in amphipathic α-helix form. However, the α-helix/β-sheet ratio of σ2 is only 0.5, characteristic of typical viral structural proteins. It should be noted that the reovirus core shell is an extremely stable structure, very resistant to proteolysis and dissociated only under strongly denaturing conditions. It is this stability that has prevented the isolation of the RNA polymerase λ3 protein from reovirus particles.

I. Protein σNS

Protein σNS is the other reovirus nonstructural protein. It has a very high α-helix content (50% α-helix; α-helix/β-sheet ratio, 1.80), particularly in its C-terminal one-third (68% α-helix). It possesses strong affinity for ssRNA (*170, 126*); removing the last traces of RNA from it when purifying it is difficult. Remarkably, it does not possess any of the known RNA-binding motifs. Like protein μNS and σ3, it binds to reovirus ssRNA very soon after it is transcribed, and is a component of the ssRNA-containing complexes that are the precursors of progeny genomes (see Section VI).

J. Protein σ3

Protein σ3, together with protein μ1C, is the major component of the outer capsid shell. It possesses strong affinity for myristoylated μ1 (126), combining with it soon after it is formed, so that most of both proteins exist in the cytoplasm of infected cells as μ1C·σ3 complexes (126, 130) that then assemble to form the outer capsid shell as multimers of σ3/μ1C–S–S–μ1C/σ3 subunits (128). As it combines with protein μ1, it causes it to cleave between residues 42 and 43, the N-terminal myristoylated peptide remaining associated with the μ1C·σ3 complex. The proteolytic site that effects the cleavage is probably on μ1; σ3 does possess a motif similar to that of picornavirus proteases, but truncated forms of σ3 that lack it can nevertheless bind μ1 and effect its cleavage (132), which is therefore thought to be autocatalytic in character, activated by a conformational change caused by binding to a specific σ3 domain.

Protein σ3 also possesses a zinc finger between positions 46 and 73 (171), which is essential both for its stability (172) and its ability to fold correctly (173). The sequence motif CGGXXXCXH also present in picornavirus proteases, referred to above, is located within it (positions 45–54).

The property of σ3 that is of particular relevance to genome segment assortment is its affinity for double-stranded RNA via an approximately 85-aminoacid domain centered around residue 265 (174). This domain has two short basic regions, each with the potential to adopt an α-helical configuration, both of which are essential for dsRNA binding ability (175, 176). The primary function of this remarkable binding activity is not clear. As isolated from reovirus-infected cells, σ3 binds dsRNA, not ssRNA (126). In infected cells, however, σ3, together with μNS and σNS, binds to nascent plus-stranded RNA to form ssRNA-containing complexes that appear to be the precursors of progeny genomes (121) (see Section VI). It has therefore been hypothesized that rather than binding reovirus dsRNA, which never exists in free form in infected cells, σ3 binds to hairpin loops of specific sequence/structure in ssRNA, stabilizing it for transport, assortment, or transcription.

The fact that σ3 binds to dsRNA, or dsRNA-like elements in ssRNA, is thought to explain its effect in stimulating translation of reporter gene mRNA during transient coexpression (177). The explanation of this phenomenon is that viral infection often represses activation of translation initiation via the inhibitory activity of an interferon-induced dsRNA-activated protein kinase, DAI (178–180). This enzyme phosphorylates the α subunit of the protein-synthesis initiation factor eIF-2, which normally, when complexed with GTP, targets initiator Met-tRNA to the 40-S ribosomal subunit, the subsequent association of which with the 60-S ribosomal subunit to form the 80-S protein synthesis initiation complex depends on the hydrolysis of the GTP. The subsequent exchange of GDP for GTP, on which continued ability to initiate protein synthesis depends, is catalyzed by a guanine-nucleotide exchange factor. However, eIF2·GDP complexes that contain a phosphorylated α subunit bind to the exchange factor with high affinity, which prevents the GDP–GTP exchange and thus inhibits initiation of translation. Protein $\sigma 3$, by binding dsRNA, prevents the activation of the DAI (*181, 182*) and thus stimulates transcription of reporter gene mRNA. Strangely enough, $\sigma 3$ fails to stimulate translation of cellular mRNA in infected cells. In fact, the s4 genome segment is implicated, through genome segment reassortment analysis, not in the stimulation, but in the inhibition, of host-cell mRNA translation in infected cells (*183*), an anomaly that may be explicable on the basis of findings presented in Section VIII. Nor is this the only anomaly in this regard. On the one hand, it has been reported that ST3 $\sigma 3$ stimulates translation of reporter gene mRNA to a much greater extent than ST1 $\sigma 3$ (*80, 184*), and the stimulating activity has indeed been traced to its dsRNA-binding region (*184*), which suggests that ST3 $\sigma 3$ binds dsRNA more strongly than ST1 $\sigma 3$. On the other hand, the replication of ST3 is inhibited much more strongly by interferon than that of ST1 (*185*), which implies that it is ST1 $\sigma 3$ that binds dsRNA more strongly. Whatever the explanation of this may be, there is no doubt that $\sigma 3$ interacts both with dsRNA, as well as with double-stranded regions in ssRNA, which makes its participation in binding nascent reovirus ssRNA species doubly intriguing.

VI. Reovirus Genome Segment Assortment Studied with Monoclonal Antibodies against Reovirus Proteins

Little is known concerning the nature of the genome segments (ss or ds, free or protein bound) that are assembled into sets of 10, or how the actual assorting process is accomplished. Largely on the basis of the fact that particles in which plus-strand RNA is transcribed into minus-strand RNA are heterogeneous in size [distributed throughout density gradients (*55*) or separable into three size classes containing l, m, and s size class RNA (*59*)], it was postulated that complexes containing several or perhaps even all 10 plus-strand RNA species are precursors of complexes that contain assembled progeny genomes. Although attractive, primarily because it is easier to imagine that ssRNA rather than dsRNA molecules display the recognizable sequence and secondary structure features that the assortment mechanism undoubtedly requires, this hypothesis has never been proved.

Recently, a detailed analysis has been carried out concerning the nature of the precursors of the complexes in which progeny dsRNA species are generated (121). The approach taken was to use a panel of MABs to identify the reovirus proteins associated with viral ss and dsRNA species as the viral multiplication cycle proceeds, using as criterion the ability of the MABs to precipitate viral RNA species. It was found that three viral proteins, μNS, σNS, and σ3, become associated with ssRNA molecules within less than 5 minutes after they are transcribed. The resultant complexes, the single-strand-RNA-containing complexes (ssRCCs), contain one RNA molecule each, based on the fact that they can be separated in density gradients into l, m, and s RNA size class containing ssRCC species, and they contain 10–30 protein molecules each, depending on the size of the RNA. Thus their RNA content is about 40% (compared with a value of about 65% for ribosomal subunits). Sequential immunoprecipitations suggest that all ssRRCs contain μNS, that about one-half contain μNS alone, that about one-half contain σ3, and that about one-quarter contain σNS. The significance of these differential protein complements is not clear. As for the relative proportions of the individual RNA species in ssRCCs, these reflect the composition of the total ssRNA population in infected cells, which differs substantially from equimolarity (159).

About 3 hours after infection, complexes that contain dsRNA rather than ssRNA (dsRCCs) appear. These complexes contain not only the same three proteins as are also present in ssRCCs, but also λ2; and, of course, they presumably contain the RNA polymerase, the catalytic component of which is protein λ3. The highly significant feature of dsRCCs is that the relative proportions of the 10 dsRNA species in them are strictly equimolar. This suggests that genome segment assortment into progeny genomes is linked to the transcription of plus strands into minus strands, and that the factor that controls assortment is the RNA polymerase.

VII. The Infectious Reovirus RNA System

Inspection of the reovirus multiplication cycle shows that the only genetic/biochemical/molecular link between parental and progeny virus particles is provided by the plus-strand forms of the 10 genome segments. Thus infectious reovirus should be formed in cells into which they are introduced. Unsuccessful attempts have been made over the years to achieve this goal, but recently conditions were devised under which reovirus RNA is indeed infectious (186). The system is complex and the precise functions of all of its components have not yet been determined, but it is highly reproducible, and it should be possible to develop it into a system for introducing new and novel genetic information into the reovirus genome, which is not possible by any other means owing to the manner in which its components are assembled. For example, options such as cloning the genome, mutating or modifying the DNA, transcribing it back into RNA, and then introducing it into cells, which have been available for plus-strand RNA viruses for some time and are now also feasible for negative-strand RNA viruses (187), are not available; nor is random assortment of genome segments feasible, such as appears to operate for influenza virus (see Section I), because of the precise assortment mechanism that assembles the reovirus genome. The only approach is to enter the reovirus multiplication cycle when all 10 species of plus strands have been transcribed, and to let the cycle take over from there. The question then is, how will the assortment mechanism respond when confronted with species other than those it has evolved to recognize, namely, the 30 species of genome segments specified by the ST1, ST2, and ST3 genome segment sets, and some of their immediate variants? This is not something than can be studied starting with infectious virus; it can be studied only starting with infectious RNA.

The conditions under which reovirus RNA is infectious are as follows. Because reovirus genome segments exist in both plus-strand and double-strand form, the first approach was to use equal amounts of both. It was thought that reovirus proteins should be introduced into cells at the same time, thus ssRNA translated for 60 minutes in a rabbit reticulocyte lysate (RRL) cell-free protein-synthesizing system was introduced at the same time; and because there was clear evidence that the outcome of reassortment was different in cells infected with virions or lipofected with cores (multi-reassortants versus monoreassortants, see Section IV), a helper virus was also used. Active helper virus turned out to be unexpectedly limited in kind: ST1 and ST2 reovirus were both active, but all 10 classes of ts mutants were inactive, even at permissive temperatures, and UV-irradiated virus was also inactive. This showed that not only are certain mutations or certain kinds of wrongly folded proteins unacceptable—it should be noted that proteins encoded by ts genes may be folded aberrantly even though they may be functional (because the functional domain may be folded correctly (188))— but also that helper virus does more than provide the components of the outer capsid shell and ISVPs. In essence, then, the system, as first developed, was structured as shown in Fig. 3. Plaques were counted on day 5, by which time ST3 virus had formed plaques whereas ST2 virus only produced plaques by day 10. If ST1, which is slightly (about twofold) more efficient, is used as helper, plaquing is done in the presence of MAB against ST1 σ1,

0.3 μ ST3 ssRNA and/or 0.25 μg ST3 dsRNA
10 μl RRL in which 0.3 μg ST3 ssRNA had been translated for 60 minutes at 30°C
50 μl BRL Lipofectin reagent in 500 μl MEM containing penicillin and streptomycin; no serum

Add to monolayers of 10^6 mouse L929 fibroblasts from which the medium had been removed
Remove after 6 hours and add 250 μl of the same medium containing 4×10^7 PFU ST2 or ST1 virus
After 1 hour add 1.75 ml MEM containing penicillin and streptomycin and 5% fetal bovine serum

Harvest 24 hours later, sonicate, and titrate virus yield. If ST1 virus was used as the helper, incorporate MAB against ST1 σ1 in medium
Count plaques on day 5

FIG. 3. The infectious reovirus RNA system.

which eliminates ST1 virus totally. The yield of ST3 virus under these conditions is $1.5 \times 10^5 \pm 50\%$ PFU (186). This yield is highly reproducible.

In addition to varying the amounts and times specified in Fig. 3, numerous variations of this scheme have been tested, among which are the following approaches:

1. The necessity for helper virus is absolute, although there is no genetic interaction between ST3 and helper virus (that is, no ST2 helper virus genome segments have ever been detected in ST3 yields).
2. The requirement for the primed RRL (PRRL) is not absolute, but in its absence virus yields are two logs lower. Further, what is supplied by the PRRL is not merely viral protein. If the nuclease in PRRLs is activated after 60 minutes so as to destroy the RNA added, the activity of the PRRL is destroyed. If the nuclease is then inactivated and another aliquot of RNA is added without permitting it to be translated, activity is not restored. Thus the active principle in PRRLs is thought to be nascent RNA–protein complexes. Further, all 10 species of RNA are required; if any of the l, m, or s size classes of RNA are omitted, PRRLs are inactive. Thus all 10 species of RNA–protein complexes are required, or at least one from each size class. However, PRRLs primed with RNA derived from monoreassortants are inactive; hence 10 homologous RNA templates are required. Attempts have been made to replace PRRLs with plasmids expressing the 10 reovirus genome segments under the control of the CMV promoter; although formation of all proteins could be demonstrated, the requirement for PRRLs was not replaced.
3. It is not necessary to use both ssRNA and dsRNA—either alone is

active. If ssRNA is used both in free form and in the RRL, the yield is, very reproducibly, two logs lower than when ssRNA plus dsRNA are used ($1.5 \times 10^3 \pm 50\%$ PFU). If dsRNA is used in free form and ssRNA in the RRL, the yield is $1.5 \times 10^4 \pm 50\%$ PFU; but if dsRNA is used in free form and melted dsRNA in the RRL, the virus yield is low and not reproducible ($10-10^3$ PFU).[4] In addition, more than 80% of the genomes of progeny virus resulting from lipofecting ssRNA plus dsRNA are derived from the ssRNA (that is, dsRNA augments the infectivity of ssRNA). Thus the optimum standard system is to use ssRNA alone (in free form and in the RRL).

To test whether the infectious RNA system would support genome segment reassortment [which was not at all certain, in view of the total absence of ST2 (helper) genome segments in ST3 yields], all 10 genome segments of ST3 and of ST2 virus were lipofected into cells as specified in Fig. 3, using ST2 virus as helper. Reassortants accounted for 14% of the yield, but all were monoreassortants. Nor were the various monoreassortants represented equally: only four genome segments of ST2 were incorporated into the ST3 genome, namely, S3, S2, M1, and L3. Thus reassortment is feasible in the system, even if it is not very efficient.

There are two possible mechanisms by which these reassortants could be generated: (1) the lipofected ssRNA species are assorted into genomes that contain genome segments from both parents; (2) the 10 ssRNA species of each parent generate homologous genomes/virus particles that then reassort as in cells infected with virions of the two serotypes. It is not known at this time which of these two mechanisms is operating.

A very important and useful property of the system is that when ssRNA genome segments are separated into their l, m, and s size classes by density gradient centrifugation, each size class alone is completely inactive, but when recombined, their ability to generate infectious virus is exactly the same as that of unresolved genome segment populations. Not only does this provide a control that no infectious virus is present, but it also permits experiments testing the reassortment of genome segments derived from different parents.

---

[4] The reason for the low yield is most probably the fact that during the time it takes to lipofect the PRRL into cells, the two strands of RNA reanneal, thereby inactivating the RNA–protein complexes. Certainly translation of melted dsRNA proceeds very efficiently in RRLs, as was shown by McCrae and Joklik (98), who translated each of the melted dsRNA genome segments in order to map the reovirus genome. Attempts to remove minus strands following translation by the addition of excess plus strands have been unsuccessful.

VIII. The Interplay of Signals Required for the Insertion of Genetic Information into the Reovirus Genome

There are many reasons for attempting to introduce novel genetic information, that is, novel genome segments, into the reovirus genome. This includes simple deletion variants and site-directed point mutants such as would be required for the identification of assortment signals; highly modified genome segments that could yield reoviruses with any set of desired characteristics, such as reduced pathogenicity or altered host range/tissue tropism; and genome segments harboring foreign genes, introduction of which could transform reovirus into an expression vector. The logical starting point for asking whether and what kinds of signals are required for introducing such novel information into the reovirus genome is to attempt to reproduce intertypic genome segment reassortment with the infectious reovirus RNA system. As discussed above, the system can accomplish this if presented with two complete genome segment sets; the question is, can it reassort if presented with one complete set of genome segments and one heterologous genome segment? When this approach was tested using the 10 ss genome segments of ST3 and the ss ST2 S1 genome segment, the result was not the reassortant containing 9 ST3 genome segments and the ST2 S1 genome segment, but viruses with 11 genome segments—that is, all 10 genome segments of ST3 plus the S1 genome segment of ST2 (88). These viruses are unstable and segregate the ST2 S1 genome segment within two replaquings.

Because a complete genome segment set will not exchange one of its own genome segments for one presented to it, the use of acceptors containing only 9 genome segments has been explored. Sets of 9 genome segments can be derived artificially (by isolating all 10 and using 9, or by destroying one of 10) or naturally, from DI particles. However, when ssRNA transcribed by ST3 DI cores is used together with wt ST3 l size class RNA, no infectious virus is generated, which suggests the presence of "incompatible" mutations in one or more of the genome segments of DI virus genomes. An analysis was therefore carried out that showed that DI m size class genome segments did not harbor such mutations, but that DI s size class genome segments are unable to assort with authentic wt l and m size class genome segments to generate infectious virus. The ST3 DI s size class genome segments were therefore sequenced; and whereas no mutations were found in the S1, S2, and S3 genome segments, two mutations were found to be present in the S4 genome segment: G74 → A (silent) and G624 → A (D → K). It should be recalled in this connection that Ahmed et al. (87) also found a variety of mutations in ST3 virus stocks passaged under conditions of high multiplicity.

Thus the S4 genome segment of DI virus contains mutations that permit acceptance of truncated genome segments into the reovirus genome, but that are incompatible with the other 9 wt genome segments.

Similar mutations exist in the sequence reassortants generated in intertypic crosses. The most basic demonstration of this is the fact that l, m, and s size class genome segments of different serotypes do not generate infectious virus even though each genome segment size class generates infectious virus when mixed with its two homologous genome segment size classes. Thus all genome segments possess the correct assortment signals, because they are able to generate functional homologous genome segment sets; but they do not possess signals permitting them to form functional heterologous genome segment sets.

This raises the question as to whether some genome segments in sequence reassortants possess mutations. Indeed, it was observed some time ago that some genome segments in sequence reassortants possessing ST1 and ST3 genome segments possess electrophoretic migration mobilities slightly different from those of wt genome segments (Herbert Virgin IV, personal communication); and the ST3 S1 genome segment in a monoreassortant containing otherwise only ST1 genome segments has been observed to migrate distinctly faster in polyacrylamide gels than the wt ST3 S1 genome segment (88). Further, if sequence reassortants do not contain certain wt genome segments but variants that are incompatible with wt genome segments, crosses of monoreassortants should not, if only one genome segment is affected, yield wt recombinants. This is in fact what is observed: when monoreassortants containing nine ST3 genome segments and one ST2 genome segment are crossed, about 5% of the progeny are double reassortants, but no wt virus is present among the yield (88). This is not true when reassortants that possess nine ST1 genome segments and one ST3 genome segment are crossed; such crosses do indeed generate wt reassortants.

The explanation of all these results was provided when a variety of intertypic mono- and multireassortants were sequenced. In brief, the S4 genome segments of all reovirus monoreassortants that contain nine ST3 genome segments and one ST2 genome segment were found to possess the G624 → A mutation and probably also the G74 → A mutation (probably, because not all monoreassortants have yet been examined: the G624 → A mutation destroys an Mnl 1 restriction site and is easily screened using PCR, but presence of the G74 → A mutation can only be established by sequencing). The same appears to be true for all multireassortants containing six or more ST3 genome segments and the rest ST2 genome segments. The question then arises as to whether the two mutations in the ST3 S4 genome segment are required in the genome segment for recognition by accepting genome segments, or in the σ3 protein, which could certainly have a function during reassortment because it is a component of ssRCCs as well as dsRCCs (see Section VI). The only information available on this point is the fact that wt ST3 σ3 protein interferes drastically with the multiplication of monoreassortant viruses under conditions when it has no effect on the multiplication of wt ST3 virus (88).

A completely different picture emerged for reassortants that contain ST3 genome segments in a background of ST1 genome segments (88). Here the pattern is that there is no mutation in any ST1 genome segment, but all heterologous, "invading" ST3 genome segments possess from one to four mutations; and in each case the effect of the mutation(s) is to change a residue where ST1 and ST3 differ, to that of ST1. The fact that the signals required for the introduction of ST2 genome segments into the ST3 genome, and of ST3 genome segments into the ST1 genome, are so different, is striking. The reason may be the fact that ST3 and ST2 genome segments are less closely related than ST1 and ST3 genome segments (see Section II). In the latter case, it seems that genome segment variants that are the most similar to those of the acceptor genome are those that are most readily accepted; in the former, it seems that special signals in the accepting genome are required, which could be thought of as "acceptance" signals.

It is remarkable that all reassortants where ST3 is the acceptor genome possess S4 genome segments with two mutations that render them incompatible with the other nine wt ST3 genome segments. What is the source of these apparently abundant double mutants in stocks of wt ST3? The answer appears to be that these two mutations accumulate rapidly during reovirus multiplication, not only under conditions of high-multiplicity passaging, but also under normal conditions of routine passaging to replenish virus stocks. Thus ST3 virus stocks differ with respect to their content of G624 (and presumably also G74) mutations: plaque isolates contain very few, and stocks that have been passaged routinely contain increasing amounts, depending on the number of passages since plaque isolation and on the multiplicity used at each passage. This scenario is confirmed experimentally in two ways. First, screening for the G624 mutation by PCR shows that "young" wt stocks contain very few mutants, whereas older stocks contain large amounts of them; and second, whereas wt ST3 stocks with large amounts of the variant S4 genome segment yield large numbers of reassortants when crossed with ST2, such amounts decrease as younger and younger stocks are used, until plaque isolates of wt ST3 virus yield essentially no reassortants (88).

The recognition of signals required for the introduction of foreign genetic information into the reovirus genome has drastically changed our thinking about genome segment reassortment as it occurs among reoviruses. For example, genome segment reassortment was simple and logical as long as only wt genome segments were thought to be involved; but the recognition of "acceptance" signals in crosses involving ST2, on the one hand, and the realization that in ST3 × ST1 crosses genome segment sequence variants are selected, on the other, has introduced an element of doubt into results obtained by the use of genome segment reassortant analysis. One possible example is the conclusion that the S4 genome segment controls inhibition of host cell protein synthesis, which it may not (see Section V). Further, the recognition that it is not wt genome segments that participate in genome segment reassortment has opened up a whole new field that may have counterparts in many other situations, possibly also involving DNA. In any case, and in practical terms, it should now be possible to move forward and identify the signals on incoming genome segments that are necessary for assortment; to carry out the sequence modifications necessary for identifying the functional domains of reovirus-encoded proteins; to assess the effects of sequence modifications on reovirus phenotype; to construct virus strains with any desired characteristic; to develop reovirus as an expression vector; and to transfer this technology to other members of the Reoviridae family, particularly the major human pathogens, the rotaviruses.

Acknowledgments

Work from the authors' laboratory was supported by Grant RO1 AI 08909 from the U.S. Public Health Service and Research Grant 3233 from the Council for Tobacco Research. We would like to thank numerous colleagues for sharing unpublished results and valuable discussions, in particular the late Bernard Fields, Herbert Virgin IV, Charles Samuel, Aaron Shatkin, Deepak Bastia, Sara Miller, Charles Lin, Igor Nepluev, and L.-P. Kong.

References

1. F. C. Robbins, J. F. Enders, T. H. Weller and G. L. Florentino, *Am. J. Hyg.* **54**, 56 (1951).
2. M. Ramos-Alvarez and A. B. Sabin, *PSEBM* **87**, 655 (1954).
3. M. Ramos-Alvarez, *Ann. N.Y. Acad. Sci.* **67**, 326 (1957).
4. M. Ramos-Alvarez and A. B. Sabin, *JAMA* **167**, 147 (1958).
5. A. B. Sabin, *Science* **130**, 1387 (1959).
6. N. F. Stanley, D. C. Dorman and J. Ponsford, *Aust. J. Exp. Biol. Med. Sci.* **31**, 147 (1953).
7. N. F. Stanley, *Nature* **189**, 687 (1961).
8. L. Rosen, *Am. J. Hyg.* **71**, 242 (1960).
9. L. Rosen, *Ann. N.Y. Acad. Sci.* **101**, 461 (1962).
10. P. J. Gomatos, I. Tamm, S. Dales and R. M. Franklin, *Virology* **17**, 441 (1962).
11. P. J. Gomatos and I. Tamm, *PNAS* **49**, 707 (1963).
12. A. R. Bellamy, L. Shapiro, J. T. August and W. K. Joklik, *JMB* **29**, 1 (1967).
13. A. J. Shatkin, *PNAS* **54**, 1721 (1965).
14. A. R. Bellamy and W. K. Joklik, *JMB* **29**, 19 (1967).
15. Y. Watanabe, L. Prevec and A. F. Graham, *PNAS* **58**, 1040 (1967).

16. A. J. Shatkin, J. D. Sipe and P. C. Loh, *J. Virol.* **2**, 986 (1968).
17. R. A. Lamb and P. W. Choppin, *ARB* **52**, 467 (1983).
18. M. Enami, G. Sharma, C. Benham and P. Palase, *Virology* **185**, 291 (1991).
19. R. S. Spendlove, M. E. McClain and E. H. Lennette, *J. Gen. Virol.* **8**, 83 (1970).
20. S. M. Larson, J. B. Antczak and W. K. Joklik, *Virology* **201**, 303 (1994).
21. B. N. Fields and W. K. Joklik, *Virology* **37**, 335 (1969).
22. R. F. Ramig, R. K. Cross and B. N. Fields, *J. Virol.* **22**, 726 (1977).
23. P. Metcalf, M. Cyrklaff and M. Adrian, *EMBO J.* **10**, 3129 (1991).
24. P. W. K. Lee, E. C. Hayes and W. K. Joklik, *Virology* **108**, 156 (1981).
25. H. L. Weiner, R. F. Ramig, T. A. Mustoe and B. N. Fields, *Virology* **86**, 581 (1978).
26. R. W. Paul and P. W. K. Lee, *Virology* **159**, 94 (1987).
27. R. W. Paul, A. H. C. Choi and P. W. K. Lee, *Virology* **172**, 382 (1989).
28. M. S. Co, G. N. Gaulton, A. Tominaga, C. J. Homcy, B. N. Fields and M. I. Greene, *PNAS* **82**, 5315 (1985).
29. A. H. C. Choi and P. W. K. Lee, *Virology* **163**, 191 (1988).
30. J. F. Strong, D. Tang and P. W. K. Lee, *Virology* **197**, 405 (1993).
31. A. H. C. Choi, R. W. Paul and P. W. K. Lee, *Virology* **178**, 316 (1990).
32. S. Dales, P. Gomatos and K. C. Hsu, *Virology* **25**, 193 (1965).
33. M. T. Tosteson, M. L. Nibert and B. N. Fields, *PNAS* **90**, 10549 (1993).
34. A. K. Jayasuriya, M. L. Nibert and B. N. Fields, *Virology* **163**, 591 (1988).
35. J. Borsa, T. P. Copps, M. D. Sargent, D. G. Long and J. D. Chapman, *J. Virol.* **11**, 552 (1973).
36. J. Borsa, D. G. Long, M. D. Sargent, T. P. Copps and J. D. Chapman, *Intervirology* **4**, 171 (1974).
37. J. Borsa, M. D. Sargent, P. A. Lievaart and T. P. Copps, *Virology* **111**, 191 (1981).
38. C.-T. Chang and H. J. Zweerink, *Virology* **46**, 544 (1971).
39. S. C. Silverstein, C. Astell, D. H. Levin, M. Schonberg and G. Acs, *Virology* **47**, 797 (1972).
40. L. J. Sturzenbecker, M. Nibert, D. Furlong and B. N. Fields, *J. Virol.* **61**, 2351 (1987).
41. K. F. H. Powell, J. D. Harvey and A. R. Bellamy, *Virology* **137**, 1 (1984).
42. A. J. Shatkin and A. J. LaFiandra, *J. Virol.* **10**, 698 (1972).
43. C. Astell, S. C. Silverstein, D. H. Levin and G. Acs, *Virology* **48**, 648 (1972).
44. M. Yamakawa, Y. Furuichi, K. Nakashima, A. J. LaFiandra and A. J. Shatkin, *JBC* **256**, 6507 (1981).
45. M. Yamakawa, Y. Furuichi and A. J. Shatkin, *Virology* **118**, 157 (1982).
46. J. B. Antczak, R. Chmelo, D. J. Pickup and W. K. Joklik, *Virology* **121**, 307 (1982).
47. A. R. Bellamy and W. K. Joklik, *PNAS* **58**, 1389 (1967).
48. A. R. Bellamy, L. V. Hole and B. C. Baguley, *Virology* **42**, 415 (1970).
49. A. R. Bellamy, J. L. Nichols and W. K. Joklik, *Nature (New Biol.)* **238**, 49 (1972).
50. A. J. Shatkin and J. D. Sipe, *PNAS* **59**, 246 (1968).
51. J. J. Skehel, and W. K. Joklik, *Virology* **39**, 822 (1969).
52. A. J. Hay and W. K. Joklik, *Virology* **44**, 450 (1971).
53. G. Acs, H. Klett, M. Schonberg, J. Christman, D. H. Levin and J. C. Silverstein, *J. Virol.* **8**, 684 (1971).
54. S. Sakuma and Y. Watanabe, *J. Virol.* **10**, 628 (1972).
55. S. Sakuma and Y. Watanabe, *J. Virol.* **8**, 190 (1971).
56. M. Schonberg, S. C. Silverstein and D. H. Levin, *PNAS* **68**, 505 (1971).
57. E. M. Morgan and H. J. Zweerink, *Virology* **77**, 421 (1977).
58. E. M. Morgan and H. J. Zweerink, *Virology* **68**, 455 (1975).
59. H. J. Zweerink, *Nature* **247**, 313 (1974).

60. L. W. Cashdollar, R. A. Chmelo, J. R. Wiener and W. K. Joklik, *PNAS* **82**, 24 (1985).
61. H. W. Virgin, IV, M. A. Mann, B. N. Fields and K. L. Tyler, *J. Virol.* **65**, 6772 (1991).
62. J. R. Wiener and W. K. Joklik, *Virology* **169**, 194 (1989).
63. J. D. Chappell, M. I. Goral, S. E. Rodgers, C. W. dePamphilis and T. S. Dermody, *J. Virol.* **68**, 750 (1994).
64. R. Kedl, S. Schmechel and L. Schiff, *J. Virol.* **69**, 552 (1995).
65. R. Bassel-Duby, A. Jayasuriya, D. Chatterjee, N. Sonenberg, J. V. Maizel, Jr. and B. N. Fields, *Nature (London)* **315**, 421 (1985).
66. L. S. Seliger, K. Zheng and A. J. Shatkin, *JBC* **262**, 16289 (1987).
67. J. A. Bartlett and W. K. Joklik, *Virology* **167**, 31 (1988).
68. J. R. Wiener, J. A. Bartlett and W. K. Joklik, *Virology* **169**, 293 (1989).
69. S. Zou and E. G. Brown, *Virus Res.* **22**, 159 (1992).
70. J. R. Wiener and W. K. Joklik, *Virology* **163**, 603 (1988).
71. L. Nagata, S. A. Masri, D. C. W. Mah and P. W. K. Lee, *NARes* **12**, 8699 (1984).
72. S. M. Munemitsu, J. A. Atwater and C. E. Samuel, *BBRC* **140**, 508 (1986).
73. R. Duncan, D. Horne, L. W. Cashdollar, W. K. Joklik and P. W. K. Lee, *Virology* **174**, 399 (1990).
74. L. W. Cashdollar, J. Esparza, G. R. Hudson, R. Chmelo, P. W. K. Lee and W. K. Joklik, *PNAS* **79**, 7644 (1982).
75. J. R. Wiener, T. MacLaughlin and W. K. Joklik, *Virology* **170**, 340 (1989).
76. T. S. Dermody, L. A. Schiff, M. L. Nibert, K. M. Coombs and B. N. Fields, *J. Virol.* **65**, 5721 (1991).
77. J. R. Wiener and W. K. Joklik, *Virology* **161**, 332 (1987).
78. M. Giantini, L. S. Seliger, Y. Furuichi and A. J. Shatkin, *J. Virol.* **52**, 984 (1984).
79. J. A. Atwater, S. M. Munemitsu and C. E. Samuel, *BBRC* **136**, 183 (1986).
80. L. S. Seliger, M. Giantini and A. J. Shatkin, *Virology* **187**, 202 (1992).
81. M. Nonoyama and A. F. Graham, *J. Virol.* **6**, 693 (1970).
82. M. Nonoyama, Y. Watanabe and A. F. Graham, *J. Virol.* **6**, 226 (1970).
83. D. A. Spandidos and A. F. Graham, *J. Virol.* **16**, 1444 (1975).
84. D. A. Spandidos and A. F. Graham, *J. Virol.* **15**, 954 (1975).
85. A. R. Schuerch, T. Matsuhisa and W. K. Joklik, *Intervirology* **3**, 36 (1974).
86. R. Ahmed and B. N. Fields, *Virology* **111**, 351 (1981).
87. R. Ahmed, P. R. Chakraborty and B. N. Fields, *J. Virol.* **34**, 285 (1980).
88. M. R. Roner, C. P. Lin, I. V. Nepluev, L.-J. Kong and W. K. Joklik, *PNAS* **92**, 12362 (1995).
89. E. G. Brown, M. L. Nibert and B. N. Fields, in "Double-Stranded RNA Viruses" (R. W. Compans and D. H. L. Bishop, eds.), p. 275. Elsevier, Amsterdam, 1983.
90. S. Zou and E. G. Brown, *Virology* **186**, 377 (1992).
91. B. N. Fields, *Virology* **46**, 142 (1971).
92. B. N. Fields, in "Virus Research" (C. F. Fox, ed.), p. 461. Academic Press, New York, 1973.
93. R. K. Cross and B. N. Fields, *Virology* **74**, 354 (1976).
94. T. A. Mustoe, R. F. Ramig, A. H. Sharpe and B. N. Fields, *Virology* **89**, 594 (1978).
94a. T. A. Mustoe, R. F. Ramig, A. H. Sharpe and B. N. Fields, *Virology* **85**, 545 (1978).
95. R. F. Ramig, R. Ahmed and B. N. Fields, *Virology* **125**, 299 (1983).
96. R. F. Ramig, T. A. Mustoe, A. H. Sharpe and B. N. Fields, *Virology* **85**, 531 (1978).
97. A. H. Sharpe, R. F. Ramig, T. A. Mustoe and B. N. Fields, *Virology* **84**, 63 (1978).
98. M. A. McCrae and W. K. Joklik, *Virology* **89**, 578 (1978).
99. H. L. Weiner, D. Drayna, D. R. Averill and B. N. Fields, *PNAS* **74**, 5744 (1977).
99a. H. L. Weiner and B. N. Fields, *J. Exp. Med.* **146**, 1305 (1977).
100. H. L. Weiner, M. L. Powers and B. N. Fields, *J. Infect. Dis.* **141**, 609 (1980).

101. A. H. Sharpe and B. N. Fields, *J. Virol.* **38**, 389 (1981).
102. D. Rubin, H. L. Weiner, B. N. Fields and M. I. Greene, *J. Immunol.* **127**, 1697 (1981).
103. A. H. Sharpe and B. N. Fields, *Virology* **122**, 381 (1982).
104. D. Drayna and B. N. Fields, *J. Gen. Virol.* **63**, 149 (1982).
105. D. Drayna and B. N. Fields, *J. Virol.* **41**, 110 (1982).
106. K. L. Tyler, W. C. Schoene and B. N. Fields, *Neurology* **34** (*Suppl. 1*), 191 (1984).
107. K. L. Tyler, D. A. McPhee and B. N. Fields, *Science* **233**, 770 (1986).
108. M. Keroack and B. N. Fields, *Science* **232**, 1635 (1986).
109. D. L. Bodkin and B. N. Fields, *J. Virol.* **63**, 1188 (1989).
110. B. Sherry and B. N. Fields, *J. Virol.* **63**, 4850 (1989).
111. P. R. Hazelton and K. M. Coombs, *Virology* **207**, 406 (1995).
112. M. D. Moody and W. K. Joklik, *Virology* **173**, 437 (1989).
113. M. C. Starnes and W. K. Joklik, *Virology* **193**, 356 (1993).
114. P. Xu, S. E. Miller and W. K. Joklik, *Virology* **197**, 726 (1993).
115. G. Lemay and C. Danis, *J. Gen. Virol.* **75**, 3261 (1994).
116. S. J. Ralph, J. D. Harvey and A. R. Bellamy, *J. Virol.* **36**, 894 (1980).
117. Z. Mao and W. K. Joklik, *Virology* **185**, 377 (1991).
118. J. Fasnaugh and A. J. Shatkin, *JBC* **265**, 7669 (1990).
119. D. R. Cleveland, H. Zarbl and S. Millward, *J. Virol.* **60**, 307 (1986).
120. J. Borsa, J. Grover and J. D. Chapman, *J. Virol.* **6**, 295 (1970).
121. J. B. Antczak and W. K. Joklik, *Virology* **187**, 760 (1992).
122. L. W. Cashdollar, *Res. Virol.* **145**, 277 (1994).
123. T. Blumenthal and G. G. Carmichael, *ARB* **48**, 525 (1979).
124. A. K. Banerjea, K. A. Brechling, C. A. Ray, H. Erikson, D. J. Pickup and W. K. Joklik, *Virology* **167**, 601 (1988).
125. M. L. Nibert, L. A. Schiff and B. N. Fields, *J. Virol.* **65**, 1960 (1991).
126. H. Huismans and W. K. Joklik, *Virology* **70**, 411 (1976).
127. H. J. Zweerink and W. K. Joklik, *Virology* **41**, 501 (1970).
128. R. E. Smith, H. J. Zweerink and W. K. Joklik, *Virology* **39**, 791 (1969).
129. M. L. Niebert and B. N. Fields, *J. Virol.* **66**, 6408 (1992).
130. P. W. K. Lee, E. C. Hayes and W. K. Joklik, *Virology* **108**, 134 (1981).
131. L. Tillotson and A. J. Shatkin, *J. Virol.* **66**, 2180 (1992).
132. T. Mabrouk and G. Lemay, *Virology* **202**, 615 (1994).
133. P. Lucia-Jandris, J. W. Hooper and B. N. Fields, *J. Virol.* **67**, 5339 (1993).
134. M. N. Rozinoff and B. N. Fields, *J. Virol.* **68**, 6667 (1994).
135. M. Mora, K. Partin, M. Bhatia, J. Partin and C. Carter, *Virology* **159**, 265 (1987).
136. B. N. Fields, *Arch. Virol.* **71**, 95 (1982).
137. B. N. Fields and M. I. Greene, *Nature* **300**, 19 (1982).
138. K. L. Tyler and B. N. Fields, in "Fields Virology" (B. N. Fields, D. M. Knipe *et al.*, eds.), 2nd Ed., p. 1307. Raven Press, New York, 1990.
139. D. H. Rubin, M. A. Eaton and A. O. Anderson, *Microb. Pathog.* **1**, 79 (1986).
140. H. L. Weiner, M. I. Greene and B. N. Fields, *J. Immunol.* **125**, 278 (1980).
141. R. Finberg, H. L. Weiner, S. J. Burakoff and B. N. Fields, *Infect. Immun.* **31**, 646 (1981).
142. R. Finberg, H. L. Weiner, B. N. Fields, B. Benacerraf and S. J. Burakoff, *PNAS* **76**, 442 (1979).
143. A. Fontana and H. L. Weiner, *J. Immunol.* **125**, 2660 (1980).
144. J. E. Strong, G. Leone, R. Duncan, R. K. Sharma and P. W. K. Lee, *Virology* **184**, 23 (1991).
145. M. L. Nibert, T. S. Dermody and B. N. Fields, *J. Virol.* **64**, 2976 (1990).

146. R. D. B. Fraser, D. B. Furlong, B. L. Trus, M. L. Nibert, B. N. Fields and A. C. Steven, *J. Virol.* **64**, 2990 (1990).
147. D. B. Furlong, M. L. Nibert and B. N. Fields, *J. Virol.* **62**, 246 (1988).
148. A. C. Banerjea and W. K. Joklik, *Virology* **179**, 460 (1990).
149. T. S. Dermody, M. L. Nibert, R. Bassel-Duby and B. N. Fields, *J. Virol.* **64**, 5173 (1990).
150. D. L. Turner, R. Duncan and P. W. K. Lee, *Virology* **186**, 219 (1992).
151. G. Leone, D. C. W. Mah and P. W. K. Lee, *Virology* **182**, 346 (1991).
152. G. Leone, L. Maybaum and P. W. K. Lee, *Cell* **71**, 479 (1992).
153. R. Duncan, D. Horne, J. E. Strong, G. Leone, R. T. Pon, M. C. Yeung and P. W. K. Lee, *Virology* **182**, 810 (1991).
154. D. C. W. Mah, G. Leone, J. M. Jankowski and P. W. K. Lee, *Virology* **179**, 95 (1990).
155. S. J. Burstin, D. R. Spriggs and B. N. Fields, *Virology* **117**, 146 (1982).
156. D. R. Spriggs, K. Kaye and B. N. Fields, *Virology* **127**, 220 (1983).
157. L. Nagata, S. A. Masri, R. T. Pon and P. W. K. Lee, *Virology* **160**, 162 (1987).
158. S. M. Larson, J. B. Antczak and W. K. Joklik, *Virology* **201**, 303 (1994).
159. R. K. Gaillard, Jr. and W. K. Joklik, *Virology* **147**, 336 (1985).
160. M. R. Roner, R. K. Gaillard, Jr. and W. K. Joklik, *Virology* **168**, 292 (1989).
161. H. Ernst and A. J. Shatkin, *PNAS* **82**, 48 (1985).
162. B. L. Jacobs and C. E. Samuel, *Virology* **143**, 63 (1985).
163. B. Sarkar, J. Pelletier, R. Bassel-Duby, A. Jayasuriya, B. N. Fields and N. Sonenberg, *J. Virol.* **54**, 720 (1985).
164. E. Fajardo and A. J. Shatkin, *Virology* **178**, 223 (1990).
165. B. A. Belli and C. E. Samuel, *Virology* **185**, 698 (1991).
166. C. X. George, A. Crowe, S. M. Munemitsu, J. A. Atwater and C. W. Samuel, *BBRC* **147**, 1153 (1987).
167. C. K. White and H. J. Zweerink, *Virology* **70**, 171 (1976).
169. M. A. Richardson and Y. Furuichi, *NARes* **11**, 6399 (1983).
170. N. M. Stomatos and P. J. Gomatos, *PNAS* **79**, 3457 (1982).
171. L. A. Schiff, M. L. Nibert, M. S. Co, E. G. Brown and B. N. Fields, *MCBiol* **8**, 273 (1988).
172. T. Mabrouk and G. Lemay, *J. Virol.* **68**, 5287 (1994).
173. C. Danis, S. Garzon and G. Lemay, *Virology* **190**, 494 (1992).
174. J. E. Miller and C. E. Samuel, *J. Virol.* **66**, 5347 (1992).
175. K. L. Denzler and B. L. Jacobs, *Virology* **204**, 190 (1994).
176. T. Mabrouk, C. Danis and G. Lemay, *Biochem. Cell Biol.* **73**, 137 (1995).
177. M. Giantini and A. J. Shatkin, *J. Virol.* **63**, 2415 (1989).
178. A. G. Hovanessian, *J. Interferon Res.* **9**, 641 (1989).
179. C. E. Samuel, *Virology* **183**, 1 (1991).
180. C. E. Samuel, *JBC* **268**, 7603 (1993).
181. F. Imani and B. L. Jacobs, *PNAS* **85**, 7887 (1988).
182. R. M. Lloyd and A. J. Shatkin, *J. Virol.* **66**, 6878 (1992).
183. A. H. Sharpe and B. N. Fields, *Virology* **122**, 381 (1982).
184. P. E. M. Martin and M. A. McCrae, *J. Gen. Virol.* **74**, 1055 (1993).
185. B. L. Jacobs and R. E. Ferguson, *J. Virol.* **65**, 5102 (1991).
186. M. R. Roner, L. A. Sutphin and W. K. Joklik, *Virology* **179**, 845 (1990).
187. N. D. Lawson, E. A. Stillman, M. A. Whitt and J. K. Rose, *PNAS* **92**, 4477 (1995).
188. D. A. Simpson and R. A. Lamb, *Virology* **185**, 477 (1991).

Alu: Structure, Origin, Evolution, Significance, and Function of One-Tenth of Human DNA

CARL W. SCHMID

Section of Molecular and Cellular
Biology and Department of Chemistry
University of California, Davis
Davis, California 95616

I. Identifying and Defining Repetitive Sequence Families 285
   A. Sequence Analysis by C_0t Renaturation Studies 285
   B. Human Repeats Are Tandemly Clustered or Dispersed 288
   C. Most Human Interspersed Repeats Have Retrotransposed 290
   D. Most Human Repeats Have Been Identified 294
II. Evolution of Alus (and Other Mammalian SINEs) 296
   A. When Alus Appeared 296
   B. Alus Accumulated during Higher Primate Evolution 296
   C. Young Alu Subfamilies Are Active 298
   D. An Alu Founder—The Simplest Possibility 299
   E. SINE Inactivation 300
III. Genetic Effects of Alus and Other Repeats 301
IV. Does Alu Have a Function? 302
   A. Selfish DNA, Junk DNA, and Cheap Genes 302
   B. Alus as *Cis*-acting Elements 304
   C. Alus as Transcription Units 307
V. Prospectives and a Speculation 313
   References 314

More than 20% of human DNA consists of repetitive sequences, provoking questions concerning the evolutionary history, genetic significance, and biological function of this redundant information. Human repeats have been largely identified and cataloged, accounting for a substantial fraction of the genome. We now know what most of these sequences are, when and how they appeared, and some of the consequences of their presence. Also, we are possibly at the threshold of learning their functions.

Repetitive sequences are ubiquitously interspersed with single-copy sequences throughout the genome. Almost all of these interspersed repeats result from retroposition, i.e., mobility, through an RNA intermediate so that the reverse flow of genetic information has essentially shaped the entire genome. Interspersed repeats are largely composed of only four distinct families (Alu, LINE 1, MIR, and MaLR), which together comprise 10 to 15% of the entire genome. These retroposed repeats are not especially mobile, the vast majority inserted prior to the divergence of humans and apes, but these four families also differ greatly in retropositional activity. At one extreme, inactive MIR family members are fossilized at orthologous positions within the genomes of different mammalian orders. At the other extreme, unfixed Alu and LINE 1 family members result in altered human genotypes as well as deleterious phenotypes.

Alu and LINE 1 subfamilies of different evolutionary ages have different degrees of retrotranspositional activity. Mechanisms for the self-perpetuation of LINEs are becoming apparent. This review features Alu repeats. Although Alus exemplify many properties of both human and nonhuman repeats, their amplification raises unsolved problems in retroposition. A young Alu subfamily appeared by simple drift of its founder, providing a model for the continual turnover of active repeat-sequence families.

Repetitive sequences, ubiquitously distributed throughout the genome, cause various genetic effects. Unequal crossing-over between interspersed repeats duplicates or deletes sequences, thereby mutating genes. Retrotransposition inserts novel sequences within or near genes leading to altered transcriptional expression as well as altered post-transcriptional events, including polyadenylylation, splicing, and the creation of new exonic sequences. Although some Alus are certainly "junk" DNA, their presence in the human genome is often deleterious, contrary to expectations for "selfish" DNA.

Despite the selective disadvantage attributable to their presence, SINEs have persistently been maintained during mammalian evolution, compelling belief in their function. Properties of Alu repeats are completely reconciled by the "cheap-gene" hypothesis, which provides a unique solution to the evolutionary problem posed by a dispersed multigene family. This hypothesis limits possible Alu functions to those requiring their dispersion and a constant recruitment of new members.

Interspersed repeats could have defined functions either as ubiquitous sequence elements or as RNA polymerase III (Pol III) transcription units encoding functional Alu RNAs. Recent findings concerning Alu repeats suggest two possibilities that also satisfy requirements of the cheap-gene hypothesis: (1) Alu repeats are methylated in a tissue-specific, locus-specific, allele-specific, and site-specific manner. Whereas Alus are fully methylated in somatic tissues and in the female germ line, they are largely unmethylated in the male germ line. Because of their genomic dispersion, Alus are uniquely suited to serve as signals for genomic imprinting. (2) The tight down-regulation of 10^6 Alu "genes" accrues at multiple levels from *cis*-acting transcriptional elements through the post-transcriptional fate of the final products, so that full-length Pol-III-directed Alu transcripts are barely detected in cultured cells. However, the abundance of Alu RNA rapidly and dramatically increases in response to cell stress and translational inhibition, suggesting a role for Alu transcripts in the cell-stress response.

I. Identifying and Defining Repetitive Sequence Families

A. Sequence Analysis by C_0t Renaturation Studies

Prior to the advent of DNA sequencing, the higher eukaryotic genomes were known to contain more DNA than necessary to encode required gene products. For example, the 3 billion base pairs in the haploid human genome could potentially encode 6.7 million genes (*1*). This coding potential is at least 17 times the number of human genes estimated from genetic load considerations (*2*). Contemporary sequence information accounts for some of this superfluous DNA in the form of introns located within genes. A major fraction of this "extra" DNA is also located between the genes. This intergenic DNA consists of sequences of largely unknown functions but includes an entire spectrum of pseudogenes. Concerning the subject of this review, eukaryotic DNAs also contain repeated copies of similar sequences, which are distributed both within and between the genes.

Beginning about 1965, DNA renaturation techniques provided one of a very few informative methods for assaying the composition and sequence organization of higher eukaryotic DNAs (*3*). Unlike the renaturation of prokaryotic DNAs, which follow homogeneous second-order kinetics, renaturation profiles of higher eukaryotic DNAs have extremely heterogeneous kinetics (Fig. 1A). In human DNA, the most rapidly renaturing sequences hybridize 50,000-fold faster than the most slowly renaturing sequences (Fig. 1A, Table I). The complexity of the most slowly renaturing sequences approximates the sequence content of human single-copy DNA, 3 billion base pairs (*1, 4, 5*). The half-time ($t_{1/2}$) for completion of a second-order reaction is inversely proportional to the molar concentration (C) of a reactant, thus the simplest interpretation of this observed kinetic heterogeneity is that the molar concentration of certain sequences is higher than that of others (Fig. 1A). Pioneering studies demonstrated that rapidly renaturing eukaryotic DNA sequences are indeed repeated (*3, 6*).

Absent other complications, the ratio of the half-times with which two sequence classes renature at a given concentration measures their relative copy-number (*3, 7*). Complications abound. One major difficulty is that homologous repetitive sequences are not identical but diverge from their com-

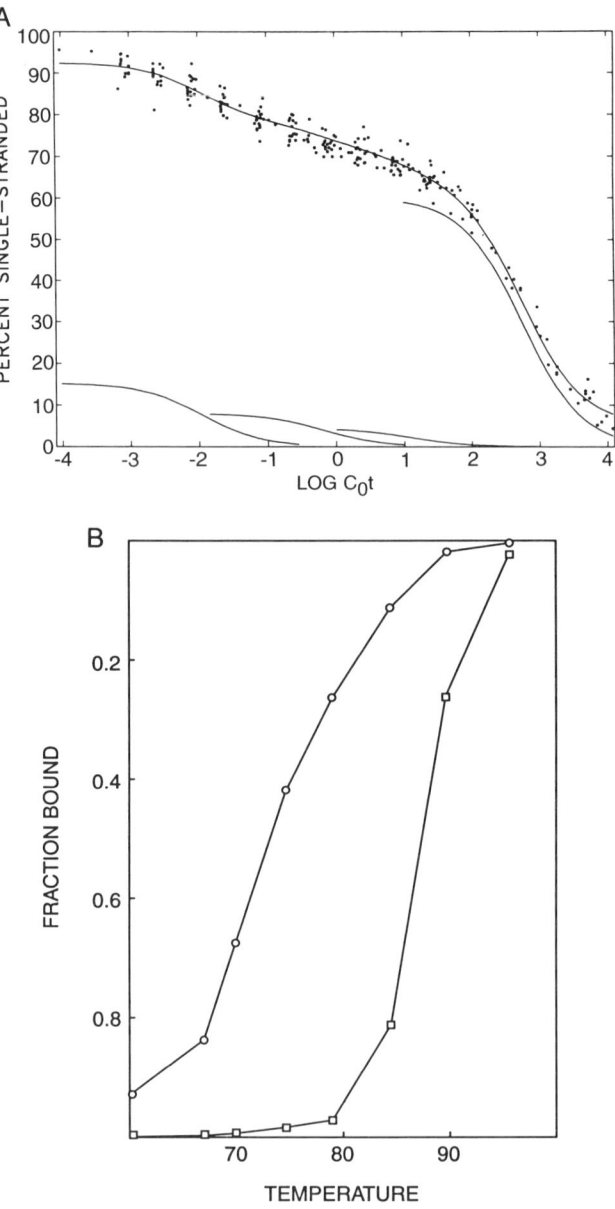

FIG. 1. (A) The renaturation of human DNA (238 data points) is assayed by binding to hydroxyapatite (12). The curve drawn through the points is a least-squares fit assuming four kinetic classes and a small amount (6.9%) of DNA renaturing with first-order kinetics and a small amount (5.5%) of nonrenaturable material. The four kinetic classes are $C_0t_{1/2} = 0.011$, 15.2%; $C_0t_{1/2} = 0.63$, 8.07%; $C_0t_{1/2} = 12.4$, 4.4%; and $C_0t_{1/2} = 550$, 60.0%. These data are analyzed further in Table I. Individual curves for these four kinetic classes are depicted. (B) The thermal elution of renatured repeats (○) from hydroxyapatite is compared to the elution of native duplex human DNA (□) (5). The difference in melting temperatures (15°C) indicates that the renatured repeats are highly mispaired, ~15%. Reprinted from Ref. 5, copyright 1975 Cell Press.

TABLE I
RENATURATION CLASSES OF HUMAN DNA[a]

$C_0t_{1/2}$	Apparent repeat frequency	Class distribution (%)	Corrected distribution (%)
0.011	50,000	20.1	11.0
0.63	870	10.6	5.8
12.4	44	5.8	3.2
550 (single copy)	1	63.5	80.0
Total	—	100.0	100.0

[a] The renaturation profile of Fig. 1A can be accurately represented by four sequence classes plus a small fraction of inverted repeated DNA and nonrenaturable material (12). The $C_0t_{1/2}$ values and their corresponding apparent repeat frequencies are reported above. The maximum amount of DNA in each repeat sequence class includes a proportional division of the inverted repeated class and the nonrenaturable material described in Fig. 1A (12). This value also includes interspersed single-copy sequences covalently attached to renatured repeats and thus overestimates the actual amount of repetitive DNA in each class. Assuming that 20% of human DNA is repetitive, these values are corrected for interspersion.

mon ancestor. This conclusion was initially achieved by investigating the thermal stability of renatured repetitive sequences (3) (Fig. 1B). Repetitive sequence hybrids are less stable than renatured single-copy sequences, suggesting that renatured repeats include mispaired bases. For example, the melting temperature of renatured human repeats is 15° lower than that of perfectly paired duplex DNA, indicating that these hybrids are 15% mispaired (Fig. 1B). Base-sequence comparisons (reviewed in Section I,C) nicely confirm the conclusions of this earlier indirect method. Mispairing coincidentally reduces the renaturation rate of complementary sequences so that the half-time of hybridization always underestimates repetition frequency of repetitive sequences (8). As a specific example, the most rapidly renaturing class of human DNA having an apparent 50,000-fold repetition frequency includes about 500,000 members of the Alu family of sequences (9, 10) (Fig. 1A).

Lowering the stringency (i.e., the temperature) of renaturation facilitates hybridization of more distantly related sequences, causing two significant changes in eukaryotic DNA renaturation profiles: the apparent repetition frequency of repetitive sequences increases, and more of the genome behaves as repetitive DNA (9, 11). Members of a repetitive sequence family are operationally recognized by their cross-hybridization under entirely arbitrary experimental conditions, such as the temperature and salt concentration used for hybridization. Renaturation conditions and methods employed in the study of Fig. 1A require at least 70% sequence identity over at least 50 nucleotides to form recognizable hybrids. As discussed above, the products of hybridization are kinetically biased toward even more closely matched sequences so that complements meeting even this minimum threshold are predictably underrepresented in the renaturation products.

Approximately 36% of the human genome, under the experimental conditions used for Fig. 1, renatures more rapidly than single-copy sequences (Fig. 1A, Table I). However, the chromatographic method employed to assay renaturation does not detect the mass of actual DNA hybrids but rather the mass of DNA in hybrids and covalently linked unrenatured single-copy sequences. This latter quantity depends on the DNA fragment size employed in a particular experiment and also on the detailed arrangement of interspersed repetitive and single-copy DNAs. Previous investigations of this additional complication suggest that at least 20% of human DNA is repetitive (*12*, and references therein) (Table I).

Finding so much repetitive DNA partially explains why the human genome contains more DNA than required for structural genes. Of human DNA, 20% or more consists of redundant information.

Repeats diverge rapidly as demonstrated by cross-hybridizing sequences of two species and examining both the extent and fidelity of cross hybridization. The basic lesson from such studies, again mostly pioneered by Britten and co-workers, is that novel repetitive sequence families appear in divergent species. For example, human and monkey repeats readily cross-hybridize, but these interspecies hybrids are more highly mispaired than intraspecies repeat-sequence hybrids (*13*). Extending this example, human and galago (a prosimian primate) repeats rarely cross-hybridize (*13*). Such comparisons raise fascinating questions (*6*). How do repeat families arise? How does the repeat-sequence composition within a genome turn over rapidly in evolution? Because repeat-sequence families evolve rapidly, do their functions, if any, have species-specific properties? As reviewed in Sections II and V, the first two questions have been answered and the answer to the third remains an exciting possibility.

B. Human Repeats Are Tandemly Clustered or Dispersed

Although as little as 20% of human DNA is repetitive, its renaturation profile shows that 36% of all 450-nt DNA fragments contain at least one repetitive element (Fig. 1A, Table I). As assayed by the same method, nearly all 20-kb human DNA fragments renature rapidly, from which we inferred that nearly all 20-kb fragments contain one or more repeats and that repetitive elements are broadly distributed throughout the genome (*5*). Early at-

TABLE II
REPETITIVE SEQUENCE COMPOSITION OF HUMAN DNA[a]

Family	Organization	Classifications	Copy number	Length	% Genome	Ref.
Satellite I–IV	Tandem repeat	Satellite	—	5-bp subunit and variations	2–5%	18, 177
α Satellite	Tandem repeat	Satellite	—	171-bp subunit	~5%	178, 179
Alu	Interspersed with single copy	SINE/retroposon	0.5×10^6 1.1×10^6	300 bp	6–13%	9, 10
MIR	Interspersed with single copy	SINE/retroposon	300,000	260 bp	1–2%	41
L1	Interspersed with single copy	LINE/transposon	~100,000	Variable 5′ truncations, ~300 bp to 6 kb; (av. = ~1000 bp?)	~4%	31
MaLR (THE 1)	Interspersed with single copy	Proretroviral transposon	10,000	2300	0.5–2%	40
Poly(CA) and other simple sequences	Tandem repeats interspersed with single copy	Minisatellite	50,000 (for poly CA)	Variable but short	1%	44
rDNA	Tandem repeat	Multigen family	200	43 kb	0.3%	180
U2 genes	Tandem repeat	Multigene family	50	6 kb	0.01%	15, 16
Total					19.8–32.3%	

[a] Technical and theoretical difficulties in determining the abundance of various repeat sequence families should be recognized. No single method is capable of revealing the total amount of any satellite DNA, and satellites I to IV are imprecisely defined. The variable length of LINE 1 elements introduces substantial uncertainty in the estimate of their mass fraction. For other repeat sequences, the cited ranges indicate experimental uncertainty or results of different methods.

tempts to quantify this sequence arrangement indicate that much of human DNA is occupied by 300-bp repeats interspersed with single-copy sequences having a modal length of 2000 bp (5, 14).

As distinct from interspersed repeats, other repetitive elements, e.g., ribosomal genes and U2 RNA genes, are arranged in long tandem clusters (15–17). Ribosomal genes alone account for about 0.3% of human DNA (Table II). "Satellite" DNAs, usually located near centrometric regions, consist of long tandem arrays of simple sequences that can be as short as five nucleotides (18). The human α satellite, as one example, consists of a 170-bp repeat-element organized in tandem blocks that can exceed several megabases in size on certain chromosomes (19). The α satellite can also be regarded as a family of sequences, but one in which different chromosomes often harbor distinct variants (20). The evolution of these chromosome-specific variants is attributed to their homogenization by unequal crossing over. As much as 10% of human DNA consists of highly repetitive satellite sequences (Table II).

Interspersed repeats must, in various ways, affect the expression of neighboring genes (Section III). Also, because of their dispersion, this class of repeats is a ubiquitous feature of human DNA. For these reasons, subsequent sections of this review exclusively concern interspersed repeats.

Interspersed repeats in some animals, in *Xenopus*, and in sea urchin belong to many different families of sequences (21–23). Because the complexity of these sequences is relevant to both their evolution and their possible functions, an estimate of the number of repeated sequence families in human is useful. As an order of magnitude calculation, 600 Mbp of human DNA is repetitive (Table I). If all of these sequences were 300 bp (which they are not), and if each family of sequences consisted of 1000 members (which they do not; Table I; Fig. 1), there could be as many as 2000 distinct families of such repeats. Surprisingly, as measured by mass, most interspersed repetitive human DNA belongs to only four sequence families, called Alu, LINE1, MER, and MaLR, which are described in the following sections (Table II). Although now taken for granted, the realization that so much of the human genome consists of two families of sequences, Alu and LINE1, and that each was generated by the reverse flow of genetic information (i.e., RNA to DNA), was extraordinary.

C. Most Human Interspersed Repeats Have Retrotransposed

1. Alu REPEATS CONTAIN AN INTERNAL PROMOTER

A detailed analysis of the physicochemical complications of DNA hybridization indicated that the repetitive sequences that renature with an apparent repetition frequency of 50,000 are actually present in about 500,000 copies per haploid genome (Fig. 1A, Table I) (24). This prediction was confirmed by the showing that a major fraction of the 300-bp repeats in human DNA belongs to a single family of sequences called Alu (25, 26). Alus constitute 5 to 10% of the mass of the human genome (Table II). A uniform dispersion of 10^6 Alus would result in one Alu occurring in every 3-kb region of the genome. Few human genes are entirely devoid of either internal or neighboring Alus, a ubiquitous feature of human DNA.

Individual Alu family members differ from each other on average by 15%, mostly in the form of point mutations, permitting the formation of a representative consensus sequence for the family (27, 28). The 282-nt consensus Alu structure has a precisely defined 5' end, shared by most members, and is a tandem repeat of two homologous sequences separated by an A-rich region (Fig. 2). The two monomer units are, in turn, homologous to 7SL RNA, also called SRP-RNA (because it is an integral component of the SRP) (Fig. 2) (29). The right monomer contains a sequence insert that is unrelated to 7SL RNA and is absent in the left monomer. Individual members have an A-rich 3' tail reminiscent of mRNA polyadenylylation (27, 29). The entire structure is surrounded by imperfect short direct repeats suggesting duplication of a preexisting genomic insertion site. This interpretation has been confirmed by comparing empty sites to sites in which an Alu has inserted in divergent species (27). Alus are mobile elements.

Alus contain internal A-box and B-box promoter elements for RNA polymerase III (Pol III) and are transcribed by Pol III *in vitro* (27, 29). The Pol-III transcription start-site corresponds precisely to the consensus 5' end

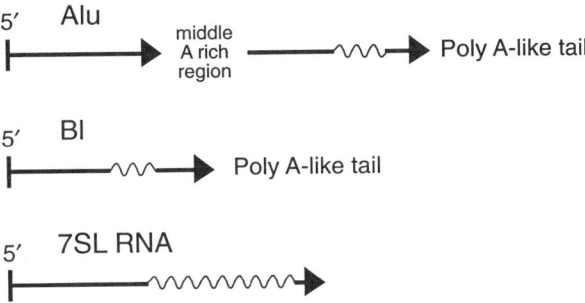

FIG. 2. Schematic depiction of the homology between Alu repeats, B1 repeats, and 7SL RNA (27, 29). The straight arrows indicate regions of homology between the left and right Alu monomers, B1 repeats, and 7SL RNA. The wavy lines indicate regions of nonhomology inserted within each of these sequences. The 5' ends of these three structures correspond precisely to their Pol III transcriptional start site.

of Alu members, suggesting their mobility via an RNA intermediate. This process is called **retroposition,** and its product is a **retroposon** (29). On insertion, a new Alu contains internal Pol-III promoter elements and is thus potentially competent to encode subsequent family members, providing a Malthusian explanation for the presence of nearly 10^6 Alu repeats in the human genome. The mobility and evolution of these elements, as well as their transcriptional expression, are examined in Sections III and IV.

2. UNLIKE SINEs, LINEs ENCODE A REVERSE TRANSCRIPTASE

7SL RNA is a structural RNA rather than an mRNA, so homologous Alu sequences do not have significant open reading frames (29). As reviewed below, other mammals also contain short interspersed elements, collectively called SINEs, which are similar to Alu in this regard (30). The *trans*-acting factors required for SINE retroposition must be encoded by other sequences. A family of long interspersed elements, called LINE1, potentially fulfills this requirement (30, 31).

LINE1 repeats are also surrounded by short direct repeats and have an A-rich 3' end reminiscent of mRNA polyadenylylation, again suggesting retroposition (31). A LINE1 consensus sequence can be derived by comparing different members to average out their divergence. However, individual LINE1 repeats are truncated at various 5' positions within this consensus as though the hypothetical reverse transcription of the RNA precursors was incomplete. The full-length LINE1 consensus sequence, which very few family members share in entirety, is 6 kb. As many as 100,000 members share 0.5–1 kb of homology to the 3' end of this consensus (31, 32). The uncertain copy-number of the different 5' truncations precludes a precise estimate of the LINE1 fraction of human DNA. Reasonable estimates suggest that this single family of sequences approximates 4% of human DNA (Table II).

Contained within the full-length LINE1 consensus sequence are several open reading frames, one of which encodes a product having reverse transcriptase activity (33, 34). LINE1 members containing this region provide at least one of the protein factors necessary for retroposition (34). Complete human and rodent LINE1 elements have Pol-II promoter activity, indicating that full-length retroposed members retain transcriptional control signals necessary for potential expression (35). The exact structure and function of this promoter remains to be determined. However, full-length LINE1s contain both *cis* elements and genes necessary for their self-perpetuation (36, 37). These discoveries account for the retropositional success of the LINE1 family.

3. MaLR: Redefining "Repeat Family"

Among many unsolved problems concerning the mobility of Alu and LINE1 repeats, the previous discussion highlights two: (1) proteins required for Alu transposition must be coordinately supplied in *trans* and (2) *cis* elements required for LINE1 expression must be included within the intermediate transcript. The sequence organization of proretroviral transposons satisfies these two requirements. Flanking long terminal repeats (LTRs) of proretroviral sequences include necessary *cis* elements for expression, and by exploiting LTR structure, the retroposition pathway ensures that reverse transcription will provide two complete LTR copies on insertion at a new site (38). The internal sequence of the transposed unit includes genes to encode products such as reverse transcriptase, required for retroposition, thereby providing new members with autonomy.

Human DNA contains thousands of copies of a 2.3-kb repeat-sequence family, formerly called THE1, which is structurally organized as a proretroviral transposon (39). Additionally, human DNA contains as many as 50,000 solitary 350-bp LTRs from the same sequence family. Subsequently, sequence analysis showed that THE1 is part of a larger superfamily, now called MaLR consisting of 40,000 to 100,000 members in rodents and primates (40) (Table II). Included in this superfamily are subfamilies that appeared in a punctuated evolutionarily manner and some among these subfamilies contain evolutionary old, highly divergent members. Collectively, this family constitutes 0.5–2% of human DNA (Table II).

Heretofore, cross-hybridization has defined families of repetitive sequences. By hybridization, a family of sequences is operationally defined by arbitrary experimental criteria such as the renaturation temperature, thereby establishing the threshold of similarity below which related sequences are not recognized (12, 40) (Section I,A). Similarity, as determined by direct sequence comparisons, provides an alternate definition of familial relatedness. Members of the MaLR super family have been repeatedly reisolated without realizing their relatedness by the criterion of hybridization (40) (Section I,D).

4. MIRs Are SINE Fossils

The evolutionary appearance of new repeat subfamilies implies inactivation and eventual disappearance of existing sequence families. The MIR family satisfies this intellectual requirement for defunct SINE families. The 260-bp MIR consensus sequence includes a tRNA-like region, so that members of this sequence family, like Alu repeats, resulted from Pol-III transcripts (41, 42). Divergence of MIR sequences from their consensus is two-to threefold greater than the divergence of Alus from theirs, implying that MIRs are evolutionarily ancient. Two lines of evidence support this interpretation: (1) Homologous MIR sequences are detected in all mammalian orders investigated, including even marsupials and monotremes. This family is ancient, perhaps appearing 130 million years ago (*41, 42*). (2) Even more significantly, some MIRs are positioned at orthologous sites in the rodent and primate genomes, indicating that those particular members inserted at their present loci prior to the ancient divergence of these lineages.

The extensive divergence of MIR elements further challenges the definition of repetitive sequence by the criterion of cross-hybridization. There may be as many as 300,000 MIR members belonging to two distinct sequence families, which correspond to 1–2% of the genome (*41*) (Table II). Despite their very high repetition frequency, as revealed by direct sequence comparison, extensive divergence would slow their rate of cross-hybridization, causing them to appear as a moderately repetitive sequence class. As discussed in Section I,A, hybridization under relaxed stringency reveals additional repetitive DNA and accelerates the hybridization of divergent homologs. Undoubtedly, MIRs contribute to such observations.

5. MICROSATELLITES EXEMPLIFY A DIFFERENT REPEAT DNA STRUCTURE

Simple sequences, such as tandem repeats of the dinucleotide CpA and other short motifs, are also broadly distributed throughout human DNA (*43, 44*) (Table II). Length polymorphisms associated with such sequences, usually called microsatellites, are useful markers in pedigree analysis and cause genetic diseases (*45–47*). Among other possibilities, slippage during DNA replication may cause these length polymorphisms (*44, 47*). Certainly, nothing in their structure suggests that microsatellites result from retroposition.

The repetition frequency of tandemly arrayed CpAs is 50,000-fold, corresponding to a highly repetitive sequence family (*43, 44*). These short arrays are heterogeneous in length, having a size range of roughly 20–50 bp. Sequence analysis indicates that tandem arrays of the possible dinucleotide repeats comprise about 1% of the human genome (Table II) (*44*).

D. Most Human Repeats Have Been Identified

Retrotransposed pseudogenes have been identified for various small RNAs, including 7SL, U1, and U2 RNAs, giving rise to corresponding moderately repetitive sequence families (*29, 32*). However, pseudogenes are also a common feature of most multigene families. For example, the seven members of the human α-globin gene family include three pseudogenes, two of which are ancient fossils in the same sense as discussed in Section I,C,4, for the MIR family (48–50). Strictly speaking, α-globin genes are a repeat sequence family, as are the more abundant immunoglobulin and histone gene families. Because members of these gene families have known functions, they are not normally considered as repeat sequences. However, they also contribute to the mass of repetitive DNA as revealed by DNA hybridization. Restricting the term "repeat" to multicopy sequences of unknown function subtly biases subsequent consideration of whether repeats have defined functions (Section IV,A).

The extraordinarily large human genome undoubtedly contains additional surprises that cannot be anticipated from current data. Imagining that all repetitive sequences have been completely catalogued would be presumptuous. Yet, useful generalizations are now possible. Allowing for considerable uncertainty, known families of repeats account for substantially all of the repetitive sequence composition of human DNA, as assayed by hybridization (Table II). This observation does not preclude the discovery of novel repetitive human structures but limits their abundance (12). Just a few examples of repeats not included in Table II are given in the preceding paragraph, and indeed there are many others. But certainly, no abundant repeat families have escaped detection, and furthermore, abundant repeat families collectively account for the mass of repetitive DNA (Table II). Supporting this contention, the MaLR and MIR families have each been independently rediscovered by many investigators, and systematic surveys for new repeat families have mostly captured additional representatives of these two known families (12, 40–42, 51).

Critical inspection of these developments, particularly as outlined by Smit, suggests that a turning point has been achieved (40, 41). As many as twenty distinct reports of human repeats are consolidated within the MaLR and MIR families. As anticipated on physicochemical principles and demonstrated by the discovery of Alu, DNA hybridization consistently overestimates the complexity of divergent sequences (Section I,A). The expansion of sequence data bases and the introduction of more powerful algorithms to identify similarity have changed the definition of repetitive DNA (12, 40–42).

Because of their ubiquity, as well as their transpositional and biochemical activities, human Alus and LINE1 sequences have been more intensely studied than other repeats. The remainder of this review concentrates on SINEs, including especially human Alu, the focus of research in my laboratory. Progress on LINEs closely parallels that of SINEs and, with few exceptions, lessons learned for one of these repeats families applies to the other (31, 52). Properties of LINEs will be examined only to supplement this description of SINEs.

II. Evolution of Alus (and Other Mammalian SINEs)

A. When Alus Appeared

The B1 repeats in rodents, although homologous to 7SL RNA and Alus, are an entirely distinct family of sequences (Fig. 2). B1 repeats resemble the left Alu monomer. Rodent DNAs do not contain Alu repeats, and although human DNA contains some left Alu monomers, these are distinct from B1 repeats (26, 53, 54). B1 and Alu repeats appeared independently in rodents and primates following divergence from their last common ancestor (Fig. 3A).

Mammalian SINEs belong to either of two superfamilies, those related to 7SL RNA or those related to tRNA. Rodents contain a second SINE family, B2, which is homologous to an ancestral tRNA and unrelated to B1 and Alu (29, 32). SINEs in most mammals (e.g., horse, rabbit, artiodactyls, pig and dog) belong exclusively to the tRNA superfamily and do not include representatives of the 7SL RNA superfamily (55–59). Plants, insects, and non-mammalian vertebrates contain SINEs belonging to the tRNA superfamily of sequences (60–62). These comparisons indicate that the ancestral mammal also had tRNA-related SINES. The existence of fossil human MIR repeats confirms this interpretation (Section I,C,4; Fig. 3A). Human B2 repeats have not been detected, implying the extinction of the tRNA SINE superfamily within the human lineage.

All primates contain recognizable Alu repeats; however, galago (a prosimian primate) Alus have diagnostic sequence differences as compared to those in higher primates (63). Distinct Alu families independently amplified in these two primate lineages (Fig. 3A). Additionally, galago contains B2-like SINEs, B1-like Alu monomers, and composites of these elements (64–66). Risking a possible oversimplification, galago SINE composition is intermediate to that in rodents and higher primates. The structure of SRP proteins that bind Alu transcripts mirrors these differences (Section IV,C,4).

B. Alus Accumulated during Higher Primate Evolution

There have been complete exchanges of SINEs in the primate genome as well as in that of other mammalian orders (Fig. 3A). New sequence families appear; previously active families disappear. These dynamics could result from simple founder effects or massive conversion of existing SINEs to a new motif.

Phylogenetic comparisons of Alus residing at particular loci show that conversion is rare and that new Alus appear by insertion of new family

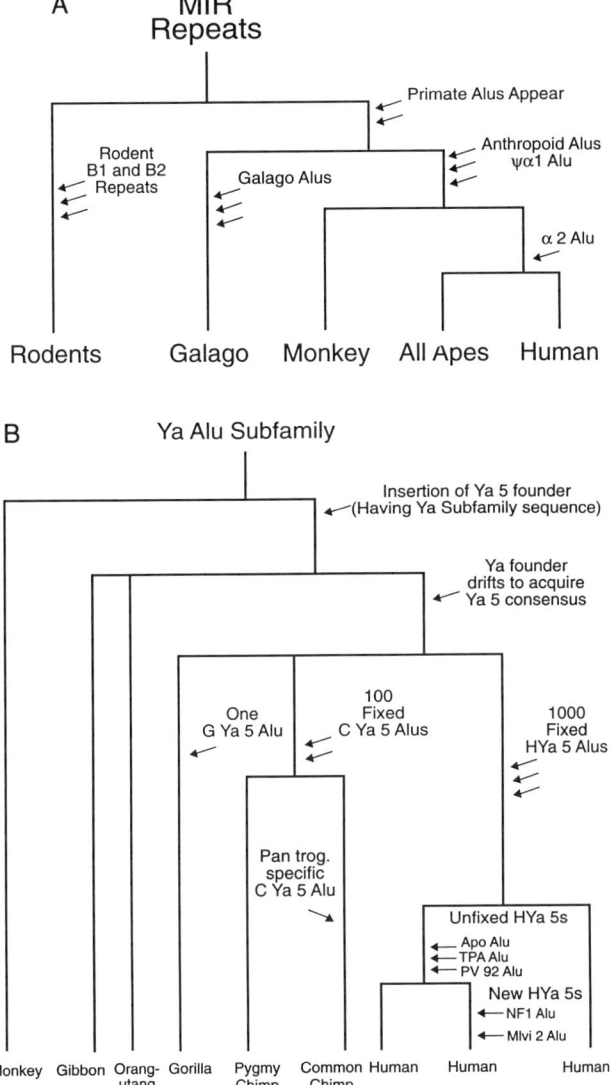

FIG. 3. (A) The succession of SINE families during rodent and primate divergence is depicted. The fossil MIR family appeared in the common ancestor of rodents and primates (41, 42). The B1 and B2 families were active in the rodent lineage, which does not contain Alus, and similarly, primate B1 and B2 repeats have not been observed. All primates have Alus, but galago Alus differ from those in higher primates (63). Alus were independently inserted in the galago and human lineages (67). Insertions of Alus at loci in the α-hemoglobin gene cluster are indicated (67, 68). (B) Appearance of the Ya5 Alu subfamily. The Ya5 founder inserted after monkey–ape divergence but initially resembled the next older Ya subfamily (81, 90). By drift, it subsequently acquired the base substitutions that identify the Ya5 subfamily prior to divergence of the human (H), chimpanzee (C), and gorilla (G). The Ya5 subfamily independently expanded following the divergence of these lineages and continued expanding during further speciation of the chimpanzee and radiation of the human population (52, 73, 80, 82, 84, 107).

members into previously empty sites. For example, humans, chimpanzees, and monkeys share an Alu repeat located 5' to the Ψ α-1 hemoglobin gene (67, 68; Fig. 3A). This Alu, allowing for sequence drift, is recognizably the same family member and was not converted to a distinct Alu sequence. This same Alu is absent from the orthologous position in galago and thus was inserted in the higher primate lineage following its divergence from the galago lineage (Fig. 3A). Similarly, independent Alu insertions occurred in monkey, ape, and human α-hemoglobin gene clusters (69, 70; Fig. 3A). Alu sequences are mobile in an evolutionary sense, but immobility is the rule. Most Alus appeared before humans diverged from chimpanzees and many appeared before the divergence from monkeys (Fig. 3A). Similarly, nearly 99.9% of the 10^6 Alus are fixed at their respective loci within the human population (31, 52).

These same phylogenetic comparisons indicate that new families appear by independent insertion of new members within a lineage (31, 52). This interpretation is confirmed by finding that the Ya5 subfamily founder acquired its identity by simple sequence drift (Section I,D). Although I discount the general importance of conversion in Alu evolution, converted Alus exist (71, 72). Simple founder effects and retropositional amplification appear to have a greater impact on Alu evolution than does conversion (Section I,D).

C. Young Alu Subfamilies Are Active

Human Alu repeats can be divided into sequence subfamilies having different levels of sequence divergence, implying their different evolutionary ages (31, 52). One subfamily, called Ya5 (also called PV, HS1, and Sb2), consists of 1000 members having five diagnostic base substitutions relative to the next older subfamily, called Ya (73–75). Members of this subfamily closely resemble their consensus sequence, some differing by only one or two nucleotides, and are rich in CpG dinucleotides (31, 52). CpGs are the target site for DNA methylation in vertebrates, resulting in 5me-C residues that have a very high mutation rate to T residues (76). As discussed in Section IV,B,2, Alus are heavily methylated in various tissues. As might then be expected, consensus CpGs in older Alus show a high transition frequency to TpG or equivalently to CpA (77, 78). Accordingly, the low level of sequence divergence and conservation of CpG residues indicates that Ya5 subfamily members are evolutionarily very young.

The youth of this subfamily is confirmed by phylogenetic and population studies. There are 1000 Ya5 Alus in humans and far fewer in chimpanzees and gorillas (73, 79–81; Fig. 3B). With one important exception (Section I,D), human Ya5 Alus are absent from orthologous positions in the chimpanzee and the gorilla. Although most Ya5 Alus are fixed in the human genome, a minority were inserted during the radiation of the population (31,

52, 82–84) (Table III). The presence or absence of an Alu at a particular site creates a readily detected sequence length dimorphism. The allele frequency of dimorphic Alus differs among racial groups (82) (Table III). Moreover, Alus are still mobile; several Alu insertions are *de novo* mutations (85–88) (Fig. 3B).

At least four distinct Alu subfamilies are young by the criterion of having members that were inserted following human and chimpanzee divergence (31, 52, 84). Just as there has been a continual turnover of mammalian SINE families, there continues to be a turnover of Alu subfamilies, raising questions concerning the mechanisms by which new families appear and existing families disappear.

D. An Alu Founder—The Simplest Possibility

As mentioned above, the Ya5 Alu subfamily was transpositionally active in the divergent chimpanzee and human lineages such that each contains multiple Ya5 Alu insertions absent from the genome of the other (73, 79–80) (Fig. 3B). There are only two copies of gorilla Ya5 Alus, of which one is gorilla specific, demonstrating that this subfamily was independently active in all three lineages, although to significantly different extents. The other, called EPL Alu, is also present as both a human and a chimpanzee ortholog (81). As revealed by this phylogenetic bottleneck, EPL Alu, which is common to all three lineages, either founded this subfamily or is one of its earliest progenies (81). The sequence of this same Alu in gibbon resembles that of the next older Ya subfamily (81) (Fig. 3B). Thus, EPL Alu was originally encoded by a member of the next older subfamily, and by drift acquired the five diagnostic base substitutions that identify the Ya5 subfamily. EPL Alu founded the Ya5 subfamily.

TABLE III
FREQUENCY OF Alu INSERTION[a]

Locus	Australian aborigine	Asian	Amerindian (Arhuaco)	Alaska native	Caucasian	Nigerian
PV92	0.152	0.813	0.975	0.645	0.141	0.091
APO	0.869	0.906	1.000	0.992	0.922	0.500
ACE	0.909	0.688	0.850	0.637	0.469	0.273

[a] The frequency of Alu insertion at three loci, PV92, APO, and ACE, is reported for various populations (82). The PV92 locus is fixed for Alu insertion in Amerindians but is almost fixed for the absence of Alu insertion in Nigerians. The APO locus is fixed for Alu insertion in North American populations but is highly dimorphic in Nigerian populations. These and other examples of Alu dimorphisms suggest an African origin for Alu dimorphisms (82).

Cis-acting elements stimulate *in vitro* transcription of EPL Alu (89). The biological significance of these elements remains to be tested, but their presence plausibly explains why this particular Alu was independently active in three lineages. No other human Ya5 Alus exactly match the ancestral EPL Alu founder sequence, which has acquired multiple base substitutions (81, 90). Also, although human DNA contains many Ya5 Alus, chimpanzee contains fewer, and gorilla contains only one additional member of this subfamily (79–81). Success of the human Ya5 subfamily is not directly attributable to the activity of its founder but rather must result from its progeny. Perhaps one such sequence happened to land in a particularly active chromosomal region.

Several Alu subfamilies are currently active and each contains a number of active source "genes" to account for the sequence heterogeneity of new subfamily members (84). Similarly, there are a small number of LINE1 source "genes" (37). Although some of the mechanistic requirements for the self-perpetuation and, correspondingly, the self-selection of successful LINE1 sources have already been demonstrated, these requirements are less obvious in the case of Alus.

Each step in the complex retroposition pathway could influence the relative success of a particular Alu (52). Among many requirements, the Alu must first be transcribed, the transcript must be protected from processing and degradation, and reverse transcriptase must be available. (Alu transcriptional and posttranscriptional expression is reviewed in Section IV,C.) For present purposes, I note merely that different Alus can be expressed in very different amounts. The source of reverse transcriptase, perhaps a LINE1 or infecting retrovirus, is not certain, but conceivably the developmental and tissue-specific regulation of this and other required products limits retropositional activity to those Alus that happened to be coordinately expressed with necessary *trans*-acting factors. According to this view, the expansion of a particular subfamily is not deterministic but rather the stochastic result of various selective advantages and chance events.

E. SINE Inactivation

All young Alus identified to date closely match one of four young subfamily consensus sequences, and none resembles divergent Alus (52, 84). Similarly, the MIR family appears to be inactive. The birth of a new Alu subfamily results from simple founder effects, but the basis by which older Alus become inactive is less certain. Even more surprising is the complete inactivation of an entire previously successful family, such as MIR.

As a minimum, active Alu source genes must be transcribed. Mutations in older Alus inactivate promoter elements, thereby accounting for the quiescence of at least some Alus (66, 91). Mutations may also decrease the post-transcriptional lifetime of an Alu transcript (92). An indirect consequence of extensive Alu methylation is a high transition frequency of CpG to TpG dinucleotides, causing older Alu transcripts to be relatively rich in A and U residues (77, 78) (Sections II,C and IV,B,2). Although not yet tested, such substitutions could destabilize RNA structure, causing transcripts from older Alus to be relatively short-lived, thereby further compromising their potential retropositional activity (52).

Evolution of primate SRP proteins hints at one possible answer to the more difficult question of how an entire SINE family is inactivated. Alu RNAs specifically bind SRP proteins, which may stabilize and process primary Alu transcripts (93–95) (Section IV,C,4). The human homolog of the rodent 14-kDa SRP protein is increased in molecular size to 18 kDa by a triplet repeat codon expansion in the primate lineage (94). This human variant SRP protein forms a more stable Alu RNP than the 14-kDa rodent protein (93). The appearance of this altered SRP protein parallels the appearance of the Alu family within primates and plausibly contributes to its selection at the expense of the preexisting MIR family (95). Galago SINE composition (Section II,A) is intermediate to that of rodents and human, as is the structure of this possibly key SRP protein (95).

III. Genetic Effects of Alus and Other Repeats

One underlying motive for studying repeats is their presumed function. Of course, repeats may be "junk" or "selfish" DNA with no physiological role (Section IV,A). Even in the absence of a defined role, abundant repeats certainly cause genetic effects simply because of their ubiquity. Two additional reviews of this topic are highly recommended (96, 97).

Ubiquitous, homologous repeats such as Alus might serve as frequent sites for unequal homologous crossing-over, consequently scrambling their flanking direct repeats. Conversely, the presence of recognizable direct repeats is evidence than an Alu has not engaged in unequal crossing-over. Unfortunately, the precise identification of the flanking repeats is somewhat ambiguous due to their short length, evolutionary divergence, and resemblance to the 3′ A-rich sequence (27). Casual inspection of individual Alu indicates that most, perhaps 80%, have retained short direct repeats. Given the residence time of these Alus within the genome (Section II,B), unequal Alu–Alu crossing-over is infrequent.

Unequal Alu–Alu crossing-over duplicates, deletes, or scrambles genetic information. Alu–Alu crossing-over within the low-density lipoprotein (LDL) receptor gene duplicates exons, thereby inactivating the gene product, leading to hypercholesterolemia (98). One case of X–X maleness resulted from unequal crossing-over between Alus on the X and Y chromosomes (99). Expansion of the human growth hormone multigene family results from unequal Alu–Alu crossing-over (100). Similarly, Alu elements at critical positions in the human and monkey α-globin gene clusters mark breakpoints in gene conversion or other sequence rearrangements (69, 101). In humans, this particular Alu does not have direct repeats and is undoubtedly a recombinant.

Alu consensus sequences do not have significant open reading frames. However, because of their ubiquitous genomic distribution, Alu elements randomly occur within Pol-II-directed transcription units, and both "sense" and "antisense" Alu fragments are abundant in pre-mRNA (26, 53). At least 10% of hnRNA consists of Alu sequences that are largely, but not completely, removed during mRNA processing, thereby altering protein products (102). The presence of SINES, including Alu B1, B2, and MIR repeats, as well as LINE1 elements, alters mRNA polyadenylylation (32, 41–42, 103). MaLRs, by virtue of sequences within their LTRs, also direct polyadenylation and 3' mRNA processing (104). Presumably, 3' ends of mRNAs are more tolerant to such mutations.

De novo insertion of an Alu within a patient's intron for the NF1 gene caused neurofibromatosis (86). Insertions of other Alus and LINE1 repeats have also resulted in gene inactivation and disease phenotypes (83, 87–88, 105–106). However, most dimorphisms for the presence of Alus do not exhibit discernable phenotypes (52, 84, 107).

Because Alu elements are transcribed by Pol III, they might not be expected to participate in the Pol-II-directed transcription of neighboring genes. However, in yeast, Pol-III transcription of tRNA genes represses Pol-II transcription of neighboring genes, and there is one case in which an Alu might have a similar effect (108, 109). Alus also interact directly with Pol-II transcription factors: Nonconsensus base substitutions confer enhancer function on an Alu interspersed within a gene's intron (110). Another Alu provides globin gene promoter elements (111).

The participation of Alus in the Pol-II expression of neighboring genes should again be considered as an effect rather than a defined role for Alu elements. The parent gene functioned prior to Alu insertion, and the activity of Alus in these examples often involves nonconsensus substitutions that appeared after insertion (112).

IV. Does Alu Have a Function?

A. Selfish DNA, Junk DNA, and Cheap Genes

Selfish DNA is construed to confer neither benefit nor harm on its host (113, 114). Based on the previous discussion, SINEs and other repeats cause a broad variety of easily detectable, deleterious effects and constitute a significant genetic load. Retropositional insertion is itself a highly mutagenic event. The insertion of an Alu within an exon would destroy function, so that the general absence of Alus within coding regions is easily understood. Even within noncoding introns, the insertion of Alus and LINE1s has been extremely deleterious.

Retroposons have been continually maintained throughout mammalian evolution. The selective disadvantage of retroposons and even the existence of the retropositional pathway suggest the corresponding existence of a compensating selective advantage, i.e., a defined function. Of course, in accordance with the selfish DNA proposal, these elements may only be imperfect parasites that have not yet fully adapted to their host's requirements. (Advancing a different point of view, Dr. Mel Green reminds me that human DNA contains 10^6 Alus with relatively few problems. Accordingly, he suggests that they might be better considered as nearly perfect parasites.)

Human DNA certainly contains evolutionary debris such as the primate pseudo-α-globin gene fossils, and such sequences are undoubtedly junk in the sense that their deletion would have no phenotypic effect (32, 67). The decrepit sequences of MIRs are also very unlikely to support even vestigal functions (reviewed in Section I,C,4). Similarly, Alus at tested genomic loci evolved at a rate approximating neutral base substitution (67, 68, 71, 115). [However, subregions of some Alus appear to be conserved, perhaps because they are required for protein binding (116) (Section IV,B).] Consequently, some ancient divergent Alus no longer promote Pol-III-directed transcription (66, 91). If pseudogenes are a form of junk DNA, the foregoing examples of highly divergent MIR and Alu repeats are probably also junk. As previously discussed, certain Alu insertions cause deleterious effects, but others have no discernible phenotype. In partial agreement with the junk DNA hypothesis, we must conclude that many individual Alus are either functionless or are dispensable.

Evolution of a multisequence family is quite different from the tightly selected evolution of a single gene. A multigene family shelters its individual members from natural selection, which normally prevents the accumulation of mutationally inactivated alleles of single-copy genes. As a solution to this problem, gene conversion efficiently maintains almost exact sequence identity among the individual members of the tandemly arrayed U2 and rDNA multigene families (15–17). For such tandemly arrayed multigene families, conversion efficiently erases the appearance of pseudogenes.

Less efficient conversion occurs within the cluster containing the somewhat more dispersed human hemoglobin genes. Conversion maintains the near identity of the two adult α-globin genes as well as the two embryonic ζ-globin genes but does not interconvert the separate identity of these two types of distinct globins and does not correct defects within the two intervening pseudo-α-globin genes (48–50, 117). Building on the example of globin genes and considering a family of genes as broadly dispersed and as numerous as human Alus, conversion simply could not effectively preserve the identity of each member. **A dispersed multigene family inevitably accumulates pseudogenes.** Unlike tandemly arrayed rRNA and U2 RNA genes, the dispersion of the α-hemoglobin genes may even be required to isolate their distinct structures and corresponding functions. The dispersion of SINEs, although more extensive than that of α-globin genes, may similarly be functionally required at the cost of accumulating junk.

As an alternative to conversion, recurrent creation of freshly selected members might maintain a sufficient number of active members in a dispersed multisequence family (118). This assumes that the genetic load associated with this profligacy is slight, so that these sequences are "cheap." As a necessary consequence of this "cheap gene" mechanism, multisequence families accumulate inactive members while always maintaining an active subset. Consistent with the cheap gene hypothesis, inactivation of SINE families and of individual SINEs has been observed, and only a subset of Alus currently serves as retropositionally active source genes (31, 52, 84).

According to this reasoning, any proposed function for Alu elements should explicitly require their broad genomic dispersion and implicitly acknowledge that, although many family members may be junk, their presence does not interfere with those that are active. Ideally, this proposed function should also account for why a select subset of Alus is active. Alus might serve as *cis*-acting elements in either DNA or RNA, or might provide necessary transcripts in the traditional sense of a multigene family. Two possible functions satisfy these criteria (Sections IV,B,2 and IV,C,5).

B. Alus as *Cis*-acting Elements

1. Arrangement with Respect to Genes and Chromosomes

No obvious correlation between transcriptional expression of genes and the presence, absence, position, or orientation of Alu elements has emerged. Also, although Alus are abundant in hnRNA, no correlation has been observed with respect to mRNA, splicing, except, of course, for those cases in which the Alu appears in mature mRNA (102) (Section III). Alu elements are enriched in chromosomal R bands, which are regions containing many housekeeping genes and which are enriched in both C and G and 5meC (119, 120). However, again there is no simple correlation between R bands and gene expression, causing us to consider other possibilities.

2. Alu METHYLATION MAY HAVE AN IMPRINTING ROLE

CpG dinucleotides are sites for DNA methylation, forming 5meC residues (76). Overall, human DNA contains 1% CpG, and depending on the tissue source, the composition of 5meC within human DNA is typically only slightly less than 1% (121). Young Alu repeats are 9% CpG, but in older Alus, consensus CpGs have frequently decayed to TpG (77, 78). These observations indicate that Alus must be methylated in the germ line and further require that source genes encoding young Alus are in some way selectively recruited in accordance with the presence of consensus CpG residues (52).

Alus are fully methylated in adult somatic tissues, e.g., spleen, accounting for one third of the 5meC residues in DNAs from these tissues (122) (Fig. 4A). In ovarian monkey oocytes, as well as human female germ-line tumors, Alus are mostly, if not entirely, methylated (123). Female germ-line methylation accounts for the rapid transition of consensus AluCpGs. In contrast, most (75%) of the young Ya5 Alu subfamily is completely unmethylated

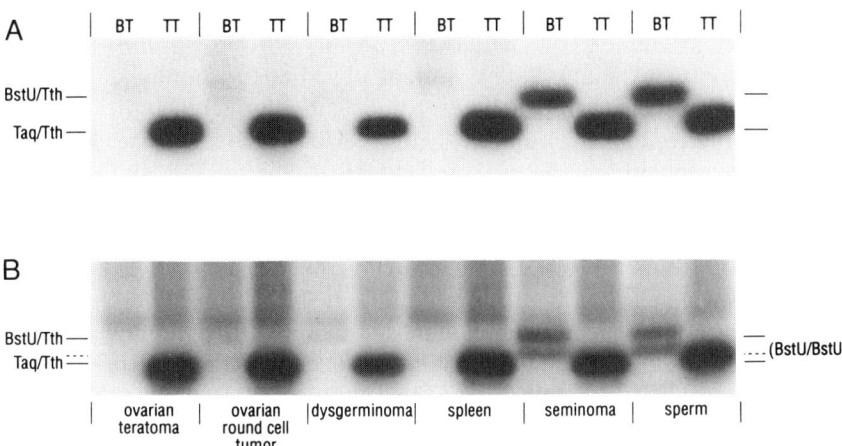

FIG. 4. (A) Sperm and seminoma Ya5 Alus are shown by Southern blot hybridization to be mostly unmethylated, but Ya5 Alus in spleen and various female germ-line tissues are mostly methylated (123). BstU/Tth (labeled BT) digestion releases a band only if the BstU1 site is unmethylated. Taq/Tth (labeled TT) digestion calibrates the total hybridization signal. The hybridization probe and experimental procedures in this experiment select for hybridization of Ya5 Alus. (B) Slightly changing the blot-hybridization conditions reveals the presence of the next older Ya Alu subfamily (123). This is apparent in the appearance of the doublet band in the BstU/Tth digest corresponding to a second consensus Bst U1 site in the Ya subfamily. Compared to the result for the younger Ya5 subfamily (A), proportionately less sperm and seminoma DNA are cleaved by BstU1 when selecting for the older Ya subfamily (see Section IV,B,2). Reproduced from Ref. 123, by permission of Oxford University Press.

at a consensus BstU1 site in sperm and seminomas (Fig. 4A). Sperm Alus unmethylated at this BstU1 site are largely unmethylated at other consensus CpGs (124).

These conclusions result from testing DNA methylation at consensus restriction enzyme cleavage sites using hybridization conditions that are specific to young Alus (Fig. 4A). As a technical note, the ability to target the sequences of young subfamilies has been instrumental to much of the recent progress concerning both Alus and LINE1 (31, 52). For example, by slightly changing hybridization conditions, the methylation status of the next older Ya Alu subfamily can be compared to that of the Ya5 subfamily (Fig. 4).

In young Alus, restriction sites are faithfully represented, whereas in older Alus these sites have been inactivated by base substitutions (123, 124) (Fig. 4B). Thus, the overall methylation level of older Alus in sperm DNA is somewhat less certain, but is apparently somewhat greater than that of younger Alus (125). The methylation status of particular sperm Alus is locus specific, showing that sperm Alus are not just randomly hypomethylated (125, 126).

Sperm chromatin contains a specific Alu binding protein, SABP, which is undetectable in HeLa cells (127). SABP specifically protects an Alu element from *in vitro* DNA methylation without blocking the methylation of *cis*-vector sequences and plausibly protects sperm Alus from methylation during germ-line development. Interestingly, the SABP binding sites are located at Alu sequence positions 25 to 33 and positions 158 to 167. Both of these regions are apparently under selection because they are conserved in some Alus (116).

Oocyte DNA is hypomethylated, but various genes and even LINE1 repeats in sperm are highly methylated (128–133). In general, mouse sperm DNA is usually regarded as being hypermethylated relative to oocyte DNA (129). The methylation patterns of Alus in sperm and oocyte are therefore sequence specific and opposite from the general methylation patterns observed in these tissues. This remarkable difference between Alu methylation in the male and female germ lines suggests possible functions. In particular, differential inheritance of Alu methylation in zygotes may have a role in genomic imprinting.

Genomic imprinting is the differential expression of the paternal and maternal genomes requiring distinguishability of the two parental genomes (134). Parent-specific differences in DNA methylation are associated with imprinted expression; however, these differences are not directly inherited but are imposed after fertilization (135–138). The gametic imprint is distinct from the functionally expressed imprint. Because of their broad dispersion, the differential methylation of Alus would effectively identify every region of the genome as being either maternal or paternal in origin, thus Alus are uniquely suited to signal the gametic imprint.

Insertion of foreign CpG-rich DNA imports methylation sites, subsequently causing methylation of surrounding sequences (*139*). Similarly, B1 and B2 repeats appear to direct methylation of the α-fetoprotein gene in cultured cells (*140*). Thus the presence of one or more CpG-rich Alus is likely to determine the local methylation pattern. At a more macroscopic level of organization, preferential insertion of Alus within R bands is likely to contribute to and possibly cause the high level of methylation within these regions (*119*). However, the rapid mutation of 5meC residues to T inevitably erases potential methylation sites on sequences that are methylated in the germ line. Although we do not know if Alus are unmethylated throughout male germ-line development or merely in its latter stages, their methylation within the female germ line alone is sufficient to ensure the loss of critical CpG residues (*123*). Recurrent recruitment of young CpG-rich Alus would be required to restore the methylation potential of critical regions.

DNA methylation represses Alu transcription *in vitro* (Section IV,C,3). Conceivably, the level of Alu methylation is *cis*-regulated by the density of neighboring Alu CpGs, providing an autoregulatory mechanism to ensure the proper level of differential germ-line methylation. When the density of CpGs falls below a critical level, germ-line transcriptional repression of Alus is relaxed, potentially initiating retroposition. Whether or not this speculation is correct, *de novo* Alu insertion occurred within the male germ line, and Pol-III-directed Alu transcripts are present in mature sperm (*86, 125*; unpublished results from this laboratory).

A role for Alus as ubiquitous DNA methylation signals virtually requires the cheap-gene model. Germ-line methylation inevitably degrades CpG methylation sites, requiring their constant replenishment.

C. Alus as Transcription Units

1. Surprisingly Low Level of Alu Expression *in Vivo*

With the exception of a few Alus having extensively mutated A box and B box promoter elements, most are readily transcribed *in vitro* by RNA Pol III (*91*). Given nearly 10^6 potential Pol-III transcription units, one might expect an abundance of corresponding transcripts. Primer extension assays show that this is not the case; in HeLa cells, Pol-III-directed Alu transcripts are barely detectable (*73, 92, 141*) (Fig. 5). Compounding this problem of detecting Alu transcripts, 10% of hnRNA consists of Alu elements inserted within Pol-II-transcription units, causing an extraordinarily high background in experiments designed to assay for authentic Pol-III-directed transcripts

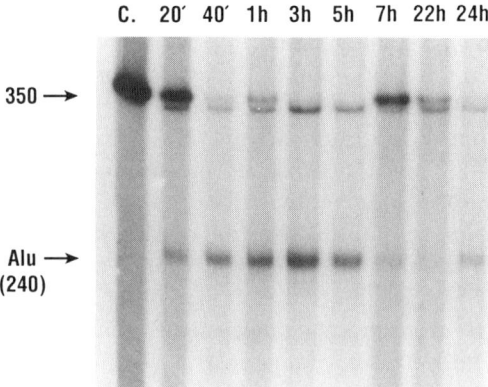

FIG. 5. Treatment of HeLa cells with cycloheximide rapidly and dramatically increases the abundance of flAlu DNA (142). The 240-nt band is the primer-extension product expected for flAlu RNA. C, Control cells; numbers indicate the exposure time to cycloheximide. Reproduced from Ref. 142, by permission of Oxford University Press.

(92, 141). Only recently has significant progress been achieved in studies of Alu RNA. In particular, the abundance of Alu RNA increases significantly in response to viral infection, cell stress, and translational inhibition (142–144) (Fig. 5). Before considering the induction of Alu RNA expression (Section IV,C,5), basal Alu transcription *in vitro* and *in vivo* is first reviewed.

2. THE Alu PROMOTER IS INHERENTLY WEAK

The Pol-III-template activity of Alus is variable, and as might be anticipated, those having the most highly conserved Pol-III-promoter elements are more actively transcribed than those having nonconsensus base substitutions in the A box and B box elements (66, 91, 145). In agreement with this finding, cDNA sequences corresponding to Pol-III-directed Alu transcripts have consensus A box and B box elements (141). Mutations in these promoter elements may silence many older Alus, partially explaining why recently inserted Alus are themselves encoded by young source genes (52).

However, in addition to an intragenic promoter, many Pol-III-directed transcription units also require flanking sequences for expression *in vivo*. The 7SL RNA gene, because of its homology to Alu, provides an especially pertinent model system to consider this possibility. Retrotransposed 7SL RNA pseudogenes are completely analogous to Alus in all structural features (146). These pseudogenes have an A-rich 3' end, are flanked by short direct repeats, and have 5' termini corresponding precisely to the Pol-III-transcription start-site for 7SL RNA (146). Also, like Alu repeats, retroposed

7SL RNA pseudogenes are transcribed *in vitro* by Pol III by virtue of their A box and B box promoter elements (*147, 147a*). However, no corresponding *in vivo* transcripts are detected, so that, like Alu repeats, they are effectively silent. In addition to its internal promoter, the authentic 7SL RNA gene has flanking 5' sequences that stimulate its expression *in vitro* and are necessary for *in vivo* expression when assayed by transient transfection (*147, 147a*). Like 7SL RNA pseudogenes, Alu elements lacking 5' control elements would be silent.

In agreement with this prediction, 5' flanking sequences of 7SL RNA significantly increase Alu template activity *in vitro* and *in vivo* as assayed by transient transfection into human 293 cells (*148*). The structure of 3' flanking sequences, specifically the transcription terminator, also affects the level of expression of transfected Alu templates and B1 RNA templates (*148, 149*). Without the benefit of these *cis* elements, an Alu having even a perfect A box and B box promoter is an inherently weak template *in vivo*. Successful source genes such as the Ya5 Alu subfamily founder may benefit from neighboring sequences that enhance their template activity (*89*).

3. Alu RNA Expression Is Regulated at Many Levels

Although the constructs described above are active on transient transfection into 293 cells, they are essentially inactive in HeLa cells, showing that Alu transcription is regulated in other ways (*148*). Given that there are potentially 10^6 Alu genes, it is difficult to imagine that their regulation is accomplished by a single mechanism operating at a single level of control (*52*).

In HeLa cells, this regulation is largely repression, and I imagine the following hierarchy: First, as noted, Alus normally lack flanking enhancer elements, and their inherently weak promoters may be further compromised by nonconsensus base substitutions (*91, 141, 145, 148*). Thus, potential expression is biased toward the Alu subset having consensus promoter elements and fortuitous combinations of flanking sequences. Second, chromatin severely represses Alu transcription *in vitro*, and alterations in chromatin structure may activate Alu templates *in vivo* (*150, 151*). Accordingly, transcription may be further restricted to a subset of those located within active chromosomal domains. Third, DNA methylation indirectly represses Alu template activity *in vitro*. A repressor present in limiting amounts binds methylated DNA (*125, 145*). Because Alus are completely methylated in somatic cells, transcription of accessible Alus might be further repressed by this mechanism. Fourth, we recently found that the TATA binding factor (TBF) binding domain of p53 represses expression of Alu templates, but that the 7SL RNA gene's promoter elements overcome p53-mediated repression (*152*). We assume that the TBF binding domain in p53 acts on a transcription factor, TFIIIB. Strong promoters, such as that in the 7SL RNA gene, may recruit an active transcriptional complex containing TBF by other pathways, thereby evading this repression (147a, 152). This additional level of repression could direct Pol III activity away from the few Alus not completely repressed by the previous levels of control. The steady-state abundance of any resulting transcripts is next regulated by their lifetimes, a topic explored below.

4. Alu RNA Exists as Three Distinct Species

Pol-III-directed transcription should read through an Alu element and its A-rich 3' tail before terminating at the first run of four or more Ts present within endogenous 3' flanking sequence (153). Such transcripts will be called full-length (fl) Alu RNA; flAlu RNA fractionates as poly(A)$^+$ cytoplasmic RNA (73, 92, 141). The half-life of an flAlu transcript resulting from a Ya5 subfamily template is approximately 0.5 hour (148). The presence of the 3' A-rich tail as well as other 3' non-Alu flanking sequences generated by this template does not affect its half-life. Although not yet tested *in vivo*, the secondary structure of flAlu RNA may affect its lifetime, thereby favoring the retropositional activity of certain Alus (154).

A 118-nt RNA called small cytoplasmic (sc) Alu RNA resembles the left Alu monomer; lacking the A-rich 3' tail, it fractionates as a poly(A)$^-$ RNA (73, 155). The scAlu RNA appears in a time-dependent manner from either flAlu RNA injected into *Xenopus* oocytes or on incubation of flAlu RNA with *in vitro* extracts, suggesting that the scAlu RNA results from processing full-length transcripts (155).

Both scAlu RNA and flAlu RNA are present, with about 1000 copies each, in HeLa cells (141). However, the abundance of scAlu is not directly dependent on the abundance of flAlu RNA but is controlled by other limiting factors (142, 148). SRP proteins p9 and p14 bind scAlu RNA, and experiments using rodent–human cell hybrids suggest that the human variants of these proteins result in higher levels of scAlu RNA (93, 155, 156). SRP protein binding may simply stabilize scAlu RNA; the half-life of scAlu RNA is at least nine times that of flAlu RNA (148). Identical observations have been made for scB1RNA (155, 156).

ScAlu RNA and flAlu RNA are each encoded by many dispersed members of the Alu family (141, 155). Like scAlu RNA, BC200 RNA, a third type of Alu RNA, resembles the left Alu monomer (157). Unlike scAlu RNA, BC200 RNA does not result from processing but is instead the primary Pol-III-directed transcript from the dedicated BC200 locus. Presumably, an Alu element that inserted at this site subsequently acquired function by a process called exaptation (158). Unlike dispersed Alus, the BC1 locus is a standard gene.

5. Cell Stress and Viral Infection Increase Full-length Alu RNA

Infection of human cells with adenovirus HIV and HSV greatly increases the abundance of flAlu RNA (143, 144). SV40 transformation of rodent cells increases the abundance of B1 and B2 RNAs, suggesting that viral induction of mammalian SINE RNA is quite general (159). Remembering that Alu retroposition requires coordinate expression of the RNA precursor and a source of reverse transcriptase, retroviral infection provides an attractive possible mechanism for Alu amplification.

Adenovirus dramatically increases flAlu RNA by increasing its transcriptional rate (144). At least four adenovirus gene products are required for this activation, which requires altering either the template or the transcription machinery. Alu transcription *in vitro* is severely repressed in chromatin isolated from control cells but not in chromatin isolated from adenovirus-infected cells (150, 151). However, factors present in chromatin isolated from virally infected cells might complicate this interpretation. As an alternative, adenovirus E1A protein increases the relative abundance of one form of TFIIIC, accounting for enhanced Pol III activity in infected cells (160). However, as yet another possibility, adenovirus gene products bind to TBF to relieve p53-mediated repression of certain genes (161). As discussed in Section IV,C,3, p53 represses Alu transcription by binding to TBF presumably in TFIIIB.

In summary, adenovirus infection might stimulate Alu transcription by altering chromatin structure, relieving p53 repression of TFIIIB and activating TFIIIC. Given the magnitude of the stimulation produced by adenovirus infection, it probably acts in several ways.

Whereas the basal level of flAlu RNA is barely detectable, cell stress, including especially heat-shock, transiently increases flAlu RNA without changing the level of scAlu RNA (148). This transient increase precedes the increase in heat-shock mRNAs, but the abundance of other Pol-III-directed RNAs, including especially scAlu RNA, is unaffected. The rapidity and magnitude of this transient increase is also evident in the effect of cycloheximide on the abundance of flAlu RNA (Fig. 5). Within 20 minutes after administration of cycloheximide, there is a noticeable increase in flAlu RNA, which reaches a maximum level after about 2 hours. The rapidity of this response leads us to believe that Alu RNA expression is tightly coupled to the translational state of the cell and is not being increased by one of the many other pleiotropic effects of this drug (Section IV,C,6). Cell stress and translational inhibition also increase the level of SINE RNA in mouse and rabbit cells, indicating that this is a universal response of all mammalian SINE RNAs (142, 162).

6. Proposed Cell-stress-response Role for flAlu RNA

We suspect that this response reflects a translational function for flAlu RNA. The cell-stress response is closely tied to the translational state of the cell, and one of the earliest effects of the cell-stress response is to alter the translational efficiency of different mRNA classes (*163, 164*). The effect of cycloheximide, which does not invoke the full heat-shock response, is interesting in this regard, because it might be considered as an upstream or downstream event in the cell-stress response (*165–167*). Viral infection also alters translation by shifting translation from host to viral messages, whereas the host cell monitoring double-strand RNA responds to infection by generally inhibiting translation (*168*).

A small 7SL-related RNA in *Tetrahymena* induced by heat shock associates with ribosomes providing heat-shock protection (*169*). In addition to their cell-stress response, all mammalian SINEs are related to small RNAs having a translational role, i.e, either 7SL RNA or various tRNAs, and it is easy to imagine that they fulfill a role similar to that of small RNA in *Tetrahymena*.

The action of adenovirus VAI RNA suggests another attractive possibility for the function of SINE RNAs. This small Pol-III-directed transcript binds eIF2 kinase, thereby serving as an antiactivator of host-cell translational inhibition (*168*). The VAI RNA gene is essential for viral propagation but can be functionally replaced by a gene for another entirely unrelated small RNA (*170*). Binding of eIF2 kinase to regions of secondary structure in small RNAs prevents its activation, thereby blocking translational inhibition (*168, 171*). There are two important points: (1) EIF2 kinase is essentially "illiterate" and will probably bind any small RNAs having sufficient secondary structure; (2) on binding this RNA, eIF2 kinase is maintained in an inactive state, thereby blocking one pathway for shutting down translational initiation.

Interestingly, eIF2 kinase is activated by cell stress and is inactivated during cell-stress recovery (*168*). Earlier literature indicates that in HeLa cells, transcription of an RNA having a half-life of about 0.5 hour is required for heat-shock recovery (*172, 173*). That transcript may be flAlu RNA, which could bind to and inactivate eIF2 kinase. This proposed mechanism is independent of the RNA sequence per se, so that nonhomologous RNAs, such as Alu or B2, might each satisfy this same function.

In summary, we find that cell stress increases the abundance of SINE RNA in all mammals, and we suspect that dissimilar SINE RNAs may have identical functions in the response of translational regulation to cell stress or viral infection.

This proposed role for flAlu RNA in translational regulation also potentially satisfies the cheap-gene requirements for Alu function identified above. As exemplified by the effect of cycloheximide, there is an extremely rapid increase in the abundance of flAlu RNA (Fig. 5). Remembering that Alus have inherently weak promoters, this response may require activation of many templates. Furthermore, since the cell-stress response must be mounted in virtually every cell type, the ubiquitous dispersion of Alus ensures that many will always be located within active chromosomal domains.

V. Prospectives and a Speculation

Thirty years ago, results from DNA hybridization produced an inexplicable finding: one-fifth of the DNA in higher eukaryotes consists of redundant sequences. This discovery raised the important issue of whether investigating structure in the absence of function might be a completely sterile exercise. Some of the puzzles produced by the initial discovery of repetitive eukaryotic DNAs have been solved. The structure, origin, and evolutionary history of human repeats have largely been described. Their functions remain enigmatic and the very existence of such functions is still debatable. We may now finally be able to propose and test meaningful hypotheses of interspersed repeat function. Alu repeats are likely to have a normal physiological role either as DNA sequence elements, perhaps in conjunction with their methylation, or as genes encoding full-length Alu RNA, in conjunction with the cell-stress response.

Another possibility deserves consideration. Repeated waves of distinct SINEs have amplified within divergent mammalian lineages, leading to a constant sequence turnover that can now be traced back to the earliest common mammalian ancestors. Phylogenetic comparisons have largely dominated this subject, and naturally we must avoid confusing methodology with conclusions. However, these changes in SINE composition, as well as other repeats, correspond to profound changes in DNA structure and correspondingly may be involved in the evolutionary divergence of these lineages. Considering, for example, the two closely related species, human and chimpanzee, human DNA contains more than 1000 Alus that are not present in chimpanzees, and chimpanzees probably contain a similar number that are not present in humans (*80*). This corresponds to 60,000 specific methylation sites in each species that are derived from Alu repeats (Section IV,B,2). Because coding sequences in chimpanzees and humans are so similar, it has long been believed that differences between these two species must result from differences in the regulated expression of their genes, particularly during development (*174, 175*). Possible roles suggested here for Alu repeats concern either translational regulation and imprinted transcriptional expression during early development. SINEs may not have been mere bystanders during evolution but may have provoked much of the speciation that has proved so useful in their study.

ACKNOWLEDGMENTS

I acknowledge and thank my students for their important contributions, which created and advanced this field. In addition to the accomplishments of these members of the nuclear family, I am pleased to review contributions from so many other labs belonging to our extended Alu family. The excellent monograph recently edited by R. M. Maraia reports the ideas of others in this increasingly exciting field (176).

This research has been generously supported since its inception by USPHS Grant GM 21346, and I greatly appreciate the confidence of my peers in this project. Also, my research has been supported by the Agricultural Experiment Station at the University of California, Davis. Ms. Sheryl Basford deserves special recognition for her expert assistance in preparing this manuscript.

REFERENCES

1. F. Vogel, Nature 201, 847 (1964).
2. T. Ohta and M. Kimura, Nature 223, 118 (1971).
3. R. J. Britten and D. E. Kohne, Science 161, 529 (1968).
4. G. F. Saunders, S. Shirakawa, P. P. Saunders, F. E. Arrighi and T. C. Hsu, JMB 63, 323 (1972).
5. C. W. Schmid and P. L. Deininger, Cell 6, 345 (1975).
6. R. J. Britten and E. H. Davidson, Qrtly. Rev. Biol. 46, 111 (1971).
7. J. G. Wetmur and N. Davidson, JMB 31, 349 (1968).
8. T. I. Bonner, D. J. Brenner, B. R. Neufeld and R. J. Britten, JMB 81, 123 (1973).
9. F. P. Rinehart, T. G. Ritch, P. L. Deininger and C. W. Schmid, Bchem 20, 3003 (1981).
10. H. R. Hwu, J. W. Roberts, E. H. Davidson, and R. J. Britten, PNAS 83, 3875 (1986).
11. A. J. Bendich and R. S. Anderson, Bchem 16, 4655 (1977).
12. C. M. Rubin, E. P. Leeflang, F. P. Rinehart and C. W. Schmid, Genomics 18, 322 (1993).
13. P. L. Deininger and C. W. Schmid, JMB 127, 437 (1979).
14. P. L. Deininger and C. W. Schmid, JMB 106, 773 (1976).
15. S. W. Van Arsdell and A. Weiner, MCBiol 4, 492 (1984).
16. G. Westin, J. Zabielski, K. Hammarstrom, H.-J. Monstein, C. Bark and U. Pettersson, PNAS 81, 3811 (1984).
17. S. Gerbi, in "Molecular Evolutionary Genetics" (R. J. MacIntyre, ed.), p. 419. Plenum, New York, 1985.
18. H. J. Cooke and J. Hindley, NARes 10, 3177 (1979).
19. R. Wevrick and H. W. Willard, PNAS 86, 9394 (1989).
20. G. M. Greig, P. E. Warburton and H. F. Willard, J. Mol. Evol. 37, 464 (1993).
21. W. H. Klein, T. L. Thomas, C. Lai, R. J. Scheller, R. J. Britten and E. H. Davidson, Cell 14, 889 (1978).
22. D. E. Graham, B. R. Neufeld, E. H. Davidson and R. J. Britten, Cell 1, 127 (1974).
23. E. H. Davidson, B. R. Hough, C. S. Amenson and R. J. Britten, JMB 77, 1 (1973).

24. C. M. Houck, F. P. Rinehart and C. W. Schmid, *BBA* **518**, 37 (1978).
25. C. M. Houck, F. P. Rinehart and C. W. Schmid, *JMB* **132**, 151 (1979).
26. C. W. Schmid and W. R. Jelinek, *Science* **216**, 1065 (1982).
27. C. W. Schmid and C.-K. J. Shen, in "Molecular Evolutionary Genetics" (R. J. MacIntire, ed.), p. 323. Plenum, New York, 1985.
28. J. Jurka and A. Milosavljevic, *J. Mol. Evol.* **32**, 105 (1991).
29. A. M. Weiner, P. L. Deininger and A. Efstratiadis, *ARB* **55**, 631 (1986).
30. M. F. Singer, *Cell* **28**, 433 (1982).
31. P. L. Deininger, M. A. Batzer, C. A. Hutchison III and M. H. Edgell, *Trends Genet.* **8**, 307 (1992).
32. C. W. Schmid, N. Deka and A. G. Matera, in "Chromosomes: Eukaryotic Prokaryotic and Viral" (K. W. Adolph, ed.), p. 323. CRC Press, Boca Raton, Florida, 1990.
33. S. L. Mathias, A. F. Scott, H. H. Kazazian, J. D. Boeke and A. Gabriel, *Science* **254**, 1808 (1991).
34. B. A. Dombroski, Q. Feng, S. L. Mathias, D. M. Sassaman, A. F. Scott, H. H. Kazazian and J. D. Boeke, *MCBiol* **14**, 4485 (1994).
35. G. D. Swergold, *MCBiol* **10**, 6718 (1990).
36. B. A. Dombroski, S. L. Mathias, E. Nanthukumar, A. F. Scott and H. H. Kazazian, *Science* **254**, 1895 (1991).
37. B. A. Dombroski, A. F. Scott and H. H. Kazazian, *PNAS* **90**, 6513 (1993).
38. H. Varmus, in "Retroviruses in Mobile Genetic Elements" (J. A. Shapiro, ed.), p. 411. Academic Press, New York, 1983.
39. K. E. Paulson, N. Deka, C. W. Schmid, R. Misra, C. W. Schindler, M. G. Rush, L. Kadyk and L. Leinwand, *Nature* **316**, 359 (1985).
40. F. A. Smit, *NARes* **21**, 1863 (1993).
41. F. A. Smit and A. D. Riggs, *NARes* **23**, 98 (1995).
42. J. Jurka, E. Zietkiewicz and D. Labuda, *NARes* **23**, 170 (1995).
43. R. Miesfeld, M. Krystal and N. Arnheim, *NARes* **9**, 5931 (1981).
44. D. Tauz and M. Renz, *NARes* **12**, 4127 (1984).
45. A. J. Jeffreys, V. Wilson and S. L. Thein, *Nature* **314**, 67 (1985).
46. A. Edwards, H. A. Hammond, L. Jin, C. T. Caskey and R. Chakraborty, *Genomics* **12**, 241 (1992).
47. P. Karran, *Science* **268**, 1857 (1995).
48. R. C. Hardison, I. Sawada, J.-F. Cheng, C.-K. J. Shen and C. W. Schmid, *NARes* **14**, 1903 (1986).
49. N. J. Proudfoot and T. Maniatis, *Cell* **21**, 537 (1980).
50. N. J. Proudfoot, A. Gil and T. Maniatis, *Cell* **31**, 553 (1982).
51. J. Jurka, D. J. Kaplan, C. H. Duncan, J. Walichiewicz, A. Milosavljevic, G. Murali and J. F. Solus, *NARes* **21**, 1273 (1993).
52. C. W. Schmid and R. Maraia, *Curr. Opin. Genet. Dev.: Genomes Evol.* **2**, 874 (1992).
53. W. R. Jelinek and C. W. Schmid, *ARB* **51**, 813 (1982).
54. Y. Quentin, *NARes* **20**, 3397 (1992).
55. M. Sakagami, K. Ohshima, H. Mukoyama, H. Yasue and N. Okada, *JMB* **239**, 731 (1994).
56. E. Frengen, P. Thomsen, T. Kristensen, S. Kran, R. Miller and W. Davis, *Genomics* **10**, 949 (1991).
57. C. H. Duncan, *NARes* **15**, 1340 (1987).
58. M. F. Minnick, L. C. Stillwell, J. M. Heineman and G. L. Stiegler, *Gene* **110**, 235 (1992).
59. D. E. Krane, A. G. Clark, J.-F. Cheng and R. C. Hardison, *Mol. Biol. Evol.* **8**, 1 (1991).
60. N. Okada, *Curr. Opin. Genet. Dev.* **1**, 498 (1991).
61. Y. Yoshioka, N. Okada and Y. Machida, *PNAS* **90**, 6562 (1991).

62. N. Okada and K. Oshima, in "The Impact of Short Interspersed Elements (SINEs) on the Host Genome" (R. M. Maraia, ed.), p. 61. Springer Publ., New York, 1995.
63. G. R. Daniels, G. M. Fox, D. Lowensteiner, C. W. Schmid and P. L. Deininger, NARes 11, 7579 (1983).
64. G. R. Daniels and P. L. Deininger, NARes 11, 7595 (1983).
65. G. R. Daniels and P. L. Deininger, Nature 317, 819 (1985).
66. G. R. Daniels and P. L. Deininger, NARes 19, 1649 (1991).
67. Sawada and C. W. Schmid, JMB 192, 693 (1986).
68. I. Sawada, C. Willard, C.-K. J. Shen, B. Chapman, A. C. Wilson and C. W. Schmid, J. Mol. Evol. 22, 316 (1985).
69. J.-P. Shaw, J. Marks and C.-K. J. Shen, J. Mol. Evol. 33, 506 (1991).
70. A. D. Bailey and C.-K. J. Shen, PNAS 90, 7205 (1993).
71. N. Maeda, C.-I. Wu, J. Blisk and J. Reneke, Mol. Biol. Evol. 5, 1 (1988).
72. D. H. Kass, M. A. Batzer and P. L. Deininger, MCBiol 15, 19 (1995).
73. A. G. Matera, U. Hellmann and C. W. Schmid, MCBiol 10, 5424–5432 (1990).
74. M. A. Batzer, G. E. Kilroy, P. E. Richard, T. H. Shaikh, T. D. Desselle, C. L. Hopkins and P. L. Deininger, NARes 18, 6793 (1990).
75. M. A. Batzer, P. L. Deininger, U. Hellmann-Blumberg, J. Jurka, D. Labuda, Y. Quentin, C. M. Rubin, C. W. Schmid, E. Zietkiewicz and E. Zuckerkandel, J. Mol. Evol. (in press) (1996).
76. A. P. Bird, NARes 8, 1499 (1980).
77. R. J. Britten, W. F. Baron, D. B. Stout and E. H. Davidson, PNAS 85, 4770 (1988).
78. J. Jurka and T. Smith, PNAS 85, 4775 (1988).
79. E. P. Leeflang, W. M. Liu, C. Hashimoto, P. V. Choudary and C. W. Schmid, J. Mol. Evol. 35, 7 (1992).
80. E. P. Leeflang, I. P. Chesnokov and C. W. Schmid, J. Mol. Evol. 37, 566 (1993).
81. E. P. Leeflang, W. M. Liu, I. N. Chesnokov and C. W. Schmid, J. Mol. Evol. 37, 559 (1993).
82. M. A. Batzer, M. Stoneking, M. Alegria-Hartman, H. Bazan, D. H. Kass, T. H. Shaikh, G. E. Novick, P. A. Ioannou, W. D. Scheer, R. J. Herrera and P. L. Deininger, PNAS 91, 12288 (1994).
83. D. H. Kass, C. Aleman, M. A. Batzer and P. L. Deininger, Genetica 94, 1 (1994).
84. M. Batzer, C. Rubin, U. Hellmann-Blumberg, M. Alegria-Hartman, E. P. Leeflang, J. Stern, H. Bazan, T. Shaikh, P. Deininger and C. Schmid, JMB 247, 418 (1995).
85. A. Economou-Pachis and P. N. Tsichlis, NARes 13, 8379 (1985).
86. M. R. Wallace, L. B. Andersen, A. M. Saulino, P. E. Gegory, T. W. Glover and F. S. Collins, Nature 353, 864 (1991).
87. D. Vidaud, M. Vidaud, B. R. Bahnak, V. Siguret, S. G. Sanchez, Y. Laurin, D. Meyer, M. Goosses and J. M. Lavergne, Eur. J. Human Genet. 1, 30 (1993).
88. K. Muratani, T. Hada, Y. Yamamoto, T. Kaneko, Y. Shigeto, T. Ohue, J. Furuyama and K. Higashino, PNAS 88, 11315 (1991).
89. I. Chesnokov and C. W. Schmid, J. Mol. Evol. (in press) (1996).
90. T. H. Shaikh and P. L. Deininger, J. Mol. Evol (in press) (1996).
91. W. M. Liu, E. P. Leeflang, and C. W. Schmid, BBA 1132, 306 (1992).
92. D. Sinnett, C. Richer, J. M. Deragon and D. Labuda, JMB 226, 689 (1992).
93. D.-Y. Chang and R. J. Maraia, JBC 268, 6423 (1993).
94. D.-Y. Chang, B. Nelson, T. Bileu, Karl Hsu, G. Darlington and R. J. Maraia, MCBiol 14, 3949 (1994).
95. D.-Y. Chang, N. Sasaki-Tozawa, L. K. Green and R. J. Maraia, MCBiol 15, 2109 (1994).
96. D. Labuda, E. Zietkiewicz and G. A. Mitchell, in "The Impact of Short Interspersed

Elements (SINEs) on the Host Genome" (R. M. Maraia, ed.), p. 1. Springer Publ., New York, 1995.
97. W. Makalowski, in "The Impact of Short Interspersed Elements (SINEs) on the Host Genome" (R. M. Maraia, ed.), p. 81. Springer Publ., New York, 1995.
98. M. A. Lehrman, J. L. Goldstein, D. Russell and M. S. Brown, Cell 48, 827 (1987).
99. F. Rouyer, M. C. Simmler, D. C. Page and J. Weissenbach, Cell 51, 417 (1987).
100. G. S. Barsh, P. H. Seeburg and R. E. Gelina, NARes 11, 3939 (1983).
101. J. F. Hess, C. W. Schmid and C.-K. J. Shen, Science 226, 67 (1984).
102. W. Makalowski, G. A. Mitchell and D. Labuda, Trends Genet. 10, 188 (1994).
103. C. J. Harendza and L. F. Johnson, PNAS 87, 2531 (1990).
104. K. E. Paulson, A. G. Matera, N. Deka and C. W. Schmid, NARes 15, 5199 (1987).
105. H. H. Kazazian, C. Wong, H. Youssoufian, A. F. Scott, D. G. Phillips and S. E. Antonarakis, Nature 332, 164 (1988).
106. B. Morse, P. G. Rothberg, V. J. South, J. M. Spandorfer and S. M. Astrin, Nature 333, 87–90 (1988).
107. P. L. Deininger and M. A. Batzer, in "Evolutionary Biology" (M. K. Hecht, .ed.), p. 157. Plenum, New York, 1993.
108. M. W. Hull, J. Erickson, J. Johnston and D. R. Engelke, MCBiol 14, 1266 (1994).
109. J. Wu, G. J. Grindlay, P. Bushel, L. Mendelsohn and M. Allan, MCBiol 10, 1209 (1990).
110. J. E. Hambor, J. Mennone, M. E. Coon, J. H. Hanke and P. Kavathas, MCBiol 13, 7056 (1993).
111. J. H. Kim, C. Y. Yu, A. Bailey, R. Hardison and C.-K. J. Shen, NARes 17, 5687 (1989).
112. F. Vidal and F. Cuzin, in "The Impact of Short Interspersed Elements (SINEs) on the Host Genome" (R. M. Maraia, ed.), p. 81. Springer Publ., New York, 1995.
113. W. F. Doolittle and C. Sapienza, Nature 284, 601 (1980).
114. L. E. Orgel and F.H.C. Crick, Nature 284, 604 (1980).
115. B. F. Koop, E. E. Miyamoto, J. E. Embury, M. Goodman, J. Czelusniak and J. L. Slightom, J. Mol. Evol. 24, 94 (1986).
116. R. J. Britten, PNAS 91, 5992–5996 (1994).
117. E. Zimmer, S. Martin, S. Beverly, Y. Kan and A. Wilson, PNAS 77, 2158 (1980).
118. E. Zuckerkandl, G. Latter and J. Jurka, J. Mol. Evol. 29, 504 (1989).
119. G. P. Holmquist, Am. J. Human Genet. 51, 17 (1992).
120. J. R. Korenberg and M. C. Rykowski, Cell 53, 391 (1988).
121. M. Ehrlich, M. A. Gama-Sosa, L. H. Huang, R. M. Midgett, K. C. Kuo, R. A. McCune and C. Gehrke, NARes 10, 2709 (1982).
122. C. W. Schmid, NARes 19, 5613 (1991).
123. C. M. Rubin, C. A. VandeVoort, R. L. Teplitz and C. W. Schmid, NARes 22, 5121 (1994).
124. U. Hellmann-Blumberg, M. F. Hintz and C. W. Schmid, MCBiol 13, 4523 (1993).
125. S. Kochanek, D. Renz and W. Doerfler, EMBO J. 12, 1141 (1993).
126. U. Hellmann-Blumberg, "Detection of Sequence Variations and Methylation Patterns of Human Alu Repeats." Ph.D. Thesis, U. C. Davis, California, 1994.
127. I. Chesnokov and C. W. Schmid, JBC 270, 18539 (1995).
128. M. Monk and M. Grant, Development (Suppl.) 55 (1990).
129. M. Monk, M. Boubelik and S. Lehnert, Development 99, 371 (1987).
130. R. Shemer, T. Kafri, A. O'Connell, S. Eisenberg, J. L. Breslow and A. Razin, PNAS 88, 11300 (1991).
131. T. Kafri, M. Ariel, M. Brandeis, R. Shemer, L. Urven, J. McCarrey, H. Cedar and A. Razin, Genes Dev. 6, 705 (1992).
132. S. K. Howlett and W. Reik, Development 113, 119 (1991).
133. D. J. Driscoll and B. R. Migeon, Som. Cell Mol. Genet. 16, 267 (1990).

134. M. A. Surani, *Nature* **366**, 302 (1993).
135. E. Li, C. Beard and R. Jaenisch, *Nature* **366**, 362 (1993).
136. R. Stoger, P. Kubicka, C. Liu, T. Kafri, A. Razin, H. Cedar and D. Barlow, *Cell* **73**, 61 (1993).
137. S. Varmuza and M. Mann, *Trends Genet.* **10**, 118 (1994).
138. D. P. Barlowe, *Trends Genet.* **10**, 194 (1994).
139. H. Heller, C. Kammer, P. Wilgenbus and W. Doerfler, *PNAS* **92**, 5515 (1995).
140. A. Haase and W. A. Schulz, *JBC* **269**, 1821–1826 (1994).
141. W. M. Liu, R. J. Maraia, C. M. Rubin and C. W. Schmid, *NARes* **22**, 1087 (1994).
142. W.-M. Liu, W.-M. Chu, P. Choudary and C. W. Schmid, *NARes* **23**, 1758 (1995).
143. K. L. Jang, M. Collins and D. Latchman, *J. Acquired Immune Defic. Syndrome.* **5**, 1142 (1992).
144. Panning and J. R. Smiley, *MCBiol* **13**, 3231 (1993).
145. W. M. Liu and C. W. Schmid, *NARes* **21**, 1351 (1993).
146. E. Ullu and A. M. Weiner, *EMBO J.* **3**, 3303 (1984).
147. E. Ullu and A. M. Weiner, *Nature* **318**, 371 (1985).
147a. S. Bredow, D. Surig, J. Muller, H. Kleinert and B. J. Benecke, *NARes* **18**, 6779 (1990).
148. W.-M. Chu, W.-M. Liu and C. W. Schmid, *NARes* **23**, 1750 (1995).
149. R. Maraia, D. Chang, A. Wolffe, R. Vorce and K. Hsu, *MCBiol* **12**, 1500 (1992).
150. E. W. Englander, A. P. Wolffe, and B. H. Howard, *JBC* **268**, 19565 (1993).
151. V. R. Russanova, C. T. Driscoll and B. H. Howard, *MCBiol* **15**, 4282 (1995).
152. I. Chesnokov, W. M. Chu and C. W. Schmid, unpublished.
153. T. Platt, *ARB* **55**, 339 (1986).
154. D. Sinnett, C. Richer, J. M. Deragon and D. Labuda, *JBC* **266**, 8675 (1991).
155. R. J. Maraia, C. Driscoll, T. Bilyeu, K. Hsu and G. J. Darlington, *MCBiol* **13**, 4233 (1993).
156. R. J. Maraia, D. J. Kenan and J. D. Keene, *MCBiol* **14**, 2147 (1994).
157. J. A. Martignetti and J. Brosius, *PNAS* **90**, 11563 (1993).
158. J. Brosius, *Science* **251**, 753 (1991).
159. M. F. Carey, K. Singh, M. Botchan and N. R. Cozzarelli, *MCBiol* **6**, 3068 (1986).
160. E. Sinn, Z. Wang, R. Kovelman and R. G. Roeder, *Genes Dev.* **9**, 675 (1995).
161. N. Horikoshi, A. Usheva, J. Chen, A. Levine, R. Weinman and T. Shenk, *MCBiol* **15**, 227 (1995).
162. M. Fornace and J. Mitchell, *NARes* **14**, 5793 (1986).
163. V. M. Pain and M. J. Clemens, in "Translation In Eukaryotes" (H. Trachsel, ed.), p. 292. CRC Press, Boca Raton, Florida, 1991.
164. C. Georgopoulos and W. J. Welch, *Annu. Rev. Cell Biol.* **9**, 601 (1993).
165. Y. J. Lee and W. C. Dewey, *J. Cell. Physiol.* **132**, 41 (1987).
166. P. Crete and J. Landry, *Radiat. Res.* **121**, 320 (1990).
167. R. Baler, W. J. Welch and R. Voellmy, *J. Cell. Biol.* **117**, 1151 (1992).
168. R. J. Jackson, in "Translation In Eukaryotes" (H. Trachsel, ed.). CRC Press, Boca Raton, Florida, 1991.
169. P. A. Fung, J. Gaertig, M. A. Gorovsky and R. L. Hallberg, *Science* **268**, 1036 (1995).
170. R. A. Bhat and B. Thimmappaya, *J. Virol.* **56**, 750 (1985).
171. P. A. Clarke, M. Schwemmle, J. Schickinger, K. Hilse and M. J. Clemens, *NARes* **19**, 243 (1991).
172. W. McCormick and S. Penman, *JMB* **39**, 315 (1968).
173. E. S. Goldstein, M. E. Reichman and S. Penman, *PNAS* **71**, 4752 (1974).
174. A. C. Wilson, L. R. Maxson and V. M. Sarich, *PNAS* **71**, 2843 (1974).
175. A. C. Wilson, V. M. Sarich and L. R. Maxson, *PNAS* **71**, 3028 (1974).

176. R. Maraia, "The Impact of Short Interspersed Elements (SINEs) on the Host Genome." Springer Publ., New York, 1995.
177. K. W. Jones, *J. Med. Genet.* **10,** 273 (1973).
178. L. Manuelidis, *NARes* **3,** 3063 (1976).
179. J. S. Waye and H. F. Willard, *NARes* **14,** 6915 (1986).
180. P. K. Wellauer and I. B. Dawid, *JMB* **128,** 289 (1979).

Recent Advances in the Molecular Biology of Vitamin D Action

HISHAM M. DARWISH AND
HECTOR F. DELUCA

Department of Biochemistry
University of Wisconsin—Madison
Madison, Wisconsin 53706

I. The Vitamin D Receptor	322
II. Vitamin D Receptor–DNA Interaction	325
III. Transcriptional Regulation Mechanism by Vitamin D	329
IV. The Molecular Biology of Vitamin D Hydroxylases	333
V. Summary	338
References	339

Vitamin D was clearly the first vitamin to be identified as a precursor of a steroid hormone. The first evidence of this was the demonstration that vitamin D_3 is normally produced in skin from 7-dehydrocholesterol by ultraviolet light originating from the sun (1, 2). Thus, this vitamin is a vitamin only when animals are not exposed to ultraviolet light. Furthermore, it is biologically inert until modified by 25-hydroxylation in liver and the subsequent 1α-hydroxylation in the proximal convoluted tubule cells of the kidney (3). Conclusive evidence that the product, 1,25-dihydroxy vitamin D_3 [1,25-$(OH)_2D_3$] is a true hormone was the demonstration that this product is not produced in anephric animals; furthermore, in the absence of kidneys, 25-hydroxy vitamin D (25-OH-D) has little biological activity, whereas 1,25-$(OH)_2D_3$ is fully active (4, 5). These results show that the active form of vitamin D is produced in the kidney and has its action on cells and organs distal from its site of production.

The action of vitamin D at the physiologic level is to elevate plasma calcium and phosphorus to levels that are required to support normal neuromuscular function and to support normal mineralization of the skeleton (6–8). Final proof that the vitamin D system is a true endocrine system with a true steroid hormone product was the finding that the production of 1,25-$(OH)_2D_3$ is strongly feedback-regulated by the need for calcium or the need for phosphorus. Thus, hypocalcemia produces a marked stimulation of the enzyme that carries out 1α-hydroxylation to produce the final hormone, and, similarly, hypophosphatemia also stimulates the production of 1,25-$(OH)_2D_3$ (9, 10).

Thus, the vitamin D system joins the other steroid hormones on the basis of its structure and in its mode of action. The vitamin D mechanism studies, therefore, benefited greatly from the massive information garnered for estrogen, progesterone, and the glucocorticoid hormones. A nuclear receptor for the vitamin D hormone was first demonstrated in the mid-1970s (*11, 12*), and since then a great deal of progress has been made in our understanding of the molecular mechanism of action of $1,25\text{-}(OH)_2D_3$ (*13–16*).

To avoid duplication of material, readers are directed to other earlier reviews of this molecular mechanism of action; it is the purpose of this review to provide the more recent advances in this research. Clearly, much remains to be learned of the molecular mechanism of action, thus this report also serves to note where the next advances might be expected.

I. The Vitamin D Receptor

Progress in our understanding of the molecular mechanism of action of vitamin D has recently been determined largely by our understanding of the function and regulation of the $1,25\text{-}(OH)_2D_3$ receptor (*13–16*). The vitamin D receptor (VDR) belongs to the superfamily of ligand-dependent transcription factors, and it mediates the activity of $1,25\text{-}(OH)_2D_3$ in the regulation of gene expression. Presumably, a tissue or cell that expresses the receptor is a potential target of vitamin D action. Using radiolabeled $1,25\text{-}(OH)_2D_3$ and autoradiography, several such tissues have been identified and confirmed by direct measurement of receptor. However, the exact role of vitamin D in many of these putative target sites remains to be defined.

The low abundance of the receptor in target tissues, including the intestine, prevented significant advances in studying its structure, function, and regulation (*17*). A major advance occurred with partial purification of the receptor, allowing production of monoclonal antibodies against the porcine (*18*) and chicken (*19*) receptors. These monoclonal antibodies provided an important tool for cloning the cDNAs, for recombinant receptor expression, purification, regulation, and phosphorylation. Using the porcine receptor antibodies, a specific immunoradiometric assay was also developed for the receptor (*20*). This allowed quantitation of total concentration of receptor protein in target tissues and cell lines (*4*), which provided a reliable approach for studying receptor regulation (*21–23*).

Perhaps the most important use of the receptor antibodies was in the cloning of receptor cDNA from different species. Thus, the amino-acid sequences were deduced for a portion of the chicken receptor (*24*) and the full-length human (*25*) and rat receptors (*26*). More recently, the full-length avian (*27*) and mouse (*28*) VDR sequences were determined. The amino-acid sequence showed that the receptor has functional domains similar to other steroid hormone receptors, namely, a well-defined amino-terminus DNA-binding domain (C domain) and a less-defined carboxy-terminus ligand-binding domain (E domain) (25–28). The availability of the receptor cDNAs also permitted recombinant receptor expression, determination of other functional domains, determination of mutations in the receptor that result in vitamin D-dependent rickets Type II, and studies of the mechanism of receptor-mediated gene expression in response to 1,25-$(OH)_2D_3$ (13–16).

Interestingly, two avian receptor protein forms have been detected in the chicken by immunoblotting (24, 27, 29), whereas only one receptor protein has been detected in the other species studied thus far. This is similar to the progesterone receptor, of which two forms were also identified in target tissues (30–32). It is not yet clear whether the two chicken receptor forms come from one or two genes or whether they are generated by differential transcription initiation or by two translational initiation sites. Initial studies suggest that the two forms might be produced by two translation sites from one mRNA (27). However, more work is required before this conclusion can be reached. Furthermore, whether the two chicken receptor forms are expressed in all tissues, linked to developmental changes, or have differential functions is still not known. Recent reports on the two forms of the progesterone receptor suggest that they have differential effects and that one form (form A) is an inhibitor of the other (form B) (33, 34). There is no information available yet on the functional significance of the two forms of the chicken receptor proteins.

The presence of the receptor in low abundance *in vivo* presented a major obstacle to the purification of adequate quantities of the protein for detailed structural and functional studies (35–37). Determination of the receptor's three-dimensional structure will be one of the most important bits of new information required for a thorough understanding the molecular mechanism of action of 1,25-$(OH)_2D_3$. In addition, the structure of the ligand-binding domain will provide a powerful tool for designing vitamin D analogs. Over the past several years, many synthetic analogs of 1,25-$(OH)_2D$ have been prepared (15). The main objective has been to generate new vitamin D compounds that are selective in their activation of specific genes or organs. Initially, the main objective was to prepare analogs that can induce differentiation without causing hypercalcemia. These analogs would be potential therapeutic agents for cancer and psoriasis without the risk of causing hypercalcemia. Several compounds meeting this criterion have been prepared, but the basis of the design of these compounds is intuitive and not based on a knowledge of the ligand binding domain of the receptor.

In order to determine the structure of the VDR, large quantities of pure receptor are needed for either crystallography or NMR. NMR has been used successfully to determine the structure of the DNA-binding domain of the estrogen receptor and the retinoic acid X receptor α (RXRα) DNA-binding domain (38, 39). More recently, the crystal structure of the human RXR ligand-binding domain has been determined (39a). These studies require substantial amounts of pure protein. The VDR has been expressed in baculovirus (40, 41), bacteria (42, 43), yeast (44), and mammalian (45) systems. The bacterial system produces high levels of the expressed receptor protein; however, its affinity for 1,25-$(OH)_2D_3$ was lower than the native receptor (43). The authors also indicated that the receptor fraction precipitated during purification. Therefore, it is difficult to evaluate the fraction of the expressed receptor that is active in ligand and DNA binding and thus possesses the native receptor structure. The VDR expressed by the baculovirus system is not only more native-like, but is produced in high yields. The baculovirus-expressed receptor can bind ligand and DNA similar to the wild-type receptor (40, 41). In addition to the full-length VDR, which may present some technical problems in the structural studies, expression and purification of the separate functional domains (DNA-binding domain and ligand-binding domain) separately provide another option, as has been done with other members of the steroid receptor family. One major advance in this field is the availability of bacterial gene-fusion expression systems, which produce high levels of fusion proteins that can be cleaved enzymatically to produce the desired protein following affinity purification. These systems produce large quantities of pure expressed proteins.

Currently, large quantities of the baculovirus-expressed rat receptor and receptor functional domains expressed as fusion proteins in bacteria are being generated and purified on affinity columns for structural and functional experiments (S. Strugnell and H. F. DeLuca, unpublished). With these ongoing efforts, it is expected that the three-dimensional structure of the VDR and its components will be available in the near future. These structures will provide clear information of the detailed molecular identity of the contact groups between the receptor, DNA, and possibly other proteins. Such information has already been provided for the glucocorticoid and estrogen receptors (46–48).

Because the VDR plays a central role in mediating 1,25-$(OH)_2D_3$ function, factors that regulate its expression in target tissues are of obvious importance. Most efforts have centered on the regulators of calcium metabolism, i.e., 1.25-$(OH)_2D_3$ calcium, phosphorus, and parathyroid hormone (13–16). Treatment with 1,25-$(OH)_2D_3$ increases VDR levels *in vivo* and *in vitro*. This appears to result from a stabilization of the receptor by its ligand (22, 23, 49). Some investigators have reported that 1,25-$(OH)_2D_3$ increases receptor mRNA (21, 24, 50). Thus, the regulation of receptor by 1,25-$(OH)_2D_3$ may result from more than one mechanism.

The VDR is rapidly phosphorylated in response to 1,25-(OH)$_2$D$_3$ (29). This phosphorylation takes place on serine residues in the ligand/hinge domains (51). The exact amino-acid residues that become phosphorylated *in vivo* in response to 1,25-(OH)$_2$D$_3$ have not been clearly determined. It seems likely that phosphorylation of the VDR plays an important role in transcriptional activity, but that has not been clearly established (52, 53). There have been several attempts to identify the sites of phosphorylation in the VDR, but the results of these studies are conflicting (3). Hilliard *et al.* reported a direct attempt to identify the active phosphorylation site(s) in the human VDR (54). These authors used labeled receptor and tryptic digestion followed by amino-acid analysis to determine that serine-205 is the primary phosphorylation site. However, when this residue was mutated, the mutant receptor was transcriptionally active, but was phosphorylated at an alternate site. Because of the different phosphorylation, it was not possible to determine whether phosphorylation is absolutely required for transcriptive activity of the receptor. Furthermore, the exact kinase(s) catalyzing phosphorylation of the receptor *in vivo* is (are) not known.

Due to the difficulty experienced by all investigators in obtaining the 5' end sequence of any of the cloned VDR cDNAs, the promoter region has not been identified for any cloned VDR genes. This represents a major deficiency in the study of the VDR regulation at the transcriptional level. Another deficiency is the lack of an animal model of vitamin-D-resistant rickets Type II. Cells that do not contain a functional VDR can be used to evaluate the proposed nongenomic action of 1,25-(OH)$_2$D$_3$. Further, cells without a functional receptor can be used to determine the activity of mutant forms of VDR.

II. Vitamin D Receptor–DNA Interaction

In the past few years, an analysis of steroid receptor binding to response elements has been at the center of the investigations of steroid hormone mechanism. Much of the work has focused on the structural and sequence determinants of the response elements that determine the specificity of their binding to the receptors. The steroid receptors are divided into two major groups based on the sequence and nature of their responsive elements. The glucocorticoid receptor group binds to the sequence AGAACA, whereas the estrogen group binds either AGGTCA or AGTTCA (55–57). In another classification, based on the nature of the half-sites, one group that includes the glucocorticoid and estrogen receptors has palindromic responsive elements, whereas the second group, which includes the thyroid, retinoid, and vitamin D receptors, binds to responsive elements that are composed of direct repeats (58, 59). The spacer region between the repeats is critical to the specificity of receptor binding to the responsive elements. Thus, VDR binding requires a 3-base intervening sequence, whereas the thyroid hormone receptor (TR) requires a 4-base separation. Retinoic acid receptors (RARs) use a 1-, 2-, or 5-base separation (60–63) and RXRα utilizes a 1-base separation (64). The molecular determinants that play a role in the specific recognition by the steroid receptors to their responsive elements have been reviewed recently (57, 65, 66). It is apparent through the gathered information that the binding of the steroid receptors to the specific elements in the target genes is influenced by a complicated array of factors that allow for flexibility and selectivity. This becomes especially important in tissues that are responsive to several steroid hormones.

Although vitamin D exerts a wide range of biological activity in several target tissues (15), only a small number of responsive genes have been described and cloned. Vitamin D response elements (VDREs) were identified and characterized in the promoter region of these genes. Table I shows the updated list of identified VDREs (67–79). Recently, new VDREs were identified in the upstream promoter regions of the mouse calbindin D28K gene (76), the chicken carbonic anhydrase II gene (77), and the human calbindin D9K gene (78). The newly identified VDRE in the mouse calbindin D28K gene (76) is different from the previously identified VDRE in the same gene (79) and lies further upstream from the transcription start site. This new VDRE is more homologous to the other previously identified VDREs, having two repeat elements separated by three nucleotides, whereas the other

TABLE I
SEQUENCES OF VITAMIN D RESPONSIVE ELEMENTS

	Sequence						Ref.
CaBP 9K			GGGTGT	CGG	AAGCCC		67
Rat osteocalcin			GGGTGA	ATG	AGGACA		68
Human osteocalcin			GGGTGA	ACG	GGGGCA		69, 70
Mouse osteopontin			GGTTCA	CGA	GGTTCA		71
Rat 24-hydroxylase, distal			GGTTCA	GCG	GGTGCG		72
Human 24-hydroxylase, distal			AGTTCA	CCG	GGTGTG		73
Rat 24-hydroxylase, proximal	GAGTCA	GCG	AGGTGA	GTG	AGGGCG		74
Human 24-hydroxylase, proximal	GAGTCA	GCG	AGGTGA	GCG	AGGGCG		73
Human PTH			GGTTCA	AAG	CAGACA		75
Mouse calbindin D28K			AGGTGA	TGA	AAGTCA		76
Chicken carbonic anhydrase-II			GGGGA	AAA	AGTCCA		77
Human calbindin D9K			TGCCCTTCCTTATGGGGTTCA				78
Mouse calbindin D28K			CTGGGGGATGTGAGGAGAAATGAGTCTGAGC				79

contains two DNA elements separated by four nucleotides. Takeda *et al.* (76) did not detect any 1,25-$(OH)_2D_3$ responses in the putative VDREs identified by Gill and Christakos (79), who did not examine the region where the new VDRE was identified. Although both groups used different cells in their studies, this does not seem to explain the difference in their findings. Among all the identified VDREs to date, the rat and human 24-hydroxylase DREs remain the most active members of these elements (72–74).

Binding of the VDR, in combination with other factors, to its responsive elements probably represents the central step in the transmission of a signal to the transcription machinery, resulting in activation or suppression of transcription. Very likely the three-dimensional structure of the DNA-bound receptor also plays a significant role in the nature and extent of the signal. Expression of the recombinant human and rat VDR in yeast and insect cells led to the discovery that a mammalian accessory factor is required for binding of the VDR to its responsive elements (80, 81). This additional factor is well recognized to be the retinoid X receptor (82, 83). Heterodimerization with RXR is also evident with the other members of the nuclear receptors, including the thyroid and retinoic acid receptors in binding to their respective response elements (82, 83). The dimerization of the VDR, RAR, or TR with RXR is mediated through specific dimerization domains located in the ligand-binding domains of these receptors (84–87). It was shown that amino-acid residues between 382 and 403 in the VDR are essential for dimerization with RXR, and that amino acid residues between 403 and 427 are required for *trans*-activation (84). Disruption of dimerization leads to the loss of the VDR *trans*-activation activity (86). The ability of the VDR to bind response elements with the correct three-nucleotide spacing is inherent in the DNA-binding domain (85–89). However, the DNA-binding domain alone cannot impart transcriptional activity, because the C-terminal region of the receptor is essential for this function (84). Apparently, VDR and RXR bind in an ordered fashion to the VDREs: RXR binds to the 5′ half-site and the VDR binds to the 3′ half-site (90). This is similar to the orientation polarity of the RAR and TR heterodimers when bound to their response elements (88, 91–93). Interestingly, the polarity of the VDR–RXR on the VDRE of the human carbonic-anhydrase gene is in reverse order: VDR binds to the 5′ half-site and RXR binds to the 3′ half-site (77). This reverse polarity is associated with an orientation-dependent activity of this VDRE (77). It should be noted that the carbonic-anhydrase II VDRE is directed in the opposite direction in comparison to the mouse osteopontin VDRE and other VDREs in relation to the transcription start site. It is not clear what role polarity plays in the mechanism of 1,25-$(OH)_2D_3$ gene *trans*-activation; however, it may allow the proper interaction of the VDR with other transcription and tissue-specific factors that are involved in the expression of vitamin D responsive genes in the different target tissues. The VDR binds palindromic sequences *in vitro* as a homodimer (*94–97*). However, the significance of this finding is uncertain, because such a VDRE has not been found in nature. Furthermore, the VDR forms heterodimers with the thyroid receptor, with the polarity of the heterodimer for the two half-sites depending on the specific VDRE (*97*). Moreover, the VDR was also reported to form heterodimers with the retinoic acid receptor (*98, 99*). Whether these *in vitro* gestures have any *in vivo* significance remains to be determined.

It is clear that all VDREs including the human preproparathyroid consist of repeat elements with three-nucleotide spacings, and none of the identified VDREs is palindromic. In addition, much of the reported binding data lack the identification of proteins in the complexes. Using specific antibodies is one direct way to obtain this information. In fact, the components comprising the VDR–VDRE complex appear to be more complicated than simply being a VDR–RXR heterodimer bound to DNA. Binding studies using semipurified RXR and VDR appear to be inadequate to produce a complex with the VDREs unless mammalian nuclear extract is added. This indicates a requirement of additional factor(s) for complex formation (C. Zierold and H. F. DeLuca, unpublished data). Furthermore, using specific antibodies, the transcription factor TFIIB was detected in these complexes (C. Zierold and H. F. DeLuca, unpublished data). Because the TFIIB transcription factor is a component of the transcriptional machinery complex, it is probably part of the final transcription complex. It is evident that the TFIIB is directly involved in the vitamin-D-mediated transcriptional mechanism, whereby it interacts directly with the VDR (*100, 101*). Similarly, TFIIB interacts with the thyroid hormone receptor (*102*), and plays a direct role in the progesterone transcriptional mechanism through direct interaction with the progesterone receptor (*103*). The TFIIB presumably acts as a bridge between these receptors and the transcription machinery complex. These new findings of polarity, receptor interactions, and transcription factor involvement illustrate the degree of complexity of how genes are regulated by the VDR and the other members of the nuclear receptor family.

Perhaps one of the most important factors in receptor-mediated transcription is the role played by the hormone. Despite years of investigation, the exact role of the ligand in the transcription activation process is not clearly understood. The VDR can bind to its responsive elements in the absence of the hormone *in vitro*, whereas the presence of the hormone increases its affinity to DNA under higher ionic-strength conditions (*104, 105*; H. M. Darwish, H. F. DeLuca, and C. Jehan-Kimmel, unpublished results). This is consistent with the receptor's nuclear localization, where it is associated with the chromatin. Although there are reports that VDR heterodimerizes with other receptors, it is clear that RXR and other unknown nuclear factors are required for formation of heterodimer or multimer complexes on the VDREs (C. Zierold and H. F. DeLuca, unpublished).

Whether these other factors in the nuclear extracts are ligand-binding proteins is still unknown. Certainly the effect of 9-*cis*-retinoic acid (the ligand for RXR) on the transcriptional activity of the VDR–RXR complex is not clear. It has been suggested that 1,25-$(OH)_2D_3$ stabilizes the VDR–RXR heterodimer complex, whereas the presence of 9-*cis*-retinoic acid disrupts heterodimer formation and stabilizes RXR homodimer formation (*106*). This apparent disruption of the heterodimer is associated with loss of transcriptional activity by the VDR (*106*). However, this effect was observed in the presence of relatively high concentrations (1 μM) of the 9-*cis* compound, which makes its *in vivo* relevance questionable.

On the other hand, other studies showed variable effects of 9-*cis*-retinoic acid on the vitamin-D-mediated gene expression. This effect varied from being inhibitory (*107*), having no effect (*97*), or causing a stimulatory effect (*77, 78, 95*). Given the available data, it is premature to reach conclusions on how the nuclear receptors interact *in vivo* in the control of gene expression. This is further complicated by whether the ligands are present and how they affect this interaction. Because phosphorylation of the receptor follows ligand binding, it does not seem to be involved in the receptor binding (*108*). However, the exact role played by phosphorylation in the VDR dimerization with RXR and binding with DNA is not known.

III. Transcriptional Regulation Mechanism by Vitamin D

Steroid hormone receptor regulation of gene expression has been thought to involve binding of a hormone to the receptor, causing its activation and binding to specific DNA sequences. The interaction with other transcription factors then causes *trans*-activation or *trans*-suppression of responsive genes. For cytosolic receptors such as the glucocorticoid and progesterone receptors, translocation to the nucleus in response to ligand precedes its effect on gene expression. This apparently does not apply to the VDR that is localized primarily in the nucleus. Further, there is no evidence that the VDR is associated with heat-shock proteins. Figure 1 presents a model of our current understanding of vitamin-D-mediated gene regulation based on experimental results. First of all, VDR can bind to VDREs in the absence of 1,25-$(OH)_2D_3$ (*81*), although the tightness of binding may be increased by the hormone (*81, 105;* H. M. Darwish, H. F. DeLuca and C. Jehan-Kimmel, unpublished results). Of some interest is the fact that the affinity of the VDR for rat osteocalcin VDRE appears lower in the presence

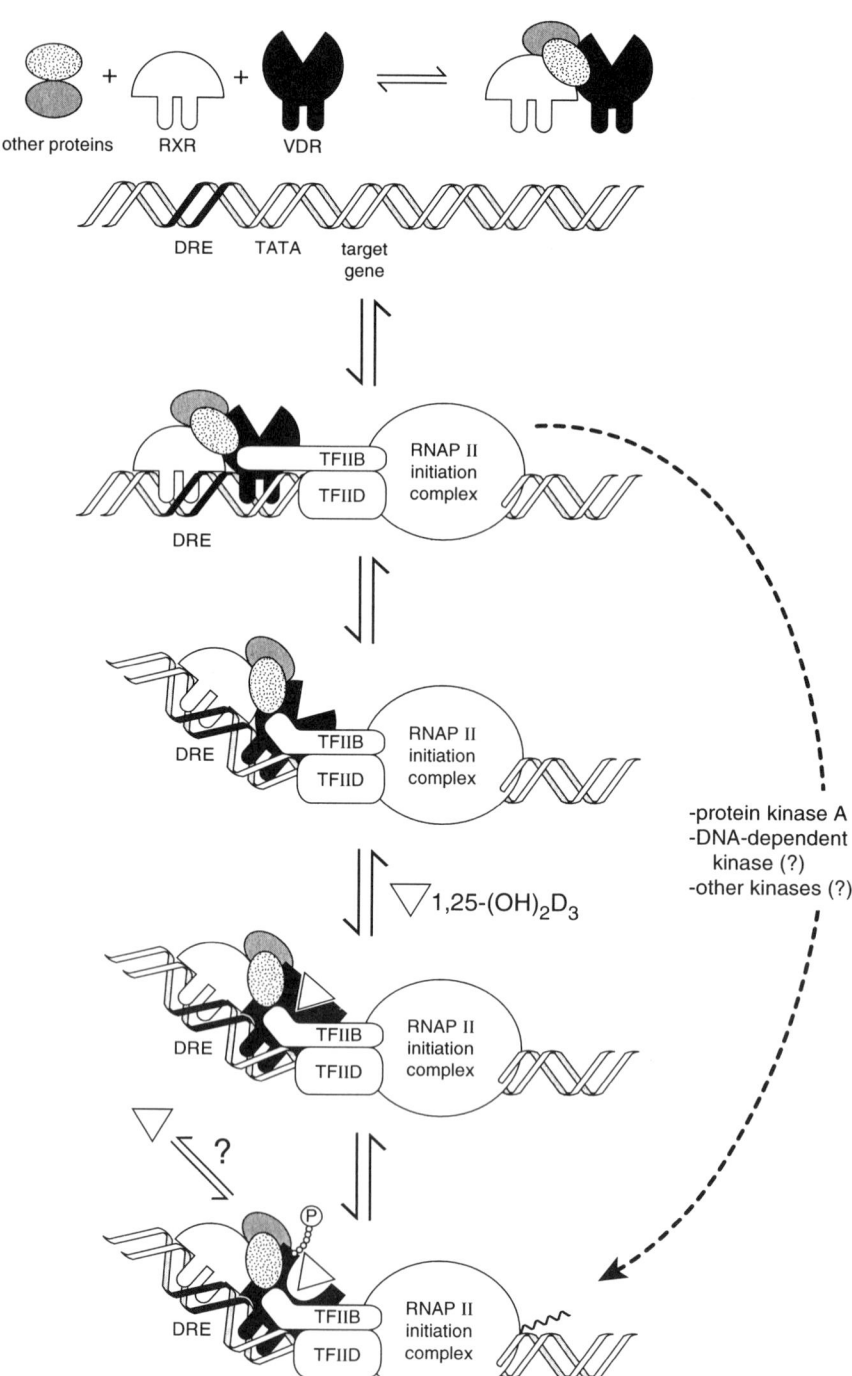

of 1,25-$(OH)_2$-24-epi-D_2 than in the presence of 1,25-$(OH)_2D_3$, even though the epi-compound is more active than 1,25-$(OH)_2D_3$ in inducing expression of the osteocalcin gene (109). Therefore, the change in receptor affinity to DNA seems to depend on the individual VDRE.

The VDR binds to the VDREs identified in the D-responsive genes as a VDR–RXR heterodimer that constitutes the transcriptionally active complex. Whether the VDR–RXR complex is formed prior to binding to DNA or after is not known. VDR and RXR form heterodimers in the absence of DNA (80); whether DNA-free heterodimers are formed in the nucleus is not known. In addition to RXR, other nuclear protein(s) seem to be required to allow the receptor to bind VDREs (C. Zierold and H. F. DeLuca; and H. M. Darwish and H. F. DeLuca, unpublished results). The exact identity of these protein(s) is not yet known, although one of these factors seems to be TFIIB (C. Zierold and H. F. DeLuca, unpublished results). This is consistent with recent studies demonstrating that VDR interacts directly with TFIIB (100, 101). A clear correlation between the VDR and TFIIB in *trans*-activation was also shown (100, 101). These studies suggest that the VDR communicates with the transcription machinery through direct interaction with TFIIB.

Next, VDR is believed to bind the hormone, followed by rapid phosphorylation in the ligand-binding domain (29, 51). As already discussed, the exact function of this phosphorylation is unknown. One of the obstacles has been the lack of exact identification of the serine residue(s) that are phosphorylated in response to 1,25-$(OH)_2D_3$. It is clear from work on the human VDR that alternate phosphorylation sites are evident (54). Previous studies showed that phosphorylation of the VDR caused by activation of protein kinase A results in a VDR-dependent increase in transcription of a reporter gene (52). Similar results were reported for the progesterone and retinoic acid receptors (53, 110). It is still not clear which kinase(s) are responsible for VDR phosphorylation *in vivo*.

The proposed model suggests that the substrate for phosphorylation is the receptor–DNA complex that may involve a DNA-dependent protein

---

FIG. 1. A model for vitamin-D-mediated gene expression. This model shows the vitamin D receptor (VDR) bound to the vitamin D response element (DRE) complexed with the retinoic acid X receptor (RXR). The RXR binds to the 5′ half site of the VDRE and the VDR binds to the 3′ half site. The RXR–VDR heterodimer binds the VDRE in the absence of 1,25-$(OH)_2D_3$. Other factor(s) seem to be required for the complex formation with VDREs, but their identity is not yet known. The VDR–RXR complex interacts with the transcription machinery through direct interaction between the VDR and the transcription factor TFIIB. The binding of 1,25-$(OH)_2D_3$ to the VDR induces receptor phosphorylation. The phosphorylated receptor bound to ligand and to the other proteins stimulates transcription.

kinase. Such a kinase was suggested to play a role in phosphorylation of the progesterone receptor (111, 112). Whether the presence of hormone serves only to induce phosphorylation of the receptor, which then becomes a hormone-independent transcription factor, requires additional investigation. Besides the activation mechanism, the molecular steps involved in the deactivation process are not known. Whether the vitamin D receptor recycles or is degraded after inducing gene transcription needs to be determined. Although phosphorylation plays some role in the activation process, it is possible that it may also serve as a signal to the receptor degradation and thereby turn down the expression of genes. It is evident that the VDR is more stable when bound to 1,25-$(OH)_2D_3$ (22, 23). However, the fate of 1,25-$(OH)_2D_3$ after inducing transcription through the receptor may be a determining factor in the transcription deactivation process.

Besides its effect on gene regulation through the nuclear receptor, another or additional effects were suggested to involve a nongenomic mechanism (113–116). Protein kinase C is suggested to mediate the effects of 1,25-$(OH)_2D_3$ on these apparently receptor-independent processes (117–119). This is supported by the observation that these effects of 1,25-$(OH)_2D_3$ are similar to the effects of other activators of protein kinase C and can be reversed by protein kinase inhibitors (116, 117, 120, 121). Also, 1,25-$(OH)_2D_3$ activates Raf kinase, which is dependent on protein kinase C (122). More recently, the effect of 1,25-$(OH)_2D_3$ on the expression of protoncogene Egr in rat hepatic Ito cells was reported to proceed through the activation of a mitogen-activated protein kinase in a protein-kinase-C-dependent pathway (123). However, it is not clear in this study if the observed effect is specific to the indicated gene. Further, no role for vitamin D in hepatic cells has ever been demonstrated.

A recent study attempted to test whether 1,25-$(OH)_2D_3$ has a direct effect on the different isoforms of protein kinase C (124). The authors used reconstituted vesicles containing 1,25-$(OH)_2D_3$ or related compounds with purified isoforms of the enzyme. They reported that enzyme activity is stimulated directly by 1,25-$(OH)_2D_3$ but not by the other metabolites of the vitamin (124). However, they tested the effect of vitamin D_3 and vitamin D_2, whereas it would have been more important and informative to test other functional metabolites of the vitamin, such as 25-OH-D_3 or 1,25-$(OH)_2D_3$.

In summary, there is some *in vitro* evidence to show that 1,25-$(OH)_2D_3$ may have effects that are not mediated through the nuclear genomic signaling pathway. However, the relevance and contribution of this pathway to the *in vivo* functions of vitamin D_3 are not clear. It is important to note that the experiments attesting to nongenomic actions usually involve large and unphysiologic amounts of vitamin D compounds, or there is a lack of specificity for the active forms of vitamin D, i.e., 1,25-$(OH)_2D_3$. In short, we are not convinced of the physiologic relevance or meaning of the reported nongenomic actions of vitamin D.

IV. The Molecular Biology of Vitamin D Hydroxylases

An obviously important determinant of biological activity of vitamin D is the production and degradation rates of $1,25\text{-}(OH)_2D_3$ (Fig. 2). Vitamin D_3 is hydroxylated at the 25-position largely in the liver by the vitamin D_3 25-hydroxylases to produce 25-OH-D_3, the major circulating form of vitamin D_3 (3). Two forms of this enzyme are present in rat liver, one in microsomes (125–127) and another in mitochondria (128, 129). This is not the case for all species, because only the mitochondrial form of the enzyme has been detected in humans (130). This 25-hydroxylase enzyme is not specific for vitamin D_3, because it can hydroxylate a variety of other steroid substrates (125–127, 131, 132).

Despite early identification of this enzyme in the metabolism of vitamin D_3, little is known about its regulation at the molecular level. Both forms of the enzyme were purified from rat liver (125, 127–129), and the microsomal form was purified from pig kidney (133). The rat mitochondrial enzyme has been cloned (134) and was expressed in yeast and mammalian cells (135). Limited data suggest a difference in expression of rat enzymes in males and females (127, 136). More studies are needed to clarify this under various physiological conditions. More importantly, it is not clear if $1,25\text{-}(OH)_2D_3$, calcium, or parathyroid hormone (PTH) exert any significant regulation on the enzyme either at the transcriptional or the post-transcriptional levels. It is possible, due to the broad specificity of the enzyme, that it may be regulated by noncalcitropic regulators. The expression of 25-hydroxylase mRNA in several tissues, including bone, has been reported (137). Bone enzyme converted 1α-hydroxy vitamin D_3 into $1,25\text{-}(OH)_2D_3$ (137). However, the contribution of the extra hepatic enzyme to the *in vivo* production of $1,25\text{-}(OH)_2D_3$ is not clear.

The 25-OH-D_3 1α-hydroxylase represents the most important enzyme in determining the level of $1,25\text{-}(OH)_2D_3$. The enzyme catalyzes the hydroxylation of 25-OH-D_3 on the 1α-position to produce $1,25\text{-}(OH)_2D_3$. It is present in kidney, and under physiologic conditions it is the only site of $1,25\text{-}(OH)_2D_3$ production (138). However, the 1α-hydroxylase may be expressed at other sites in disease conditions such as sarcoidosis or lymphoma (139). The 1α-hydroxylase is tightly regulated by PTH, $1,25\text{-}(OH)_2D_3$, calcium, and phosphorus. Hypocalcemia causes an increase in 1-hydroxylase

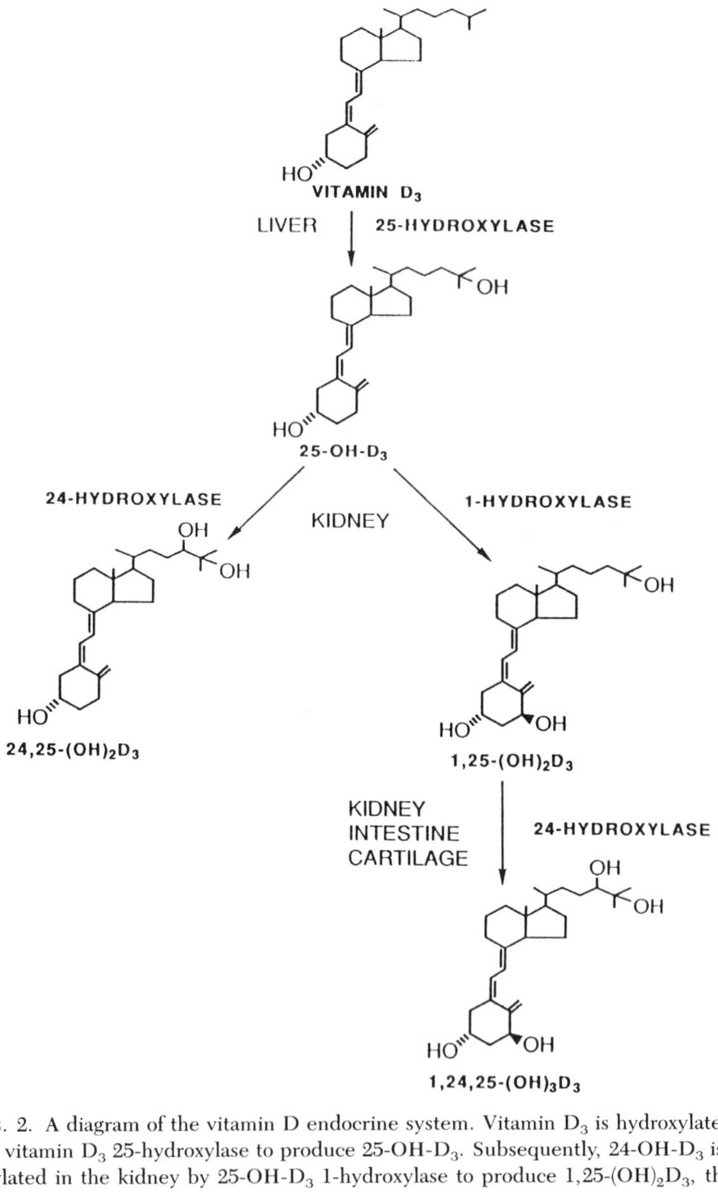

FIG. 2. A diagram of the vitamin D endocrine system. Vitamin D_3 is hydroxylated in the liver by vitamin D_3 25-hydroxylase to produce 25-OH-D_3. Subsequently, 24-OH-D_3 is further hydroxylated in the kidney by 25-OH-D_3 1-hydroxylase to produce 1,25-$(OH)_2D_3$, the active hormonal form of vitamin D_3. Consequently, 25-OH-D_3 and 1,25-$(OH)_2D_3$ are hydroxylated at the 24 position to produce 24,25-$(OH)_2D_3$ and 1,24,25-$(OH)_3D_3$, respectively, by the 24-hydroxylase in the kidney and other tissues. This latter hydroxylation is the initial step in the degradative pathway of vitamin D_3.

activity, probably as a result of increased PTH levels. This was shown by an increase in serum levels of 1,25-$(OH)_2D_3$ *(140, 141)* and directly by an increase in enzyme activity in kidney homogenates *(142, 143)*. The effect of hypocalcemia is eliminated by parathyroidectomy *(140)*.

However, a direct effect of calcium cannot be completely excluded. PTH clearly causes an increase in the activity of 1-hydroxylase *(143–145)*. The effect of PTH on 1-hydroxylase activity was clearly demonstrated in rats during lactation, because parathyroidectomy of lactating rats blocked the increase in 1-hydroxylase activity in these animals *(146)*. The effect of PTH on the activation of 1-hydroxylase is mimicked by cAMP. This was shown both *in vivo* by infusion *(147)* and *in vitro* by treatment of kidney slices and cells *(148–150)*. In addition to cAMP, protein kinase C apparently mediates the activation of 1-hydroxylase by PTH *(151, 152)*.

One of the most important regulators of 1-hydroxylase activity is 1,25-$(OH)_2D_3$. The activity of 1-hydroxylase is clearly diminished by administration of 1,25-$(OH)_2D_3$ *(143, 153–155)*. Plasma phosphorus levels also regulate 1-hydroxylase activity. Hypophosphatemia increases 1α-hydroxylase, as shown by increased serum levels of 1,25-$(OH)_2D_3$ *(140, 141)* and by increased 1α-hydroxylase levels *(143, 156, 157)*. The mechanism by which phosphorus regulates 1-hydroxylase activity is not known, although it has been suggested that growth hormone may be involved in this mechanism *(158)*.

In avian species there is a profound regulation of 1α-hydroxylase by the sex hormones during the egg-laying cycle. Injection of estradiol increases 1α-hydroxylase activity *(159)*. Testosterone *(159)* and progesterone *(160)* also increase estrogen stimulation. As yet, the mechanism by which these hormones increase 1-hydroxylase activity is not known.

Unfortunately, isolation of this enzyme or cloning of the cDNA has evaded investigators, preventing any advances on the molecular events involved in its regulation. This has also prevented elucidating the molecular basis of vitamin-D-dependent rickets Type I. It seems that the lack of success in isolating this enzyme is due to its instability; a major loss of activity is observed after each purification step and all attempts thus far to solve this problem have met with limited success (R. A. Ettinger and H. F. DeLuca, unpublished results).

Among the vitamin D family of hydroxylases, 25-OH-D_3 24-hydroxylase is the best characterized and studied at the molecular level. This enzyme hydroxylates both 25-OH-D_3 and 1,25-$(OH)_2D_3$, producing 24,25-$(OH)_2D_3$ and 1,24,25-$(OH)_3D_3$, respectively. The 24-hydroxylase was first detected in kidney *(161, 162)* and later was found in all target tissues *(137)*. The 24-hydroxylase is localized in the inner membrane of mitochondria *(163)*. Regulation of this enzyme is almost dramatically opposite to the regulation of the 1α-hydroxylase by the same factors *(164)* (Fig. 3). The 24-hydroxylase from

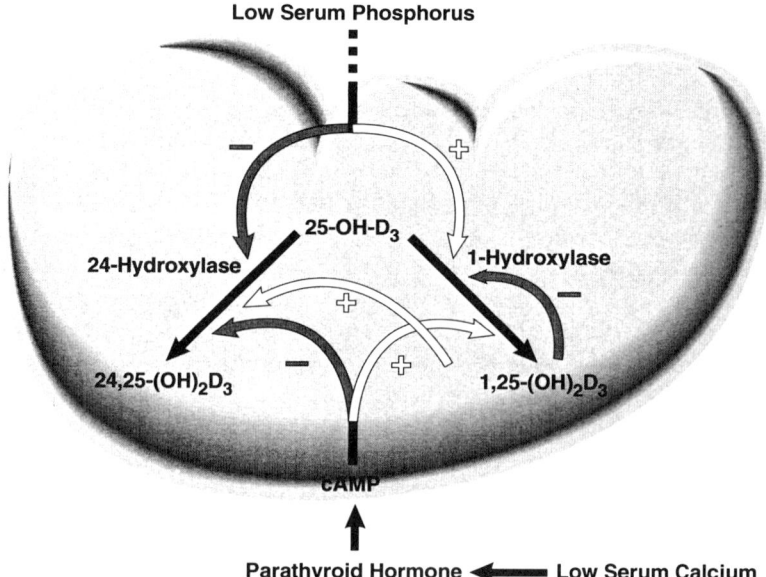

FIG. 3. Reciprocal regulation of 1-hydroxylase and 24-hydroxylase activities in the kidney by the same factors. Low serum phosphorus and low serum calcium induce 1-hydroxylase activity and suppress 24-hydroxylase activity. Similarly, high 1,25-$(OH)_2D_3$ inhibits 1-hydroxylase activity and induces 24-hydroxylase activity. The effect of low serum calcium on both enzymes is mediated through PTH and intracellular cAMP.

rat has been purified to homogeneity (165). This allowed production of antibodies for the enzyme and eventually led to cloning of the rat 24-hydroxylase cDNA (166).

Cloning the 24-hydroxylase was the key event that led to our understanding of the molecular regulation of this enzyme. Using the rat clone, it was shown that 1,25-$(OH)_2D_3$ markedly increases the mRNA levels for the 24-hydroxylase in rat kidney, intestine (167, 168), and osteoblasts (169). These effects parallel the enzymatic activity of 24-hydroxylase in these tissues. PTH down-regulates the mRNA levels for this enzyme in rat kidney but not in intestine (170). This is not surprising because the intestines do not have receptors for PTH. Interestingly, PTH did not result in a decrease in 24-hydroxylase mRNA in rat osteoblasts, despite the presence of PTH receptors in these cells (169). When rats were fed a low-calcium diet, kidney 24-hydroxylase was suppressed due to increased PTH secretion, but no change was observed in expression of the bone enzyme in the same animals (169). These results suggest a tissue-specific differential regulation of the 24-hydroxylase gene in response to PTH.

Besides 1,25-$(OH)_2D_3$ and PTH, it seems that protein kinase C is involved in the regulation of 24-hydroxylase. Treatment of rat intestinal epithelial cells with H-7 (a protein kinase C inhibitor) eliminates the induction of 24-hydroxylase mRNA by 1,25-$(OH)_2D_3$; treatment of cells with TPA, an activator of protein kinase C, facilitates the 1,25-$(OH)_2D_3$ induction (170). Similarly, TPA has similar effects on 24-hydroxylase induction in primary rat renal cell cultures (171). These results agree with previous studies showing that TPA treatment increases 24-hydroxylase activity in cultured chick kidney cells (172). The response of these cells to TPA was suggested to be correlated with phosphorylation of mitochondrial matrix proteins (172).

The 24-hydroxylase cDNA clone allowed cloning and characterization of the promoter and upstream region of the rat 24-hydroxylase gene (72, 74). Two active VDREs were identified in this region, the first between nucleotides -262 and -238 (72) and the second between nucleotides -151 and -137 (74). The two VDREs are thought to function cooperatively, resulting in maximum induction by 1,25-$(OH)_2D_3$ (105). The cDNA and promoter region of the human 24-hydroxylase were also cloned (73, 173). The human and rat enzymes have a large degree of homology, both at the nucleotide level and the amino-acid sequence level (173). Similar to the rat gene, the human 24-hydroxylase gene contains two VDREs in the promoter upstream region (73).

The two 24-hydroxylase genes are the first examples among the vitamin-D-dependent genes shown to have two active VDREs. This probably explains the strong responsiveness of this gene to treatment with 1,25-$(OH)_2D_3$. Interestingly, the proximal VDRE in both the rat and the human genes consists of three repeat elements, whereas the distal VDRE is similar to the other identified VDREs with two repeat elements. The cDNA and promoter region of chicken 24-hydroxylase has been cloned and characterized (F. Jehan, R. Ismail and H. F. DeLuca, unpublished results). The data show that the chicken cDNA is homologous to the rat and human cDNAs, and at least three putative VDREs are located in the 5' upstream region of the chicken gene. The human gene has been localized to chromosome 20q13.2–q13.3 by Hahn *et al.* (174). These clones will permit a wide range of studies on the expression and regulation of the 24-hydroxylase in different tissues. Ettinger *et al.* reported the cloning and characterization of a kidney-specific protein termed vitamin D_3 hydroxylase-associated protein (VDHAP) (175). No regulation of this protein by 1,25-$(OH)_2D_3$ was observed (176). This protein was later identified as an amidase (176). It is not clear if this enzyme plays any role in regulation of either the 24-hydroxylase or the 1α-hydroxylase in kidney. Beckman *et al.* reported a possible role for calcitonin in suppression of 24-hydroxylase mRNA expression in intestine (177). Whether this is a direct or indirect effect by the hormone on the 24-hydroxylase gene awaits further investigation.

Interestingly, when the rat 24-hydroxylase is expressed in bacteria, it has 23-hydroxylase activity (178). This activity was also detected with the balculovirus-expressed rat enzyme (M. Beckman and H. F. DeLuca, unpublished results). Because 23-hydroxylase is part of the *in vivo* metabolism of 25-OH-D_3 (179–181), these results suggest that 24- and 23-hydroxylations of the vitamin D_3 compounds are catalyzed by the same enzyme.

V. Summary

Following the cloning and deletion analysis of the vitamin D receptor, most recent advances have been in the isolation and characterization of the DNA response elements found in the promoter region of target genes of vitamin D. Vitamin D, like the thyroid and retinoid hormones, binds to repeat sequences, but the repeats are separated by three nonspecified bases. The action of the VDR requires the presence of the RXR proteins and evidently other proteins that are involved in regulating transcription.

A possible role of phosphorylation of the ligand binding domain of the VDR in transcription has also appeared. Very likely, the molecular events involved in vitamin D stimulation or suppression of a target gene will include its interaction with a number of transcription factors, both in the regulation of transcription and in the actual machinery involved in the transcription process through polymerase II. Although likely, it is not entirely clear whether the genomic action of vitamin D can account for all of its biological activities. Nongenomic actions of the vitamin D hormone have been reported, but convincing evidence that this is of biological importance *in vivo* is lacking.

Advances in our understanding of the vitamin D mechanism of action can clearly be expected from physical studies of cloned and expressed vitamin D receptor and its subdomains, elucidation of the transcription factors in vitamin D-modulated transcription of target genes, elucidation of the role of phosphorylation in the transcription process, and the identification of important genes that are regulated in the specific target tissues responsive to vitamin D. This will definitely remain as a very active field of investigation well into the future.

Acknowledgments

This work was supported in part by program project Grant No. DK14881, from the National Institutes of Health, a fund from the National Foundation for Cancer Research, and a fund from the Wisconsin Alumni Research Foundation.

REFERENCES

1. R. P. Esvelt, H. K. Schnoes and H. F. DeLuca, *ABB* **188**, 282 (1978).
2. M. F. Holick, J. A. MacLaughlin, M. B. Clark, S. A. Holick, J. T. Potts, Jr., R. R. Anderson, I. H. Blank, J. A. Parrish and P. Elias, *Science* **210**, 203 (1980).
3. R. Brommage and H. F. DeLuca, *Endocrine Rev.* **6**, 491 (1985).
4. I. T. Boyle, L. Miravet, R. W. Gray, M. F. Holick and H. F. DeLuca. *Endocrinology* **90**, 605 (1972).
5. M. F. Holick, M. Garabedian and H. F. DeLuca, *Science* **176**, 1146 (1972).
6. M. Lamm and W. F. Neuman, *Arch. Pathol.* **66**, 204 (1958).
7. H. F. DeLuca, *Vitam. Horm.* **25**, 315 (1967).
8. J. L. Underwood and H. F. DeLuca, *Am. J. Physiol.* **246**, E493 (1984).
9. M. R. Hughes, P. F. Brumbaugh, M. R. Haussler, J. E. Wergedal and D. J. Baylink, *Science* **190**, 578 (1975).
10. J. I. Radar, D. J. Baylink, M. R. Hughes, E. G. Safilian and M. R. Haussler, *Am. J. Physiol.* **236**, E118 (1979).
11. P. F. Brumbaugh and M. R. Haussler, *Life Sci.* **16**, 353 (1975).
12. B. E. Kream, R. D. Reynolds, J. C. Knutson, J. A. Eisman and H. F. DeLuca, *ABB* **176**, 779 (1976).
13. J. W. Pike, *Annu. Rev. Nutr.* **11**, 189 (1991).
14. H. Darwish and H. F. DeLuca, *Crit. Rev. Euk. Gene Exp.* **3**, 89 (1993).
15. T. K. Ross, H. M. Darwish and H. F. DeLuca, *Vit. Horm.* **49**, 281 (1994).
16. D. N. MacDonald, D. R. Dowd and M. R. Haussler, *Sem. Nephrol.* **14**, 101 (1994).
17. M. E. Sandgren, M. Bronnegard and H. F. DeLuca, *BBRC* **181**, 611 (1991).
18. M. C. Dame, E. A. Pierce, J. M. Prahl, C. E. Hayes and H. F. DeLuca, *Bchem* **25**, 4523 (1986).
19. J. W. Pike, N. M. Sleator and M. R. Haussler, *J. Cell Biol.* **111**, 2385 (1983).
20. M. E. Sandgren and H. F. DeLuca, *Anal. Biochem.* **183**, 57 (1989).
21. M. Strom, M. E. Sandgren, T. A. Brown and H. F. DeLuca, *PNAS* **86**, 9770 (1989).
22. R.J.A. Wiese, T. K. Ross, J. M. Prahl and H. F. DeLuca, *JBC* **267**, 20082 (1992).
23. N. C. Arbour, J. M. Prahl and H. F. DeLuca, *Mol Endocrinol.* **7**, 1307 (1993).
24. D. P. McDonnell, D. J. Mangelsdorf, J. W. Pike, M. R. Haussler and B. W. O'Malley, *Science* **235**, 1214 (1987).
25. A. R. Baker, D. P. McDonnell, M. Hughes, T. M. Crisp, D. J. Mangelsdorf, M. R. Haussler, J. W. Pike, J. Shine and B. W. O'Malley, *PNAS* **85**, 3294 (1988).
26. J. K. Burmester, R. J. Wiese, N. Maeda and H. F. DeLuca, *PNAS* **85**, 9499 (1988).
27. M. A. Elaroussi, J. M. Prahl and H. F. DeLuca, *PNAS* **91**, 11596 (1994).
28. Y. Kamei, T. Kawada, T. Fukuwatari, T. Ono, S. Kato and E. Sugimoto, *Gene* **152**, 281 (1995).
29. T. A. Brown and H. F. DeLuca, *JBC* **265**, 10025 (1990).
30. W. T. Schrader and B. W. O'Malley, *JBC* **247**, 51 (1972).
31. H. Gronemeyer, P. Harry and P. Chambon, *FEBS Lett.* **156**, 287 (1983).
32. K. B. Horwitz and P. S. Alexander, *Endocrinology* **113**, 2195 (1983).
33. C. Dany and G. Florence, *JBC* **269**, 23007 (1994).
34. D. X. Wen, Y.-F. Xu, D. E. Mais, M. E. Goldman and D. P. McDonnell, *MCBiol* **14**, 8356 (1994).
35. J. W. Pike and M. R. Haussler, *PNAS* **76**, 5485 (1979).
36. R. U. Simpson and H. F. DeLuca, *PNAS* **79**, 16 (1982).
37. R. U. Simpson, A. Hamstra, N. C. Kendrick and H. F. DeLuca, *Bchem* **22**, 2586 (1983).

38. M. S. Lee, S. A. Kliewer, J. Provencal, P. E. Wright and R. M. Evans, *Science* **260**, 1117 (1993).
39. M. S. Lee, D. S. Sem, S. A. Kliewer, J. Provencal, R. M. Evans and P. E. Wright, *EJB* **224**, 639 (1994).
39a. W. Bourquet, M. Ruff, P. Chambon, H. Gronemeyer and D. Moras, *Nature* **375**, 377 (1995).
40. T. K. Ross, J. M. Prahl and H. F. DeLuca, *PNAS* **88**, 6555 (1991).
41. P. N. MacDonald, C. A. Haussler, C. M. Terpening, M. A. Galligan, M. C. Reeder, G. K. Whitfield and M. R. Haussler, *JBC* **266**, 18808 (1991).
42. J. Schaefer-Klein, J. M. Londowski and R. Kumar, *BBRC* **196**, 167 (1993).
43. R. Kumar, J. Schaefer and E. Wieben, *BBRC* **139**, 1414 (1992).
44. T. Sone, D. P. McDonnell, B. W. O'Malley and J. W. Pike, *JBC* **265**, 21997 (1990).
45. C. L. Smith, G. L. Hager, J. W. Pike and S. J. Marx, *Mol. Endocrinol.* **5**, 867 (1991).
46. T. Härd, E. Kellenbach, R. Boelens, B. A. Maler, K. Dahlman, L. P. Freedman, J. Carlstedt-Duke, K. R. Yamamoto, J.-A. Gustafsson and R. Kaptein, *Science* **249**, 157 (1990).
47. B. F. Luisi, W. X. Xu, Z. Otwinowski, L. P. Freedman, K. R. Yamamoto and P. B. Sigler, *Nature* **352**, 497 (1991).
48. J. W. R. Schwabe, L. Chapman, J. T. Finch and D. Rhodes, *Cell* **75**, 267 (1993).
49. D. Santiso-Mere, T. Sone, G. M. Hilliard, J. W. Pike and D. P. McDonnell, *Mol. Endocrinol.* **7**, 833 (1993).
50. E. M. Costa and D. Feldman, *BBRC* **137**, 742 (1986).
51. T. A. Brown and H. F. DeLuca, *ABB* **286**, 466 (1991).
52. H. M. Darwish, J. K. Burmester, V. E. Moss and H. F. DeLuca, *BBA* **1167**, 29 (1993).
53. T. Matkovitas and S. Christakos, *Mol. Endocrinol.* **9**, 232 (1995).
54. G. M. Hilliard, R. G. Cook, N. L. Weigel and J. W. Pike, *Bchem* **33**, 4300 (1994).
55. E. Martinez, F. Givel and W. Wahli, *EMBO J.* **6**, 3719 (1987).
56. G. Klock, U. Strahle and G. Schutz, *Nature* **329**, 734 (1987).
57. C. K. Glass, *Endocrine Rev.* **15**, 391 (1994).
58. M. Leid, P. Kastner, R. Lyons, H. Nakashatri, M. Saunders, T. Zacharewski, J. Chen, A. Staub, J. Garnier, S. Mader and P. Chambon, *Cell* **68**, 377 (1992).
59. V. C. Yu, C. Delsert, B. Anderson, J. M. Holloway, O. V. Devary, A. M. Näär, S. Y. Kim, J. M. Boutin, C. K. Glass and M. G. Rosenfeld, *Cell* **67**, 1251 (1991).
60. K. Umesono, K. K. Murakami, C. C. Thompson and R. M. Evans, *Cell* **65**, 1255 (1991).
61. A. M. Näär, J. M. Boutin, S. M. Lipkin, V. C. Yu, J. M. Holloway, C. K. Glass and M. G. Rosenfeld, *Cell* **65**, 1267 (1991).
62. J. N. Rottman, R. L. Widom, B. Nadal-Ginard, V. Mahdavi and S. K. Karathanasis, *MCBiol* **11**, 8814 (1991).
63. B. Durand, M. Saunders, P. Leroy, M. Leid and P. Chambon, *Cell* **71**, 73 (1992).
64. D. J. Mangelsdorf, K. Umesono, S. A. Kliewer, U. Borgmeyer, E. S. Ong and R. M. Evans, *Cell* **66**, 555 (1991).
65. P. Chambon, *Sem. Cell Biol.* **5**, 115 (1994).
66. D. J. Mangelsdorf, K. Umesono and R. M. Evans, in "The Retinoids: Biology and Medicine" (M. B. Sporn, A. B. Roberts and D. S. Goodman, eds.), 2nd Ed., p. 319. Raven Press, New York, 1994.
67. H. M. Darwish and H. F. DeLuca, *PNAS* **89**, 603 (1992).
68. M. B. Demay, J. M. Gerardi, H. F. DeLuca and H. M. Kronenberg, *PNAS* **87**, 369 (1990).
69. N. A. Morrison, J. Shine, J.-C. Fragonas, V. Verkest, M. L. McMenemy and J. A. Eisman, *Science* **246**, 1158 (1989).
70. K. Ozono, J. Liao, S. A. Kerner, R. A. Scott and J. W. Pike, *JBC* **265**(35), 21881 (1990).

71. M. Noda, R. L. Vogel, A. M. Craig, J. Prahl and H. F. DeLuca, *PNAS* **87**, 9995 (1990).
72. C. Zierold, H. M. Darwish and H. F. DeLuca, *PNAS* **91**, 900 (1994).
73. K. S. Chen and H. F. DeLuca, *BBA* **1263**, 1 (1995).
74. Y. Ohyama, K. Ozono, M. Uchida, T. Shinki, S. Kato, T. Suda, O. Yamamoto, M. Noshiro and Y. Kato, *JBC* **269**(14), 10545 (1994).
75. M. B. Demay, M. S. Kiernan, H. F. DeLuca and H. M. Kronenberg, *PNAS* **89**, 8097 (1994).
76. T. Takeda, M. Arkawa and R. Kuwano, *BBRC* **204**, 889 (1994).
77. I. Quelo, J.-P. Kahlen, A. Rascle, P. Jurdic and C. Carlberg, *DNA Cell. Biol.* **13**, 1181 (1994).
78. M. Schräder, S. Mayeri, J.-P. Kahlen, K. M. Müller and C. Carlberg, *MCBiol* **15**, 1154 (1995).
79. R. K. Gill and S. Christakos, *PNAS* **90**, 2984 (1994).
80. T. Sone, K. Ozono and J. W. Pike, *Mol. Endocrinol.* **5**, 1578 (1990).
81. T. K. Ross, V. E. Moss, J. M. Prahl and H. F. DeLuca, *PNAS* **89**, 256 (1992).
82. S. A. Kliewer, K. Umesono, D. J. Mangelsdorf and R. M. Evans, *Nature* **355**, 446 (1992).
83. V. C. Yu, C. Delsert, B. Andersen, J. M. Holloway, O. V. Devary, A. M. Näär, S. Y. Kim, J.-M. Boutin, C. K. Glass and M. G. Rosenfeld, *Cell* **67**, 1251 (1991).
84. S. Nakajima, J.-C. Hsieh, P. N. MacDonald, M. A. Galligan, C. A. Haussler, G. K. Whitfield and M. R. Haussler, *Mol. Endocrinol.* **8**, 159 (1994).
85. J.-I. Nishikawa, M. Kitaura, M. Imagawa and T. Nishihara, *NARes* **23**, 606 (1995).
86. E. Rosen, E. G. Beninghof and R. Koenig, *JBC* **268**, 11534 (1993).
87. X.-K. Zhang, G. Salbert, M.-O. Lee and M. Pfahl, *MCBiol* **14**, 4311 (1994).
88. J.-I. Nishikawa, M. Matsumoto, K. Sakoda, M. Kitaura, M. Imagawa and T. Nishihara, *JBC* **268**, 19739 (1993).
89. T. L. Towers, B. F. Luisi, A. Asiano and L. P. Freedman, *PNAS* **90**, 6310 (1993).
90. R. Kurokawa, V. C. Yu, A. Näär, S. Kyakumoto, Z. Han, S. Silverman, M. G. Rosenfeld and C. K. Glass, *Genes Dev.* **7**, 1423 (1993).
91. S. Mader, J.-Y. Chen, Z. Chen, J. White, P. Chambon and H. Gronemeyer, *EMBO J.* **12**, 5029 (1993).
92. T. Perlmann, P. N. Rangavajan, K. Umesono and R. M. Evans, *Genes Dev.* **7**, 1411 (1993).
93. P. F. Predki, D. Zamble, B. Sarkar and V. Giguere, *Mol. Endocrinol.* **8**, 31 (1994).
94. C. Carlberg, *BBRC* **195**, 1345 (1993).
95. C. Carlberg, I. Bendik, A. Wyss, E. Meier, L. J. Sturzenbecker, J. F. Grippo and W. Hunziker, *Nature* **361**, 657 (1993).
96. M. Schräder, K. M. Müller, M. Becker-André and C. Carlberg, *J. Mol. Endocrinol.* **12**, 327 (1994).
97. M. Schräder, K. M. Müller, S. Nayeri, J.-P. Kahlen and C. Carlberg, *Nature* **370**, 382 (1994).
98. M. Schrader, I. Bendik, M. Becker-André and C. Carlberg, *JBC* **268**, 17830 (1993).
99. J.-P. Kahlen and C. Carlberg, *BBRC* **202**, 1366 (1994).
100. J. C. G. Blanco, I.-M. Wang, S. Y. Tsai, M.-J. Tsai, B. W. O'Malley, P. W. Jurutka, M. R. Haussler and K. Ozato, *PNAS* **92**, 1535 (1995).
101. P. N. MacDonald, D. R. Sherman, D. R. Dowd, S. C. Jefcoate, and R. K. Delisle, *JBC* **270**, 4748 (1995).
102. A. Baniahmad, I. Ha, D. Reinberg, S. Tsai, M.-J. Tsai and B. W. O'Malley, *PNAS* **90**, 8832 (1993).
103. N. H. Ing, J. M. Beckman, S. Y. Tsai, M.-J. Tsai and B. W. O'Malley, *JBC* **267**, 17617 (1992).

104. T. K. Ross, H. M. Darwish, V. E. Moss and H. F. DeLuca, *PNAS* **90**, 9257 (1993).
105. C. Zierold, H. M. Darwish and H. F. DeLuca, *JBC* **270**, 1675 (1995).
106. B. Cheskis and L. P. Freedman, *MCBiol* **14**, 3329 (1994).
107. P. N. McDonald, D. R. Dowd, S. Nakajima, M. A. Galligan, M. C. Reeder, C. A. Haussler, K. Ozato and M. R. Haussler, *MCBiol* **13**, 5907 (1993).
108. F. Auricchio, *J. Steroid Biochem.* **32**, 613 (1989).
109. N. Arbour, H. M. Darwish and H. F. DeLuca, *BBA* **1263**, 147 (1995).
110. L. A. Denner, N. L. Weigel, B. Maxwell, W. T. Schrader and B. W. O'Malley, *Science* **250**, 1740 (1990).
111. N. L. Weigel, T. H. Carter, W. T. Schrader and B. W. O'Malley, *Mol. Endocrinol.* **6**, 8 (1992).
112. M. K. Bagchi, S. Y. Tsai, M.-J. Tsai and B. W. O'Malley, *PNAS* **89**, 2664 (1992).
113. A. R. deBoland, S. Movelli and R. Boland, *JBC* **269**, 8675 (1994).
114. S. Morelli, A. R. deBoland and R. L. Boland, *BJ* **289**, 675 (1993).
115. R. Civitelli, Y. S. Kim, S. L. Gunsten, A. Fujimori, M. Huskey, L. V. Avioli and K. A. Hruska, *Endocrinology* **127**, 2253 (1990).
116. A. Bourdeau, F. Atmani, B. Grosse and N. Lieberherr, *Endocrinology* **127**, 2738 (1990).
117. R. U. Simpson, T. Hsu, M. D. Wendt and J. M. Taylor, *JBC* **264**, 19710 (1989).
118. A. R. deBoland and A. Norman, *Endocrinology* **127**, 39 (1990).
119. M. Simboli-Campbell, A. Gagnon, D. J. Franks and J. Welsh, *JBC* **269**, 3257 (1994).
120. J. P. van Leeuwen, J. C. Birkenhager, G. J. van den Bemd, C. J. Buurman, A. Staal, M. P. Bos and H. A. Pols, *JBC* **267**, 12562 (1992).
121. S. Khave, D. M. Wilson, X. Y. Then, P. K. Dudeja, R. K. Wali, M. D. Sitrin and T. A. Brasitus, *Endocrinology* **133**, 2213 (1993).
122. D. W. A. Beno, L. M. Brady, M. Bissonnette and B. H. Davis, *JBC* **268**, 25132 (1993).
123. T. W. Lissoos, D. W. A. Beno and B. H. Davis, *JBC* **268**, 25132 (1993).
124. S. J. Slater, M. B. Kelly, F. J. Taddeo, J. D. Larkin, M. D. Yeager, J. A. McLane, C. Ho and C. D. Stubbs, *JBC* **270**, 6639 (1995).
125. S. Hayashi, M. Noshiro and K. Okuda, *J. Biochem.* **99**, 1753 (1986).
125. S. Andersson, I. Holmberg and K. Wikvall, *JBC* **258**, 6777 (1983).
127. S. Andersson and H. Jornvall, *JBC* **261**, 16932 (1986).
128. O. Masumoto, Y. Ohyama and K. Okuda, *JBC* **263**, 14256 (1988).
129. I. Björkhem, I. Holmberg, H. Oftebro and J. I. Pederson, *JBC* **255**, 5244 (1980).
130. K. Saarem, S. Bergseth, H. Oftebro and J. I. Pederson, *JBC* **259**, 10936 (1989).
131. E. Usui, M. Noshiro, Y. Ohyama and K. Okuda, *FEBS Lett.* **274**, 175 (1990).
132. M. Akiyoshi-Shibata, E. Usui, T. Sakaki, Y. Yabusaki, M. Noshiro, K. Okuda and H. Ohkawa, *FEBS Lett.* **280**, 367 (1991).
133. H. Postlind and K. Wikvall, *BJ* **253**, 549 (1988).
134. E. Usui, M. Noshiro and K. Okuda, *FEBS Lett.* **262**, 135 (1990).
135. M. Akiyoshi-Shibata, E. Usui, T. Sakaki, Y. Yabusaki, M. Noshiro, K. Okuda and H. Ohkawa, *FEBS Lett.* **280**, 367 (1991).
136. K. Saarem and J. I. Pedersen, *BJ* **247**, 73 (1987).
137. F. Ichikawa, K. Sato, M. Nanjo, Y. Nishii, T. Shinki, N. Takahashi and T. Suda, *Bone* **16**, 129 (1995).
138. H. F. DeLuca, *FASEB J.* **2**, 224 (1988).
139. H. J. Armbrecht, K. Okuda, N. Wongsurawat, R. Nemani, M. L. Chen and M. A. Boltz, *J. Steroid Biochem. Mol. Biol.* **43**, 1073 (1992).
140. M. R. Hughes, P. F. Brumbaugh, M. R. Haussler, J. E. Wegedal and D. J. Baylink, *Science* **190**, 578 (1975).

141. J. I. Radar, D. J. Baylink, M. R. Hughes, E. F. Safilian and M. R. Haussler, *Am. J. Physiol.* **236**, E118 (1979).
142. H. L. Henry and A. W. Norman, *JBC* **249**, 7529 (1974).
143. Y. Tanaka and H. F. DeLuca, *Am. J. Physiol.* **246**, E168 (1984).
144. D. R. Fraser and E. Kodicek, *Nature (New Biol.)* **241**, 163 (1973).
145. B. E. Booth, H. C. Tsai and R. C. Morris, *J. Clin. Invest.* **60**, 1314 (1977).
146. B. Lobaugh, S. C. Garner, J. A. Lovdal, A. Boass and S. U. Toverud, *Am. J. Physiol.* **264**, E981 (1993).
147. N. Horiuchi, T. Suda, H. Takahashi, E. Shimazawa and E. Ogata, *Endocrinology* **101**, 969 (1977).
148. H. Rasmussen, M. Wong, D. Bikle and D. B. P. Goodman, *J. Clin. Invest.* **51**, 2502 (1972).
149. C. R. Rost, D. D. Bikle and R. A. Kaplan, *Endocrinology* **108**, 1002 (1981).
150. H. L. Henry, *Endocrinology* **108**, 733 (1981).
151. H. K. Ro, V. Tembe and M. J. Favus, *Endocrinology* **131**, 1424 (1992).
152. M. Janalis, M.-S. Wong and M. J. Favus, *Endocrinology* **133**, 713 (1993).
153. A. Trechsel, J.-P. Bonjour and H. Fleisch, *J. Clin. Invest.* **64**, 206 (1979).
154. K. W. Colston, I.M.A. Evans, T. C. Spelsberg and I. MacIntyre, *BJ* **164**, 83 (1977).
155. H. I. Henry, *JBC* **254**, 2722 (1979).
156. R. W. Gray and J. L. Napoli, *JBC* **258**, 1152 (1983).
157. L. A. Baxter and H. F. DeLuca, *JBC* **251**, 3158 (1976).
158. R. W. Gray and T. L. Garthwaite, *Endocrinology* **116**, 189 (1985).
159. Y. Tanaka, L. Castillo and H. F. DeLuca, *PNAS* **73**, 2701 (1976).
160. Y. Tanaka, L. Castillo, M. J. Wineland and H. F. DeLuca, *Endocrinology* **103**, 2035 (1978).
161. J. L. Omdahl, R. W. Gray, I. T. Boyle, J. Knutson and H. F. DeLuca, *Nature* **237**, 63 (1972).
162. Y. Tanaka, L. Castillo and H. F. DeLuca, *JBC* **252**, 1421 (1977).
163. J. I. Pedersen, H. H. Shobaki, I. Holmberg, S. Bergseth and I. Björkhem, *JBC* **258**, 742 (1983).
164. H. F. DeLuca, in "Vitamin D" (R. Kumar, ed.), p. 1. Nijhoff, Boston, Massachusetts, 1984.
165. Y. Ohyama and K. Okuda, *JBC* **266**, 8690 (1991).
166. Y. Ohyama, M. Noshiro and K. Okuda, *FEBS Lett.* **278**, 195 (1991).
167. H. J. Armbrecht and M. A. Boltz, *FEBS Lett.* **292**, 17 (1991).
168. T. Shinki, C. H. Jin, A. Nishimura, Y. Nagai, Y. Ohyama, M. Noshiro, K. Okuda and T. Suda, *JBC* **267**, 13757 (1992).
169. A. Nishimura, T. Shinki, C. H. Jin, Y. Ohyama, M. Noshiro, K. Okuda and T. Suda, *Endocrinology* **134**, 1794 (1994).
170. H. Koyama, M. Inaba, Y. Nishizawa, S. Ohno and H. Morii, *J. Cell. Biochem.* **55**, 230 (1994).
171. M. L. Chen, M. A. Boltz and H. J. Armbrecht, *Endocrinology* **132**, 1782 (1993).
172. C. Tang, S. R. Kain and H. L. Henry, *Endocrinology* **133**, 1823 (1993).
173. K.-S. Chen and H. F. DeLuca, *BBA* **1219**, 26 (1994).
174. C. N. Hahn, E. Baker, B. K. May, J. L. Omdahl, and G. R. Sutherland, *Cytogenet. Cell. Genet.* **62**, 192 (1993).
175. R. A. Ettinger, R. Ismail and H. F. DeLuca, *JBC* **269**, 176 (1994).
176. R. A. Ettinger and H. F. DeLuca, *ABB* **316**, 14 (1995).
177. M. J. Bechman, J. P. Goff, T. A. Reinhardt, D. C. Beitz and R. L. Horst, *Endocrinology* **135**, 1951 (1994).

178. M. Akiyoshi-Shibata, T. Sakaki, Y. Ohyama, M. Noshiro, K. Okuda and Y. Yabuski, *EJB* **224,** 335 (1994).
179. H. F. DeLuca and H. K. Schnoes, *ARB* **52,** 411 (1983).
180. S. Yamada, M. Ohmori, H. Takayama, Y. Takasaki and T. Suda, *JBC* **258,** 457 (1983).
181. G. W. Engstrom, T. A. Reinhardt and R. L. Horst, *ABB* **250,** 86 (1986).

Regulation of Synthesis of Ribonucleotide Reductase and Relationship to DNA Replication in Various Systems

G. ROBERT GREENBERG* AND
JOHN M. HILFINGER†

*Department of Biological Chemistry
University of Michigan
Ann Arbor, Michigan 48109
†Department of Internal Medicine
University of Michigan
and Therapeutic Systems Research
Laboratories
Ann Arbor, Michigan 48108

I.	Nature of Ribonucleotide Reductase Enzyme and Reaction	345
II.	Time of Initiation of Bacteriophage T4 DNA Replication	349
III.	Expression of T4 *nrdA* and *nrdB* Genes	351
IV.	Expression of *Escherichia coli nrdA* and *nrdB* Genes	356
V.	Phospholipid Synthesis and mRNA Breakdown Provide CMP for Synthesis of CDP and dCDP	359
VI.	Control of Ribonucleotide Reductase Synthesis in Eukaryotes and Activity in Cell Division Cycle	361
VII.	Effect of DNA Damage on *nrdA* and *nrdB* Expression	371
VIII.	Ribonucleotide Reductase Regulation during Oogenesis	375
IX.	Eukaryotic Animal DNA Viruses Encoding Ribonucleotide Reductases	379
X.	Kinetics of Synthesis of T4-encoded Ribonucleotide Reductase and Relationship to T4 DNA Replication	383
XI.	General Discussion	390
	References	391

I. Nature of Ribonucleotide Reductase Enzyme and Reaction

A. Early History

The ever growing interest in the ribonucleotide reductases and their functions is a tribute to the creative and unrelenting work of Peter Reichard of the Karolinska Institute in Stockholm and his many able students. In 1950,

Hammarsten, Reichard, and Saluste found that an ^{15}N-labeled pyrimidine ribonucleoside administered to rats was readily converted to the deoxyribonucleotide residues of DNA, though the free base was poorly utilized. Therefore, they proposed that the N-glycosyl bond was preserved in the reduction of ribonucleosides (and ribonucleotides) to precursors of the deoxyribonucleotides of DNA (1). In 1953, Rose and Schweigert placed this proposal on firm ground by the use of randomly ^{14}C-labeled cytidine (2). In 1960–1961, Reichard et al. demonstrated that ribonucleoside diphosphates are converted to dNDPs by both *Escherichia coli* and animal cell extracts, and thenceforth, the studies bubbled over (3–6). [For a review of the early findings and the basic phenomena, see Ref. 5.]

B. Ribonucleotide Reductase

Ribonucleotide reductase is a central and crucial enzyme in all cells. It catalyzes an irreversible step, the first in the only *de novo* pathway to deoxyribonucleotides, and it follows a large number of enzymes that form the pyrimidine and purine ribonucleotide precursors. The classical ribonucleotide reductase is ribonucleoside diphosphate reductase of aerobically grown *E. coli*.

Escherichia coli ribonucleotide reductase (RnR)[1] is an $\alpha_2\beta_2$ enzyme that catalyzes the difficult reduction of a secondary hydroxyl group (2'-OH) to a methylene group (5, 6). Its α_2 subunit has an M_r of $2 \times 86{,}000$; the β_2 subunit has an M_r of $2 \times 43{,}500$. The overall reaction may be written as two oxidation–reduction steps:

RnR:

$$\text{rNDP} + \text{thioredoxin(SH)}_2 \rightarrow \text{dNDP} + \text{thioredoxin(S)}_2$$

---

[1] Abbreviations and definitions: NDP signifies ribonucleoside diphosphate, but rNDP is sometimes employed to contrast it with dNDP; P_E, T4 early promoter of the *nrdB* gene; P_M, T4 middle or intermediate promoter; Hm, 5-hydroxymethyl; Hmase, hydroxymethylase; HmC, hydroxymethyl cytosine; RnR, ribonucleotide reductase; p75, a protein of M_r 75,000. In the context of phage T4, p62 is the protein product of gene 62, gp*nrdB* is the gene product of the *nrdB* gene, and NrdB protein is the protein product of *nrdB*. NrdB is also commonly used to designate the corresponding protein; R1, the protein subunit of RnR formed by the *nrdA* gene; R2, the protein subunit of RnR formed by the *nrdB* gene; 3' UTR, 3' untranslated region of a mRNA; TIR, translational initiation region; PKC, protein kinase C; TGF-β1, transforming growth factor-β1; TPA, a phorbol ester; DAG, diacylglycerol; PI, phosphatidyl inositol; PC, phosphatidylcholine; PL, phospholipid; NDK, nucleoside diphosphate kinase; P_{nrdB}, promoter for *nrdB*; nt, nucleotide(s); kb, kilobasepairs; HSV-1, HSV-2, human simplex virus 1 and 2; α, β, and γ, immediate-early, early, and late transcripts, respectively, in herpes virus infection; ICP, infected cell peptides in herpes infection; R1-PK, large subunit of herpes RnR with protein kinase domain; mRNP, ribonucleoprotein particles; RBS, ribosome binding site; pRb, retinoblastoma susceptibility protein.

Thioredoxin reductase:

thioredoxin(S)$_2$ + NADPH + H$^+$ $\rightleftharpoons$ thioredoxin(SH)$_2$ + NADP

Thioredoxin is a small carrier protein with two specific cysteine sulfhydryl groups at its active site, and it is regenerated by the flavoprotein, thioredoxin reductase, and NADPH + H$^+$. In a recent paper Holmgren and co-workers have demonstrated that the primary carrier system in *E. coli* is glutaredoxin III/glutathione reductase (7).

RnRs are allosteric enzymes under the control of dNTPs that bind to specific sites on the α chains, and function as positive and negative effectors. Effectors determine the K_m values of specific rNDPs, which compete for the two identical substrate sites. Both the α$_2$ and β$_2$ subunits are required to create the active sites (5, 6, 8). The precise actions of the effectors and their binding sites vary with the biological origin of the enzyme (5, 6).

In the α$_2$β$_2$ enzyme, the accepted terminology for the α$_2$ subunit, no matter its biological source, is R1, and for the β$_2$ subunit, R2 (9). The R2 protein consists of one dinuclear iron center for each chain with the antiferromagnetic ferric ions connected by a μ-oxobridge. Its most structurally exciting characteristic is an unusually stable tyrosyl free radical (located at residue 122 in the *E. coli* enzyme and in comparable positions in other species). The tyrosyl hydroxyl oxygen is 5.3 Å from its iron center, and the free radical is apparently protected and stabilized by its hydrophobic environment (*10*).

Hydroxyurea, a radical scavenger, acts by reducing the tyrosyl radical. The sensitivity to hydroxyurea depends on the origin of the enzyme and the availability of the target, but the susceptibility of the radical also varies with the regulatory state of the enzyme. It is far less labile in the R1R2 enzyme complex than in R2 alone, presumably because R1 blocks accessibility. Positive allosteric effectors increase accessibility and negative effector conditions cause a decrease (*11*). The variation in accessibility of the free radical to such scavengers may be related to the lability of the iron in R2. The nature of the intricate assembly of the important R2 active structure and its function are the continuing targets of a number of laboratories (9, 10, 12–18).

C. Goals

It is our purpose to examine the biosynthesis of the subunits of ribonucleotide reductases in several diverse systems and to describe and compare their controls. Primarily, we have attempted to gather pertinent information from bacteriophage T4, *E. coli*, the mouse and other mammals, *Saccharomyces cerevisiae*, a few animal DNA viruses that encode ribonucleotide reductases, and oocyte systems, especially the clam, which displays an intriguing block in the translation of its *nrdB* message.

We pay special attention to phage T4-encoded ribonucleotide reductase. This enzyme plays a central role in the T4-induced dNTP synthetase complex, which possibly is unique to this phage, forming dNTPs in a channeled synthesis and funneling them to the associated T4 DNA replication complex. We do not, however, dwell on the detailed nature of the dNTP synthetase complex, which is admirably reviewed by Mathews in Volume 44 of this series (19). Readers may also refer to a review from this laboratory (20). We have not covered the status of the existence or role of such complexes in eukaryotes, which remains unresolved (19). Our intention is to review the state of the expression of the T4-encoded *nrdA* and *nrdB* genes, including studies from this laboratory, some unpublished. We are particularly interested in reviewing the kinetics of formation and utilization of the dNTPs, primarily from *in vivo* analysis. When we felt that additional data or emphasis on the dNTP synthetase complex was needed, we have presented our results and views.

This review arbitrarily has not included a number of important systems, such as the 5'-deoxyadenosylcobalamin-dependent ribonucleotide reductases (21) or the recently discovered anaerobic ribonucleotide reductases of *E. coli* and from phage T4 infection (22) or controls in many other bacterial and eukaryotic sources.

D. Role of RnR Activity in Initiation of DNA Replication

It is axiomatic that the initiation and continuation of DNA replication is dependent on a supply of the four deoxyribonucleotide substrates (6, 23). Yet, with few exceptions, the literature and textbooks seem unaware of any possible regulatory role of deoxyribonucleotides. Instead, merely the pathways to their syntheses may be described. The student is left with an ostensibly established concept that dNTPs are formed in excess, flooding the hungry DNA replication apparatus and marking time until replication switches on. Only the control of initiation of DNA replication and the exponential increase in DNA growing points (24) might be considered in describing the kinetics of the process.

To indicate the contrast in interpretation that prevails, in a recent study it was proposed that RnR subunits may be produced only after DNA origins have been activated (25). As discussed in Sections II and III, in T4 infection the observed initial exponential character of the increase in DNA replication is controlled by the kinetics of formation of the dNTPs, apparently at the level of assembly of the T4-induced dNTP synthetase complexes. The notions of "exclusive control" by DNA replication may have been reinforced by the studies of Arthur Kornberg, an early advocate of the important role played by dNTPs, and co-workers, who painstakingly unraveled a set of exquisite initiation mechanisms for the *E. coli* DNA replication apparatus (23). The T4 DNA replication fork has been reviewed recently (23a).

These initiation controls cannot be effective in replication until dNTPs become available from ribonucleotide reductase and associated deoxyribonucleotide enzymes. However, the function and control of the complex of enzymes initiating DNA replication are in no way disavowed by the factors regulating the initiation of dNTP synthesis. In our view, these are systems initiating in series and functioning in parallel.

II. Time of Initiation of Bacteriophage T4 DNA Replication

After phage T4 infection, an amazing number of events, involving the formation of many messengers, proteins, and other products, take place in a few minutes: i.e., immediately, at early, and at delayed-early times. "Late" proteins are formed only after a round of replication. But T4 DNA replication is detected only after about 5 minutes. Let us then consider the chronology of some of these events. First, when does T4 DNA replication initiate?

A. *In Vivo* Analysis of Pyrimidine Deoxynucleotide Exponential Synthesis and DNA Replication

To analyze these relationships kinetically and to develop tools to examine, *in vivo*, our earlier suggestions of a phage-induced complex synthesizing dNTPs (26, 27), Tomich measured both deoxyribonucleotide synthesis and DNA replication by administering a mixture of [5-^{3}H]uridine and [6-^{3}H]uridine to an *E. coli* culture shortly after infection by bacteriophage T4. The assay is based on the conversion of the labeled uridine to 5' UMP, to UTP, and to CTP, and ultimately, through the RnR pathway, to become substrates for dTMP synthase and dCMP hydroxymethylase (28). The substrates, [5-^{3}H]dUMP and [5-^{3}H]dCMP, are converted by these enzymes to dTMP and 5-HmdCMP, respectively, with an obligatory displacement of the 5-^{3}H atoms into the water phase, and ^{3}HOH is assayed. The 6-^{3}H-labeled substrates, however, form labeled dNTPs that enter into DNA replication. The two administered labeled substrates share common intermediate pools through to the formation of dUMP and dCMP.

Figure 1 shows the results of such a simultaneous measurement of deoxyribonucleotide synthesis and DNA replication *in vivo*. Because RnR is the limiting step in the series, the rate of ^{3}H release is taken to be a measure of the rate of RnR activity within the dNTP synthetase complex structure, at

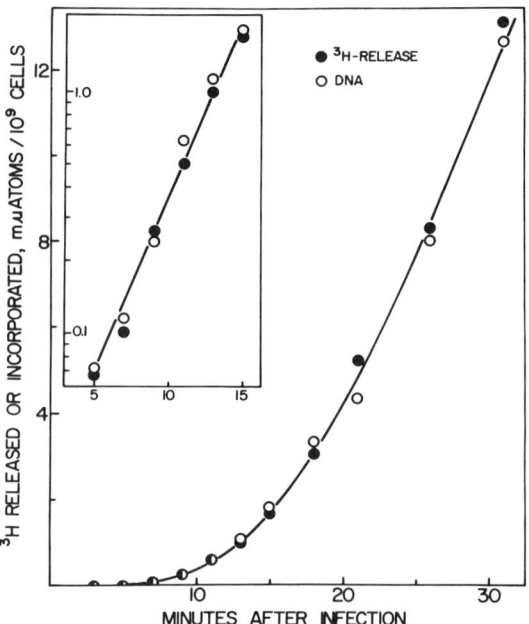

FIG. 1. Simultaneous initiation *in vivo* of tritium release (pyrimidine deoxyribonucleotide synthesis) and DNA replication after phage T4 infection. Note that the curves of the two processes are superimposable. The inset shows the ^{3}H release and DNA replication plotted on a semilogarithmic scale. From Ref. 28, with permission.

physiological concentrations of the RnR substrates, CDP and UDP, and the substrates for dTMP synthase and dCMP Hmase. Two major points were demonstrated: (1) the rate of deoxyribonucleotide synthesis and the rate of DNA replication were equal, and (2) the two curves were initially exponential, apparently beginning at about 5 minutes after infection and coinciding exactly. Because DNA replication depends on dNTP synthesis, the curves will be parallel or superimposable.

Concurrency of the two curves means that the concentrations of free deoxyribonucleotide intermediates are very low. Note that the ^{3}H-release values do not represent concentrations of HmdCMP plus dTMP derivatives, but rather a trail reflecting their cumulative syntheses versus time.

To determine the times of initiation of dNTP synthesis and of T4 DNA replication, chloramphenicol was added at various times after infection to decrease the concentrations of the enzymes involved in dNTP synthesis (28, 29). Semilogarithmic graphical presentation of the log phases of several resulting curves gives 4.8 minutes as the consistently reproducible point in time that the process begins after infection at 30°C (Fig. 2).

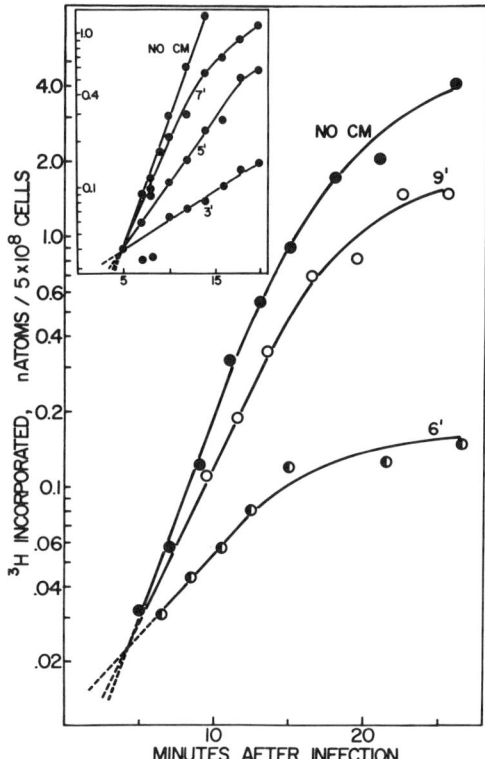

FIG. 2. Time of initiation of T4 DNA replication. At the indicated times after infection of a culture at 30°C, chloramphenicol (CM) was added. The inset shows the time of initiation of tritium release activity (HmdCMP and dTMP synthesis) run at a different time (28). The time of initiation of the two processes was 4.8 minutes with an estimated maximum error of ±0.5 minutes. From Ref. 29.

The kinetic relationships of dNTP synthesis to the dNTP synthetase complex and other aspects of this complex, as well as the quantification of the tritium release pathways and the role of dCMP deaminase, are returned to in Section X.

III. Expression of T4 nrdA and nrdB Genes

A. Events Leading to Expression

In 1969, phage T4 RnR activity was described by Berglund *et al.* (30) and the *nrdA* and *nrdB* genes were identified as well by Yeh and Tessman (31).

Earlier, Cohen had shown an increased RnR activity in T6-infected cultures (32). There followed a series of papers by Berglund on the purified enzyme (see Ref. 33), on the T4-encoded thioredoxin by Berglund and Sjöberg (32), and on the T4 thioredoxin gene, *nrdC*, by Yeh and Tessman (35).

Figure 3 shows an event line of our current information on the initiation of expression and translation of the phage T4 *nrdA* and *nrdB* genes. *nrdA* lies about 142.7 to 140.4 kb on the T4 168.9-kb genetically circular genome (36, 37), 700 nt upstream of *nrdB*. *nrdA* is positioned just downstream of *td* (thymidylate synthetase), which is downstream of *frd* (dihydrofolate reductase). The *frd*, *td*, *nrdA*, and *nrdB* genes are expressed on a strong "early" promoter, P_E, at about 146.8 kb, and its message continues through the downstream *denA* (endonuclease II) gene and apparently beyond (38, 39).

nrdB also has its own middle promoter that requires the T4-encoded MotA protein for activity. The latter protein derives from the *motA* gene and regulates middle gene expression (Fig. 3). The *nrdB* promoter initiates the *nrdB* messenger at about 2 minutes after infection, but the P_E messenger

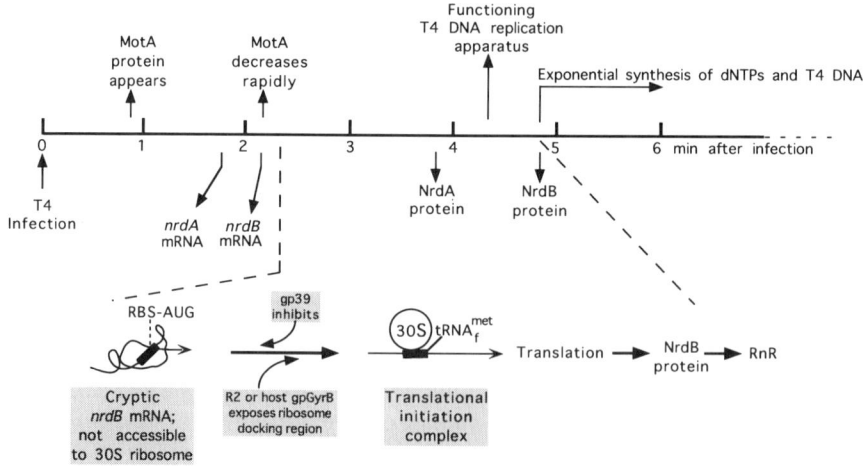

FIG. 3. Chronology of appearance of components participating in the synthesis of phage T4 RnR and thence replication of T4 DNA; cryptic control of the initiation of *nrdB* mRNA translation. The figure shows that the time of initiation of T4 DNA replication depends both on the activation of the *nrdB* promoter by T4 MotA protein and the effect of a translationally cryptic *nrdB* mRNA structure on the appearance of R2 protein. RnR enters into the formation of the dNTP synthetase complex, and dNTPs are funneled to the replication complex, already in place. Not to scale.

does not reach the *nrdB* gene until about 5 minutes, *i.e.*, after transcription and translation of the *nrdB* mRNA have initiated. Therefore, P_{nrdB} provides the mRNA during this crucial period, both for transcriptional and translational initiation (*39*).

In summary *nrdA* is transcribed by a multicistronic mRNA, and *nrdB* is transcribed *both* coordinately, from the *nrdA* message, and separately, from the *nrdB* promoter. Transcription of *nrdB* from its own promoter occurs long before the multicistronic message reaches it. A clear delay in the appearance of the NrdB protein indicates that transcription and translation are dissociated.

B. The T4 *nrdB* mRNA Translational Delay: *nrdB93*, a Fortuitous Mutation

Arising from the chronology of these events is the question: why does it take more than 2.5 minutes after *nrdB* mRNA initiates at the *nrdB* promoter to initiate the synthesis of the NrdB protein—even considering a brief delay to excise the *nrdB* intron? Commonly, transcription and translation occur concurrently, but in this case the two processes are separated. The P_{nrdB} messenger, measured through the *nrdB* translational frame, is 1.92 kb. By contrast, the mRNA initiating from the P_E promoter through the *nrdA* gene is 6.42 kb and may be expected to take some three times longer to be transcribed. Accordingly, the R2 protein ought to appear before the R1 protein. In fact, R2 protein appears up to 1 minute after R1, *i.e.*, at about 4.8 minutes (Fig. 3).

We were first directed to the regulation of *nrdB* by a temperature-sensitive missense mutation, *nrdB93* (*40, 41*), having a Gly-253-to-Glu switch. This mutation leads to a great lability of the protein even at lower temperatures, and at 41°C both a rapid loss of activity and an inability of the α_2 subunit to bind to the β_2 unit. *nrdB93* was discovered as a second mutation in a well-studied gene-39 mutant, now named 39-01 (DNA topoisomerase). *nrdB93* grows and forms its product poorly. However, 39-01 acts as a suppressor of the *nrdB93* defect, restoring wild-type growth, except that in its initial stages the infection is temperature sensitive. The gradual conversion of the suppressed system to temperature stability may result from the incorporation of the $\alpha_2\beta_2^{93}$ enzyme into the tightly structured dNTP synthetase (*42*). The suppression by gene-39 mutants appears to be site specific. Gene-39 amber mutants also cause suppression. Therefore, the defect in *nrdB93* infection results from the inhibition by gene-39 protein. Most enticingly, the suppression did not occur in the presence of a mutation in the host gene, *gyrB* (DNA gyrase; DNA topoisomerase II) (*40, 43*).

C. Block in T4 nrdB mRNA Translation: Linear, but Cryptic, 30-S Ribosome Binding Sites

nrdB93 is blocked post-transcriptionally (44). Because translation was implicated, the wild-type system was examined[2] for the binding of 30-S ribosomes to the nrdB mRNA ribosome docking region, an approximately 34-nt segment covered by the ribosome and including the RBS and AUG start-codon binding sites (45). The binding process is the initiation of translation and the most common step for a defect. For this study, a phage T7 promoter plasmid was constructed to synthesize an nrdB mRNA fragment that included a 159-nt segment upstream of the encoding region, constituting the 5' terminus as well as the translational initiation region and extending 78 nt into the translational frame.

The binding of 30-S ribosomes was measured by the primer-extension inhibition "toeprint" analysis (45). The movement of the reverse transcriptase was blocked predominately at nucleotide +16 (numbering the A of the AUG start codon as +1), but only when the very large 30-S/tRNA^{fMet} initiation complex was assembled at the mRNA ribosome docking region.

The 30-S ribosomes (plus tRNA^{fMet}) were unable to bind to the mRNA, even though the ribosome concentration was raised eight-fold over the usual values in the literature. When the temperature was raised to 41°C, or when even higher concentrations of 30-S ribosomes were used, binding occurred. Therefore, the process was reversible and could be driven by mass action.

When pure NrdB protein (β_2 or β_2^{93}) or GyrB protein was added to the system, binding of 30-S ribosomes to the ribosome docking region occurred readily. On the other hand, gene-39 protein inhibited the binding process. On retardation gels, these proteins showed specific complexes with the mRNA and titration curves of the mRNA fragment with the several proteins showed binding constants ranging between about 35 and 200 nM.

An analysis of the TIR by single- and double-strand RNases (46, 47) provided the structure shown in Fig. 4. Additional analysis of the ribosome docking region verified that it was single-stranded; i.e., its inaccessibility could not be explained by the more common finding that RBS or AUG segments may be secluded within secondary structures (48). For this reason and because of the effects of temperature and of high ribosome concentrations and some aspects of the structure of the TIR, we propose that the TIR has a tertiary structure that renders the linear RBS and AUG segments inaccessible (cryptic) to 30-S ribosomes. RNase protection studies showed that the β_2 and GyrB proteins bound individually, but in an overlapping manner, to the TIR in

---

[2] P. He and G. R. Greenberg, unpublished. Presented at May, 1994 meeting of the American Society of Biological Chemistry, and Molecular Biology, Washington, D.C.

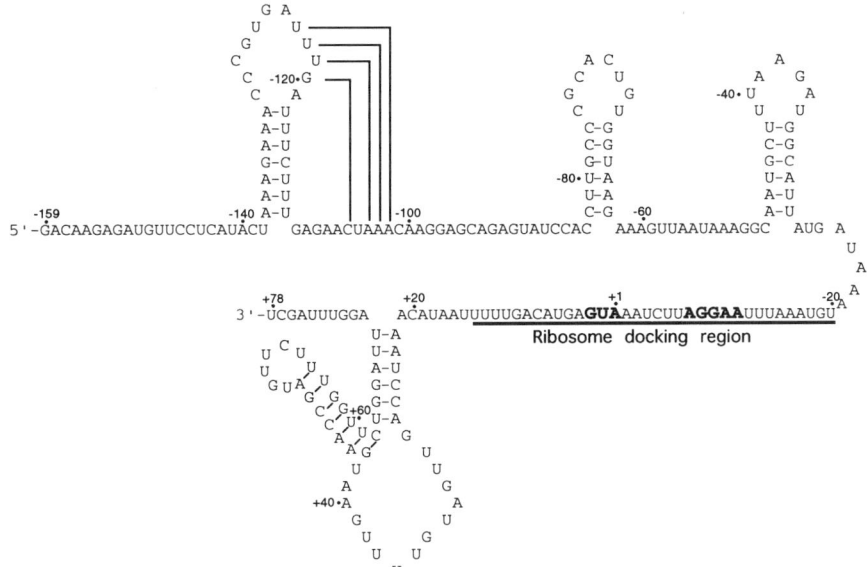

FIG. 4. Proposed secondary structure of translational initiation region of phage T4 *nrdB* mRNA. The positions of the Shine and Dalgarno ribosome binding site and the AUG start codon are shown in bold letters, and the ribosome docking region, the sequence considered to be covered by the 30-S ribosome (45), is indicated. This structure was determined by analysis with single- and double-stranded RNases, but the sites of cleavage are omitted for clarity.

the region from about -49 to -152 nt from the AUG start codon. In initial experiments, gp39 appeared to protect the AUG codon.

By our interpretation, GyrB or β_2 (or β_2^{93}) protein causes a reversible unfolding of the TIR, and the ribosome binding region becomes accessible to 30-S ribosomes. At 30°C the process may be considered as a series of reversible reactions (shown with GyrB protein as the activator, where c is closed and o is open):

$$mRNA^c \rightleftharpoons mRNA^o$$
$$mRNA^c + GyrB \rightleftharpoons mRNA^o \cdot GyrB$$
$$mRNA^o \cdot GyrB + 30S\ ribosome/tRNA^{fMet}$$
$$\rightleftharpoons mRNA^o \cdot 30S\ ribosome \cdot tRNA^{fMet} + GyrB$$

The equilibrium of the first reaction lies far to the left, whereas that of the second lies far to the right and is called the "open-sesame reaction." At 41°C, the first reaction is also forced to its open form.

gp39 is an immediate-early product, and it would inhibit the opening of the *nrdB* mRNA 30-S ribosome binding region. However, gp39 enters into the tightly associated T4 DNA topoisomerase complex with gp60 and gp52 (*49*), and the question may boil down to the excess or availability of gp39. Neither gp39 nor the NrdB protein has been tested *in situ* in the whole enzymes. By contrast, gpGyrB is expected to be available immediately because DNA gyrase is readily dissociated to its two subunit proteins (*43*). The delay in the time of the putative opening of the TIR appears to depend on a competition between the action of gp39 on the one hand and the R2 and GyrB proteins on the other.

D. Summary

The *nrdB93* mutation has helped to give the following insight into T4 *nrdB* expression. In the expression of phage T4-encoded *nrdB*, translation is about 2 minutes behind transcription. In *in vitro* studies, the formation of the ribosome/tRNA^{fMet} · mRNA translation initiation complex at the *nrdB* mRNA ribosome docking region is blocked, but is overcome by the NrdB or GyrB protein, which binds to the mRNA translational initiation region. T4-encoded p39 inhibited the formation of the 30-S translation initiation complex. The complexity of this process plus the time to initiate *nrdB* transcription is considered to be part of the clock mechanism determining the initiation of RnR activity and of DNA replication from *de novo* sources of dNTPs (Fig. 4). By our proposal, the ribosome binding sites within the mRNA TIR three-dimensional structure is cryptic and inaccessible to 30-S ribosomes, but the activating proteins unveil the binding site.

IV. Expression of *Escherichia coli* nrdA and nrdB Genes

Much of the insight into the nature and regulation of expression of the *E. coli nrdA* and *nrdB* genes has come from studies over some 20 years by Fuchs and his students at the University of Minnesota. The two genes, lying at about 48 minutes on the *E. coli* chromosome, are transcribed on a polycistronic message in a clockwise manner in the order, AB. Figure 5 shows the region upstream of the promoter. The message begins 110 nucleotides upstream of the *nrdA* translational frame.

Beginning with the observation that inhibition of DNA replication increases RnR synthesis (*50*), these workers demonstrated that the regulation of *nrdA* and *nrdB* expression is cell-cycle regulated, that its synthesis parallels the initiation of DNA replication, and that *nrd* expression is coupled to the point of initiation of DNA replication at the *oriC* site on the chromosome.

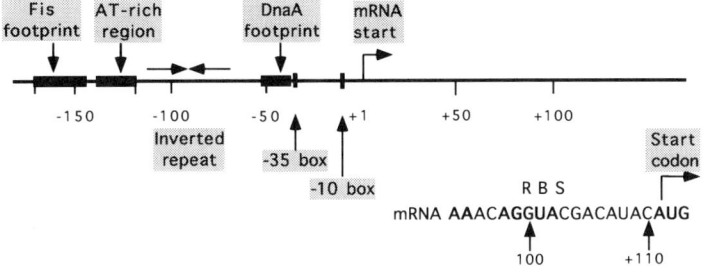

FIG. 5. Promoter region of *Escherichia coli nrd* mRNA and regulation of expression (see Table I for sequences). The figure shows the segments that carry the binding sites of the Fis and DnaA proteins, important in the expression of the gene. An AT-rich region is required to couple the expression of the *nrd* promoter to the cell cycle, and for coupling to DNA replication at the *oriC* site. The functioning of the −35 and −10 boxes (bold) in RNA polymerase binding and the action of Fis protein on expression require DNA supercoiling.

A. Roles of *fis* and *dnaA* Genes in *nrd* Promoter

The Fis and DnaA proteins bind to separate specific sites in the *nrd* promoter region (*51*). Fis protein is known to be involved in widely varied DNA interactions, including the initiation of replication at the *oriC* origin, inversion stimulation, and recombination reactions, and DnaA protein is a basic player in the initiation of DNA replication (*52*). *nrd–lac* fusion experiments demonstrated that homogeneous Fis protein binds to the segment between 142 and 171 bp, upstream from the site of initiation of the transcript, and causes an increase in the rate of expression (Fig. 5). In turn, DnaA protein activates the *nrd* promoter by binding to a 22-bp segment, immediately upstream of the −35 RNA polymerase binding site.

Accordingly, the expression of the *E. coli nrd* system requires two proteins, DnaA and Fis, both appearing to have connections to DNA replication. Notwithstanding, Sun *et al.* (*53*) reported that although *nrd* expression was cell-cycle regulated and the DnaA and Fis proteins were required for promoter activity, neither of the proteins appeared to be necessary for the control of *nrd* expression by the cell cycle. For the time, the problem was compounded.

B. Relationship of *nrd* Expression and the Cell Cycle: Superhelicity in the Promoter Region

Instead of the DnaA and Fis proteins being involved, connection to the cell cycle depends on a *cis*-acting region between −139 and −124 bp. The functionality of this segment appears to be related to its AT richness rather than to its sequence (54). Just downstream of the AT-rich region (Table I) is another possible component of the *cis*-acting region, a set of long inverted repeats, presumably able to form a C-type cruciform (citations in Ref. 54). The positive action of the Fis protein on the expression of the *nrd* promoter in an *in vitro nrd–lac* fusion system depends on the supercoiling of the DNA

TABLE I
nrdA AND *nrdB* CONTROL SEQUENCES FROM DIFFERENT SOURCES

Source	Sequence	Identification	Ref.
E. coli nrd 5' UTR	-55 GAG**TTATCCACA**AAGTTATGC -35 DnaA consensus box	DnaA footprint	51
E. coli nrd 5' UTR	-171 ATTATGCCGTTCAAGAAATCGCCGAACAGT -142	Fis footprint region, centered on -156	51
E. coli nrd 5' UTR	-139 TTTTTAACAAATTTTT -124	AT-rich region	53
E. coli nrd 5'UTR	-113 TGACTTTCCCGGACACCTTGTC(T) -91	One of the inverted repeats (5')	53
E. coli nrd 5' UTR	-32 **TTGCAA**GAGGGTCATTTT**CACACTATCTT**GCAG +1, mRNA start ⟵ 17 nt ⟶	-10 (Pribnow) and -35 boxes	54
Mouse nrdA 5'UTR	β -189 TTGCCCACAC**C**CAATATGGCGGC -167 α -98 TTGCCCACACTCAATATGGCGGC -76	23-nt segments bold bases differ	72
Mouse nrdA mRNA 3'UTR	5' CUAGACAAACUUCUAUAAGUCAUUUUG- AAAUAAACAUUUCUAAGUGAUA-; Bold: nrdA mRNA polyA signal	49-nt p57 binding site	75
Mouse nrdB mRNA 3' UTR	+1861 AAUUCACAAGUGAGUUUGAGCCCAGUGGU GGGUACACCCGUGGGACUCUCUUACAAACCAA AACAGGAAAAGCAAGUGUUCCC +1943	p75 binding segment	81
Mouse nrdB mRNA 3' UTR	+1234 AAGUAAAUGAUCGUGUGCUC +1253	p45 binding site	80
Mouse nrdB intron 1	Exon/intron boundary ⟵—87nt—⟶ ACTGGCAGGTCGTGGGTTC CCGGACCC**GGGAGGG**TGGAGGGTGCCTACCG*, *block (text)	Potential Me1al protein site (bold)	78
Human c-myc Human C2 Consensus	**GGGAGGGG**AGGG **GGGAGGGG**GAGG **GGGAGGGG**RRGG	Me1al protein binding site; human complement	79
Yeast nrdB 5' upstream	-489 ACC**ACACCCACGCGCG**ATCGCCATGGCAACGAGGTC GCACACGCCCC**ACACCCAGACCTC**CCTGCGAGCGGG CATGGGTACAATGT-403 Bold: RAP1 consensus sites	DNA-damage responsive element (DRE). Binds several yeast proteins	100, 101
HSV-2 nrdA OCT-1/GARAT	-137 ATGCAAATGGGATTCA -122	Potential Oct-1 and VP-16 binding site	132
HSV-1 nrdA OCT-1/GARAT	-135 ATGCAAATGGGATACA -120	As above	135
HSV-1 nrdA TATA box	-39 GCGCGATGTGG**ATAAAAA**GCCAGCGCGGGTGGTTTAGGG -1	ICP-O-responsive site	135
S. solidissima nrdB 3'UTR	+1281 ATACTGGAAGCGTTGTTGTTTCATTGGTTT TAAAACATTTTAATAATAGTGCAAATTATA ATTTTACTTTTTTATTAGTGTGCAATAATG CATGAATTATGATTTTACCGGTTATTTATT TTTGTAAATTTGGT +1414	134-bp translational masking region	111

and the presence of the AT-rich region. To illustrate how sensitive and accurate the control is, if constructs contained inserts between the Fis binding site and the promoter, the system did not show positive activation by Fis. Such constructs were also insensitive to supercoiling and were not regulated by the cell cycle *in vivo*.

The Pribnow and the -35 boxes of *nrd* are separated by 17 bp. At a separation of 17 bp, the two hexameric boxes lie on the same face of the DNA helix, and this orientation has been accepted as a specific landing dock for RNA polymerase. The concept of twist-sensitive promoters (55) relates to DNA superhelicity and the spatial orientation of the two components, but conceivably could involve the nearest-neighbor composition of the spacer region, among other variables.

C. Summary

Escherichia coli presents a sophisticated combination of systems regulating the initiation of ribonucleotide reductase and tying its activity to the cell cycle. As yet, the exact fit of these systems remains to be established, but workers in this area know that (1) activation of the *nrd* promoter requires that both the DnaA and Fis proteins bind to their respective specific sites, (2) an AT-rich region upstream of the site of initiation of the mRNA of the *nrd* operon is required for coupling to the cell cycle, (3) DNA supercoiling is required for positive regulation of the promoter by Fis protein, and (4) a synergistic action of Fis protein and DNA gyrase (type II DNA topoisomerase) appears to be functioning in the initiation of DNA replication at the *oriC* site. Indeed, Fis protein has been shown to bind to the *oriC* site (56).

V. Phospholipid Synthesis and mRNA Breakdown Provide CMP for Synthesis of CDP and dCDP

It has long been accepted that CTP, formed by the amination of UTP, is converted to CDP by the ubiquitous nucleoside diphosphate kinase: CTP + ADP ↔ CDP + ATP, which shows an equilibrium constant of about 1. CDP is then converted to dCDP by the irreversible RnR reaction.

Fricke *et al.* have now challenged this picture by the discovery of the primary source(s) of CDP in *E. coli* (57). CMP kinase converts CMP to CDP. CMP, though not formed by a *de novo* pathway, is the normal product from the lipo-CDP intermediate in phospholipid synthesis, which, in turn, comes from CTP (Fig. 6). CMP also is formed by the degradation of mRNA [recalling the finding of Cohen and Barner (32) that labeled mRNA appears to be an efficient precursor of dNMPs in phage T6 infection].

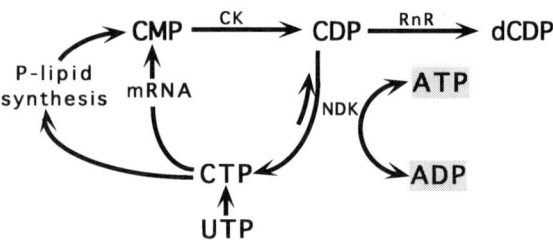

FIG. 6. A major role of cytidine monophosphate (CMP) kinase in dCDP formation. Nucleoside diphosphate kinase (NDK) does not provide sufficient CDP for dCDP synthesis. CMP, formed in the synthesis of phospholipids and in mRNA degradation, is converted by CMP kinase to CDP.

A mutation in the CMP kinase gene greatly decreases DNA replication. Therefore, CDP does not appear to come primarily from the NDK reaction, but from CMP kinase, as shown in Fig. 6.

On first thought, this phenomenon might result merely from the prevailing high ATP-vs.-CTP concentrations. However, by such an argument, CDP derived from CMP would also be driven by ATP to CTP through NDK. Because the levels of the four diphosphates appear to be adequate in wild-type organisms (58, 59), this is an unlikely notion. GDP, ADP, and UDP do not require NDK in their synthesis, but can be derived directly from their monophosphates. We are unaware of any evidence with purified NDK of a kinetic quirk that could explain the apparent inability of the enzyme to carry out the CTP → CDP reaction *in vivo*.

We see no obvious argument against the same CMP bypass prevailing in eukaryotes, which also show the UTP → CTP and CMP → CDP pathways and possess the NDK enzymes.

It is possible that compartmentation can be implicated to explain the NDK anomaly, though the CMP → CDP reaction also requires ATP. The idea would require that the kinase be in a "compartment," ostensibly associated with phospholipid synthesis, and that CDP would enter a channel to dCDP and to dCTP, and thence to DNA, *i.e.*, a dNTP synthetase-like complex. Conceptually, the proposal is made tortuous by the need for NDK in the complex at the dNDP → dNTP level, but not in the phospholipid synthesis compartment, and by the inherent danger in interpreting *in vivo* data in the realm of compartmentation. Still, several studies presenting convincing arguments for such compartmentation in eukaryotes may be consulted (60–64).

Evidence has been presented that 3T3 mouse fibroblasts show compartmentation of deoxycytidine nucleotides during the S phase of the cell cycle (63). One pool was labeled preferentially from cytidine and then entered into

DNA replication. The other pool derived from deoxycytidine, which enters into dCDP liponucleotide synthesis. The investigators concluded that DNA and deoxyliponucleotides are synthesized from separate dCTP pools. Labeled cytidine forms CDP liponucleotides as well as leading to dCTP for DNA replication. From comparisons of nucleotide pools and from a consideration of the pools in the G0 and S phases of the cell cycle, proposals have been made (63) that dance close to the idea of a sequestered "channel" from CMP to DNA. These findings are in keeping with the earlier suggestion that liponucleotides are formed in a membrane fraction of the cell (64).

VI. Control of Ribonucleotide Reductase Synthesis in Eukaryotes and Activity in Cell Division Cycle

For some years, it has been recognized that RnR activity occurs in conjunction with the appearance of the S phase of the cell cycle of eukaryotic systems (65), and that the activity represents newly synthesized RnR (66). Ultimately, the activity rises just prior to and merging into the time of replication of DNA and then falls. By the use of specific inhibitors and activators of the cell cycle, RnR has been shown to be formed in the same time period without DNA replication. Recently, these kinetic observations became more intriguing. In mouse cell cultures, RnR activity reflects the synthesis and degradation of the R2 subunit of the enzyme (67). By contrast, the R1 subunit appeared to be synthesized throughout the cycle and remained relatively constant (67, 68). These properties may, in part, reflect a half-life of greater than 24 hours for R1 (69) and 3 hours for R2 (67).

An obvious question arises: is it possible that the time of appearance of DNA replication activity during cell division and its definition of the S phase may in some instances be reflections of the time of appearance of RnR activity, even though inhibitors allow RnR synthesis without concomitant DNA replication and without entry into the S phase? Such a proposal will, at least, require evidence that factors ascribed to the cell division cycle and the initiation of DNA replication are intimately related to or physically interact with the messenger RNAs forming the R1 and/or R2 subunits. Is RnR merely an end product in the regulation of the cell cycle, *i.e.*, the fuel injector for the replicating machine, or does it also have a more direct role in the timing mechanism, the phasing of the cycle?

Table II is a summary of the controls operating in the normal cells and also the effects of cell-cycle mitogens and other agents on the formation and stability of the mRNAs of the *nrdA* and *nrdB* genes. We discuss part of this table. [A major portion of the work has come from the laboratories of James A. Wright and Lars Thelander.]

TABLE II

EFFECTS OF CELL-CYCLE FACTORS AND OTHER AGENTS ON SYNTHESIS OF nrdA AND nrdB mRNAs AND R1 AND R2 SUBUNITS OF RIBONUCLEOTIDE REDUCTASE IN MOUSE AND MONKEY CELLS

Added factor or agent	nrd	Intermediate factor[a]	Action[b]	Cell line	Comments	Ref.
1. None	A	Normal p57 R1BP	p57 destabilizes nrdA mRNA by binding to a 49-nt segment of 3' UTR	COS-7 monkey line and BALB/c 3T3	Protein kinase C is vector	75
2. None	A	Promoter control	3 protein-DNA complexes formed upstream of the TATA-less promoter(See text)	BALB/3T3	One complex may be with Sp1 protein, another S-phase-specific	72
3. None	B	Block is in first intron	A G_0/G_1 transcriptional block is released by binding S-phase-specific nuclear protein	BALB/3T3	Cell-cycle regulated gene expression	78
4. None	B	p45(R2BP) (Cytosol)	Binding of p45 to 3' UTR of nrdB mRNA apparently required for degradation	BALB/c 3T3	p45 binds to 20-nt sequence	80
5. Phorbol ester	A	p57 (R1BP)	Half life of nrdA mRNA ▲ 5-fold. Phorbol ester decreases binding of p57 to 49 nt (No. 1).	Monkey COS-7 and BALB/c 3T3	Protein kinase C vector. Stabilizes mRNA.	75
6. Staurosporine	A	p57	▲ R1BP-mRNA complex (49-nt segment). Agent inhibits protein kinase C.	Monkey COS-7; BALB/c 3T3	Decreases mRNA stability	76
7. Phorbol ester	B	p45 (R2BP)	▲ in half life of nrdB mRNA by 3-fold, apparently by down-regulating p45 binding (No. 4)	BALB/c 3T3	Stabilizes mRNA	80
8. cAMP signal	A, B	Agents such as forskolin, cholera toxin, 8-Br-cAMP, and IBMX	▲ levels of both nrdA and B mRNAs via cAMP-dependent kinase A. Not dependent on S phase.	Malignant H-ras-transformed mouse cells	Effects not observed in parental cell line	88
9. $p34^{cdc2}$ and CDK2 protein kinases	B	Cell-cycle regulatory factors	Phosphorylation of purified R2 protein may be at Ser-20; presumably ▲ enzymatic activity.	Mouse L-cells	Other kinases not active. Also occurs in vivo.	91
10. TGF-β1	B	Induces 75kD cytoplasmic protein	Blocks degradation. nrdB mRNA ▲ >6 times. p75 binds to 83-nt segment of 3' UTR.	BALB/c 3T3 mouse	Ras-transformed cells uncouple TGF-β1 signal	81, 89
11. Okadaic acid	A, B	Protein phosphatase inhibitor; non-phorbol ester tumor promoter	Marked ▲ in nrdA & B mRNAs and R2 protein. RNR ▲ ca 4-fold.	BALB/c 3T3 mouse	▲ in enzyme activity may be via phospho-R2 level	77
12. Calyculin A	A, B	As in No. 11	Rapid rise in nrdA and B mRNAs	BALB/c 3T3 mouse	R2 increases	77
13. Chorambucil	A, B	Damages DNA	Transient ▲ in nrdA & B mRNAs and ▲ only in R2 protein	BALB/c 3T3 mouse	May involve protein kinase C	90

[a] IBMX, 3-Isobutyl-1-methylxanthine.
[b] ▲, Increase(s).

A. Mouse nrdA and nrdB Genes

The synthesis of eukaryotic ribonucleotide reductases presented in this discussion was studied in the cell lines indicated in Table II. The full-length cDNA carrying the mouse nrdA gene has been mapped to chromosome 7, and *nrdB* was found in chromosome 12 (70, 71), ruling out multigenic relationships comparable to those of prokaryotic operons. *nrdA* has been cloned and sequenced. Its gene is 26 kb long and is made up of 19 exons; all of the exon/intron splice junctions have been sequenced (72). R1 protein is an 180-kDa homodimer consisting of 792 amino-acid residues (73) and carries binding sites for dNTPs and ATP allosteric effectors and the sites for the rNDP substrates.

Figure 7 depicts the known controls of the expression of the *nrdA* gene and the stability of its messenger. Because this gene lacks a TATA box, it has evolved an elaborate set of controls that, among others, links it to the S phase of the cell cycle. Table II summarizes some of these relationships and the effects of a number of agents discussed below.

Upstream of the first of multiple mRNA starts lie two 23-nt segments, α and β, 68 bp apart and differing by only 1 bp; α extends from −76 to −98 and β from −167 to −189 (Table I). In footprinting studies, the regions α and β were exclusively protected against DNase-I digestion by proteins from a nuclear extract.

To examine the nature of the protection, factors from the nuclear extract were allowed to bind to a synthetic double-stranded oligodeoxynucleotide corresponding to the β segment. Three protein–DNA complexes, A, B, and C, were identified by gel electrophoresis retardation (72). One protein was specific to the S phase. Another had properties of Sp1, a well-known promoter-specific transcription factor (74). Using a slightly different synthetic oligodeoxynucleotide, Sp1 successfully competed with protein B to form the complex.

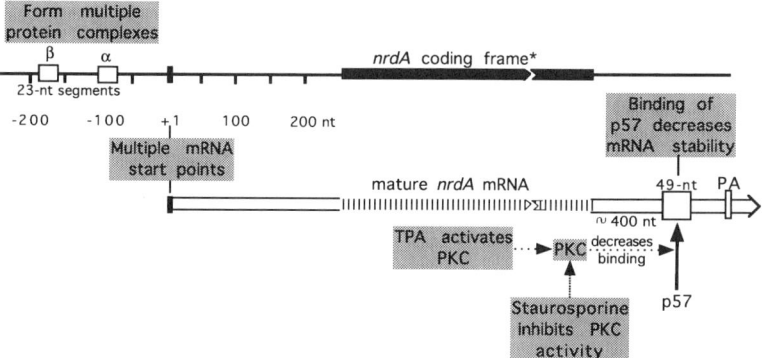

FIG. 7. Mouse *nrdA* gene, TATA-less promoter, and control of mRNA degradation (parts of the figure are not to scale). The scheme for the actions of phorbol ester and staurosporine on the binding of p57 to the 49-nt 3′ UTR segment represents our interpretation of the studies.

B. Factors Controlling Mouse nrdA mRNA Stability

Not only does the mouse *nrdA* gene exhibit a remarkable control of its transcription, but it has developed an elegant set of controls to regulate the stability of its mRNA. About 400 nt downstream from the carboxyl terminus of the coding region lies a 49-nt segment that binds a cytosolic protein with an M_r of approximately 57,000 as determined by gel electrophoresis (Fig. 7).

The action of the 57-kDa protein on the stability of the mRNA is mediated by protein kinase C (Section VI,E). Figure 7 summarizes the relationships between p57, the 49-nt piece, and PKC. The binding of p57 to *nrdA* mRNA increases the rate of degradation (75). Intriguingly, this supports the idea that this binding is decreased by a PKC-mediated pathway. In turn, TPA activates PKC and therefore would be expected to stabilize the messenger. Indeed, it did, increasing the half-life about fivefold. As would then be anticipated, staurosporine, a known inhibitor of PKC activity, though not completely specific, increased the rate of degradation of the *nrdA* mRNA. In further support of PKC involvement, these research workers tested okadaic acid, which is a nonphorbol ester tumor promoter and an inhibitor of protein phosphoserine/phosphothreonine phosphatases 1 and 2A, and which augments the action of PKC. To be sure, okadaic acid also stabilized the mRNA (75–77). In fact, its effect on mRNA synthesis was detected by 10 minutes.

Thus, three protein factors appear to control the *nrdA* promoter, one arising in the S phase and the second identified as Sp1. *nrdA* mRNA degradation is tied to a protein, p57, which binds to the 3′ UTR region and is regulated, at least in part, by the PKC pathway.

C. Mouse nrdB Expression: Truncation of mRNA in the G1 Phase of the Cell Cycle

Bjorklund *et al.* provided an important inroad into the control of synthesis of the R2 protein and its relationship to the cell cycle (78). Previously it was thought that *nrdB* transcription was initiated in the S phase. In mouse cell cultures, the *nrdB* transcription is, in fact, initiated during the G1 phase. However, the transcript is blocked in the first intron, about 87 nt downstream of the first exon–intron boundary (Fig. 8); as a result, a greatly truncated mRNA product accumulates. On further examination, still another control surfaced. Initial experiments suggested that as the S phase of the cycle is entered, a nuclear protein becomes available and binds to the gene just upstream of the block. Presumably, the bound protein overcomes the transcription block and allows continuation of RNA polymerase II and formation of the mature mRNA.

Just 17 nt upstream of the transcriptional block is the sequence GGG-

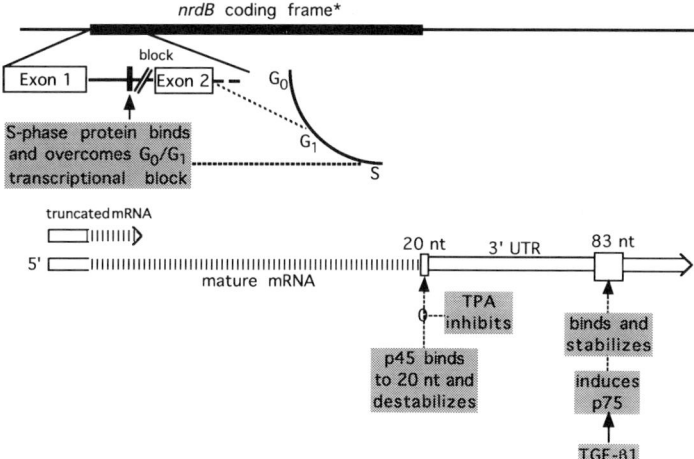

FIG. 8. Composite effects on expression of the mouse *nrdB* gene and stabilities of its mRNA (*, not drawn to scale). Three major controls are shown. One is at the level of a transcriptional block in the G0/G1 interface and is overcome by a S-phase protein; the others involve two newly discovered proteins that have opposing effects on mRNA stability and degradation in the 3' UTR. TPA acts by activating PKC as in Fig. 7, which inhibits the binding of p45. Both PKC and TGF-β1 act to stabilize the mRNA.

AGGG (Table I), which is similar to the binding site of the ME1a1 protein. This protein appears to induce the bending of DNA and to play a role in the transcription of human complement gene, *C2*, and also of c-*myc* (79).

D. nrdB mRNA: Stabilization and Destabilization in the Same Message; Roles of PKC and TGF-β1

Many mechanisms have developed in the control of the degradation of mouse *nrdB* mRNA. In the one case, a newly uncovered 45-kDa cytoplasmic protein specifically binds to the mRNA (Tables I and II) at a 20-nt segment that begins at the second nucleotide of the UAA translation terminator (80). Formation of the mRNA · p45 complex increases the rate of degradation of the messenger.

The binding of p45 to the 20-nt region in the mRNA is inhibited by the action of TPA (Section VI,B), thus stabilizing the message, in a manner comparable to its inhibition of the binding of p57 to the 49-nt region of the *nrdA* mRNA (Fig. 8). It should be pointed out that okadaic acid or calyculin A increases the levels of both the *nrdA* and *nrdB* mRNAs (Section VI,B and Table II).

In a second case, binding of a 75-kDa cytoplasmic protein to a site within an 83-nt region about 400 nt further downstream has the opposite effect on messenger stability; it causes greater stability. p75 appears to be present at low concentrations in unstimulated cells until TGF-β1 is added to the cultures (81).

TGF-β1 is the prototype of a superfamily (82) of protein cytokines. These factors not only transform, but, in different lines, actually inhibit proliferation. β TGFs interact with and activate two distantly related (by sequence) serine/threonine protein-kinase transmembrane receptors, I and II. Thus, they function through phosphorylated intermediates (83).

TGF-β1 (Table II), added to BALB/c 3T3 mouse line cultures at levels in the range of 10 ng/ml, causes a sharp increase in the level of the R2 protein of RnR. The effect derives from an approximately fourfold increase in the level of *nrdB* mRNA. In turn, the higher level results from an increased stability of the messenger. This stability is explained by the binding of a protein to the 3' untranslated region of the mRNA, a mechanism of controlling degradation that is frequently being found in eukaryotic genes.

In this instance, the evidence supports the idea that TGF-β1 action induces the formation of a cytoplasmic protein, p75, which in turn binds specifically to an 83-nt segment of the *nrdB* mRNA (Fig. 8).

E. Relationships between RnR Synthesis and G-protein Receptors, Phosphatidylinositol and Phosphatidylcholine Breakdown, PC-specific Phospholipase C, PKC, and TGF-β1

PKC has a multitude of actions as a transducing agent in cell regulation, functioning by phosphorylating specific protein serine/threonine sites (84, 85).

Figure 9 shows the relationships between phospholipid synthesis and RnR synthesis. Induction of G-protein-linked receptors activates the two phospholipase C enzymes that hydrolyze PI and PC to DAG. In the process, PI also forms phosphatidylinositol 4,5-bisphosphate, which is phosphorylated and hydrolyzed to yield inositol 1,4,5-trisphosphate, the important second messenger that acts on the endoplasmic reticulum and releases Ca^{2+}. After interaction with Ca^{2+}, PKC is believed to translocate to the membrane, binding to the negatively charged phosphatidylserine residues at the cytoplasmic interface. In this state, PKC is activated by DAG and Ca^{2+}. These relationships are described clearly by Alberts *et al.* (86). Phorbol ester appears to activate PKC by mimicking part of the structure of DAG. PKC then stabilizes the mRNAs of *nrdA* and *nrdB* (Figs. 7 and 8), and more RnR and DNA are formed. At the same time, phospholipid synthesis, especially from

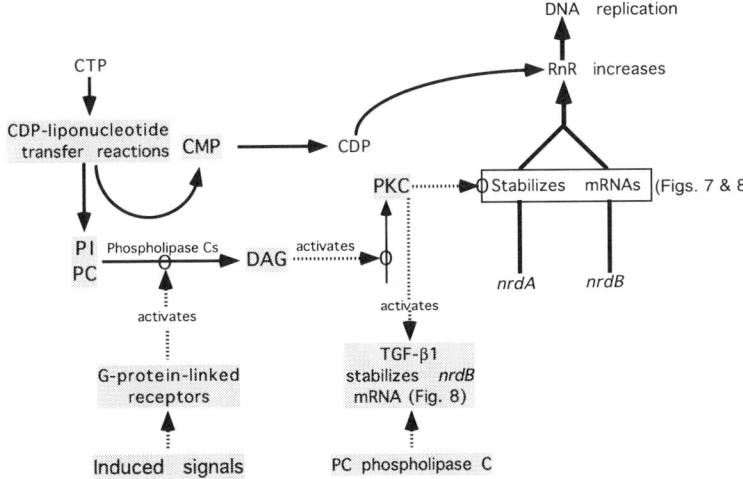

FIG. 9. A scheme showing the relationships in the mouse cell among *nrdA* and *nrdB* mRNAs, PKC, TGF-β1, G-protein-linked receptors, phosphatidyl phospholipase Cs, DAG, CDP-liponucleotides, and PC-specific phospholipase C. The interactions represent controls of the stabilities of *nrd* mRNAs by induced signals and other factors, and the relations to the synthesis and degradation of the membrane.

the far more abundant PC (and from mRNA breakdown), supplies CDP for dCDP synthesis (Fig. 6). Clearly, cyclic relationships have developed.

After these relationships became clearer to us, it was gratifying to see evidence that both phosphatidylcholine-specific phospholipase C and protein kinase C have roles in the activation of TGF-β1 (87). The involvement of PKC in the pathway was also suggested (88). These new findings have been incorporated into Fig. 9 (see also Figs. 7 and 8). Accordingly, evidence appears to be accruing for an intimate, perhaps tight, relationship between the synthesis and degradation of the membrane and RnR synthesis. The growing argument for a causal relationship of RnR synthesis with the cell cycle extends the possibilities.

F. Eukaryotic Cells: Apparent Coordinated Synthesis of *nrdA* and *nrdB* mRNAs or of R1 and R2 Proteins

In prokaryotic cells, many genes may be expressed in a coordinated manner on single, multigenic transcripts. Although mouse *nrdA* and *nrdB* genes are on separate chromosomes, in some conditions expression of the two genes or the formation of their protein products appears to be coordinated. Indeed, "coordinated" synthesis may be necessary in mammalian cells, e.g., to control the levels of both R1 and R2 to meet different biological requirements. On exposure of H-*ras*-transformed mouse 10 ½ T cells to TGF-β1, the messages of both *nrd* genes are elevated and in certain cell lines both R1 and R2 proteins are elevated (89). Chlorambucil treatment of mouse cell lines increases both messages as well, but only R2 protein is elevated (90). R1 was already present at higher levels, so that RnR activity increased.

G. Other Agents: nrdB

The activity of mouse R2 protein also appears to be regulated by cell cycle factors. CDK2 and p34^{cdc2}, both cyclin-dependent protein-serine/threonine kinases, bring about phosphorylation of purified R2 (91). The reaction appears to be limited to one amino-acid site, based on comparing the *in vivo* phosphorylation of R2 to *in vitro* experiments employing the two kinases. *In vivo*, only the R2 subunit is phosphorylated, not R1. Preliminary analysis narrowed the site to the Ser-20 positions of the dimeric subunit. Ostensibly, phosphorylation increases the activity, though at present this is not clear.

The tertiary structure of *E. coli* R2 has been determined to 2.2 Å (10), and R1 to 2.6 Å (8), but no obvious role has been described for Ser-20 of R2. However, the 33 amino acids at the N-terminal end of the β-chain do not form a regular secondary structure capable of giving a diffraction pattern, but essentially fill the gap between the two chains of the R2 protein. Although the R2 subunits of the *E. coli* and mouse enzymes show only a 24% amino-acid identity, most likely they have quite similar three-dimensional structures. For example, the mechanisms of the mouse and *E. coli* RnRs are basically the same, in terms of the placements of their iron atoms and tyrosyl radicals, and the nature of their dNTP effector functions.

A possible role for phosphorylation of Ser-20 could be at the level of a physical interaction of RnR with other enzymes.

H. Fe Requirement in R2 Synthesis

Fe(III) atoms are an integral part of the R2 active structure. From the intricate structure of the tyrosine free-radical/dioxygen (μ-oxo)-Fe(III) active site, with its iron atoms coordinately linked to many different amino acids, it follows that the incorporation of four Fe(II) atoms into the $β_2$ subunit and their oxidation to Fe(III) is unlikely to occur until the β protein chains are completed and have coalesced to their dimeric structure (9, 10).

Naturally, during the S phase at the time of maximum R2 synthesis,

Fe(II) atoms must be available in sufficient concentrations. In eukaryotes, iron is carried as transferrin (92). *Escherichia coli* contains a bacterioferritin–Fe(III) storage system (93). Because iron is needed by many enzymes (and heme proteins), the Fe(II) of the transferrin carrier pool would be expected to be distributed to each according to its K_D value and rate constant. Cytosolic aconitase, an iron–sulfur cluster protein, has two states. In the one, the [4Fe–S] state, it is fully active. In the second, in the absence of iron, aconitase is inactive but binds avidly to "iron-responsive elements" in the 5' UTRs of certain mRNAs. In the absence of sufficient Fe(II), aconitase inhibits the translation of ferritin mRNA and a number of other mRNAs, but at the same time stabilizes the mRNAs (92). A nonaconitase protein also has such properties (94). Apparently, *nrdB* mRNAs have not been examined for such binding sequences.

I. Mutational Studies Suggesting a Relationship between Mouse *nrd* Genes and Ferritin

Hurta and Wright (95), in approaching this problem in mice cell cultures, found first that resistance to hydroxyurea is associated with increased expression of both *nrdA* and *nrdB* and increased levels of the heavy and light mRNA chains of ferritin. To pursue this finding, they expanded the genetic study by employing hydroxyurea-sensitive, -resistant, and -revertant mouse L-cell lines and examining the levels of expression of the *nrdB* and ferritin genes. Hydroxyurea-resistant cells showed *nrdB* gene amplification, and increased *nrdB* message, R2 protein concentration, and RnR activity. Furthermore, resistant cells exhibited increases in the levels of the ferritin light chain mRNA and the heavy protein chain. On the other hand, revertants to sensitivity to hydroxyurea behaved like wild-type cells in the several characteristics that were changed in resistant cells. Chinese hamster ovary cells showed similar relationships. These experiments appear to support a relationship between the controls of the synthesis of ferritin and R2 protein.

Although hydroxyurea reduces the tyrosyl radical in the *E. coli* R2 protein, it does not affect the state of the iron center; *i.e.*, the latter is still in the ferric state connected in pairs (per each β chain) by μ-oxo bridges. But hydroxyurea reacts differently with mouse R2 protein. It causes the loss of both the free radical and the iron center. Furthermore, the iron is released in the ferrous form. For some time it has been known that both iron and oxygen are required continuously to convert mammalian RnR to an active state (Section I). Nyholm *et al.* (96) have wondered about the conditions of enzyme renewal, because transient reduction of the iron center could occur. Such a process may be expected to activate ferritin synthesis.

J. Yeast Ferric Reductase: Relation to RAP1 Gene

In *S. cerevisiae*, an iron requirement is associated with a membrane-spanning enzyme, ferric reductase, which converts external Fe(III) to Fe(II) for delivery to a transport system (97). Increasing the iron in the medium represses the mRNA levels of the ferric reductase *FRE1* gene, and decreases the level of ferric reductase. This repression is dependent on an 85-bp 5'-noncoding sequence, carrying a binding site for the RAP1 gene product, GRF1 (see Section VII,B), and a repeated sequence, TTTTTGCTCAYC. The binding of RAP1 protein, which has a broad range of interactions with DNA and may be part of the DNA-damage-sensing system of S. *cerevisiae* (Section VII,B), appears to point to a relationship between the regulatory structures of the promoter region of *nrdB* and the ferric reductase gene. Higher eukaryotes appear to have similar externally oriented ferric reductases. *Escherichia coli*, instead, secretes specific sideropheres that carry out the process of solubilizing external Fe(III) and delivering the iron to specific receptors for utilization by the cell.

K. Summary

Numerous links between cell-cycle specific factors and the regulatory sequences of the *nrd* genes are evident. In mouse *nrdA* expression, an S-phase-specific factor, in conjunction with the well-studied Sp1 promoter-activating protein and one other protein, interacts with *nrdA* promoter sequences to stimulate transcription. The degradation of the mRNA is under the control of p57, which binds to the 3' UTR and is itself regulated by the PKC pathway (Fig. 7).

A unique cell-cycle-dependent block in mouse *nrdB* mRNA transcriptional elongation, possibly under control of the protein factor, ME1a1, ensures that the full-length mRNA is formed only on entry of the cell into S phase. Once the complete message is formed, two proteins, p45 and p75, mediate mRNA stability by binding to sites within the 3' UTR. p45 binding destabilizes the message and is under the control of the PKC pathway. In contrast, p75 binding stabilizes *nrdB* mRNA and is controlled by TGF-β1 (Fig. 8).

Additional studies show that *nrd* expression can be coordinated, even though the *nrd* genes are on separate chromosomes, and that two cyclin-dependent protein serine/threonine kinases can phosphorylate purified R2. Finally, numerous studies link the expression of *nrdB* and proteins involved in iron metabolism. Because iron plays a crucial role in the structure of the RnR active site, this linkage would seem necessary. Taken together, the range and breadth of controls employed by eukaryotic systems in the regulation of *nrdA* and *nrdB* clearly point to the unique and critical role held by RnR in DNA replication and the cell cycle.

VII. Effect of DNA Damage on *nrdA* and *nrdB* Expression

Saccharomyces cerevisiae has provided an unusual insight into the regulation of the expression of the subunits of RnR. In 1987, the sequence of yeast *nrdB* was described (98, 99) and it was reported that DNA damage by chemical and physical agents induced the expression of the gene. This induction led to an increase in R2 synthesis and thence of ribonucleotide reductase. *nrdB* mRNA levels increased independently of protein synthesis (100), for example, by as much as 18-fold on exposure to the DNA-damaging agent, 4-nitroquinoline 1-oxide (98). R2 protein increased comparably, about 8-fold after treatment with methyl methanesulfonate (99).

A. Cell-cycle-independent Induction of Yeast *nrdB* by DNA Damage

Further study demonstrated that *nrdB* mRNA can be induced by DNA damage even when the cell cycle is blocked by treatment with yeast α-factor (102). α-Factor is a secretible, diffusible peptide hormone that blocks the cell cycle at the G1 phase and that appears to be involved in transmitting the status of mating type from one cell to another (103).

B. DNA Damage-responsive Element of *Saccharomyces cerevisiae nrdB*: The RAP1 (GRF1) Protein

In DNA damage, a number of protein factors bind to the *nrdB* promoter region (100). This binding occurs at an upstream *cis*-activating sequence that is 60% G + C and lies between −270 and −120 bp relative to the TATA segment. By deletion analysis, it is apparent that the segment has a potential repressor element and upstream activator sequences. Of this region, 87 bp constitute a DNA damage-responsive element that binds four proteins from yeast extracts; the sites have been more accurately defined (Table I).

These studies led to some intriguing new relationships. One of the four sites binds the product of the *RAP1* gene (100, 101). The *RAP1* gene produces GRF1, an abundant protein that binds to the promoter regions of many yeast genes: e.g., certain ribosomal proteins, pyruvate kinase, enolase, and the *MAT* α gene. GRF1 binds both to silencer and activator elements. It apparently acts as an activator in the *nrdB* promoter. Although deletion analysis suggests that GRF1 does not play a direct role in the DNA damage response, it may be part of a repressor system, and its concurrent role in the Fe utilization system indicates that it has a pertinent function.

C. Yeast *RNR3*: a New "*nrdA*" Gene Especially Responsive to DNA Damage

The studies of Elledge and Davis (103) uncovered a new yeast gene that has an 80% amino-acid identity to *nrdA;* it was named *RNR3*. At high concentrations, *RNR3* suppresses lethal mutations in the *nrdA* gene. *nrdA* is essential for mitotic activity, but *RNR3* is not. Both *nrdA* and *RNR3* are induced by DNA damage, but *RNR3* shows a 20- to 30-fold greater response. Induction of these genes also occurs by a block in DNA replication with hydroxyurea, which inactivates the R2 subunit, and by nitroquinoline 1-oxide and methyl methanesulfonate. Clearly, *S. cerevisiae* has developed a second defense against DNA damage.

Other relationships of the *nrd* genes have been discovered. In contrast to the findings in the mouse, the yeast *nrdA* message changes drastically during the cell cycle, whereas the *nrdB* transcript remains relatively constant. It also became apparent that, in yeast, the expressions of the *nrdA* and *POL1* genes are coordinately regulated, a relationship that is in keeping with the simultaneous appearance of the two systems.

On treatment with hydroxyurea, the cells are arrested in the S phase and form large, budded, uninucleate cells. *RAD9* is a gene that controls the response of the cell cycle to DNA damage (104). In yeast, DNA damage causes arrest of cell division at a checkpoint in the G2 phase by a process that is dependent on the *RAD9* gene product. After the damage is repaired, the cell cycle resumes. But the cell-cycle arrest resulting in a budding response does not involve *RAD9*. Therefore, it was proposed that a new pathway had been unearthed for coordination of DNA replication and the control of the cell cycle, a new checkpoint in the regulation of the cycle.

D. A Yeast Protein Kinase That Controls DNA Damage by Protein Phosphorylation: The *DUN1* Gene, an SOS Sensor?

Recently, mutations that did not allow induction of *RNR3* by DNA damage uncovered *DUN1*, a gene encoding a nuclear protein kinase (105). In the region from amino-acid 194 to 480, the *DUN1*-encoded protein shows significant identity with a number of Ser/Thr protein kinases, including *Dictyostelium discoideum* myosin light-chain kinase and mouse calmodulin-dependent protein kinase IV. The segment contains all 15 of the invariant residues and 11 subdomains conserved by the known protein kinases.

Autophosphorylation increases on DNA damage, perhaps mainly in the N-terminal segment of Dun1 protein, which contains a higher proportion of Thr/Ser residues. The authors propose that the *DUN1* gene (one of five complementary groups) is a transducer of DNA damage. In fact, Dun1 protein has two levels of phosphorylation, its normal state and an increased phosphorylation state after DNA damage. The authors also speculate that in the state of increased phosphorylation, Dun1 kinase activity on cellular kinase substrates is elevated. The Dun1 pathway provides a firm argument for a eukaryotic SOS response, and the group has proposed that it is a sensor of DNA damage.

More recently, a new mutant, *dun2*, defective both in its ability to induce *RNR3* synthesis and its entry into mitosis, was isolated (*106*). Dun2 protein is, in fact, Pol2 protein, the DNA polymerase epsilon, which is required for the S-phase checkpoint and the activation of Dun1 kinase. Navas *et al.* proposed that DNA Pol-epsilon coordinate the responses of transcription and the cell cycle to blocks in replication. Clearly, these workers are approaching the nature of the interreactive controls that appear to be operating among DNA replication, the cell cycle, and the synthesis and function of RnR.

E. Effect of DNA Damage on Expression of Mouse *nrdA* and *nrdB* Genes

Mouse *nrd* expression is also affected by DNA damage. Exposure of BALB/c 3T3 mouse cell cultures to the alkylating agent, chlorambucil, for 2 hours causes increases of about 10-fold in the *nrdA* and *nrdB* mRNAs and a concomitant rise in RnR of about 4.5-fold (Table II). The change in RnR activity may be assigned to increases in the R2 protein of up to sixfold (*90*). In extracts, chlorambucil inactivates RnR in a dose-dependent manner.

None of these *in vivo* effects changed the rate of DNA replication. In such cases, the increase in RnR activity normally is taken as that required for DNA repair (*90*).

From TPA studies (Figs. 7 and 8), it appears that chlorambucil acts through the PKC pathway. Thus the expression of mouse *nrdA* and *nrdB* genes responds to DNA damage, apparently as in *S. cerevisiae*.

F. DNA Damage in *Escherichia coli* Causes Increased RnR Activity

Early studies reported that induction of DNA damage in thymine-starved *E. coli thyA* organisms causes an increase in RnR activity (50). Such findings led to investigations of the expression of the *nrd* genes.

G. Comments on RnR and Cancer

Although the relation of RnR biosynthesis and regulation to the control of cancer proliferation is not within the scope of this review and lies far from our experience, a few cursory remarks may be made. RnR activity has had a long association with cancer cell treatment. Howard Elford and colleagues were among the first to suggest that RnR had a central role in cancer cell proliferation and that the enzyme was part of a complex of enzymes forming DNA precursors (66, 107). The enzymes on the pathway to dNTPs have been exhaustively employed to design competitive anticancer agents. RnR is a particularly attractive target, inasmuch as it is a unique enzyme catalyzing an unusual reaction. An understanding of the nature and control of the synthetic pathway to the R1 and R2 proteins and their assembly into the enzyme, and ostensibly into a complex of enzymes, should provide a different approach to developing therapeutic agents.

Because DNA damage has profound effects on the expression of the *nrdB* genes in eukaryotic cells and because the p53 oncogenic protein also brings about DNA damage, it is conceivable that the two systems may be related, perhaps more than merely because RnR is necessary for DNA repair or replication.

It has been suggested that *nrdA* and *nrdB* genes (and the genes of other enzymes that increase during the S phase of the cell cycle) should show increased transcription in the absence of pRb or its active phosphorylated form. pRb, a widespread tumor suppressor protein and the product of the retinoblastoma susceptibility gene, appears to be a central cog in the cell cycle clock (108). pRb binds to and reduces the activity of E2F, a promoter-activating factor required for the adenovirus E2 promoter (109). This kind of modulating mechanism contributes to the control of the G1/S boundary. It may be added that TGF-β factors markedly decrease the levels of the kinase, CDK4, which functions in the phosphorylation of Rb protein.

Hartwell and Kastan briefly discuss possible relationships of RnR and related enzymes to the cell cycle and cancer induction (110).

We have alluded to the role of TGF-β1 in the expression of the *nrdB* gene of mouse BALB/c 3T3 fibroblast line (Table II). In *ras*-transformed mouse $10\frac{1}{2}$ T cells, TGF-β1 signaling is uncoupled (89). Significantly, the early increases in *nrdA* and *nrdB* messages induced by TGF-β1 were found only in malignant cell lines. Such results suggest an autocrine activation of malignant cells by TGF-β1, and early changes in the control of RnR synthesis are implicated.

In H-*ras* transformed mouse cell lines, cyclic-AMP signals, induced by stable cAMP analogs or agents stimulating protein kinase A, greatly increased the levels of *nrdA* and *nrdB* messages as well as R1 and R2 proteins (*88*). The effects appeared to be S-phase independent because no changes in DNA formation were detected. Provocatively, these analogs and agents show no effects in the normal parental lines (Table II). The authors have speculated on the significance of this profound difference.

H. Summary

From the lowly yeast and *E. coli* to the mighty mouse, and beyond, organisms have developed mechanisms for repair of DNA damage. Basic to these systems is RnR. In yeast, *nrdB* induction by DNA damage is independent of the cell cycle. GRF1 protein, which is important in the iron utilization system, is also found to interact with the DNA damage response element found upstream of the yeast *nrdB* gene. Further, a newly discovered yeast *nrdA*-type gene, termed RNR3, is hyperresponsive to DNA damage. At least in part, its expression is regulated by Dun1, an autophosphorylating nuclear protein kinase. A second important control of *RNR3* occurs via the Dun2 protein, the DNA polymerase epsilon, which also connects the system to mitosis. In the mouse, DNA damage induces both the *nrdA* and *nrdB* genes in a pathway mediated by PKC.

VIII. Ribonucleotide Reductase Regulation during Oogenesis

In the developing clam oocyte, the mRNA of the *nrdB* gene accumulates to unusually high concentrations and is one of the major messenger species (*111–113*). This enticing finding may provide clues to a general mechanism of regulation of RnR biosynthesis; thus we summarize briefly the biological events in oogenesis. (For a more detailed development, see Refs. 86 and *114*).

During oogenesis, the developing diploid germ cells or oogonia undergo numerous mitotic divisions before beginning their first meiotic division. During the prophase of the first meiotic division, the progression of the primary oocytes through the first meiotic division is arrested, the length of this arrest being dependent on the species. At this stage, the cell, now termed a primary oocyte, increases in volume as it accumulates many of the proteins and the RNA required for its future development. It can then respond to cellular signals (*i.e.*, hormones) that induce it to mature and complete its first meiotic division. The resulting "secondary" oocyte undergoes a second meiotic division, in which the chromatids are separated and give rise to a mature haploid oocyte or egg (*86, 114*).

A. mRNA Masking and Polyadenylation

Throughout oogenesis, the processes of transcription and translation continue. However, during the extended arrest period, a whole class of mRNAs is not translated, but simply stored (*115–118*). Storage and subsequent release of maternal mRNA is one of the hallmarks of development and is found throughout the animal kingdom. Spirin first proposed in 1966 that specific maternal mRNAs are "masked" and that proteins found in the cytoplasm of the developing oocyte bind to the maternal mRNAs and prevent their translation. According to this model, the untranslated mRNA is later released and translated at the appropriate stage in the development of the embryo (Fig. 10). The mRNAs are stored during oogenesis in ribonucleoprotein particles (mRNPs) (*114, 119*). Storage of the maternal mRNA within the mRNPs is thought to prevent ribosome binding until the appropriate developmental time.

In *Xenopus laevis*, two oocyte-specific proteins, mRNP3 and mRNP4, have been identified as the major protein components of the mRNPs and are thought to regulate the translation of the mRNAs to which they are bound. mRNP4 is identical to the Y-box factor, FRGY2, whereas the mRNP3 has about 85% sequence identity. The Y-box proteins are highly conserved transcription factors and also bear a striking resemblance (55% identity) in their DNA binding domains to prokaryotic cold-shock proteins (*120–122*). FRGY2 then has a dual function: as a transcription factor and as an mRNA sequestering factor.

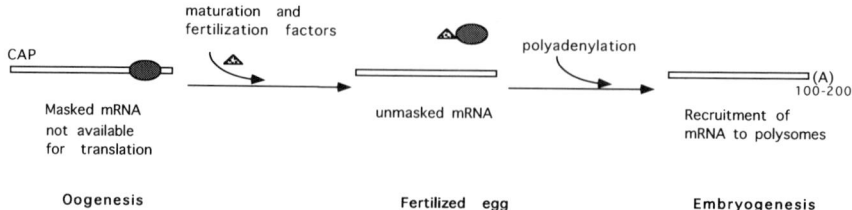

FIG. 10. Translational regulation of *nrdB* in *Spisula solidissima*. The figure indicates the relationship between the masking, unmasking, and polyadenylation of the mRNA encoding the small subunit of ribonucleotide reductase and the developmental stages of the oocyte. Note that the protein masking and unmasking factors have not been identified but are implied from the currently available data.

The mechanism of masking is separate from the polyadenylation/deadenylation mechanisms described for maternal mRNA (115, 116, 118). In X. laevis, two classes of mRNAs are found in the maturing oocyte. Those messages that are sequestered and stored are underadenylated, carrying between 15 and 90 A residues (115). A second class of mRNAs carries long poly(A) tails (>150) and is actively translated. Polyadenylation of the stored short poly(A) transcripts is under the control of specific sequences within the 3′ UTR. A nuclear polyadenylation signal, AAUAAA, directs poly(A) formation in the nucleus. By contrast, the cytoplasmic polyadenylation element (CPE), $U_{4-6}A_{1-2}U$, directs polyadenylation in the cytoplasm.

B. Regulation of Translation of the Small Subunit of Ribonucleotide Reductase

In the surf clam, *Spisula solidissima*, the most abundant stored, but masked, mRNA encode the small subunit of ribonucleotide reductase and cyclins A and B. They are not translated in oocytes, but are actively translated within 20 to 30 minutes after fertilization. The mRNAs retain their masked state in cell-free extracts of oocytes, but they can be activated or unmasked by gel filtration in high salt (113, 123, 124). Presumably, this process removes specific repressor proteins that bind to the 3′ UTR. For *nrdB*, the site of binding of the putative repressor protein was defined by "competitive unmasking experiments." In these assays, short segments of antisense RNA complementary to specific regions of the mRNA were annealed to the message, and its ability to undergo translation was assessed. Thus, a 134-nt segment in the 3′ UTR starting 80 nucleotides downstream of the coding region is responsible for the masking of the message (Fig. 11).

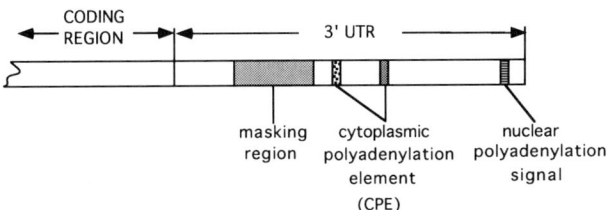

FIG. 11. Masking region and polyadenylation signals associated with the 3′ UTR of the *nrdB* mRNA of *Spisula solidissima*. The masking region is thought to bind a protein that functions in the prevention of translation of the message. After fertilization, the message is "unmasked" and sequences within the 3′ UTR direct the formation of the poly(A) tail, with a subsequent initiation of translation. The mechanism is addressed in Section VIII,B.

An important feature in this system is the relationship between masking of the mRNA and polyadenylation. In S. *solidissima*, the 3' UTR of *nrdB* contains two uridine-rich sites similar to the cytoplasmic polyadenylation elements described in X. *laevis*, along with the AAUAAA nuclear polyadenylation sequence (Fig. 11). These are located approximately 150 nt downstream of the masking region. Unmasking of the message can occur *in vitro* in the absence of polyadenylation, indicating that the two processes can be functionally separated. Further, fertilized egg extracts, but not oocyte extracts, have the ability to polyadenylate unmasked *nrdB* 3' UTR mRNA.

Finally, efficient polyadenylation by egg extracts does not require the AAUAAA nuclear polyadenylation sequence, but requires sequences upstream of the identified U-rich motifs, possibly extending into the masking region. Therefore, these experiments suggest that, in the case of the small subunit of ribonucleotide reductase, a timed sequence of events occurs in the activation of the stored mRNAs. In the process, the masking repressor is first removed from a site within the mRNA 3' UTR, allowing the polyadenylation of the mRNA to be directed by sequences within the 3' UTR.

C. Regulation of Synthesis of the Large Subunit of Ribonucleotide Reductase

In sharp contrast to the small subunit of RnR in S. *solidissima*, the large subunit has no storage mechanism for its mRNA (*112*). Instead, the R1 protein is found in high concentrations in the oocyte, and its level remains constant for at least 24 hours postfertilization. Little or no synthesis of the R1 occurs after fertilization.

The two subunits are thus stored in a completely different manner in the oocyte, the small subunit as masked maternal mRNA and the large subunit as the fully functional protein. In pioneering work on sea urchins, it was found that RnR lost activity when protein synthesis was inhibited (*125*). Therein, perhaps, lies the reason behind the differential storage. The small subunit contains a tyrosyl radical. The half-life of the tyrosyl radical, though remarkably stable, may be on the order of hours or days, depending on temperature. However, the lifetime of the primary oocyte can be very long, on the order of months or years in higher eukaroytes. It seems reasonable, then, that a mechanism developed to store the corresponding, relatively stable mRNA of the small subunit, ready to be translated into R2 after fertilization, when large quantities of dNTPs are required.

D. Summary

RnR plays an important role in the developing embryo, because large amounts of dNTPs are required for the embryonic cell divisions. During oogenesis, the two subunits of RnR are stored in different manners. The small subunit is stored as a masked form of maternal mRNA unavailable for ribosome binding and subsequent translation. This masking mechanism is mediated by a 134-nucleotide sequence located in the 3′ UTR. In contrast, the large subunit is stored as a functional protein that, on fertilization, combines with the newly synthesized small subunit to form active RnR.

IX. Eukaryotic Animal DNA Viruses Encoding Ribonucleotide Reductases

Few eukaryotic viruses encode ribonucleotide reductases. Most DNA viruses rely primarily on the host enzymes for the synthesis of their deoxyribonucleotide precursors. The prominent exceptions are the herpes viruses, with close to a hundred distinct members throughout the animal world (126), and the orthopoxviruses, typified by vaccinia virus. Six herpes viruses have been found in human beings: herpes simplex viruses (HSV-1 and HSV-2), human herpes virus, varicella-zoster virus, Epstein–Barr virus, and human herpes virus 6. Both herpes viruses and orthopoxviruses encode many of the enzymes required for *de novo* dNTP synthesis and DNA replication. This capability is vital for viral replication in fully differentiated cells such as neurons, which do not divide during most of the life of the host.

A. Herpes Viruses

Three separate subfamilies of herpes viruses have been identified. The Alphaherpesvirinae (HSVs and varicella-zoster) are characterized by rapid growth, lysis of infected cells, and formation of latent infections in sensory nerve ganglia. Betaherpesvirinae (cytomegalovirus and human herpes virus 6) have a relatively slow replication cycle, form multinucleated cells (cytomegalia), and establish latent infections in lymphoreticular tissue, secretory glands, kidneys, and other tissues. The Gammaherpesvirinae (Epstein–Barr virus) actively and latently infect lymphoid cells (25). All three subfamilies of herpes viruses are characterized by (1) a linear, double-stranded DNA, approximately 155 kb in HSV, (2) an icosadeltahedral capsid approximately 100-nm in diameter, (3) an amorphous region, called the tegument, surrounding the capsid and consisting of a number of viral proteins, including the virion host shut-off (VHS) protein and a number of viral *trans*-activating factors, and (4) an envelope containing viral glycoproteins. (For reviews, see Refs. 126 and 127.

Transcriptional control of herpes virus genes is tightly regulated, a process resulting in the formation of an ordered cascade of gene products as they are required in the herpes life-cycle. In HSV-1 and -2, the most studied herpes viruses, transcriptional patterns fall into three discernible groups; immediate-early, early, and late (also referred to as α, β, and γ). Herpes viruses encode a large number of proteins intimately involved in the DNA synthetic process. These proteins fall into two categories, those involved in *de novo* deoxyribonucleotide synthesis and those making up the DNA replication apparatus. The enzymes of the *de novo* synthetic pathway are referred to as "nonessential" (*126*); *i.e.*, productive infection can occur in cultured cells in their absence (*128, 129*). However, in the absence of ribonucleotide reductase, both acute replication and latent reactivation of viral replication are markedly reduced (*130*).

1. HSV TRANSCRIPTIONAL CONTROL

On infection by HSV-1 and -2, the first genes expressed are five immediate-early genes encoding the infected cell polypeptides (ICP): 0, 4, 22, 27, 47. This expression occurs in the absence of viral protein synthesis. The immediate-early genes are activated by viral protein 16 (VP-16), present in large concentration in the tegument of the viral particle. Immediate-early protein synthesis peaks at approximately 2 to 4 hours postinfection. However, these proteins continue to accumulate over the course of the entire infective process.

Transcription of the early genes is dependent on the successful expression of the immediate-early genes. The proteins translated from these transcripts peak at approximately 5 to 7 hours postinfection. The late genes, $\gamma 1$ and $\gamma 2$, form the structural proteins that make up the progeny virus. $\gamma 2$ gene expression is differentiated from $\gamma 1$ expression by its requirement for DNA synthesis.

2. TRANSCRIPTIONAL CONTROL OF HSV-1 *nrdA*: R1-PROTEIN KINASE

The large subunit R1-PK of RnR in HSV-1 and -2 carries an added protein kinase domain located in the first 340 amino-acid residues (*131–134*). This kinase domain is not found in the RnR of any other herpes, nor, indeed, in RnRs from any source. Because the transformation potential was mapped to the protein kinase domain, its expression has been the subject of intense scrutiny. Perhaps because of this added domain, the transcriptional regulation of the R1-PK subunit of HSV-1 and -2 does not fit the normal transcriptional pattern of other early herpes genes.

The transcription of early genes is normally regulated by one or more of the immediate-early gene products. However, in the case of HSV-1, VP-16, together with the Oct-1 cellular transcription factor, *trans*-activate the gene coding for the R1-PK (*132, 135*). Here, *trans*-activation is mediated by an octamer/TAATGARAT site (GARAT, in Table I) located 120 nucleotides upstream of the transcriptional start site (Fig. 12). Additionally, in the absence of VP-16 transactivation, i.e., during infection with HSV-1 containing a deleted octamer/TAATGARAT site, the immediate-early gene product, ICP-0, can mediate activation of the *nrdA* gene through interaction at the TATA box region. Thus, the HSV-1 *nrdA* gene expression shows characteristics of both immediate-early and early genes.

The herpes *nrdB* gene is under the control of the immediate-early ICP-4 protein and therefore follows the more established *trans*-activation pattern of herpes early genes (Fig. 12).

3. TRANSCRIPTIONAL CONTROL OF HSV-2 *nrdA*

In general, regulation of the synthesis of *nrdA* in HSV-2 follows the same pattern of transcriptional activation as seen in HSV-1 (*132*). Thus, the presence of the octamer/TAATGARAT site at nucleotide −122, ATGCAAATG-GGATTCA, suggests that the gene is regulated as an immediate-early gene and that VP-16 mediates *trans*-activation of the gene. This prediction was substantiated by cotransfection experiments using a VP-16 expression vector and a construct containing the HSV-2 *nrdA* promoter enhancer fused to the chloramphenicol acetyltransferase gene. The VP-16 expression vector stimulated transcription from the HSV-2 *nrdA* promoter nearly 20-fold. Further, gel-shift mobility assays suggested that VP-16 bound to a DNA fragment containing the octamer/TAATGARAT binding site in lysates made from puri-

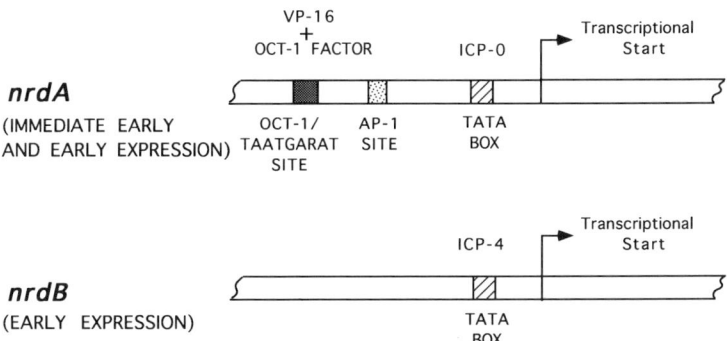

FIG. 12. The 5′ upstream regions of both *nrdA* and *nrdB* genes of HSV-1. Factors crucial to *trans*-activation of the RnR genes are indicated above their respective sites.

fied virions. The HSV-2 RnR-1 gene was also *trans*-activated by ICP-0, but this *trans*-activation was found to be cell-type specific. Thus, ICP-0 activates transcription of HSV-2 RnR in Vero cells, but has no effect in primary astrocytes and in 293 cells, an adenovirus-transformed human kidney line (132).

The added complexity seen in transcriptional regulation of the large subunit of HSV-1 and -2 reductases is largely a result of the Ser/Thr-specific kinase domain found in the N-terminal portion of the protein. In HSV-2, the PK region promotes both auto- and transphosphorylating activity (131, 134, 136), but in HSV-1, only autophosphorylating activity has been demonstrated. Further, only the HSV-2 PK domain causes neoplastic transformation (133, 137). Nevertheless, the role of the PK domain in normal HSV infection remains largely unknown. It has been suggested that it may play a critical role in reactivation of viral replication from latency (135, 138). Activation of R1-PK would then serve two purposes: the formation of the protein kinase, which could regulate the activity of essential viral and cellular gene products, and the activation of the R1 component domain for the assembly of RnR and the synthesis of dNTPs for subsequent viral DNA synthesis.

B. Vaccinia Virus

Vaccinia virus is a member of the orthopoxvirus family and contains a 187-kb duplex DNA genome. This group of viruses is unique in its ability to express and replicate its DNA within the cytoplasm of the host cell (139). Host enzymes catalyzing the *de novo* dNTP synthesis pathway are cytoplasmic, whereas DNA replication is localized in the nucleus. The vaccinia virus encodes most of the enzymes necessary for its transcription and replication, including DNA polymerase, thymidine kinase, DNA topoisomerase, and RnR (140–146).

RnR Regulation in Vaccinia

Numerous reports indicate that the RnR is not essential for wild-type growth in cultured cells (142); however, as in the HSV studies cited above, fully active RnR is required for optimal virulence *in vivo*. Presumably, the RnR dependence of vaccinia as well as herpes is a function of the type of cell it infects. For example, in fully differentiated, nondividing cells, (e.g., neurons), in which the amount of host RnR is virtually zero, viral RnR is essential for viral *de novo* synthesis and DNA replication.

Transcriptional and functional data on vaccinia RnR have been gathered in large part by Mathews and co-workers (140, 142–144). Vaccinia *nrdA* and *nrdB* genes are not expressed coordinately, being separated by approximately 35 kb. Both genes are maximally transcribed between 1 and 3 hours after infection, either in the presence or in the absence of cyclohexamide, and are therefore classified as immediate-early genes. In keeping with the transcriptional data, both RnR subunits were detected 1 to 2 hours postinfection, with maximal rates of protein synthesis between 3 and 4 hours. In comparison, vaccinia DNA replication can be first detected between 1 and 2 hours postinfection. These kinetic studies indicate that vaccinia virus RnR transcription and translation precede (as they must) viral DNA synthesis.

Recent studies established that an unusual situation holds in vaccinia infection (*140*). The rate of the encoded RnR enzyme is such that dNTPs are not limiting in DNA replication. Actually, of course, even in this case dNTP synthesis is still an absolute requirement for DNA replication.

C. Summary

Unlike other eukaryotic viruses, the herpes and pox viruses carry the genes for the two subunits of ribonucleotide reductase along with many of the genes required for *de novo* dNTP synthesis and DNA replication. The presence of these genes appears to reflect the unique life cycles of the viruses, wherein DNA replication occurs outside of the normal controls of the cell cycle. In HSV infection, an added protein kinase domain is carried by the large subunit of the RnR. Transcriptional analysis indicates that the large subunit is expressed from an immediate-early gene, whereas the small subunit is formed at a time typical of early gene expression. By contrast, both subunits of the vaccinia RnR are expressed as products of immediate-early genes. It is important to note that, in both cases, RnR is required for optimal virulence.

X. Kinetics of Synthesis of T4-encoded Ribonucleotide Reductase and Relationship to T4 DNA Replication

A. What Is the Trigger Initiating DNA Replication? The Phage T4 Caper

In Section II, we described the tritium-release procedure, a technique for simultaneously comparing deoxyribonucleotide synthesis and DNA replication *in vivo*. Now, we examine the kinds of systems that T4 has evolved to control a complex of enzymes synthesizing dNTPs, and inquire into the nature of the interactions of this complex with the DNA replication machinery.

On infection by phage T4 carrying amber mutations in the *dna* genes 41, 44, or 45, T4 DNA replication does not occur (*28, 29*). Yet, the synthesis of deoxyribonucleotides, as measured by ^{3}H release, exhibits the same kinetics; *i.e.*, it shows an exponential curve superimposable with that of wild-type infection. Thus, these observations provide strong argument that the exponential increase in DNA replication from the *de novo* pathway is related to dNTP synthesis and not to the increase in DNA growing-points (24). Of course, an increase in DNA growing-points occurs, but in our view, only by the limiting behest of the exponentially increasing rate of *de novo* dNTP synthesis. In phage T4 infection, it is apparent that at 30°C a functional DNA replication complex is in place as much as 1 minute before deoxyribonucleotides appear by the *de novo* pathway (29). Naturally, the turn-on of T4 DNA replication is ultimately determined by the formation of active ribonucleotide reductase, more specifically by its inclusion into the dNTP synthetase complex.

It is apparent that the exponential increase derives from the kinetics of formation of the dNTP synthetase complexes (see Ref. 20) or that an exponential increase in the number of complexes occurs as the concentrations of enzyme precursors of the complex (at least of the limiting enzyme) increase.

B. Ratio of Synthesis of Deoxyribonucleotides Reflects Ratio of Bases in T4 DNA

Two decades ago, it was found that, after T4 infection, the synthesis of dTMP and HmdCMP occurs in a ratio close to 2.1:1, *i.e.*, at the ratio of the *average* occurrence of Thy and HmC in T4 DNA. But the same ratio of synthesis of the deoxyribonucleotides was maintained for 25 to 30 minutes after infection, even when infection was by a phage carrying an amber mutation in a *dna* gene and DNA replication could not occur (147). Therefore, the ratio was not controlled by feedback, but appeared to result from an intrinsic property of the combined set of enzymes we now know as the dNTP synthetase complex. We guess, though without analysis, that the average rate of synthesis of dAMP:dGMP is also 2.1:1, so that the DNA polymerase is not flooded with unfavorable, mutation-prone ratios of dNTPs. Furthermore, analyses demonstrate that dATP and dGTP do not accumulate in wild-type infection (20, 58).

Recently, evidence for the physical interaction of the dTMP synthase and dCMP hydroxymethylase enzymes in the core dNTP synthetase complex (19, 148) and the partial identities in the amino-acid sequences (149) of the two enzymes have been reported. As a possible explanation of the finding that dTMP and HmdCMP derivatives are formed at a ratio of 2.1:1, it was suggested that the two enzymes may control each other's rate by physical interaction. This idea certainly merits further study.

C. Ratio of Synthesis of dTMP to HmdCMP Is Regulated by Precursor Levels

The ratio of the activities of dTMP synthase and dCMP hydroxymethylase appears to be determined, at least in great part, by the rates of formation of their respective substrates, dUMP and dCMP, which, in turn, depend on the relative rates of reduction of UDP and CDP by ribonucleotide reductase, the K_m values for the different rNDPs being controlled by the dNTP (and ATP) effectors. Important to the control of the ratio is the rate of the allosterically regulated T4-encoded dCMP deaminase reaction (150, 151). In the latter, dTTP is a negative effector and HmdCTP, an activator. Figure 13 shows the relative rates of the several reactions involved. The major role played by dCMP deaminase is shown on infection by a phage T4 mutated in the *cd* gene (dCMP deaminase); the ratio of synthesis of dTMP:HmdCMP drops to 0.6:1. dUTPase/dCTPase, converting dUDP and dCDP to dUMP and dCMP, respectively, is a very high turnover enzyme, apparently the most rapid in the complex. Even so, the activities of dCMP hydroxymethylase and dTMP synthase must be equal to the task of turning over their substrates, or more accurately, they are being fed by the limiting RnR reactions, because dCMP and dUMP are found at very low concentrations (20, 28). To compound the problem, it should be mentioned that dTMP synthase is product-inhibited by dTMP, and dCMP hydroxymethylase by HmdCMP (20).

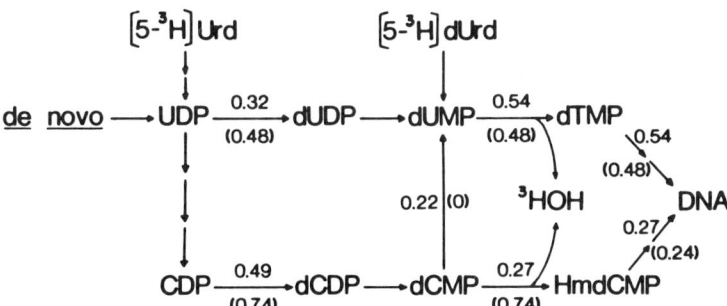

FIG. 13. Synthesis of pyrimidine deoxyribonucleotides from UDP and CDP, and replication of DNA in bacteriophage T4 infection. The labeled precursor was [5-³H]uridine. The figure shows the sites of release of ³H into the aqueous phase. The numbers are calculated relative rates at each step in wild-type infection. Figures in parentheses are the values after infection by a phage *cd* mutant lacking dCMP deaminase. Thy:HmC base ratios were determined by labeling simultaneously with administered [6-³H]uracil (which gives identical results to labeled uridine.) From Ref. 152, with permission.

Phage T4 RnR, while requiring dNTPs for its activation, does not exhibit significant feedback by dNTPs as in the case of the *E. coli* and mammalian enzymes. In fact, concentrations of dATP that shut off *E. coli* RnR completely activate the T4 enzyme.

D. Effect of Damage to the T4 DNA Template on Deoxyribonucleotide Synthesis: Role of Template

Because the 2.1:1 ratio could not entirely be explained kinetically, Flanegan *et al.* turned to the possibility that the T4 template might have a role in the control (153). Exposure of phage to ultraviolet radiation or methyl methanesulfonate treatment had no effect on the formation of the early enzymes, but the rate of ^{3}H release was greatly decreased. These effects required a mutation in the *uvsX* gene, which plays a critical role in DNA repair. Thus, wild-type phage showed little or no response to radiation or the methylation reaction. Therefore, the effect of these agents is on DNA. It is important to understand that these results were also obtained with a phage T4 *dna* mutant, that is, unable to replicate DNA, providing strong argument that the effects were on the T4 template DNA.

It appears, then, that T4 template DNA has a structural role in the dNTP synthetase complex, probably at RnR. In initial studies, we found no effect of DNA or oligodeoxyribonucleotides on the activity of pure RnR preparations in the presence and absence of dATP activator (20). These studies need to be repeated and extended, especially considering the role of dATP in maintaining the association of the R1 and R2 proteins (154).

The idea that the template DNA could have a structural role is not new. DNAs play a structural role in a number of well-studied enzymes (23). Nevertheless, one is compelled to consider the possibility that RnR or the dNTP synthetase complex monitors or scans the template DNA and somehow directs the choice of the incoming rNDPs to form the precise dNTPs required at each of the growing 3' termini of the DNA chains. The bypasses necessary in the synthesis of HmdCMP and dTMP complicate this attractive fantasy. Furthermore, it is obvious that no direct evidence for such a mechanisms exists. Finally, we should mention, if not emphasize, that preparations of an extended T4 dNTP synthetase complex contained no DNA (155) nor membrane (by absence of NADH oxidase), nor are these components apparent in the more highly purified preparations of dNTP synthetase (19, 157).

Comparing these findings with the effects of DNA damage agents on *E. coli*, mouse, and yeast, it is apparent that T4 presents a different system of control.

E. Comments on the T4 dNTP Synthetase Extended Complex

T4 DNA polymerase, p43, and many ancillary enzymes associated with the T4 DNA replication are found in the dNTP synthetase complex fraction at concentrations that appear to be approximately equivalent to those of the enzymes of the core fraction (155). The number of proteins in the complex has been increased (20) as a result of the identification of additional T4-encoded proteins by two-dimensional electrophoretic gels (Table III). A question has been posed as to whether the DNA polymerase and ancillary

TABLE III

PROTEINS OF PHAGE T4 dNTP SYNTHETASE COMPLEX AND THE EXTENDED COMPLEX, AND PROTEINS BINDING TO dCMP HYDROXYMETHYLASE COLUMNS[a]

Gene	Gene product	Bound to dCMP Hmase	Extended complex	dNTP synthetase core
42	dCMP Hmase	+	+	+
td	dTMP synthase	+	+	+
frd	Dihydrofolate reductase	+	-[a]	+
56	dCTPase/dUTPase	+	+	+
nrdA	rNDP reductase R1 subunit	+	+	+
nrdB	rNDP reductase R2 subunit	+	+	+
32	ss DNA binding protein	+[b]	+	-
43	T4 DNA polymerase	+/-[c]	+	-
44	Replication accessory protein	+[d]	-	-
45	Replication accessory protein	+[b]	-	-
61	DNA Primase	+[b]	+	-
62	Replication accessory protein	+[d]	-	-
46	Host DNA degradation	+/-[e]	+	-
47	Host DNA degradation	-	+	-
uvsX	RecA analogue, DNA repair	+	+	-
uvsY	DNA repair	+[b]	+	-
pseT	Phosphatase/kinase	+	+	-
βgt/RNase H	β glucose transferase/RNase H	+[d]	+	-
ddA	DNA helicase	-	+	-
regA	RegA protein	-	+	g
lp III	lp III protein	-	+	-
41	gp41 helicase/primase	-	+	-
39, 52, 60	T4 DNA topoisomerase proteins	-	+	-
rIIA	RIIA protein	-	+	-
rIIB	RIIB protein	-	+	-
endonuclease IV	DenB protein	-	+	-
T4 gene 1	dGMP/dTMP/HmdCMP kinase	-	+	+
cd	dCMP deaminase	-	+	+
host ndk	Nucleoside diphosphate kinase	f	f	+
host dAMP kinase	dAMP kinase	f	f	+

[a] Notes: Binding to columns was determined from extracts of cells infected in presence of [35S]methionine. (a) Not detected in 2-D electrophoretic gels; see text (155). (b) Bound to NDK column also. (c) Bound weakly to both columns. Normal binding means elution by 0.6 M NaCl. Weak binding is elution by 0.2 M NaCl. (d) Bound to dCMP Hmase only in absence of p32 or ptd. (e) Bound weakly. (f) Not measurable. (g) See Ref. 157.

DNA replication enzymes are contaminants of the core complex. The lack of DNA polymerase in the more extensively purified dNTP synthetase complex preparations was explained as having been stripped by the more stringent purification (19). By the same reasoning, our mild procedure would not strip off the DNA polymerase and other factors. At this stage, we do not believe that p43 and the ancillary DNA replication proteins in our preparations were contaminants.

F. Proteins Binding to dCMP Hydroxymethylase and to NDK Columns

A study by Wheeler and co-workers (156) may address parts of the question in Section X,E, but perhaps introduces new questions. Many of the dNTP synthetase core enzymes in an extract of T4-infected cells bind to a dCMP hydroxymethylase column. Very interestingly, a number of enzymes concerned with DNA replication or ancillary enzymes were also bound to the column. In Table III we list the proteins binding to the dCMP hydroxymethylase and host nucleoside diphosphate kinase columns, compared to proteins that we found associated in an extended complex. In view of the overlap in Table III of a number of replication-related proteins in these two studies, it is possible that the column procedure approaches the more gentle method employed by us to isolate the extended complex.

Because it did not appear reasonable that all proteins could bind to Hmase directly, possible bridging proteins were examined (156). By the use of extracts of cells infected by amber mutants, gene-32 protein was required to bind gp*nrdA*, gp*nrdB*, gp32, gp*uvsX*, and gp*uvsY*. Further, gp*td* was necessary to bind gp*td*, gp*nrdA*, gp*nrdB*, gp32, gp*uvsX*, and gp*uvsY*. These findings provide strong clues to the relationships between some proteins in the complex. gp44, gp62, and gpβt/RNase-H are bound to dCMP columns only in the absence of gp32, or gp*td*, whereas gp43 was bound weakly (see Table III footnotes) from all the extracts, and also was bound weakly to NDK columns.

One enzyme, dCMP deaminase, is missing on the dCMP Hmase columns, whether extracts of cells infected by wild-type T4, gp32 *am*, or gp*td am* phage were employed. Yet this enzyme is at the juncture metabolically with dCTPase/dUTPase, dCMP Hmase, and dTMP synthase. One would expect that if this were a tightly locked interactive complex, dCMP deaminase would interact physically with one or all of the three enzymes. This question becomes particularly vexing because dCMP deaminase is one of only a few enzymes that is required for the structural integrity of the dNTP synthetase core complex (19, 157). It seems that this apparent anomaly represents another clue either to the structure of the complex or to the fraction accumulating on the dCMP Hmase column. Possibly dCMP deaminase requires the presence of one of its effectors.

The T4-encoded kinase, acting only on dGMP, dTMP, and HmdCMP, also was not adsorbed to these columns.

Table III indicates that the extended complex lacks gp44–gp62 and gp45, well-studied accessory DNA proteins in replication (23). It seems more than chance that infection by phage T4 carrying the amber mutants of genes 41, 44, and 45 (Fig. 14) had no effect on the rate and kinetics of deoxyribonucleotide synthesis *in vivo* (as measured by ^{3}H release from administered [5-^{3}H]uridine), compared to the rates shown in wild-type infection (Fig. 1). By contrast, all five amber mutants of gene 43 that were tested showed a

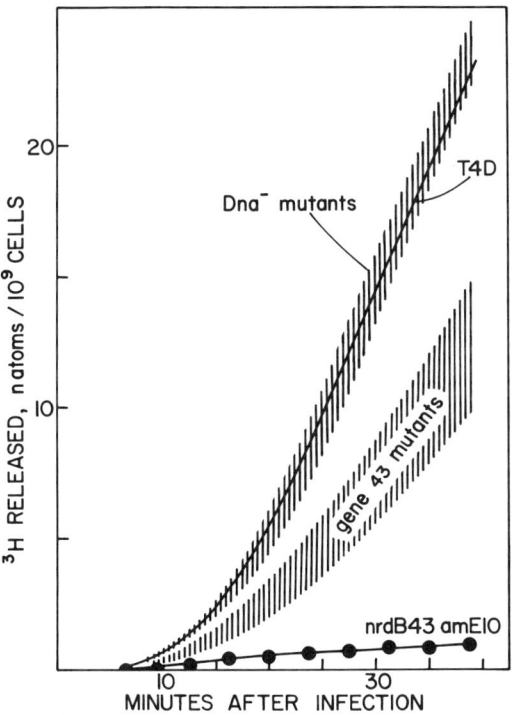

FIG. 14. *In vivo* evidence that the dNTP synthetase complex and T4 DNA polymerase interact. The ^{3}H release *in vivo* on infection by amber mutants of gene 43 (DNA polymerase) is greatly diminished compared to wild-type phage. But amber mutations in other *dna* genes (41, 44, and 45) and in gene 1 (HmdCMP kinase, Dna$^-$ phenotypes) showed the same rate of deoxyribonucleotide synthesis. Our interpretation: DNA polymerase interacts with the dNTP synthetase complex. An *nrdB* amber/gene-45 amber double mutant shows only slight activity. From Ref. 158, with permission.

clear decrease in the rate of ^{3}H release (29, 158). Gene-43 mutants block DNA replication, but mutants of genes 41, 44, and 45 also block replication. Therefore, the gene-43 effect is not caused by the *dna* block. We have proposed that this is *in vivo* evidence for the interaction of the dNTP synthetase complex and T4 DNA polymerase (20, 29, 158).

p44 and p62 participate together in the T4 replication complex and are isolated as a rather tight couple. In addition, p45 interacts with 44–62, in the complex (23). In our view, the absence of these several proteins strengthens the idea that the other proteins, mostly associated with replication (found in the extended complex), are not present by happenstance or occlusion. Furthermore, a number of other proteins were not found in the extended complex. A few were found only in the complex (155).

To further elaborate on the gene-43 amber-mutant effect, the lack of inhibition of ^{3}H release by amber mutants of genes 41, 44, and 45 might suggest that these components are not in the direct line of deoxyribonucleotide incorporation in DNA replication. p61 is known to be an RNA-primase protein functioning together in a complex (primosome) with the helicase/primase, p41, and is directly in the rNTP/primer pathway (23a). The suggestion that the accessory proteins are not in a direct line of the polymerization reaction is made with the understanding that these several gene products certainly have an absolute role in replication. Nevertheless, it is obvious that only p43 actually catalyzes the transfers of the dNTPs into the growing DNA chain.

G. Summary

The distinctive controls timing the initiation of the R1 and R2 subunits assure the entry of T4 RnR, probably as the last component, into its keystone position in the dNTP synthetase complex. We believe that the T4 DNA template has a fundamental role in the structure and/or function of dNTP complex. From our *in vivo* kinetic studies and from genetic studies in other laboratories, we propose that the dNTP synthetase complex interacts directly with T4 DNA polymerase.

XI. General Discussion

We asked whether the syntheses of the RnR subunits in different biological systems show comparable controls and whether the controls are intimately related to the protein factors determining the course and timing of the cell cycle.

In both instances, the response is positive, but the picture is yet to be completed and the uniqueness of RnR established. Some of the obvious similarities, their usefulness as tools, and their possible biological roles have been mentioned. The binding and reactive sequences of the different RNAs and DNAs and the binding sites of the proteins still need to be examined systematically. Phage T4 is emphasized as a basic system to study the control of the formation of RnR and its function and to identify the factors that determine the initiation of T4 DNA replication.

All of the regulations of gene expression appear to have elements in common, but that may not make them a family. Nevertheless, all of the systems show a relation to DNA replication. Most organisms, including humans, have evolved elaborate, though not exclusive, *nrd* expression mechanisms to respond to DNA damage. The viruses excepted, all of the pathways synthesizing RnR are tied to the cell cycle, in some cases intimately. Little question can be raised about the control of R1 and R2 synthesis by a number of established cell-cycle protein factors, and insight into the system is expanding. The newly evolving relationships between membrane synthesis and degradation and the cell-cycle factors and inducers of G-protein-linked receptors offer an exciting direction of study (Fig. 9).

Acknowledgments

Supported by The University of Michigan and its Department of Biological Chemistry. We are most grateful to Susan J. Greenberg for typing the manuscript and creating most of the figures. We thank Elizabeth Leibold, University of Utah, for sending a reprint of her paper.

References

1. E. Hammarsten, P. Reichard and P. Saluste, *JBC* **183**, 105 (1950).
2. I. A. Rose and B. S. Schweigert, *JBC* **202**, 635 (1953).
3. P. Reichard and L. Rutberg, *BBA* **37**, 554 (1960).
4. P. Reichard, Z. N. Canellakis and E. S. Canellakis, *JBC* **236**, 2514 (1961).
5. L. Thelander and P. Reichard, *ARB* **48**, 133 (1979).
6. P. Reichard, *ARB* **57**, 349 (1988).
7. F. Åslund, B. Ehn, A. Miranda-Vizuete, C. Pueyo and A. Holmgren, *PNAS* **91**, 9813 (1994).
8. U. Uhlin and H. Eklund, *Nature* **370**, 533 (1994).
9. M. Fontecave, P. Nordlund, H. Eklund and P. Reichard, in "Advances in Enzymology" (A. Meister, ed.), Vol. 65, p. 147. Academic Press, San Diego, 1992.
10. P. Nordlund and H. Eklund, *JMB* **232**, 133 (1993).
11. M. Karlsson, M. Sahlin and B.-M. Sjöberg, *JBC* **267**, 12622 (1992).
12. B. M. Sjöberg, M. Karlsson and H. Jörnval, *JBC* **262**, 9736 (1987).
13. A. Åberg, M. Ormö, P. Nordlund and B.-M. Sjöberg, *Bchem* **32**, 9845 (1993).
14. K. Regnström, A. Åberg, M. Ormö, M. Sahlin and B.-M. Sjöberg, *JBC* **269**, 6355 (1994).
15. J. Ling, M. Sahlin, B.-M. Sjöberg, T. M. Loehr and J. Sanders-Loehr, *JBC* **269**, 5595 (1994).

16. J. M. Bollinger, D. E. Edmondson, B. H. Huynh, J. Filley, J. R. Norton and J. Stubbe, *Science* **253**, 292 (1991).
17. J. Stubbe, *Nature* **370**, 502 (1994).
18. S. S. Mao, T. P. Holler, J. M. Bollinger, Jr., G. X. Yu, M. I. Johnston and J. Stubbe, *Bchem* **31**, 9744 (1992).
19. C. K. Mathews, *This Series* **44**, 167 (1993).
20. G. R. Greenberg, P. He, J. Hilfinger and M.-J. Tseng, in "Molecular Biology of Bacteriophage T4" (J. D. Karam, ed.), p. 14. Am. Soc. for Microbiology, Washington, D.C., 1994.
21. J. Stubbe, G. Smith and R. L. Blakley, *JBC* **258**, 1619 (1983).
22. P. Reichard, *JBC* **268**, 8383 (1995).
23. A. Kornberg and T. Baker, "DNA Replication," 2nd Ed. W. H. Freeman, San Francisco, California, 1992.
23a. N. G. Nossal, in "Molecular Biology of Bacteriophage T4" (J. D. Karam, ed.), p. 43. Am. Soc. for Microbiology, Washington, D.C., 1994.
24. R. Werner, *CSHSQB* **33**, 501 (1969).
25. J. H. Toyn, W. M. Toone, B. A. Morgan and L. H. Johnston, *Trends Biochem. Sci.* **20**, 70 (1995).
26. C.-S. Chiu and G. R. Greenberg, *CSHSQB* **33**, 351 (1968).
27. M. G. Wovcha, P. K. Tomich, C.-S. Chiu and G. R. Greenberg, *PNAS* **70**, 2196 (1973).
28. P. Tomich, C.-C. Chiu, M. G. Wovcha and G. R. Greenberg, *JBC* **249**, 7613 (1974).
29. C.-S. Chiu, P. K. Tomich and G. R. Greenberg, *PNAS* **73**, 757 (1976).
30. O. Berglund, O. Karlstrom and P. Reichard, *PNAS* **62**, 2735 (1969).
31. Y.-C. Yeh, E. J. Dubovi and I. Tessman, *Virology* **37**, 615 (1969).
32. S. S. Cohen and H. D. Barner, *JBC* **237**, 1376 (1962).
33. O. Berglund, *JBC* **250**, 7550 (1975).
34. O. Berglund and B.-M. Sjöberg, *JBC* **245**, 6030 (1970).
35. Y.-C. Yeh and I. Tessman, *Virology* **47**, 767 (1972).
36. M.-J. Tseng, J. M. Hilfinger, A. Walsh and G. R. Greenberg, *JBC* **263**, 16242 (1988).
37. E. Kutter *et al.*, in "Molecular Biology of Bacteriophage T4" (J. D. Karam, ed.), p. 491. Am. Soc. for Microbiology, Washington, D.C., 1994.
38. B.-M. Sjoberg, S. Hahne, C. Z. Mathews, C. K. Mathews, K. N. Rand and M. J. Gait, *EMBO J.* **5**, 2031 (1986).
39. M.-J. Tseng, P. He, J. M. Hilfinger and G. R. Greenberg, *J. Bact.* **172**, 6323 (1990).
40. D. O. Wirak, K. S. Cook and G. R. Greenberg, *JBC* **263**, 6193 (1988).
41. K. S. Cook, D. O. Wirak, A. F. Seasholtz and G. R. Greenberg, *JBC* **263**, 6202 (1988).
42. D. O. Wirak and G. R. Greenberg, *JBC* **255**, 1896 (1980).
43. K. Mizuuchi, M. Mizuuchi, M. H. O'Dea and M. Gellert, *JBC* **259**, 9199 (1984).
44. J. M. Hilfinger and P. He, *Gene* **141**, 55 (1994).
45. D. Hartz, D. S. McPheeters, R. Traut and L. Gold, in "Methods in Enzymology" (H. F. Noller, Jr. and K. Moldave, eds.), Vol. 164, p. 419. Academic Press, San Diego, 1994.
46. G. Knapp, in "Methods in Enzymology" (J. E. Dahlberg and J. N. Abelson, eds.), Vol. 180, p. 192. Academic Press, San Diego, 1989.
47. D. S. McPheeters, G. D. Stormo and L. Gold, *JMB* **201**, 507 (1988).
48. S. Altuvia, S. Kornitzer, K. Simi and A. B. Oppenheim, *JMB* **218**, (1991).
49. A. F. Seasholtz and G. R. Greenberg, *JBC* **258**, 1221 (1983).
50. D. F. Filpula and J. A. Fuchs, *J. Bact.* **130**, 107 (1977).
51. L. B. Augustin, B. A. Jacobson and J. A. Fuchs, *J. Bact.* **176**, 378 (1994).
52. R. S. Fuller, B. E. Funnell and A. Kornberg, *Cell* **38**, 889 (1984).
53. L. Sun, B. A. Jacobson, B. S. Dien, F. Srienc and J. A. Fuchs, *J. Bact.* **176**, 2415 (1994).

54. L. Sun and J. A. Fuchs, *J. Bact.* **176,** 4617 (1994).
55. J. Y. Wang and M. Syvanen, *Mol. Microbiol.* **6,** 1861 (1992).
56. M. Filutowicz, W. Ross, J. Wild and R. L. Gourse, *J. Bact.* **174,** 398 (1992).
57. J. Fricke, J. Neuhard, R. A. Kelln and S. Pedersen, *J. Bact.* **177,** 517 (1995).
58. C. K. Mathews, *JBC* **247,** 7430 (1972).
59. J. Neuhard and P. Nygaard in *"Escherichia coli* and *Salmonella typhimurium,* Cellular and Molecular Biology" (F. C. Neidhardt, ed.), p. 445. Am. Soc. for Microbiology, Washington, D.C., 1987.
60. Y.-Z. Xu, P. Huang and W. Plunkett, *JBC* **270,** 631 (1995).
61. B. Nicander and P. Reichard, *JBC* **260,** 5376 (1984).
62. G. Spyrou and P. Reichard, *JBC* **262,** 16425 (1987).
63. G. Spyrou and P. Reichard, *JBC* **264,** 960 (1989).
64. R. Sleight and C. Kent, *JBC* **258,** 831 (1983).
65. M. K. Turner, R. Abrams and I. Lieberman, *JBC* **243,** 3725 (1968).
66. H. L. Elford, *Adv. Enzyme Reg.* **10,** 19 (1972).
67. S. Eriksson, A. Gräslund, S. Skog, L. Thelander and B. Tribukait, *JBC* **259,** 11695 (1984).
68. Y. Engström, S. Eriksson, I. Jildevik, S. Skog and B. Tribukait, *JBC* **260,** 9114 (1985).
69. G. J. Mann, E. A. Musgrove, R. M. Fox and L. Thelander, *Cancer Res.* **48,** 5151 (1988).
70. M. Thelander and L. Thelander, *EMBO J.* **8,** 2475 (1989).
71. J. E. Brussenden, I. Carasi, L. Thelander and U. F. Francke, *Exp. Cell Res.* **174,** 302 (1988).
72. J. Bjorklund, K. Hjortsberg, E. Johansson and L. Thelander, *PNAS* **90,** 11322 (1993).
73. I. W. Caras, B. B. Levinson, M. Fabry, S. R. Williams and D. W. Martin, Jr., *JBC* **260,** 7015 (1985).
74. M. R. Briggs, J. T. Kadonaga, S. P. Bell and R. Tijian, *Science* **234,** 47 (1986).
75. F. Y. Chen, F. M. Amara and J. A. Wright, *EMBO J.* **12,** 3917 (1993).
76. F. Y. Chen, F. M. Amara and J. A. Wright, *BJ* **302,** 125 (1994).
77. R.A.R. Hurta and J. A. Wright, *Biochem. Cell Biol.* **70,** 1081 (1992).
78. S. Bjorklund, E. Skogman and L. Thelander, *EMBO J.* **11,** 4953 (1992).
79. R. Ashfield, P. Enriquez-Harris and N. J. Proudfoot, *EMBO J.* **10,** 4197 (1991).
80. F. M. Amara, F. Y. Chen and J. A. Wright, *JBC* **269,** 6709 (1994).
81. F. M. Amara, F. Y. Chen and J. A. Wright, *NARes* **21,** 4803 (1993).
82. D. M. Kingsley, *Genes Dev.* **8,** 133 (1994).
83. F. Ventura, J. Doody, F. Liu, J. L. Wrana and J. Massague, *EMBO J.* **13,** 5581 (1994).
84. Y. Asaoka, S.-I. Nakamura, K. Yoshida and Y. Nishizuka, *Trends Biochem. Sci.* **17,** 414 (1992).
85. Y. Nishizuka, *Science* **258,** 607 (1992).
86. B. Alberts, D. Bray, J. Lewis, M. Faff, K. Roberts and J. D. Watson, "Molecular Biology of the Cell," 3rd Ed., Garland Publ., New York, 1994.
87. J. Halstead, K. Kemp and R. A. Ignotz, *JBC* **270,** 13600 (1995).
88. R. A. R. Hurta and J. A. Wright, *J. Cell. Physiol.* **158,** 187 (1994).
89. R. A. R. Hurta, S. K. Samuel, A. H. Greenberg and J. A. Wright, *JBC* **266,** 24097 (1991).
90. R. A. R. Hurta and J. A. Wright, *JBC* **267,** 7066 (1992).
91. A. K. Chan, D. W. Litchfield and J. A. Wright, *Bchem* **32,** 12835 (1993).
92. R. D. Klausner, T. A. Rouault and J. B. Harford, *Cell* **72,** 19 (1993).
93. E. R. Bauminger, A. Teffry, A. J. Hudson, D. Hechel, N. W. Hodson, S. C. Andrews, S. Levi, I. Nowik, P. Arosio, J. R. Guest and P. W. Harrison, *BJ* **302,** 813 (1994).
94. B. Guo, Y. Yu and E. A. Leibold, *JBC* **269,** 24252 (1994).
95. R. A. R. Hurta and J. A. Wright, *Biochem. Cell Biol.* **69,** 635 (1991).
96. S. Nyholm, L. Thelander and A. Gräslund, *Bchem* **32,** 11569 (1993).

97. A. Dancis, D. G. Roman, G. J. Anderson, A. G. Hinnebusch and R. D. Klausner, *PNAS* **89**, 3869 (1992).
98. S. J. Elledge and R. W. Davis, *MCBiol* **7**, 2783 (1987).
99. H. K. Hurd, C. W. Roberts and J. W. Roberts, *MCBiol* **7**, 3673 (1987).
100. S. J. Elledge and R. W. Davis, *MCBiol* **9**, 5373 (1989).
101. H. K. Hurd and J. W. Roberts, *MCBiol* **9**, 5359 (1989).
102. E. Bucking-Throm, W. Duntze, L. H. Hartwell and T. Manney, *Exp. Cell Res.* **76**, 99 (1973).
103. S. J. Elledge and R. W. Davis, *Genes Dev.* **4**, 740 (1990).
104. T. A. Weinert and L. H. Hartwell, *Science* **241**, 317 (1988).
105. Z. Zhou and S. J. Elledge, *Cell* **75**, 1119 (1993).
106. T. A. Navas, Z. Zhou and S. J. Elledge, *Cell* **80**, 29 (1995).
107. H. L. Elford, M. Freese, E. Passamani and H. P. Morris, *JBC* **245**, 5228 (1970).
108. R. A. Weinberg, *Cell* **81**, 323 (1995).
109. J. R. Nevins, *Science* **258**, 424 (1992).
110. L. H. Hartwell and M. B. Kastan, *Science* **266**, 182 (1994).
111. N. M. Standart, S. J. Bray, E. L. George, T. Hunt and J. V. Ruderman, *J. Cell Biol.* **100**, 1968 (1985).
112. N. Standart, T. Hunt and J. Ruderman, *J. Cell Biol.* **103**, 2129 (1986).
113. N. Standart, M. Dale, E. Stewart and T. Hunt, *Genes Dev.* **4**, 2157 (1990).
114. E. H. Davidson, "Gene Activity in Early Development." Academic Press, Orlando, Florida, 1986.
115. R. F. Bachvarova, *Cell* **69**, 895 (1992).
116. M. Wormington, *Curr. Opin. Cell Biol.* **5**, 950 (1993).
117. N. Standart and R. Jackson, *Curr. Biol.* **4**, 939 (1994).
118. M. Wormington, *BioEssays* **16**, 533 (1994).
119. A. S. Spirin, in "Current Topics in Developmental Biology" (A. A. Moscona and A. Monroy, eds.), Vol. 1, p. 1. Academic Press, New York, 1966.
120. D. H. Darnbrough and P. J. Ford, *EJB* **113**, 415 (1981).
121. M. Ranjan, S. R. Tafuri and A. P. Wolffe, *Genes Dev.* **7**, 1725 (1993).
122. M. T. Murray, D. L. Shiller and W. W. Franks, *PNAS* **89**, 11 (1991).
123. N. Standart and T. Hunt, *Enzyme* **44**, 106 (1990).
124. N. Standart and M. Dale, *Dev. Genet.* **14**, 492 (1993).
125. J. M. Noronha, G. H. Sheys and J. M. Buchanan, *PNAS* **69**, 2006 (1972).
126. B. Roizman, in "Fundamental Virology" (B. N. Fields and D. M. Knipe, eds.), 2nd Ed., p. 849. Raven Press, Ltd., New York, 1991.
127. "Medical Virology" (D. O. White and F. J. Fenner, eds.), 3rd Ed., p. 317. Academic Press, San Diego, 1986.
128. D. J. Goldstein and S. K. Weller, *J. Virol.* **62**, 196 (1988).
129. V. G. Preston, A. J. Darling and I. M. McDougall, *Virology* **167**, 458 (1988).
130. J. G. Jacobson, D. A. Leib, D. J. Goldstein, C. L. Bogard, P. A. Schaffer, S. K. Weller and D. M. Coen, *Virology* **173**, 276 (1989).
131. T. D. Chung, J. P. Wymer, M. Kulka, C. C. Smith and L. Aurelian, *Virology* **179**, 168 (1990).
132. J. P. Wymer, T. D. Chung, Y.-N. Chang, G. S. Hayward and L. Aurelian, *J. Virol.* **63**, 2773 (1989).
133. C. Jones, F. Zhu and K. R. Dhanwada, *DNA and Cell Biology* **12**, 127 (1993).
134. T. D. Chung, J. P. Wymer, M. Kulka, C. C. Smith and L. Aurelian, *J. Virol.* **63**, 3389 (1989).

135. P. Desai, R. Ramakrishnan, Z. W. Lin, B. Osak, J. Glorioso and M. Levine, *J. Virol.* **67**, 6125 (1993).
136. J.-H. Luo and L. Aurelian, *JBC* **267**, 9645 (1992).
137. R. J. Jariwalla, B. Tanzos, C. Jones, J. Ortiz and S. Salimi-Lopez, *PNAS* **83**, 1738 (1986).
138. W. Cai and P. A. Shaffer, *J. Virol.* **66**, 2904 (1992).
139. B. Moss, *in* "Virology" (B. Fields, ed.), p. 685. Raven Press, New York, 1985.
140. M. L. Howell, N. A. Roseman, M. B. Slaubaugh and C. K. Mathews, *JBC* **268**, 7155 (1993).
141. S. J. Child, G. J. Palumbo, R.M.L. Buller and D. E. Hruby, *Virology* **174**, 625 (1990).
142. M. L. Howell, J. Sanders-Loehr, T. M. Loehr, N. A. Roseman, C. K. Mathews and M. B. Slaubaugh, *JBC* **267**, 1705 (1992).
143. M. B. Slaubaugh, R. E. Davis, N. A. Roseman and C. K. Mathews, *JBC* **268**, 17803 (1993).
144. M. B. Slaubaugh, M. L. Howell, Y. Wang and C. K. Mathews, *JBC* **265**, 2290 (1991).
145. P. Traktman, P. Sridhar, C. Condit and B. E. Roberts, *J. Virol.* **49**, 125 (1984).
146. D. E. Hruby and L. A. Ball, *Virology* **113**, 594 (1981).
147. J. B. Flanegan and G. R. Greenberg, *JBC* **252**, 3019 (1977).
148. J. P. Young and C. K. Mathews, *JBC* **267**, 10786 (1992).
149. N. Lamm, J. Tomaschewski and W. Rüger, *NARes* **15**, 3920 (1987).
150. J. J. Scocca, S. R. Panny and M. J. Bessman, *JBC* **244**, 3698 (1969).
151. G. F. Maley, B. W. Duceman, A. M. Wang, J. Martinez and F. Maley, *JBC* **265**, 47 (1990).
152. C.-S. Chiu, T. Ruettinger, J. B. Flanegan and G. R. Greenberg, *JBC* **252**, 8603 (1977).
153. J. B. Flanegan, C.-S. Chiu and G. R. Greenberg, *JBC* **252**, 6031 (1977).
154. E. Hanson and C. K. Mathews, *JBC* **269**, 30999 (1994).
155. C.-S. Chiu, K. S. Cook and G. R. Greenberg, *JBC* **257**, 15087 (1982).
156. L. J. Wheeler, Y. Wang and C. K. Mathews, *JBC* **267**, 7664 (1992).
157. L. Moen, M. L. Howell, G. W. Lasser and C. K. Mathews, *J. Mol. Recogn.* **1**, 48 (1988).
158. C.-S. Chiu, S. M. Cox and G. R. Greenberg, *JBC* **255**, 2747 (1980).

Index

A

β-Alanine synthase
 allosteric regulation by oligomer dissociation, 13–15, 52–57
 complementary DNA from rat liver, 57–59
 5-fluorouracil degradation, 51–52
 importance, 49–51
 sequence homology with other enzymes, 57–59
Allosteric regulation
 conformational change stabilized by ligand, 16–17
 enzyme oligomer dissociation, 13–15
 ligand binding between subunits, 17–18
Alu
 arrangement in genome, 304
 cheap-gene hypothesis, 284–285, 302–303
 EPL Alu, 299–300
 evolution, 296, 298–301
 genetic effects, 284, 301–302
 internal promoter, 291–292
 methylation and imprinting role, 305–307, 313–314
 prevalence in human genome, 291
 RNA
  cell stress response, 311–313
  expression *in vivo*, 307–308
  promoter strength, 308–309
  regulation of expression, 309–310
  species, 310
 SINE inactivation, 300–301
 subfamilies, 298–299
Amino acid, control of gene expression
 asparagine synthetase, 232–237
 bacteria, 219–220
 plasma membrane transport, adaptive regulation, 226–232
 ribosomal proteins, 238–239, 241–245
 yeast, 220–221, 223–228
Asparagine synthetase, amino acid control of gene expression, 232–237

ATP-binding site
 kinase motifs, 60
 types, 60–63, 71

B

Bacteriophage T4
 deoxyribonucleotide synthetase extended complex, 387–390
 expression of ribonucleotide reductase, 351–356
 initiation of replication, 349–351, 383–384
 synthesis of deoxyribonucleotides and ratio of bases, 384–386
 template damage and deoxyribonucleotide synthesis, 386
Bisulfite, probing of RNA structure, 143, 146

C

Cell cycle
 cancer, 198, 214
 ribonucleotide reductase expression, 361–367
 signal transduction and G0 to S transition, 198–201
 thymidine kinase, transcriptional regulation, 202–214
Cyclin A-dependent kinase, cell cycle regulation, 200–201
Cyclin D-dependent kinase, cell cycle regulation, 199–200
Cyclin E-dependent kinase, cell cycle regulation, 200
1-Cyclohexyl-3-(2-morpholinoethyl)carbodiimide metho-*p*-toluene sulfate, probing of RNA structure, 143, 149
Cytidine nucleotide, sources in *Escherichia coli*, 359–361

D

Diethylpyrocarbonate, probing of RNA structure, 143, 149, 154
Dihydrouridine, conformation and dynamics, 116–117
Dimethyl sulfate, probing of RNA structure, 142
Dissociating enzyme, allosteric regulation, 13–15

E

E2F, cell cycle regulation, 201, 204–208
Enzyme evolution
  ATP-binding sites, kinase motifs, types, 60–63, 71
  gene fusion and multifunctional proteins, 66–70
  nucleoside monophosphate-binding sites, 63–66
Ethyl nitrosourea, probing of RNA structure, 142

F

5-Fluorouracil, degradation, 51–52

G

GCN4, amino acid control of gene expression, 221, 223–226
Genomic imprinting, Alu role, 305–307, 313–314

H

Herpes virus
  ribonucleotide reductase, control of expression, 380–382
  subfamilies, 379
  transcriptional control, 380

I

Imidazole, probing of RNA structure, 147
Iron(II)–EDTA, probing of RNA structure, 147, 153–154, 159

K

Kethoxal, probing of RNA structure, 143

M

Magnesium, transfer RNA modification effect on binding, 113–114, 119–120

N

Nucleoside, modified
  biological functions in transfer RNA, 81–84
  discovery, 79–80
  drug therapy, 121
  modification effect on nucleoside conformation and dynamics, 116–119
  modifying enzymes, 80
  physicochemical contributions
    base stacking, 109
    disruption of canonical base-pairing, 114–116, 120
    magnesium binding, 113–114, 119–120
    reordering of water, 112
    types, 81
  site-selective positioning of modifications, 101–104
  structures and chemistry, 84–101
  symbols and common names, 122–123

O

Orotate phosphoribosyltransferase, see also UMP synthase
  amino acid sequence homology between species, 20–26
  developmental changes of enzyme activity, 9–10
  dimerization, 34–35
  stability, 35–36
  structure and function, 26–29
Orotic aciduria, etiology, 19–20
Orotidylate decarboxylase, see also UMP synthase
  abzyme model, 39–40
  amino acid sequence homology between species, 20–26

developmental changes of enzyme activity, 9–10
dimerization, 34–35
stability, 35–36
structure and function, 29–32

P

Pseudogene, accumulation, 303–304
Pseudoknot, prediction in RNA, 180–186
Psoralen, cross-linking of RNA, 155–158
Pyrimidine metabolism
  compartmentation of nucleotide pools, 12–13
  liver function, 10–12
  overview, 2–4
  salvage versus *de novo* pathways
    developmental changes of UMP synthesis enzyme activities, 9–10
    enzyme activity variation, 4–6
    incorporation of exogenous orotate and uridine, 6–9

R

Reovirus
  assembly of segmented genome
    assortment mechanism, 250–251
    functions of reovirus proteins
      λ1, 261–262
      λ2, 262–263
      λ3, 263–264
      μ1, 264–265
      μ2, 263–264
      μNS, 265–266
      σ1, 266–267
      σ1S, 267
      σ2, 267
      σ3, 268–269
      σNS, 267–268
    monoclonal antibodies against reovirus proteins, 269–270
    nature of genome segments, 255, 257–259
    reassortments, 260–261
    signal interplay requirements for genetic information insertion, 274–277
  defective interfering particles, 259–260
  infectious RNA system, 270–273
  multiplication cycle, 252–253
  taxonomy, 249–250
Repetitive sequences, *see also* Alu
  evolution, 296, 298–301, 313
  families, 284
  genetic effects, 284, 301–302
  identification, 294–295
  LINEs, 292
  MaLR, 293
  microsatellites, 294
  MIRs, 293–294
  prevalence in human genome, 283–284, 288, 313
  retroposition, 290–294
  sequence analysis by DNA renaturation studies, 285, 287–288
  tandem clusters, 288, 290
Retinoblastoma protein, cell cycle regulation, 201, 210
Ribonucleotide reductase
  activity in eukaryotic cell cycle, 361
  bacteriophage T4 enzyme
    deoxyribonucleotide synthetase extended complex, 387–390
    events leading to expression, 351–353
    synthesis of deoxyribonucleotides and ratio of bases, 384–386
    template damage and deoxyribonucleotide synthesis, 386
    translational delay, 353–356
  cancer role, 374
  clam oogenesis, regulation of expression
    large subunit, 378
    mRNA masking and polyadenylation, 376–377
    small subunit translation, 377–378
  early history, 345–346
  Escherichia coli enzyme subunits
    expression control, 356–359
    induction by DNA damage, 374
    types, 346–347
  herpes virus, transcriptional control, 380–382
  initiation of replication, role, 348–349
  mouse enzyme subunits
    coordinated synthesis, 367–368
    factors controlling mRNA stability, 364–366, 370
    ferritin relationship in expression, 369
    gene structure, 362–363
    induction by DNA damage, 373

iron requirement in R2 synthesis, 368–369
mRNA truncation during G1, 364–365
phosphorylation of R2, 368
relationship to synthesis of other proteins, 366–367
vaccinia virus enzyme regulation, 382–383
yeast enzyme subunits
  ferric reductase relation to RAP1 gene, 370
  induction by DNA damage, 371–373
Ribosomal protein, amino acid control of gene expression, 238–239, 241–245
RNA, see also Transfer RNA
  biophysical investigation, 108–109
  computer analysis of secondary structure
    algorithms, 164, 168–170
    comparative approach, 174–180
    prediction of pseudoknots, 180–186
    simulation of folding, 172–174
    statistical analysis, 186–190
    thermodynamic approach, 166–170
  gel sequencing, 137–138
  modification, see Nucleoside, modified; transfer RNA
  radiolabeling, 137–138
  structure determination
    chemical probing of modified RNA, 133–140, 142–143, 146–147, 149, 153–159
    cross-linking and intramolecular modifications, 155–161, 191
    enzymatic probing, 133–142, 150–151, 159–161
    nucleotide interactions and folding, 148–152
    physical methods, 132, 191
    scission by complexed metals, 161–164
    tertiary structure, 152–155
  thermodynamic parameters of secondary structure, 165–166
R-point protein, cell cycle regulation, 198–199, 202, 214

S

Substrate channeling, UMP synthase, 69–71
System A
  amino acid regulation, 230–232

amino acid transport, 227–228
repressor protein, 228–230

T

T4, see Bacteriophage T4
2-Thiouridine, modification effect on conformation and dynamics, 117–119
Thymidine kinase
  cloning of MT3-binding proteins, 210–212
  promoter, MT2 region
    binding
      E2F complexes, 204–208
      Yi complexes, 208–210, 213–214
      DNase I footprinting, 203
    transcriptional regulation and late G1 events, 202–204
Transfer RNA
  biological functions of modified nucleosides, 81–84
  factors affecting chemical reactivity of groups, 135–136
  physicochemical contributions of nucleoside modification
    base stacking, 109
    disruption of canonical base-pairing, 114–116, 120
    magnesium binding, 113–114, 119–120
    reordering of water, 112
  types, 81
  reverse transcription and analysis of modified RNA, 138–140
  site-selective positioning of modifications
    biophysical investigation, 108–109
    synthesis, 105–110
    transfer RNA, 101–104
  structure determination
    chemical probing of modified RNA, 133–140, 142–143, 146–147, 149, 153–159
    cross-linking and intramolecular modifications, 155–161, 191
    enzymatic probing, 133–142, 150–151, 159–161
    nucleotide interactions and folding, 148–152
    physical methods, 132, 191
    scission by complexed metals, 161–164
    tertiary structure, 152–155

INDEX

tertiary structure, 133–134
Tryptophan operon, attenuation, 220

U

UMP synthase, *see also* Orotate phosphoribosyltransferase; Orotidylate decarboxylase
  allosteric regulation by oligomer dissociation, 13–15
  amino acid sequence homology between species, 20–26
  defects in orotic aciduria, 18–20
  dimerization, 34–35
  gene fusion, 67–70
  human mutants, 38–39
  kinetics of orotidylate decarboxylase, 37–38
  proteolysis, 36–37
  stability, 35–36
  substrate channeling, 69–71
Uridine kinase
  allosteric regulation by oligomer dissociation, 13–15
  complementary DNA from mouse brain, 46–48
  gene copy number, 48–49, 71
  gene expression, 40–42
  importance, 40
  isozymes, 42–44
  pseudogene, 48
  regulation, 45–46
  structure, 44–45

V

Vitamin D
  forms, 321
  genes, transcriptional regulation, 329, 331–333
  hydroxylases, molecular biology, 333, 335–338
  synthesis, 321, 333, 335–338
Vitamin D receptor
  cloning, 322–323, 338
  genes, transcriptional regulation, 329, 331–333
  immunoradioassay, 322
  phosphorylation, 325, 338
  responsive elements, 325–328
  structure determination, 323–324
  tissue expression, 322–324

Y

Yi complex, cell cycle regulation, 208–210, 213–214

ISBN 0-12-540053-5